Philosophy of Engineering and Technology

Volume 20

More information about this series at http://www.springer.com/series/8657

Steen Hyldgaard Christensen
Christelle Didier • Andrew Jamison
Martin Meganck • Carl Mitcham
Byron Newberry
Editors

International Perspectives on Engineering Education

Engineering Education and Practice in Context, Volume 1

Springer

Editors
Steen Hyldgaard Christensen
Aalborg University
Aalborg, Denmark

Christelle Didier
Université Charles de Gaulle-Lille3
Lille, France

Andrew Jamison
Aalborg University
Aalborg, Denmark

Martin Meganck
KU Leuven
Ghent, Belgium

Carl Mitcham
Colorado School of Mines
Golden, Colorado, USA

Renmin University
Beijing, China

Byron Newberry
Baylor University
Waco, Texas, USA

ISSN 1879-7202 ISSN 1879-7210 (electronic)
Philosophy of Engineering and Technology
ISBN 978-3-319-16168-6 ISBN 978-3-319-16169-3 (eBook)
DOI 10.1007/978-3-319-16169-3

Library of Congress Control Number: 2015939348

Springer Cham Heidelberg New York Dordrecht London

Printed on acid-free paper

Springer International Publishing AG Switzerland is part of Springer Science+Business Media (www.springer.com)

Preface

And some time make the time to drive out west
Into County Clare, along the Flaggy Shore,
[T]he ocean on one side is wild
With foam and glitter, and inland among stones
The surface of a slate-grey lake....
Useless to think you'll park and capture it
More thoroughly. You are neither here nor there,
A hurry through which known and strange things pass
As big soft buffetings come at the car sideways
And catch the heart off guard and blow it open.

– Seamus Heaney, "Postscript" (1996)

We live today in a world progressively in the process of becoming an engineered artifact. We engineer not only roads and buildings but communication systems and biologies. In such a world, thinking about engineering is increasingly important – and yet incredibly difficult.

Among themselves, engineers are continuously trying to figure out what and who they are: skilled workers, project managers, applied scientists, designers, entrepreneurs, and more. Additionally, there are a host of competing interests that would enroll engineering for their purposes: military interests, nation-building interests, commercial interests, social interests, environmental interests, and more. Finally, multiple disciplines attempt to take the measure of engineers and engineering: history, sociology, philosophy, and more.

There is no simple resolution to the tensions inherent in this complexity of contextualizations for the engineered constructions in which we progressively live and move and have our being. The best we can do is take an intellectual drive through diverse intellectual landscapes, with a willingness to let what poet Seamus Heaney calls "big soft buffetings" come at us sideways, opening the mind. Open to its contexts, the mind is at once:

- More reflective in negotiating the pressures that enfold it
- Better at spanning different engineering visions and practices

- More insightful when conciliating the corporeal powers of engineering with the ethereal truths of poetry or art
- More resistant to commercial, political, and military distortions of human and professional responsibilities
- Better at constructing a more just world – one in which lives well-lived and well-examined transcend mere existence

To contribute to this opening up, not so much of the black box of what takes place behind the scenes in engineering, but of our own thinking about engineering, is the central effort of our collective reflection.

The two books we offer – *International Perspectives on Engineering Education: Engineering Education and Practice in Context. Volume 1* and *Engineering Identities, Epistemologies and Values: Engineering Education and Practice in Context. Volume 2* – are the result of an extended dialogue or bridge-building between humanists and engineers with whom we have been involved both individually and more recently as a group. Steen Hyldgaard Christensen, the editor-in-chief, studied literature and history of ideas at Aarhus University in the 1970s, and since 1987 has taught humanities for engineering and business students at what was originally a technical vocational college in Herning, Denmark (which in 1995 became the Institute of Business and Technology, and in 2006 Aarhus University). Since 2003, Christensen has been facilitating processes of collaboration between engineers, social scientists, and humanists in a series of book projects. The first, with coeditors Martin Meganck and Bernard Delahousse, was on *Profession, Culture and Communication: An Interdisciplinary Challenge to Business and Engineering* (2003); the second, with the same coeditors, was *Philosophy in Engineering* (2007); a third, again with coeditors Meganck and Delahousse, was *Engineering in Context* (2009), the precursor of the present two volumes.

Martin Meganck has a doctorate in chemical engineering and is a former Dominican friar who studied theology and currently teaches ethics for engineering students at KU Leuven in Ghent, Belgium. Bernard Delahousse was an English language scholar who taught at an engineering college in Lille, France, and served as head of the school's international office. Delahousse has retired, but participates now as coauthor of one of the chapters in Volume I. Christensen got to know them while serving as the international officer at his institution, which is now part of Aarhus University.

For each book, Christensen and his coeditors organized a gathering of potential authors. Two days of deliberations by participants lead to a table of contents, after which Christensen and his coeditors orchestrated the logistics of book production: first draft submission, final draft submission, index submission, proofreading, etc.

In 2008, Andrew Jamison was drawn into the process, as a contributor to the project that became *Engineering in Context*. But even before that book was published, Jamison, with Christensen and several other contributors to these volumes, asked the Danish Strategic Research Council to fund a four-year Program of Research on Opportunities and Challenges in Engineering Education in Denmark (PROCEED). This ambitious, interdisciplinary project took place between 2010 and 2013.

PROCEED was organized as a strategic research alliance between four universities: Aalborg University, Aarhus University (including the former engineering college in Herning), Roskilde University, and the Danish Technical University. The research was divided into five thematic projects: "Challenges and Responses in Historical Perspective," "Curriculum Design and Learning Outcomes," "Modeling and Simulation in Engineering," "Engineering Practice and Design Competence," and "Integrating Contextual Knowledge into Engineering Education" (cf. PROCEED 2010). The alliance included engineers, social scientists, philosophers, and historians; numerous chapters in these books are based on research and teaching activities that were part of the program.

Prior to the initiation of PROCEED, another project took shape that has also influenced the present two volumes. Christensen, Jamison, and Carl Mitcham teamed up to organize an interdisciplinary reflection on relationships between "engineering and development" that involved American, Chinese, and European perspectives. Christensen invited ten Europeans, Mitcham ten Americans, and Li Bocong, from the Graduate University of the Chinese Academy of Sciences (with whom Mitcham had been working since the early 1990s), ten Chinese scholars. Together these scholars met at the Colorado School of Mines in April 2010 in a workshop supported by the CSM Hennebach Program in the Humanities for an exercise in reflective, cross-cultural learning. PROCEED served as a cosponsor of the workshop by funding travel by some of the European participants.

Mitcham – a key node in the Christensen network from 2006 on – organized the CSM workshop around a series of "tutorials" designed to stimulate dialogue. Mitcham and his colleague Juan Lucena led tutorials on engineering and development from an American perspective (Mitcham for the North and Lucena, originally from Colombia, from the South); Christensen and Jamison offered a tutorial on engineering and development from a European perspective; while Li Bocong and Yanming An introduced a Chinese perspective. By the end of the meeting in Golden, CO, a table of contents was developed for a book that was eventually published in 2012 under the title, *Engineering, Development and Philosophy: American, Chinese and European Perspectives*, edited by Steen Hyldgaard Christensen, Carl Mitcham, Li Bocong, and Yanming An. The book appeared in the Springer series *Philosophy of Engineering and Technology*.

As a further contribution to the American-Chinese-European collaboration project, Li Bocong arranged another workshop on "Engineering and Sociology" in Beijing, China, in the fall of 2011. Li had long been concerned that engineering in the West was too focused on an individualistic professionalism, and he sought to stimulate reflections that would broaden the contexts of understanding in both the West and the East. It was thus in Beijing, around the pleasures of extended Chinese meals, and in a country undergoing a historically unique engineering construction, that there emerged the germ of an idea that has grown into these two volumes on *Engineering in Context*.

Another contributory linkage to these publications can be found in the European Ethics Network (EEN) from the 1990s. The EEN brought together ethicists from 40 European universities and had a broad set of objectives. One of these was creating a book series with core materials for professional ethics in the fields of biomedicine, business, press, and engineering. A kick-off conference in Barcelona, under the title "Rethinking Professional Ethics," was the starting point of a series of collaborations among ethicists involved in engineering and technology from mainly Western European countries. An immediate and tangible result was the publication of Philippe Goujon and Bertrand Hériard-Dubreuil's edited volume, *Technology and Ethics: A European Quest for Responsible Engineering* (2001). The engineering ethics team of the Catholic University of Lille (France) was the motor and the pivoting centre behind the book, and Christelle Didier and Martin Meganck were members of the editorial team. Mitcham contributed an afterword comparing American and European efforts in this area. The ethics journal *Ethical Perspectives* served for some time as the official organ of EEN and is the only *ad extra* visible remainder of that EEN period. A less visible outgrowth, however, is a continuing set of ties among ethicists in different professional fields. When the *Profession, Culture and Communication* project sought a continuation in *Philosophy in Engineering*, the ties between research groups and individual researches resulting from the EEN experience were useful in identifying new partners. The presence of Christelle Didier in the current editorial team has its basis there.

Still one more contributing stream to our collaborative effort, one that draws again on the work of Li Bocong, among others, is the 2012 Forum on Philosophy, Engineering, and Technology (fPET) held in Beijing, China. fPET-2012 was a follow-on to an earlier fPET-2010 hosted at CSM in Colorado. The fPET conferences grew out of previous workshops held in 2007 and 2008 known as the Workshops on Philosophy and Engineering (WPE). The fPET conferences, like the WPE workshops before them, have provided opportunities to bring together scholars from a variety of cultures and disciplines, all sharing a common interest in trying to better understand the human activities we call engineering, the people we call engineers, and the creations we call technology. At the latest meeting in Beijing, approximately 15 countries and 5 continents were represented. Philosophers, historians, and other humanists, along with social scientists and engineers, participated. The range of presentations included philosophical, historical, cultural, and ethical analyses of engineers, engineering, and technology. These events have proved invaluable as catalysts for ideas, scholarly exchanges, and collaborations. In fact, almost half the contributors to the present volumes have been participants in one or more of these events. Byron Newberry, another member of the current editorial team, whose background is in aerospace and mechanical engineering, served as cochair, along with Li Bocong, of the fPET-2012 meeting. Newberry also contributed to the earlier *Engineering in Context* book.

These different strands come together in the current set of two books. An international editorial kick-off workshop was initiated by Christensen and organized with the help of Louis L. Bucciarelli at MIT in May 2012. The main purpose was to define the objectives, structure, and content of the volumes. After introductory presentations by workshop host Bucciarelli, Gary Downey, and Jamison, an intensive process of discussions began. And, as the French say, *Du choc des idées jaillit la lumière*: at first confrontational ideas finally result in understanding and constructive proposals.

We hereby present the final result of a long writing and editorial process. We trust that our readers will find the work worthwhile and they may be inspired by it to do even more to think and rethink engineering contexts so as to transform engineering into a truly humanizing enterprise.

As those two books are meant to be a contribution to furthering the dialogue between engineering and philosophy in order to explore ways in which the humanities can contribute to self-development in engineering education through appreciation of the multiple contexts within which engineers increasingly work, these groups of academics are the primary audience for our books. Moreover, we believe that the very process of creating these volumes, bringing together as it has a host of scholars from a diversity of disciplinary and cultural perspectives, marks a major milestone on the path toward creating a sense of identity and shared culture, while recognizing the value of differences, and building a vibrant community of scholars dedicated to bridging the gaps between engineers, humanists, and social scientists.

However, the book is also addressing a wider academic audience and may actually function as a means to achieve greater self-understanding for both teachers in engineering disciplines and for practitioners. Educational policy makers, both on a political and an institutional level, may also find valuable matter for reflection and inspiration in this book. We believe that, not least, the process of globalization compels engineering educators to rethink and recontextualize engineering education in order to educate a better and more rounded type of engineer. We finally hope that the book may inspire students of engineering as well as students of the humanities and social sciences who are interested in the challenges and complexities that a rapidly changing and globalized world pose for higher education in general and for engineering education in particular.

Herning, Denmark — Steen Hyldgaard Christensen
Lille, France — Christelle Didier
Aalborg, Denmark — Andrew Jamison
Ghent, Belgium — Martin Meganck
Golden, Colorado, USA — Carl Mitcham
Waco, Texas, USA — Byron Newberry
1 October 2014

Acknowledgment

The editors would like to express our heartfelt gratitude to the two anonymous reviewers who provided thorough assessments of our two volumes, respectively. The comments, suggestions, and criticisms provided by these two scholars were both detailed and insightful. As a result of their feedback, we added new material on topics that deserved more attention (particularly with respect to issues of gender, race, and class), made significant improvements to several chapters, reorganized some of the chapters for better coherence and flow, and have tightened up some of our introductory sections. Our manuscript has been made stronger due to the care and diligence of these reviewers.

Contents

Author Biographies

Steen Hyldgaard Christensen M.A. in Scandinavian Language and Literature and the History of Ideas, Aarhus University. Ph.D. in Educational Studies, Aalborg University. Senior lecturer at Aarhus University, School of Business and Social Sciences, Herning, Denmark, until 2014. From 2014, adjunct associate professor at Aalborg University, Denmark. He has initiated six big international inter- and meta-disciplinary research projects on engineering including PROCEED and coordinated five of them. He has acted in roles of editor-in-chief and coauthor of four books: *Profession, Culture, and Communication: An Interdisciplinary Challenge to Business and Engineering* (Institute of Business Administration and Technology Press 2003); *Philosophy in Engineering* (Academica 2007); *Engineering in Context* (Academica 2009); and *Engineering, Development and Philosophy: American, Chinese, and European Perspectives* (Springer 2012). Besides, he has coauthored *A Hybrid Imagination: Science and Technology in Cultural Perspective* (Morgan & Claypool Publishers 2011) together with Andrew Jamison and Lars Botin. In addition, he has published a number of articles on engineering epistemology, culture, and education. Current research interest includes academic drift in engineering education and structural dynamics in higher education.

Christelle Didier B.S. in Electrochemistry Engineering, M.A. in Education, Ph.D. in Sociology from Ecole des Hautes Etudes en Sciences Sociales (EHESS), Paris. From 1993 to 2013, Assistant Professor, Lille University, France, Ethics Department. Assistant Professor, Charles de Gaulle University of Lille, Education Department. Member of CIREL (EA 4354). Coauthor of *Ethique industrielle* (DeBoeck, Brussels, 1998) and author of *Penser l'éthique des ingénieurs* (PUF, Paris 2008) and *Les ingénieurs et l'éthique: Pour un regard sociologique* (Hermes 2008). She has published many articles on ethics and social responsibility in the engineering profession and education and on the engineering profession's values (from interviews and extensive surveys). Her research areas are engineering ethics and values, including historical, cultural, and gender perspective, sustainable development and corporate social responsibility, social responsibility.

Andrew Jamison B.A. in History and Science from Harvard University, Ph.D. in Theory of Science from University of Gothenburg (Göteborg). Docent in Theory of Science, University of Gothenburg. Professor of Technology, Environment, and Society at Aalborg University. Coordinator of Program of Research on Opportunities and Challenges in Engineering Education in Denmark (PROCEED), 2010–2013, and author, most recently, of *The Making of Green Knowledge: Environmental Politics and Cultural Transformation* (Cambridge 2001), *Hubris and Hybrids: A Cultural History of Technology and Science*, with Mikael Hård (Routledge 2005), *A Hybrid Imagination: Science and Technology in Cultural Perspective* (Morgan & Claypool Publishers 2011) together with Steen Hyldgaard Christensen and Lars Botin, and *The Making of Green Engineers: Sustainable Development and the Hybrid Imagination* (Morgan and Claypool 2013).

Martin Meganck M.Sc. in Chemical Engineering from Ghent University, Ph.D. in Chemical Engineering, and M.A. Moral Theology both from KU Leuven. Lecturer in Philosophy and Ethics in the Faculty of Engineering Technology, researcher at the Center for Science, Technology, and Ethics, and the Theological Faculty of the KU Leuven. Teaching areas include Philosophy of Science, Philosophy of Technology, and Professional and Business Ethics. He was coauthor and coeditor of *Philosophy in Engineering* (Academica 2007) and *Engineering in Context* (Academica 2009).

Carl Mitcham B.A. and M.A. in Philosophy from University of Colorado, Boulder. Ph.D. in Philosophy from Fordham University. Professor of Philosophy of Science and Technology, Renmin University of China, Beijing; Liberal Arts and International Studies, Colorado School of Mines, Golden, Colorado. Scholarly contributions have been directed toward the philosophy and ethics of science, technology, engineering, and medicine and to science, technology, and society (STS) studies. Teaching areas include ethics, STS, and science and technology policy.

Byron Newberry B.S. in Aerospace Engineering, University of Alabama. M.S. in Aerospace Engineering and Ph.D. in Engineering Mechanics, both from Iowa State University. Professor of Mechanical Engineering, Baylor University. Baylor Fellow for teaching. Professional Engineer (PE) Texas, USA. Research interests include engineering design, engineering ethics, and philosophy of engineering and technology. Aircraft structural engineering consultant. Executive board member, National Institute for Engineering Ethics. Editor of the Springer *Philosophy of Engineering and Technology* book series.

General Introduction
The Engineering-Context Nexus
A Perennial Discourse

Steen Hyldgaard Christensen, Christelle Didier, Andrew Jamison, Martin Meganck, Carl Mitcham, and Byron Newberry

In 1982, Barry Barnes and David Edge published *Science in Context: Readings in the Sociology of Science*, which significantly influenced the sociology of science. The volume collected 18 previously published articles from the 20-year period 1961 to 1981 – articles almost exclusively by social scientists – to promote reflection on relationships between the subculture of science and the wider culture that surrounds it. Although the editors did not present it as such, the program for understanding

S.H. Christensen
Department of Development & Planning, Aalborg University,
Vestre Havnepromenade 5, Aalborg 9000, Denmark
e-mail: steenhc@plan.aau.dk

C. Didier
Département des sciences de l'éducation UFR DECCID,
Université Charles de Gaulle-Lille3, 3 Rue du Barreau,
Villeneuve-d'Ascq, Lille F-59650, France
e-mail: christelle.didier@univ-lille3.fr

A. Jamison
Department of Development and Planning, Aalborg University,
Erik Dahlbergsgatan 22, Malmö S-211 48, Sweden
e-mail: andy@plan.aau.dk

M. Meganck
KU Leuven, Faculty of Engineering Technology, Technologiecampus Ghent,
Gebroeders De Smetstraat 1, Ghent B-9000, Belgium
e-mail: martin.meganck@kuleuven.be

C. Mitcham
Colorado School of Mines, Golden, Colorado, USA

Renmin University, Beijing, China
e-mail: cmitcham@mines.edu

B. Newberry
Department of Mechanical Engineering, Baylor University,
One Bear Place 97356, Waco, TX 76798, USA
e-mail: Byron_Newberry@Baylor.edu

"science in context" can be read as responding to the challenge of C.P. Snow's 1959 "two cultures" lecture, which identified a debilitating split between scientific and literary intellectuals. For Snow, there really were two cultures that approached the world from antagonistic perspectives. For the social scientists collected by Barnes and Edge, however, scientific culture is always part of culture in a more expansive sense. The two cultures are really one, and science needs to be understood precisely as an aspect of what it may indeed partially oppose.

In the spirit of that earlier title, the present two companion volumes focus on *Engineering Education and Practice in Context* (EEPiC, read as "epic"). This project differs, however, not only in its concern with engineering instead of science but also in being composed of more than 40 original articles contributed by a much more interdisciplinary group: social scientists, yes, but also engineers, philosophers, historians, and even scholars from the fields of classics, communication, and film studies. Additionally, among the more than 60 contributors are representatives from 16 countries on the 6 inhabited continents. The volumes direct attention to four primary contexts of engineering: formal education, the design process, workplace and institutional experience, and civil society. Yet like Barnes and Edge, these new volumes postulate an integral if sometimes contentious relationship between engineering cultures and their larger cultural contexts.

Comparing work on science with the present work on engineering, there emerges what may be termed a contextualization-decontextualization paradox. Scientists qua scientists think of their work as decontextualized and, therefore, have trouble recognizing the ways in which it is also contextualized. Engineers qua engineers think of their work as contextual and, therefore, tend to overlook the ways in which it is decontextualized. Scientists, for example, see formulas such as $F = ma$ and $E = mc^2$ as independent of context and universally true, failing to appreciate their knowledge production can reflect particular cultures (as, in these cases, a mathematical rhetoric enacted in distinctive social institutions). By contrast, engineers engage with contexts in which they deploy those same formulas in particular projects. But it is precisely because they think of themselves as so context dependent and sensitive that engineers also so often presume they can go into any situation and provide appropriate solutions; they often too readily believe all their solutions are inherently contextual, even when this fails to be the case. The existence of such a paradox suggests the need to use the *Science in Context* project as defined by Barnes and Edge as a foil with which to exploit difference.

Beyond Science in Context

The science in context argument is in an important respect nihilistic. The significance of natural science, which the sociology of science aims to disclose, is that natural science has no special significance. Its reputed claims to significance are unmasked, demythologized, and demystified. The sociological argument, as succinctly summarized by Barnes and Edge, is that "There is no way in which [natural scientific] expertise can be guaranteed by reference to reason rather than habitual inference, nature rather than culture" (p. 11). Natural science is a social institution like any other; it rests on purely social foundations and its reasons are no more privileged than those of politics, economics, or the military.

Yet as Barnes and Edge also admit, "to conceive of expert knowledge solely in terms of advocacy" is to ignore the normative question concerning which advocates are most credible or authoritative. The normative question is not one that can be "reduced to a matter of what beliefs are immediately expedient, or immediately relevant to vested interests" (p. 10). Among natural scientists and nonscientists alike, the problem of credibility has customarily been resolved by granting natural science a measure of rational authority – although a rational authority that social scientific analysis questions.

The social scientific analysis of science in context is nevertheless faced with three problems. First, social science is not generally granted the same social recognition as natural science – that is, as the astronomical, physical, geological, and biological sciences. So its claims with regard to the natural sciences often carry little weight. It is not clear what influence the analysis of science in context can ever really have.

Second, even if the social sciences were magically to acquire social prestige and power, it is not clear how more careful and detailed sociological studies – which are repeatedly recommended by Barnes, Edge, and others, in order to give a better understanding of what really happens with science – would escape the acidic analysis that they apply to the natural sciences. That is, the sociological analysis addressed to the natural sciences would seem necessarily to apply as well to the social sciences. The social sciences, too, would have to be conceived as social constructions.

As a result, third, the social sciences can "offer no obvious solutions to the normative problems involved in the evaluation of [scientific] expertise" (p. 12). It is not just that the normative question is, as Barnes and Edge later claim, "of no sociological interest" (p. 194); normativity is not an issue that it is even possible in principle for sociology to address. The sociology of science reveals science to be without distinctive authority and thus at the mercy of political, economic, and military powers – powers that are not troubled, in their real-world exercise of power, by any alleged lack of authoritative rationality. This is what Barnes and Edge refer to as "the tragedy of the expert" (p. 237). Experts can never deploy the methods of expertise, which exist within a community of experts, to legitimate such expertise to the wider public. "If science itself is called into question, then the scientific expert can only retire gracefully" (p. 234). Scientific experts appear dependent on irrational acceptance by the public, with an irrationality that can at most and only on occasion be meliorated by programs of public participation – although Barnes and Edge acknowledge the "power" present in science, especially as reflected by the close linkages of science "with 'the higher levels' of government and industry" (p. 248).

The science in context project is thus fraught with implications the engineering in context project seeks as much as possible to avoid. To this end, we offer three observations. First, by way of a brief historicophilosophical gloss, note that while the idea of the social construction of science can be manifest among scholars without serious immediate harm, the idea has been applied elsewhere with quite harmful results. Insofar as the administration of U.S. President George W. Bush refused, when making decisions about how best to reduce teenage pregnancy, respond to climate change, and the invasion of Iraq to grant any privileged status to scientific knowledge, both natural and social, he adopted a social constructivist stance. To the realist objection that one needs to respect reality, one of Bush's senior advisers is reported simply to have replied, "When we act, we create our own reality" (Suskind 2004). Such application and its results surely provide a good reason to revisit the normative question and defend the rationality of engineering as well as of science.

Second, and more positively, as if offering a means for addressing the normative question, our engineering in context project, is inherently more interdisciplinary. It involves not just sociologists and historians but also engineers and philosophers – along with scholars in the further reaches of the humanities and the social sciences. Indeed, while science in context sought to broaden the reach of science, the broadening went no further than to describe science as not just a "source of knowledge and competence [but as] a repository of theories, findings, procedures and techniques which it makes generally available both directly, via expert intervention and consultation, and indirectly, via its interaction with technology and with specialized institutions in the economic and political structure" (p. 2). What is lacking is recognition of science as a font of social, ethical, and even environmental problems.

To recognize science or engineering as a source of problems – especially environmental problems – is not to deny that it can also contribute to solutions or, better, responses. Indeed, to adopt and adapt the naturalistic pragmatism of John Dewey and to recognize something as a problem is implicitly to imagine a better state of affairs. For Dewey, engineering is ultimately and properly subordinate to the enhancement of life and the qualitative enlargement of human experience. Insofar as science and its sibling engineering fail to accord with this transcendent end – an end that is subject to continuous reimagination and reinstitutionalization in culture – it calls forth its own reconceptualization, regulation, or delimitation along with parallel and complementary extensions and expansions.

It is precisely this that best functions as our own context for the study of engineering. We are studying engineering not simply to promote sociological understanding but in pursuit of better engagement between engineering and society – and the better education of engineers. Moreover, although to some degree a socially constructed or contingent end, it is an end for which we are willing and able to develop rational arguments. Only insofar as we can give good reasons for such ends – not just insofar as such ends are popularly accepted – should we wish to defend and built upon or toward them.

Thus the EEPiC project includes a strongly reflexive element. In the Barnes and Edge volume, for instance, there was no discussion of the meaning of context. By contrast, our two volumes both explicitly and implicitly address different meanings of context. On the explicit side, some chapters grapple overtly with the issue of context, whether trying to elucidate its meaning, to highlight its importance, or in at least one case to reject it. On the implicit side, ideas about contexts were built in via the selection of authors and topics, along with the organization of the volume sections. For example, while Barnes and Edge relied heavily on the problematic concept of culture, for which it assumes an anthropological meaning (see p. 193), here the question of culture is itself placed in context by the presence of contributions from multiple cultures and cultural perspectives, not to mention disciplines and disciplinary perspectives. In addition, chapters in the two volumes are organized in sections designed to explore particular contextual facets, whether historical, ideological, or institutional.

We should note, however, that it is not the objective of these volumes to definitively demarcate the meaning of context in engineering. For our purposes, context is not an end-in-itself but rather a means to an end. In the spirit of further reflexivity, the contingent but nonetheless rationally defensible (and inherently normative) end of the engineering in context project is to foster a better understanding of and engagement with engineering. This engagement will be intentionally provocative and argue for an end that is not explicitly given but implicitly found embedded within it: the transcendence of engineering, what has been called postengineering (see Mitcham 2009).

Remaining for the present in the European tradition, there exists a long-standing or sedimented distinction between liberal and professional education. From the perspective of liberal studies, the contrast is one between education and training, even vocational or technical training. From the perspective of professional studies, the contrast is between useless discussion or mere theory and useful or practical learning. It seems clear that engineering education accords primarily with professional or practical studies. Yet this is not to deny its possible involvement with liberal or even useless studies. We need to move beyond simple dependence on engineering. We must not become so effective at and engrossed with engineering that we forget that engineering is not everything. We need to exercise again the classical humanities disciplines of self-moderation.

Two Volumes and Their Complementarities

In summary, in relation to science in context, which it references as an ancestor, the two EEPiC volumes aim to be more interdisciplinary and original, more critical and reflexive, and more openly normative. Taken as a whole, this collection of original scholarly work is unique in its broad, multidisciplinary consideration of the changing character of engineering education and engineering practice in and from the perspective of multiple contexts.

Volume 1 on engineering education includes analyses of the history, structure, and ideologies of engineering education, challenges and critical perspectives, along with discussions of new pathways in 25 contributions by 50 authors from engineering, social sciences, and humanities. Key overlapping questions examine such issues as:

- What are the different approaches to engineering education?
- Are differences competitive or complementary?
- What special challenges are emerging for engineering from concerns for sustainable community development, energy ethics, sustainability, and demands for innovative design?

- What new efforts are being made to reform engineering education from the perspectives of design, engineering education research, and case-based learning?
- What is the role of the social sciences and the humanities in engineering education?

The chapters of Volume 1 are grouped into four sections, roughly following a see-judge-act logic. Part I historically frames engineering education in the United States, Western Europe, and a selection of locations elsewhere (India, Brazil, Slavic Europe). What appears initially simply descriptive is interwoven with a reflexive/interpretative layer. Part II groups a series of more fundamental reflections on the hidden and overt ideologies in engineering and engineering education. Parts III and IV collect contributions on experiences and approaches for reform and innovations in engineering education.

In Part I, the institutional history and evolution of engineering education in different geographical/cultural contexts is the carrying canvas. Regional, cultural, and historically bound aspects form one approach. Although these historiographical descriptions focus on regional and cultural differences, some common themes emerge. One is “academic drift”: vocational-oriented training programs tend to be swept into more academic structures, inducing changes in professional profile and educational culture. A shift of focus from local toward more global perspectives can also be observed throughout the contributions. Insertion in the global economy seems to induce more pragmatic and neoliberal entrepreneurial tendencies in engineering education.

Part II shifts from institutional history to the asking of critical questions regarding theory and practice in engineering education. Like all institutionalized programs of education, engineering schools explicitly or implicitly assume and promote beliefs about how engineers should behave, not just in technical terms but in their social relationships. As previous scholars have noted, there are deeply ingrained ideas in the American context about positive relationships between engineering and business. The chapters in this section invite consideration of some alternative perspectives by calling attention to how engineering education functions differently in China and how the engineering-business nexus may not be experienced as unquestionably rational by members of nondominate social groups.

The framework of Part III extends an exploration of the limitations of received ideologies in engineering education by considering specific cases in the emergence of alternative futures. Hence the majority of chapters in Part III contribute to the construction of a counter-hegemonic discourse or “heterotopia,” to use a term of Baillie et al. (2012). Some themes that come into view are engineering mindsets that get in the way of engineers seeing social justice, social justice in the context of global energy consumption and use, critique of the prevailing “weed out” culture in undergraduate programs as an impediment to diversity, developing a hybrid imagination in prospective engineering students, and questioning the ideology and codes of knowledge behind the dominant construction of the epistemological core in engineering education and more.

The chapters in Part IV focus on the renovation of engineering education. Different in their structures and approaches, the innovations that are discussed in this section have in common to refuse reducing education to a mere transmission of knowledge from a master to passive students. Instead, they rely on the active participation of the students and their personal experiences. Most importantly, rather than discussing which content should be added to enrich engineering education, some chapters focus on how to teach with pedagogical methods such as problem-based learning, and how to combine engineering teaching and engineering education research. Others propose a more radical transformation of engineering education through a definition of engineering not only as problem solution but also a contribution to problem definition or a new understanding of engineering knowledge, as the products of contextualized experience.

Volume 2 on engineering practice advances contextual analyses of engineering identity, epistemologies, and values in 23 contributions by more than 30 authors from engineering, social sciences, and humanities. Key overlapping questions examine such issues as:

- What does it mean to be an engineer?
- How are engineering self-understandings enacted in the professional world?
- What is the distinctive character of engineering knowledge?
- How do engineering science and engineering design interact in practice?
- What are the prominent norms of engineering?
- How do they interact with the values of efficiency or environmental sustainability?

The reflection on engineering identities in Part I fans out in the following sections: Is there anything like "engineering knowledge" (Part II)? Is there an inherent normativity in engineering, and how does it connect with the norms and values of the surrounding world (Part III)? The concluding Part IV gives a further exploration of the idea of context itself: in practice, a sharp delineation between "text" and "context" may appear difficult if not impossible. This can either lead to fundamentally questioning the very concept of context or to the vision that engineers can make their own context.

How do engineers distinguish themselves from scientists? From business people? From technologists? How do engineers define themselves professionally, and how are those professional identities uniquely shaped within particular national contexts. How do those outside of engineering perceive engineers? Is there a common unifying element between the diverse types of engineers? And how do gender-based stereotypes of and within engineering serve to limit equitable participation in the field? These are the types of questions that are grappled with by the chapters in Part I of Volume 2, in an effort to gain a clearer understanding of the *identities* of engineers. In addition, a final chapter provides a statistical overview of the scope of the engineering occupation worldwide.

Another field – expounded in the chapters in Part II – where the contextuality of engineering appears, is in the epistemology of engineering: the knowledge engineers need or use in their work cannot be clearly defined and demarcated. There are many uncertainties, as well in the available knowledge itself as in the evaluation of possible outcomes. Data may be lacking or hidden in an overload of information of indistinct relevance. And the boundaries within which engineering projects are to be solved are subject to negotiation with economic or political instances and societal groups and stakeholders of many kinds. Part II of this volume gathers reflections on engineering epistemology. What kinds of efficiencies are pursued by engineers? How do they situate themselves in the tension between pure science and design practice? And how can the many layers of engineering knowledge be reflected in modern curricula of engineering education?

In Part III, the central issue is the values that carry engineers and engineering (which is nowadays our common world) and cultural norms that are or should be at work in professional practice engineers. Some authors question the ambiguous influence of professional associations on the consideration by engineers of ethical issues. Others wonder how the culture of the engineers, the way they look at the world, shapes and is shaped by their relationship with the world of politics. Still others discuss the influence of social values on the attitudes of engineers and those of economic and political issues on how the problems they are asked to solve are formulated.

Do engineers create their own contexts or are they created by contexts? The authors in Part IV, the final section of Volume 2, all take explicit aim at the notion of context. Aptly titled "Competing Contexts in Engineering," the chapters present contrasting views of what context might mean or even how important the concept might be. One author argues that engineers create their own contexts. Another argues that the very idea of context is too static and should be abandoned in favor of more dynamic ways of characterizing engineering. Other chapters seek useful ways to differentiate context, whether by scale (from the micro to the macro) or by vantage point (internal versus external to the engineering activity). A final chapter explores the challenge faced by engineering practitioners with respect to reflexively incorporating an understanding of context in their work.

Contexts, Challenges, and Paths to Transformation

The notion of context in engineering education and practice is an object of heated debate. On the one hand, claims are made that context is an artificial construct, reifying a distinction between context and content and producing the sense of an inside and an outside. On the other hand, claims are made that the distinction between technical context and social context (a) reflects real tensions in engineering education and practice, (b) is constantly being re-negotiated, and, most importantly, (c) the outcome of such negotiations has real world consequences. Positions that adopt the context approach often focus on social justice, and more broadly empirical studies of engineering students' engagement with context, have been reflected in a number of path breaking works. Among these are: Cindy Atman and colleagues (1996, 2008), Caroline Baillie (2006), Donna Riley (2008), Baillie and colleagues (2011, 2012), and Juan Lucena (2013). Baillie et al. (2012), Most recently Bill Williams, José Figueiredo, and James Trevelyan in a collection on *Engineering Practice in a Global Context* (2014) have made another significant contribution.

The position taken here is that context matters and has practical consequences. The relevance of context is related to at least three different meanings of context and to inherent tensions that result:

- The embedding of institutions of engineering education into higher education systems,
- The breadth of problem scoping in engineering problem solving
- Contextual knowledge

Context, however, is an inherently dialectical concept, since contextualizing is itself dependent on definitions of what are perceived to be the relevant boundaries regarding both the education and the practice of engineers. Contextualizing unfolds its inherent dialectics in the terrain between "is" and "ought," fact and value. In this way, the quest for a recontextualizing of engineering education and practice inevitably is a value-laden enterprise and thus not without a certain degree of controversy. It is concerned with both what engineering "is" and what it "ought" to be. Ultimately a greater awareness and understanding of context should result in better preparation of engineers to render those contexts visible in their work, and consequently enable them to contribute to more socially robust and responsible endeavors.

When thinking about how far context can influence engineering and engineering education, one rapidly discovers challenges or even crises that can be roughly categorized into a number of ideal typical arguments:

- The captivity argument
- The cultural change argument
- The identity crisis argument
- The weak profession argument
- The convergence argument

This list of arguments, most of which have been developed in one form or another over recent years, should be understood as neither complete nor definitive, although it provides a useful point of departure for anyone interested in understanding and innovating with respect to engineering and engineering education. Despite overlaps between these arguments, the merit of distinguishing them is that each emphasizes a specific aspect of engineering and/or engineering education that poses challenges – and opportunities – for the engineering profession.

In many chapters of these two volumes, the ideas and analyses aim to further identify, characterize, and explicate one or more of these challenges. Other chapters, drawing on such analyses, propose responses in hopes of transforming engineering and engineering education in ways that will sustain the profession as a vital, constructive, and responsive social institution. A brief summary of relevant arguments follows.

The *captivity argument* is that the engineering profession, in regard to both education and practice, has been locked in a number of social and intellectual captivities that may be interpreted as a "fundamental usurpation of the intellectual and social dimensions of engineering as an autonomous discipline" (Goldman 1991,

p. 121). An "intellectual captivity" consists of engineering being considered subordinated to science. Engineering education requires students to master large doses of mathematics and physical sciences. Engineers in turn tend to believe that science and engineering are objective and able to exclude human values from influencing the esoteric work taking place in engineering disciplines. Engineers become overly concerned with order and certainty and adverse to ambiguity. Issues of meaning and social impact are marginalized because scientific methodology, the structure of hypothesis, proof, validation, publication, and critique are embedded in a scientific culture to which engineers find themselves attached. A "social captivity" lies with engineering practice being subordinated to a managerial agenda driven by economics and the market. Engineers exercise their power only within that mandate, which raises questions about the idea of engineers as the primary agents of technological change. According to Johnston et al. (1996), the result has been a serious limitation in engineers' capacity to examine the social meanings and effects of their work and to self-consciously reflect on their practices and professional identities.

Captivity arguments surface throughout these volumes. For example, in Volume I, Chap. 1, Atsushi Akera and Bruce Seely provide a historical account of the American system of engineering education. In it they highlight the rise to dominance of the *engineering science* paradigm, as well as the influences of "neoliberal economic doctrine." Similarly, in Volume II, Chap. 10, Stig Andur Pedersen delves into the intellectual tensions between science and engineering. Other chapters present ideas for moving beyond such intellectual and social captivities. For example, Tony Marjoram argues in Volume I, Chap. 16, for a problem-based, as opposed to science-based, education, with an emphasis on addressing human and social development goals. And in Volume II, Chap. 17, Carl Mitcham and Wang Nan advocate an expansion of engineering ethics into the political arena, so that "taking a global perspective on investing in a new technological innovation, for instance, would involve going beyond economics to include assessments of multiple risks and benefits at the social and environmental levels."

The *cultural change argument* concerns an alleged lack of diversity in engineering. In one version of this argument, feminist research criticizes the social norms of engineering culture as overly masculine. How could female students feel attracted to engineering faculties that are not only demographically dominated by men but also culturally emphasizing of male interests? Research has shown that male students go for engineering because they like to tinker; the choice of female students seems more inspired by a general interest in mathematics and physics. Even without giving in to the caricature of the pragmatic and performance-oriented male vs. the more caring and relation-oriented woman, bridging these "two cultures" is far from evident. But this is only one aspect of the cultural change argument. In Volume I, Chap. 8, Amy Slaton describes the "less-than–democratic character" of engineering and other science, technology, engineering, and mathematics (STEM) occupations and the weak influence of many inclusive efforts made in the United States to address diversity issues (gender issues, but also social diversity). Wendy Faulkner in Volume II, Chap. 2, highlights how gender operates alongside professional and organizational to produce engineering culture and proposes to disseminate "heterogeneous" images of engineering in order to create space for a more diverse range of people.

In a context where the global engineering competency becomes "a problem of engaging people from different cultures" (Downey et al. 2006), another aspect of cultural change has to do with cross cultural and globalization issues. In Volume I, Chap. 7, Qin Zhu and Brent Jesiek highlight the need to develop a better understanding of the history and cultural context of engineering education and profession in other countries and regions. They propose three key intellectual concepts enabling understanding Chinese culture: Confucianism, Marxism, and pragmatism.

A further aspect of cultural change involves preparing engineers to deal with environmental issues. In Volume II, Chap. 13, Christelle Didier and Kristoff Talin highlight French engineers' attitudes toward the environment and how they differ from those of their fellow citizens; "ecoskepticism" is the norm even among the younger generation of engineers. In Volume II, Chap. 15, Jen Schneider, Abraham Tidwell, and Savannah Fitzwater describe the tremendous difficulty of reforming nuclear science and engineering education in the United States to better integrate environmental issues. Encouraged by physics and engineering educators, student skepticism toward climate change research constitutes a cultural value and contributes to constructing an "insular culture." Rather than simply objecting to their opinions, the authors invite nuclear engineers to make their voices better heard at the "table of discussion."

The *identity crisis argument* has several manifestations, ranging from how engineering is understood – or misunderstood – by the public, to uncertainties in the roles engineers play, or will continue to play in the future, in technology development. The latter issue, for example, was developed forcefully by Rosalind Williams (2002). In a reflection that grew out of her service as Dean for Undergraduate Education and Student Affairs at MIT, she analyzes how a division of labor has eroded the identity of the engineering profession.

> What engineers are being asked to learn keeps expanding along with the scope and complexity of the hybrid world. Engineering has evolved into an open-ended Profession of Everything in a world where technology shades into society, into art, and into management, with no strong institutions to define an overarching mission. All the forces that are pulling engineering in different directions – toward science, toward the market, toward design, toward systems, towards socialization – add logs to the curricular jam. (Williams 2002, p. 70)

The challenge for engineering education is complex: it can lead to cramming more and more into the curriculum. It can lead to hyper-specialization, with a set of narrowly defined skills and competencies for preestablished jobs. But this contrasts with future demands for "educating active, rigorous and flexible individuals, rather than skilled workers for pre-established jobs." For Williams, the curricular response should be a convergence between the technological and liberal arts, educating the engineering student both for life and flexible employment.

> Only a hybrid educational environment will … prepare students for handling … life in a hybrid world. Students need to be prepared for life in a world where technological, scientific, humanistic, and the social issues are all mixed together. Such mixing will not take place if students have to decide from the outset that they are attending an "engineering school" as opposed to a "non-engineering school." (Williams 2003, p. 4)

Elements of the identity crisis argument are apparent in many chapters here. Byron Newberry, in Volume II, Chap. 1, discusses what he terms the *dialectics of identity*, which is created by ambiguities in the understanding who engineers are and what they do, ambiguities that exist both internally (engineers' self-identity) and externally (engineers as viewed by others). A detailed example of ambiguous self-identity is provided, for example, in Volume II, Chap. 3, where Mike Murphy, Shannon Chance, and Eddie Conlon present empirical results of engineering students' self-conceptions. Looking toward engineering's future Andrew Jamison, Niels Mejlgaard, and Jette Egelund Holgaard, in Volume I, Chap. 14, reimagine engineering by advocating development of what they call a *hybrid identity*:

> Fostering hybridity or a hybrid imagination involves a mixing of scientific education and training in technical skills with an appreciation of the broader cultural implications of science and technology in general and one's own role as an engineer, in particular.

The *weak profession argument* deals with the professional status of engineers. Mitcham (2009) distinguished between *strong* and *weak* professions. According to his argument, strong professions (such as medicine and law) rest on the formulations of ideals that are well embedded in the professional curriculum and practice. Weak professions (such as military and business) either lack such ideals or only weakly include the relevant specialized knowledge in a professional curriculum and practice. Somewhat provocatively he argues that engineering has more in common with weak than with strong professions.

This overlaps with the captivity argument in that engineers themselves may see their job as executing what others have decided: clients or patrons, sponsors, government, the market; decisions about the ultimate end-use of engineering work seem removed from engineers themselves. Seeing engineering as a weak profession is nevertheless at odds with the aspiration to have "engineers who will assume leadership positions from which they can serve as positive influences in the making of public policy and in the administration of government and industry" (National Academy of Engineering 2004). There is a call for engineers who would not just be technocrats, but public intellectuals, who would accompany society in dealing with a technological culture, and

> show to a broad array of audiences – politicians, engineers, scientists, and the general public – that science and technology are value laden, that all aspects of modern culture are infused with science and technology, that science and technology do play key roles in keeping society together, and that they are equally central in all events that threaten its stability. It is therefore necessary that science and technology, in their explicit and implicit forms, be subject to political debate. (Bijker 2003, p. 444)

This argument can be seen as part of the choices university education has to make in general, and not only for engineering. Will universities be training camps for professionals, under a regime run by "academic capitalism and managerialism" (Slaughter and Leslie 1997)? Or should universities be places of intellectual critique and cultural citizenship?

Especially in the second volume of this diptych, several chapters deal with the disputed professional status of engineering, either as part of a main line of discussion or at least as an aside. It is part of Newberry's consideration of the "dialectics of engineering." The "engineering-label" covers a wide range of specializations and occupational activities, and the boundaries between professionals and other educational backgrounds are blurred. This makes it difficult for engineers to gather in one recognizable group and to speak with an authoritative voice, even concerning topics that are within their realms of competence. Michael Davis has a long record of publications on professionalism and engineering. In Volume II, Chap. 4, he enters into discussion with some comments and objections his publications have raised and deals with methodological and conceptual misunderstandings that blur the vision of engineering as a profession. Martin Meganck in Volume II, Chap. 12, questions why a professionalism label should be important at all and discusses whether a professionalism-based ethics cannot be reduced to principles of ordinary morality.

Finally, the educational consequences of the above-mentioned arguments are related to a *convergence argument*, which focuses on relatively recent evolutions in higher education across many countries. Democratization of education, homogenization (e.g., through the Bologna process in Europe), political decisions, and the application of new management styles seem to lead to an academic drift – or convergence in mission – in and of nonuniversity institutions, and vocational drift in universities or institutions similar to universities. For engineering, some fear that this will lead to a gradual loss of the practice-oriented nature of engineering. Curricula will become more theoretical. Teaching staff will be evaluated more on their research activity than on their teaching or contacts with industry. The blurring of boundaries between "noble" and "less noble" institutions is a tendency that seems to occur spontaneously and organically; yet it solicits further fundamental reflection.

In Volume I, Chap. 2, Steen Hyldgaard Christensen and Newberry zoom in on major differences between and dynamics of change in European and American higher education. They examine two European examples of academic and research drift in nonuniversity institutions – Irish *Institutes of Technology* (IoTs) and Dutch *Hogescholen* (HBOs) – and three American examples – a public technical institute (Southern Polytechnic State University in Georgia), a state teacher's college (Western Kentucky University), and a sectarian liberal arts university (Baylor University). They argue that convergence in mission between universities and former vocationally oriented designated teaching institutions both in Europe and the United States are likely to create a number of tensions and dilemmas as well as winners and losers. Shifting emphases in engineering degree programs from teaching based on practical experience derived from engineering work to research-informed and research-led education creates crisis for many faculty members whose values and identities embody the core of a teaching culture. Many of these practically experienced teachers are likely to be one obvious group of losers in this process of institutional transformation.

Bernard Delahousse and Wilhelm Bomke in Volume I, Chap. 3, further substantiate the convergence argument in presenting a comparative study of two more profession-oriented institutions in Europe – the French *Instituts Universitaires de Technologie* (IUTs) and the German *Fachhochschulen* (FHs). In their study, the focus is on the historical evolution of the two types of institution in terms of degree of autonomy, creation or adaptation of curricula, pedagogical methods, student standing, personnel status, and research opportunities. The two institutions have a number of traits in common: a strong focus on teaching rather than research, fixed curricula oriented toward practice including internships, close links with companies, academic staff recruitment, a particular stand with regard to universities, insistence on graduate operational skills, and more. The authors argue that academic drift should be regarded as a natural and irreversible process: "natural" because it interacts with the inevitable evolutions of society in its economic, political, social, cultural, and technological dimensions; and "irreversible" as it constitutes a never-ending trajectory. Generally transformations take place in moments of opportunity provided by external state, public, private, or transnational agencies. Yet the void after the transformation of institutions may need filling by a new type of short-cycle institution and the process can go on once again.

Conclusion

These two EEPiC volumes thus aim to stimulate critical reflection on the past, the present, and the future of engineering in both education and practice. They offer no final answers or even a well-formed methodology. Instead, their programmatic character invites readers themselves to reflect on the engineering-context nexus and contribute their own insights to a perennial discourse – a discourse that can help us all, engineers and nonengineers alike, live more consciously and carefully in our increasingly engineered world.

With regard to issues addressed in Volume 1, engineering education in all its dimensions – histories and structures, ideologies, reforms, and innovations – can be expected to be continuing subjects for empirical research and critical reflection. More empirical research on the institutional contexts of engineering education with respect to ongoing institutional transformations both locally and globally will be a priority. Given the increased blurring of boundaries between university and nonuniversity engineering educational programs, there are ongoing needs to explore what does and does not work under what conditions to achieve diverse goals. A related issue for research and reflection is the engineering-business nexus, a multilayered relationship with implications for both engineering education and practice. More systematic empirical research along the lines of Cindy Atman and colleagues (1996 and 2008) on student engagement would also be important.

With regard to Volume 2, engineering practice as reflected in identifies, epistemologies, and values calls as well for further research and reflection. Here recent (and no doubt future) analyses of the normativity in engineering and technology are (and will become more) relevant; see, for example, the work by Ibo van de Poel and Peter Kroes (2006) and Sergei Gepshtein (2009). Additionally, the relationship between engineering, social sciences, and humanities has implications not only for education but for engineering identity, knowledge, and ethics. The need to integrate these three perspectives on engineering practice has been pointed toward by Mitcham (2014) as well as many others trying to assess large-scale social problems that have emerged in conjunction with the engineering transformations of human ways of life (see, e.g., Mike Hulme 2014).

More generally, our introduction began by referencing and criticizing a previous "science in culture" project. But we should also acknowledge the extent to which this project has received its own criticism in the science studies field. One extension of the science in context program argued that since there is no reason to grant scientific expertise any special cognitive privilege, everyone is justified in claiming expertise. In an insightful response to this developmental trajectory, Harry Collins argues at length that although everyone may be some kind of expert, we are not all scientific experts "because we do not [all] belong to the scientific community and we do not necessarily make our judgments from the platform of the norms and aspirations that drive that community" (2014, p. 131). For Collins, "If we start to believe we are all scientific experts, society will change: it will be those with the power to enforce their ideas or those with the most media appeal who will make our truths, according to whatever set of interests they are pursuing" (ibid.).

Adopting Collins' framework, we can note that there has been little temptation for any social critic to argue that "we are all engineers now." Additionally, despite Snow's blurring of any science-engineering distinction in his famous two-culture argument, engineering intellectuals are probably something different than either scientific or literary intellectuals. At the same time, there is some sense in which even literary intellectuals would have to admit their dependency on engineers much more than on scientists. This is the case, first, insofar as engineering is conceived as attempting to satisfy human needs and, second, insofar as engineering has been argued by engineers themselves to be a more refined form of that making and using that permeates all human activities (see, e.g., Koen 2003). To the extent that either of these theses is even partially true, it is all the more incumbent on us to struggle to examine engineering in context.

References

Atman, C. J., & Nair, J. (1996). Engineering in context: An empirical study of freshmen students' conceptual frameworks. *Journal of Engineering Education, 85*(4), 317–326.

Atman, C. J., Kilgore, D., Yasuhara, K., & Morozov, A. (2008). Considering context over time: Emerging findings from a longitudinal study of engineering students. *Proceedings of the Research in Engineering Education Symposium*, Davos, Switzerland, 7–10 July 2008.

Baillie, C., Pawley, A. L., & Riley, D. (2011). *Engineering and social justice: In the university and beyond*. West Lafayette: Purdue University Press.

Baillie, C., Kabo, J., & Reader, J. (2012). *Heterotopia. Alternative pathways to social justice*. Hants: Zero Books, John Hunt Publishing Ltd.

Barnes, B., & Edge, D. (1982). *Science in context. Readings in the sociology of science*. Cambridge, MA: MIT Press.

Bijker, W. (2003). The need for public intellectuals. A space for STS. Pre-presidential address, annual meeting 2001, Cambridge MA. *Science, Technology & Human Values, 28*(4), 443–450.

Collins, H. (2014). *Are we all scientific experts now?* Cambridge, UK: Polity Press.
Downey, G. L., Lucena, J. C., Moskal, B. M., Parkhurst, R., Bigley, T., Hays, C., Jesiek, B. K., Kelly, L., Miller, J., Ruff, S., Lehr, J. L., & Nichols-Belo, A. (2006). The globally competent engineer: Working effectively with people who define problems differently. *Journal of Engineering Education, 95*(2), 1–17.
Gepshtein, S. (2009). Closing the gap between ideal and real behavior: Scientific vs. engineering approaches to normativity. *Philosophical Psychology, 22*(1), 61–75.
Goldman, S. L. (1991). The social captivity of engineering. In P. T. Durbin (Ed.), *Critical perspectives on nonacademic science and engineering* (pp. 121–145). London: Research in Technology Studies Series, Lehigh University Press.
Hulme, M. (2014). *Can science fix climate change?* Cambridge, UK: Polity Press.
Johnston, S., Lee, A., & McGregor, H. (1996). Engineering as captive discourse. *Society for Philosophy and Technology, 1*, 3–4. http://scholar.lib.vt.edu/ejournals/SPT/v1n3n4/Johnston.html.
Koen, B. V. (2003). *Discussion of the method: Conducting the engineer's approach to problem solving*. New York: Oxford University Press.
Lucena, J. (Ed.). (2013). *Engineering education for social justice: Critical explorations and opportunities*. Dordrecht: Springer.
Mitcham, C. (2009a). A historico-ethical perspective on engineering education: From use and convenience to policy engagement. *Engineering Studies, 1*(1), 35–53.
Mitcham, C. (2009b). A philosophical inadequacy of engineering. *The Monist, 92*(3), 339–356.
Mitcham, C. (2014). The true grand challenge for engineering: Self-knowledge. *Issues in Science and Technology, 31*(1), 19–22.
Riley, D. (2008). *Engineering and social justice*. San Rafael: Morgan & Claypool Publishers.
Slaughter, S., & Leslie, L. L. (1997). *Academic capitalism: Politics, policies, and the entrepreneurial university*. Baltimore: Johns Hopkins University Press.
Suskind, R. (2004). Faith, certainty and the presidency of George W. Bush. *New York Times Magazine*, October 17, 2004.
Van de Poel, I., & Kroes, P. (2006). Technology and normativity. *Techne, 10*(1), 1–6.
Williams, R. H. (2002). *Retooling. A historian confronts technological change*. Cambridge, MA: MIT Press.
Williams, R. H. (2003). Education for the profession formerly known as engineering. *The Chronicle of Higher Education, 49*, 20. 24 January.
Williams, B., Figueiredo, J., & Trevelyan, J. (2014). *Engineering practice in a global context: Understanding the technical and the social*. Leiden: CRC Press.

Part I
Histories, Dynamics and Structures in Engineering Education

Introduction

Atsushi Akera and Erin A. Cech

The chapters in this section focus on the structural transformation of engineering education in historical and contemporary perspective, with a focus on processes around the globe. Based largely on country and region-specific case studies, these chapters discuss developments in the United States, Western and Eastern Europe, Brazil, India, and Latin America, respectively. They also provide the broad canvas upon which the subsequent sections of this two-volume edited compilation can be interpreted.

As with all introductions of this sort, our goal is to build coherence across the chapters by pointing to common themes and threads, and use the material to point to new questions and research directions not addressed by the individual authors. In other words, our hope is to direct the reader toward their own reading of what is, after all, a still new and emerging literature. We do so in three parts: first by discussing what we see to be the interesting interplay between the historical and contemporary material in these chapters; second, by focusing on the global dimensions of the current changes in engineering education; and third by discussing several institutional processes that appear to be at work across the case studies. Following a brief introduction of the individual chapters, we then conclude this introduction with some broad observations about the overall structure, or *institutional ecology,* for the production (research) and reproduction (education) of engineering knowledge.

One of the first things that comes to focus in this section is the productive tension that exists between the historical and contemporary material presented both within

A. Akera (✉)
Department of Science and Technology Studies—Sage 5206,
Rensselaer Polytechnic Institute, 110 8th Street, Troy, NY 12180, USA
e-mail: akeraa@rpi.edu

E.A. Cech
Department of Sociology, Rice University, 6100 Main St, MS28, 77005 Houston, TX, USA
e-mail: ecech@rice.edu

and across this collection of chapters. None of the chapters are simply historical or contemporary in their design. The "historical" chapters, which are built around a broad history of structural transformations in engineering education in a specific country or region, and the stories that they tell, are nevertheless haunted by neoliberalism and the ghost of contemporary global economic transformations. Meanwhile, the two "contemporary" articles, both of which focus on the academic drift that has occurred among occupationally-centered, non-PhD granting engineering schools and degree programs (e.g. the German *Fachhochschule* and U.S. Engineering Technology degree programs) are nevertheless firmly grounded by institutional histories and a deep concern for a changing historical context. To the extent to which the interplay between historical and contemporary issues are not spelled out by individual authors – whether with regard to a Cold War legacy of engineering science and research or how the particular demands of an "innovation economy" shape the author's rendering of history – there are rich opportunities to read across the chapters to arrive at one's own insights and conclusion.

We see other ways in which the historical and contemporary chapters appear in a productive juxtaposition. For example, the "contemporary" studies by Christensen and Newberry (Chap. 2), and Delahousse and Bomke (Chap. 3), who express interest in the unfolding relationship between universities and occupationally-oriented institutions, point to the tendency of the historical chapters to deal, preferentially, with university-level education and engineering curricula. As noted early on by Ken Alder (1997) with regard to the system of engineering education in France, any attempt to analyze a national system of engineering education needs to give balanced treatment to the different facets of technical education rather than dealing exclusively with the level most closely associated with engineering professional identities.

Conversely, the "historical" studies (Chaps. 1, 4, 5, and 6) help shed light on the two contemporary studies by hinting at how the process of academic drift might be a part of a broader historical transformation in engineering education. As noted by Akera and Seely, engineers claim a cultivated responsibility for adapting engineering education to changing times and needs. It is interesting, in this respect, that the authors of the two contemporary studies appear to be ambivalent about the process of academic drift that they describe: on the one hand they remain sympathetic to the older, vocationally-oriented curricula offered by the former engineering colleges and technical institutes, and yet at other points, they appear critical of the neoliberal reforms that are designed to bring greater accountability to educational institutions and bring them into greater alignment with industrial interests. This is not a critique. Rather, we regard the authors as having tapped into a tension endemic to engineering education that merits further study. For instance, that the Indian Institutes of Technology described by Balasundaram (Chap. 5) has embraced – and continues to embrace – a logic of academic drift points to our need to understand how this logic operates in the vastly different parts the world's educational systems.

This may also be a useful segue into the "global" dimension of the chapters in this section. The editors selected this set of chapters in order to describe the "structure" of engineering education with a global scope. As typical of such volumes, the essays are arranged with Euro-American (U.S.) perspectives presented first, followed by developments in India, Brazil, and Latin America. There is a historic logic for this: the IITs, for instance, were a specific attempt to emulate Western models. However, given that our *contemporary* conversations about engineering education are being driven by economic transformations that are occurring outside of the West, it might be equally interesting to read these chapters in a different order, beginning with the chapter on India (Chap. 5) and the material on China to be found elsewhere in this two-volume series; proceeding to the frustrated efforts in Brazil and Latin America; and concluding with the developments that are occurring in Europe and the United States. This might lend additional insights into how recent events such as the Bologna Declaration and ABET's Engineering Criterion 2000, and their practical manifestations, are (or are not) effective responses to the actual changes in engineering education resulting from economic globalization.

There are other ways in which cross-national comparisons can be made. Indeed, several of the chapters already make effective use of comparative study, both within and across regions. Two interesting examples of difference that we can point to include the greater persistence of a still partially separate, occupationally-centered curriculum within the German *Fachhochschulen* in spite of, or perhaps even as a result of, the Bologna Declaration; and the stronger historic articulation of a tripartite system (research universities, undergraduate colleges, and junior colleges) in the United States as opposed to the university-centric model that had been dominant in Europe. Similar comparisons made across the chapters could yield additional insights into the national contexts and institutional change processes that generate and perpetuate differences, despite common global causes and trends.

This last observation brings us to a set of common institutional processes that appear to be at work across the individual chapters of this Part. First among these are institutional isomorphism and academic drift introduced by Christensen and Newberry in Chap. 2 and Delahousse and Bomke in Chap. 3. Drawing on the insights of the New Institutionalists in sociology, Christensen and Newberry explain a mimetic process by which diverse institutional structures within engineering education come to resemble one another by mimicking the practices of the most prestigious institutions. Within academia, isomorphism often takes the form of "academic drift," whereby less-prestigious academic institutions shift their organizational structures to look more like prestigious 4 year institutions. Academic drift, a feature of academe explicitly or implicitly described within the United States (Chap. 2), Western Europe (Chaps. 2 and 3), Eastern Europe (Chap. 6), Brazil (Chap. 4) and India (Chap. 5), largely results in institutional forms that emphasize research and resemble 4 year universities. Several authors point to the "Bologna process," a series of agreements reached by many nations in the European Union in 1999, to ensure comparability and quality standards across each nation's higher education system, as a powerful driver (or accelerator) of academic drift in Europe and beyond.

Second, partly due to this academic drift and partly a result of socio-cultural and demographic changes, the chapters in this Part tell a story of institutional change in engineering education that promotes social mobility. Academic drift, combined with the "engineering [person] power crisis" of the 1960s and 1970s, opened up 4-

and 5-year degrees – and the salaries and professional status that accompanies them – to working-class students who might otherwise have obtained lower-status vocational degrees. Although the efficaciousness of this social mobility to undermine broad social inequality may be overstated, Chaps. 1, 2, 3 and 6 clearly illustrate how these globalized changes in engineering education opened up new pathways to the middle-class.

The third process that comes to light in this section is the tension between the global and the local in these shifts in engineering education. Such tensions, best illuminated by the chapters on Brazil (Silva et al., Chap. 4), India (Subramanian, Chap. 5) and Slavic-language countries (Kostyszak et al., Chap. 6), exist around the structure and content of engineering education – the tension between promoting local engineering knowledge, directly applicable to regional or national needs (e.g. what Silva et al. call "Brazilian Engineering"), versus globalizing pressures such as academic drift and market competition. While engineering has long been understood by scholars as a tool of economic development and national identity projects, as Akera and Seely discuss in Chap. 1, recent institutional changes have moved engineering away from local concerns toward more global concerns.

Since full abstracts accompany each chapter, we do not provide extensive summaries, but merely list the chapters in this abbreviated form: Chap. 1, by Akera and Seely, is a historical article on the structural transformation of the U.S. system of engineering education, with a focus on its origins, its transformation during the Cold War, and the subsequent changes associated with liberalism during the 1970s and then the neoliberal turn during the 1980s and beyond; Chap. 2, by Christensen and Newberry focus on the process of academic drift and the disappearing middle-tier, both in Europe and the United States. In Chap. 3, Delahousse and Bomke build on Christensen and Newberry's insights by providing a more detailed, historically grounded analysis of academic drift in the IoT in France and FHS in Germany, with specific attention to associated changes in institutional autonomy, curricula, pedagogy, the student body, and faculty research. In Chap. 4, Silva, Bartholo and Proença historically examine the case of Brazil's struggle to develop "Brazilian Engineering," or a more localized knowledge and application of engineering techniques, out from under the dominance of colonial "engineering in Brazil" and the threatening of this more localized approach by globalizing trends. Taking a similar single-country historical approach, Subramanian (Chap. 5) describes the emergence and reform of engineering education in India, tracing similar shifts from more localized, diversified approaches to engineering education to more standardized, research-oriented arrangements. Finally, Chap. 6 explores the shift in engineering education within formerly communist contexts in Slavic-language countries.

In closing, we wish to take note of what the chapters in this section makes evident, namely that engineering education continues to unfold within a rather complex institutional ecology (Star [ed.], 1995; Akera 2007). This is clearly an ecology of knowledge characterized by national differences in the relationship between engineering education, national workforce needs, cultural values, and the state; the specific mix of public and private institutions that itself is influenced by economic context and the growth rate of the economy; regional and national differences in industrial capacity, and placement within the global economy; and the long shadow cast by Cold War engineering science ideology and its redeployment within the new "innovation economy." The process of academic drift that is central to two of the chapters and implied in the rest points to tensions between research and education that remain endemic to engineering education institutions. How this is being resolved in different ways in different national contexts, points to the diverse structure of engineering education that persists around the world. Thus, although no single picture emerges out of these chapters, we invite readers to read across the chapters, as well as across the sections in order to yield new insights about current and past transformations in engineering education.

References

Akera, A. (2007). Constructing a representation for an ecology of knowledge: Methodological advances in the integration of knowledge and its social context. *Social Studies of Science, 37*(3), 413–441.

Alder, K. (1997). *Engineering the revolution: Arms and enlightenment in France, 1763–1815.* Princeton: Princeton University Press.

Star, S. L. (Ed.). (1995). *Ecologies of knowledge: Work and politics in science and technology.* Albany: State University of New York Press.

Atsushi Akera M.A. and Ph.D. in the History and Sociology of Science, University of Pennsylvania. Associate Professor, Department of Science and Technology Studies, and Director, First Year Studies at Rensselaer Polytechnic Institute, Troy, New York. As Director of First Year Studies he has been active in the development of new pedagogic strategies for engineering education. Associate Editor of *Engineering Studies,* officer for the Liberal Education/Engineering & Society Division, American Society for Engineering Education; and elected member of the Executive Council to the Society for the History of Technology. His current research, with co-author Bruce Seely, is on the history of engineering education reform in the United States (1945-present). Publications include *Calculating a Natural World: Scientists, Engineers and Computers during the Rise of U.S. Cold War Research* (MIT Press, 2006).

Erin A. Cech B.S. in Electrical Engineering and B.S. in Sociology from Montana State University-Bozeman; M.A. and Ph.D. in Sociology from the University of California, San Diego. Assistant Professor in the Department of Sociology at Rice University. Cech's research seeks to uncover cultural mechanisms of inequality reproduction – particularly gender, sexual identity and racial/ethnic inequality within science and engineering professions. Her research has appeared in the *American Sociological Review, Social Problems,* and *Engineering Studies.*

Chapter 1
A Historical Survey of the Structural Changes in the American System of Engineering Education

Atsushi Akera and Bruce Seely

Abstract This chapter provides a historical overview of the U.S. system of engineering education from its origins in the nineteenth Century until the present. It is organized chronologically, describing the early institutional formation of the U.S. system of engineering education; the post World War II ascent of engineering science ideology; late- and post Cold War changes in engineering education. As a broad brush stroke history, this text does not attempt to be comprehensive, nor does it touch on every major historical development. Instead, the chapter adopts a more analytic view of the structural features of the U.S. system of engineering education and it transformation over time. The primary intent of the chapter is to provide background historical knowledge for the other chapters in his volume, but it also closes with several observations of broader interest.

Keywords Engineering education • U.S. • Educational reform • Educational structure • Engineering science • Theory and practice • Professional structure • General education • American Society for Engineering Education (ASEE) • Ethnomethodology

Introduction

This chapter provides a historical overview of the U.S. system of engineering education from its origins in the nineteenth Century to the present. Written in broad strokes, this text is not comprehensive, nor does it even touch upon every major

A. Akera (✉)
Department of Science and Technology Studies – Sage 5206,
Rensselaer Polytechnic Institute, 110 8th Street, Troy, NY 12180, USA
e-mail: akeraa@rpi.edu

B. Seely
College of Sciences and Arts, Michigan Technological University,
1400 Townsend Drive, Houghton, MI 49931-1295, USA
e-mail: bseely@mtu.edu

S.H. Christensen et al. (eds.), *International Perspectives on Engineering Education*,
Philosophy of Engineering and Technology 20, DOI 10.1007/978-3-319-16169-3_1

development. Rather, this chapter dons a more analytic lens, focusing on the structural features of the U.S. system of engineering education and its transformation over time. Even then, many important aspects of engineering education as practiced in the United States will be bracketed out of this account. Some are addressed elsewhere in this edited compilation, while for yet others we merely refer to appropriate literatures.

Institutional Formation of the U.S. System of Engineering Education

Most historians consider the U.S. system of engineering education to be a hybrid that combined elements from the British and European continental traditions for engineering education. As summarized by Terry Reynolds in his review of nineteenth Century developments in American engineering, U.S. approaches to formal engineering education varied from the mathematical approaches established at the U.S. Military Academy at West Point to the hands-on, apprentice-based tradition adopted at schools such as Worchester Polytechnic Institute (Reynolds 1991, pp. 16–23). It is perhaps more important to realize that this variation in U.S. approaches resulted not from the eclectic vision of each institution's founder, but from a unique institutional ecology of American institutions and cities during the mid-nineteenth Century that produced competing visions about the knowledge and labor requirements of a young and growing nation. Most notable was the distinct role that science and Enlightenment ideals played within the cultural imaginary of the American republican experiment. While notions of American exceptionalism have been discredited as an accurate description of the U.S. historical condition, at least through entire nineteenth century republican rhetoric and ideology held substantial sway in national politics, as well as among state and philanthropic institutions that invested substantially in technical training and education. At the same time, a theory of manufactures, and republican notions about artisanal labor and the proper development of a new working class, gave rise to a set of early technical schools. From the point of view of many states governments, however, the dominant concern remained that of promoting scientific approaches to agriculture. Meanwhile, at military schools, fortifications, military ordnance, and geography, and the underlying mathematics behind them, played a greater part in the imperial ambitions of a new nation. As the century wore on, the rise of substantial urban centers along the Eastern seaboard created the accumulations of capital and a demand for an augmented, skilled labor force that produced institutions more closely modeled after British mechanic's institutes (Sinclair 1974; Lerman 1997).

The early history of Rensselaer Polytechnic Institute (1824), among the earliest of civilian school to offer engineering education in the United States, illustrates the changing demand for engineering knowledge in the United States during the first half of the nineteenth Century. Originally the Rensselaer School, Rensselaer was established to apply science to the problems of agriculture and "the common purposes of life," but soon introduced a curriculum in civil engineering after the completion of the Erie Canal launched a growing demand for surveying and civil works that accompanied Western expansion (Phelan et al. 1995). Rensselaer was fairly late in adopting a mechanical engineering curriculum, but the Stevens Institute of Technology was set up in 1871 specifically to set up a more formal curriculum in this field in direct response to the growth in U.S. manufacturing capabilities (Reynolds 1991, p. 21; Calvert 1967).

Ken Alder, among others, has made the point that the actual approach to engineering education in France, and by extension, other European countries, were more diverse than has often appeared in many historical studies, especially once we look beyond the most elite institutions (Alder 1997; also Kranakis 1997). This is not to say that politics and ideology could not skew the allocation of resources or define the structural patterns of engineering education as practiced within different countries. More to the point, each nation's approach to engineering education matched the particular institutional ecology of each country and the specific emphasis they placed on engineering knowledge and skills relevant to their military, civic, commercial and industrial enterprises. But if the U.S. case was not exceptional in this respect, early patterns of engineering education within the United States matched the relative immaturity of the American economic system, especially where republican rhetoric was invoked. In such settings, an individual founder's vision could indeed produce original and usually eclectic institutional formulations, as reflected in the diverse array of early U.S. engineering schools that included Michigan, Harvard, Yale, Union College, and Dartmouth (Reynolds 1991, p. 20).

The seminal event that brought greater uniformity to the U.S. system of engineering education was the 1862 Morrill Act, which transferred title to federal lands to the states to support colleges "to teach such branches of learning as are related to agriculture and the mechanic arts, in such manner as the legislatures of the States may respectively prescribe, in order to promote the liberal and practical education of the industrial classes in the several pursuits and professions in life."[1] While the political and ideological origins of the land-grant act also merits discussion (Williams 1991; also Eddy 1956, pp. 1–46; Geiger 2000), from the standpoint of the professional and disciplinary formation of engineering, a more novel insight can be gleaned from how this act reshaped engineering in relation to other professions. The resources and political impetus created by the Morrill Act led to the rapid expansion of engineering professional training through an expanded system of state colleges. While private engineering schools and technical institutes were not eliminated by this move – indeed, they continued to proliferate under private philanthropy into the early part of the twentieth century – the substantial and continuous state financing of the state colleges allowed engineering curricula finally to develop several common features, most notably a 4-year undergraduate degree program that varied in form from traditional liberal arts degrees. Thus engineering professional training during the latter part of the nineteenth century became both more uniform and rigor-

[1] The original text of the Morrill Act is available at this site: http://www.loc.gov/rr/program/bib/ourdocs/ Morrill.html. Accessed Aug. 2012.

ous compared to many other professional education endeavors, despite the delivery of engineering education by a very eclectic array of proprietary colleges and universities. Engineering also remained a broadly accessible profession in that professional standing could be achieved upon completion of an undergraduate degree.[2]

Yet this general pattern also was flexible enough – or lacked the rigidity to prevent – significant variation. From the 1870s onwards, both new engineering schools – and even more importantly newer and more specialized degree programs – proliferated. These developments paralleled the rise of new industries and of major engineering professional societies. Thus a significant degree of regional difference and specialization remained among engineering schools. Strong degree programs in mechanical engineering, beginning with the Stevens Institute of Technology and confirmed by the school of mechanical engineering at Cornell proliferated in the East and the mid-west, even as schools with a strong emphasis in mining and metallurgy, and agriculture and agricultural engineering expanded in the Midwest and West. Many engineering schools – both private and public – aligned themselves to the specific needs of regional industries, ranging from an interest in metals, mining, and manufacturing at the Carnegie Institute of Technology, to the chemical engineering needs of the pulp and paper industry at the University of Maine, Orono. Other subtle variations in agricultural and manufacturing interests emerged at the other land-grant institutions.

These regional differences contributed to the disciplinary fragmentation of engineering, yet as noted above, curricula also began to converge. A key to this countervailing trend within the nation's engineering schools was the development of a primary occupational identification among engineering faculty. Unlike most other professional schools, strong state patronage enabled large numbers of engineering professors to secure fulltime academic positions. Their occupational identification as educators came in addition to, and remained distinct, from, their disciplinary professional identities as, for example, civil or mechanical engineers.

Largely for this reason, in 1893 engineering educators convened at a special section of the Engineering Congress organized in conjunction with the World's Columbian Exposition in Chicago. This meeting produced, in turn, an independent Society for the Promotion of Engineering Education (SPEE). While the early goal of this Society centered on providing a collegial forum for the exchange of knowledge and ideas about engineering teaching and curricula, by the early twentieth century the leading members of SPEE actively worked to bring even greater standardization to engineering curricula.

Against this backdrop of curricular and professional convergence despite continued institutional variation and pressure for disciplinary specialization, perhaps the earliest systematic review of American professional education carried out by Carnegie Foundation for the Advancement of Teaching during early 1900s. The famous, or infamous Flexner Report (1910), which is widely regarded as having catalyzed a major transformation in medical education was only the most widely

[2] Illustrative in this respect is a comparison of the Mann Report (1918) and Flexner Report (1910), as noted below.

known report – Bulletin #4 – in a series of studies produced by the Foundation. Bulletin #11 (1918), by the University of Chicago applied scientist, Charles Riborg Mann, examined engineering education.[3] While Mann's report clearly reflected a professional bias towards his chosen field of applied science, it nevertheless accurately noted the existence of a number of important trends and patterns: substantial curricular standardization across educational institutions; the profession's commitment to a liberal-professional program of study where practical and professional training occurred alongside basic scientific education; and the introduction of laboratory methods of instruction in science into the education of engineers.

As noted by Terry Reynolds and Bruce Seely in their institutional history of SPEE (American Society for Engineering Education [ASEE] since 1946), the Mann Report opened a period of roughly five decades during which this organization served as "the voice of engineering education" (Reynolds and Seely 1993). SPEE learned a great deal about how to structure and conduct an investigation through its affiliation with the Carnegie Foundation. SPEE came to adopt a grand investigative tradition in which the organization at intervals convened a board of investigation, hired an independent director of investigation, and assembled one or more investigative teams. Drawing on broader Progressive Era educational reform practices, they convened and met with defined interests groups, adopted fieldwork practices, and most importantly proceeded to survey best practices that could be projected to the membership as a way of establishing a regime of accountability. Beginning with the Mann Report, SPEE or ASEE produced four major and two lesser studies on engineering education between the 1920s and 1960s. Lesser variants of this investigative tradition were practiced by specific ASEE Divisions, as seen by the three studies produced by its Humanistic-Social (later Liberal Studies) Division.[4]

It is significant that each of these reform efforts sought to realign the epistemological basis of engineering to match "changing times and needs," namely the changing social and economic context within which engineers advanced their knowledge claims and professional identities. Through World War I, the main emphasis was rooted in a rhetoric of applied science, as befitting the ascendant stature of the sciences within American society (Kline 1995). During the 1920s, engineering educators also began placing greater emphasis on business, personnel management, and engineering economics in a classic jurisdictional response to the rapid ascent of a new managerial profession. The Great Depression brought about a shift in emphasis away from the narrow instrumental skills of the 1920s towards more fundamental knowledge in economics and the social sciences necessary for upholding a broader vision of professional responsibility. The educational reforms during World War II were less about any basic curricular change as opposed to the administrative challenges of accelerating degree programs and producing specialized war training programs. However, both the exhilaration and horrors of wartime

[3]The so-called Mann Report originated with a conversation that began within SPEE in 1907, and the Society contributed both content and guidance to Mann (Grayson 1993).

[4]All of these studies were published or summarized in the society's *Journal of Engineering Education*. Most are specifically cited in the text below.

technological accomplishments, and the further ascent of science and scientists during the war, prompted more significant responses and additional ASEE studies during the postwar era, as will be discussed further below (Akera 2012a; Seely 1995, pp. 742–749).

However, some additional observations are needed before proceeding to these postwar developments. The first is to note that whatever SPEE's role as the "voice" of engineering education, actual practice and curricula in engineering education often veered significantly away from the stated ideal. While the SPEE investigations produced articulations about what would be ideal in terms of advancing engineering as a profession, each engineering school remained accountable to regional interests and the disciplinary interests of academic departments and faculties at a time when many degree program operated as semi-autonomous schools within an engineering college or a university. Private consulting and the production of industrially-relevant knowledge remained more important to the reward structure for faculty, resulting in greater emphasis on specialization as opposed to general education. Rampant specialization was in fact the backdrop against which leading engineering educators – increasingly engineering deans and senior administrators who began to hold disproportionate influence within SPEE – expressed their desire for a broader and more fundamental curriculum for engineers.

This did not mean that pockets of science-based approaches to engineering could not be found in the U.S. during the interwar years. While industry affiliation and obligations provided one set of institutional ecologies for engineering education, the continued growth of an applied science ideology, the ascendant reputation of European scientific institutions, and the emergence of scientifically-oriented industrial and applied science laboratories in the United States as well as Germany, prompted selected faculty members and schools to focus on more scientific approaches to engineering. As pointed out by W. Bernard Carlson (1988), educational institutions could pursue objectives that were distinct from the needs of the regional industries. This could be true with regards to an industrial internship program, as described by Carlson (1988), but could also involve a more direct attempt to align a department or a major laboratory within an engineering school along "European" lines, often as a result of the arrival of émigré engineers with a stronger scientific orientation and background. There was, moreover, often a technical basis for this turn towards applied science, whether with respect to heat transfer in the mechanical industries; electrical circuits and phenomena for the electrical industries; physical chemistry as relevant not only to the chemical industries, but electrical firms such as General Electric; and aerodynamics, structural mechanics, and other fields are relevant to a nascent aviation industry. The large-scale civil works projects of the 1920s and 1930s, as represented by the Hoover Dam, created other opportunities for science-based engineering. From Robert Thurston's work on thermodynamics at Cornell, to Von Karman's work on aerodynamics at Cal Tech, and institutional-level decisions to hire someone like Stephan Timoshenko (Michigan and Stanford), Robert E. Doherty (Carnegie Tech) and Solomon Cady Hollister (Cornell University), all this represented early attempts to establish more science-based engineering programs within United States (Seely 1993, 1999a, b).

Still, it is important to note that at no institution, except perhaps Cal Tech, were these transformations substantially complete before World War II. Events at MIT highlight this point. MIT's trustees decided to hire physicist Karl Compton, as its 12th president in 1930 in order to claim a more fundamental and science-based approach to engineering. Compton created a School of Science at MIT, and successful built up its physics and math departments to become senior academic entities (Lecuyer 1993; Leslie 1993). Several of MIT's engineering departments, notably the aeronautical, chemical, and mechanical engineering departments, followed Compton's lead and reoriented their curricula to include a stronger scientific foundation. The most notable changes occurred in the Electrical Engineering, where Vannevar Bush and his colleagues introduced a rigorous, math-based curriculum, expanded their graduate program, and introduced new laboratory practices centered on scientific and mathematical instruments designed to analyze and model electrical phenomena (Wildes and Lindgren 1985, pp. 82–95; Mindell 2002). Even so, other engineering departments at MIT, including civil engineering and naval architecture, saw only partial changes. Developments were even more uneven at most other engineering colleges, as isolated faculty – either born or educated in Europe introduced mathematically-rigorous and analytical approaches to engineering problem solving.

The Postwar Ascent of Engineering Science

If science-based approaches to engineering represented an ideal during the interwar period, World War II provided the compelling rationale for making it the dominant approach during the Cold War years. The mobilization of "science" during World War II is an often told story, although certain misperceptions persist that are worth correcting. As historian David Mindell (2002) observed, despite the unprecedented efforts to harness science to the war, most of the wartime mobilization of the nation's technical capacity, including its engineering schools, occurred within well-worn patterns. Engineering schools across the nation developed accelerated and specialized war training programs, even as individual laboratories and faculty members were drawn directly into the war effort through their affiliations with regional industries and military laboratories. The U.S. science mobilization differed from these other initiatives in that it established new scientific laboratories and development facilities that remained, by design, outside the already established channels for administration and direction.

In other words, the most significant aspect of the interwar changes at MIT was not the teleological reading of an institution that already practiced science-based engineering before the Cold War era, but the more direct historical fact that these changes placed this particular engineering institution in the direct path of the science mobilization effort. Central to this happenstance was the entrepreneurial energies of Vannevar Bush, who after becoming Vice President of MIT, the President of the Carnegie Institution of Washington, and the chair of the National Advisory

Committee for Aeronautics, became the chief architect of the U.S. science mobilization effort (Kevles 1977, pp. 294–301). While MIT was clearly not the only academic institution drawn into the civilian science mobilization effort – Harvard, Cal Tech, Johns Hopkins, and Chicago, along with Bell Labs, were among the other major academic or quasi academic laboratories connected to the National Defense Research Committee (NDRC) and its parent, the Office of Scientific Research and Development (OSRD) by large contracts – MIT nevertheless garnered a lion's share of wartime civilian military research expenditures. The primary project was the Radiation Laboratory, the central laboratory for the wartime work on radar (Leslie 1993). The success of the science mobilization effort profoundly affected how faculty and administrators within U.S. engineering schools subsequently positioned their knowledge claims. The accelerated war training programs also had an important effect on the nation's engineering colleges, insofar as they gave engineering deans the experience of orchestrating major curricular changes through top-down administrative guidance.

During the period from the end of World War II until the end of the 1960s, "engineering science" and, more generally, a science-and mathematically-based approach to engineering and engineering research became ensconced within most U.S. engineering schools. The phrase, "engineering science," is probably best understood as a rhetorical construct, the etymology of which can be traced to the desire since the latter part of the nineteenth century of many professions to root their knowledge claims on a scientific footing. Nevertheless, the term "engineering science" came to encapsulate an identifiable reform movement during the 1950s and 1960s. Many contemporaries felt the phrase lacked precise meaning (a particularly irksome issue for many engineers educators), but it nevertheless linked a loosely coordinated coalition that embraced several components including curricular changes introducing more science-based, analytic subjects in the place of skills-based and hands-on, experiential modes of learning (such as machine shop, engineering drawing, and survey camps); expectation faculty should possess a doctorate; the related faculty commitment to research and graduate education, along with the associated rhetorical claims about a necessary synergy between research and the quality of undergraduate instruction; and the associated changes in engineering professional identity that resulted from the adoption of distinct scientific methods including greater reliance on analytic methods, apparatus and facilities.[5]

Before describing this transition, it is important to recognize that a handful of institutions that pursued other directions in engineering education during the immediate postwar years. Indeed, in 1950 a number of leaders in engineering education, including those who saw the value of engineering science, nonetheless did not envision completely transforming all undergraduate educational experiences into engineering science programs. Many also assumed (at first) that the focus upon graduate

[5] For one commentary on engineering science, see Ferguson (1992), pp. 160–161. But the concept also attracted the attention of historians of technology, most notably Edwin Layton ad David Channel, who in the 1960s and 1970s were exploring the nature and historical development of the engineering profession, as distinct from science (Layton 1971, 1976; Channell 1989).

education and research would best fit a handful of large universities. Thus variation in educational goals remained apparent, as at the Carnegie Institute of Technology under Robert E. Doherty, and the Case Institute of Technology under William E. Wickenden and his successor, T. Keith Glennan. During the postwar period, both institutions pursued a liberal-professional vision for engineering education that they had already embraced prior to World War II. Both engineering programs focused significant attention upon the humanities and social sciences components of engineering curricula; both remained focused on their undergraduate program. It is significant that Carnegie Tech and Case were private institutions which, after World War II, experienced specific pressures to distinguish their graduates from those of rapidly expanding state college programs, bolstered by new state and federal commitments to education, including the G.I. Bill. The fact that both schools turned their attention away from this approach and firmly embraced engineering science and sponsored research by 1960 indicates the compelling effect that the paradigm, backed by unprecedented levels of federal research spending, had on engineering institutions (Akera 2010).

A full description of the Cold War transformation of the U.S. system of engineering education would require a longer and more wide-ranging account – what anthropologists would refer to as a multi-site, multi-scale analysis – in order to explore the continued complexity of the ever-changing institutional ecology for engineering education in the United States. But to provide a sense of the dynamic process of change, we focus here upon three major developments and two additional institutions that at least initially existed on the periphery. What follows will be an account of the changes at MIT beginning with the 1949 Lewis Survey and tracing its implementation into the early 1960s; two key ASEE studies published in 1955 and 1968 and associated developments within the U.S. engineering accreditation organization, the Engineers' Council for Professional Development (ECPD); parallel developments in California that culminated in the *Master Plan for Higher Education in California* (1960). The section concludes with a brief review of events at the University of Texas and the University of Maine.

Not surprisingly, MIT was widely associated with the turn to engineering science, although Frederick Terman, dean of engineering and then provost at Stanford was as influential in shaping the course of postwar reform in engineering education. In 1947, the MIT faculty, which operated as a single body, convened a five-member Committee on Educational Survey and charged it to produce a future vision for the Institute. Usually known as the Lewis Survey (1949), after the committee's chair, Warren K. Lewis, succeeding generations of MIT faculty and administrators considered the report the Cold War blueprint for MIT. As an indication of the seriousness with which these five faculty members approached this task, the committee met no less than 118 times between 1947 and 1949. Less obviously, the Lewis committee also placed heavy emphasis on MIT's undergraduate program, for into the postwar years most MIT faculty still regarded MIT primarily as an educational institution devoted to the production of engineers and other graduates, not new knowledge.

The MIT report also bore the clear imprint of the interwar dialogue within SPEE – MIT President Karl Compton was himself the President of SPEE back in

1930. Thus, the primary conclusions of the Lewis Survey were that MIT's undergraduate degree programs had to place greater emphasis on science and fundamentals; and that MIT, as a private institution offering a distinctive education needed to embrace a broader professional vision based on further increases in, and a more coherent and integrated program of humanistic and social scientific study. While presented as an original vision of the MIT faculty, this position followed, very closely, the recommendations of two lesser reports of the SPEE produced during the 1940s. The latter of these, published in 1944, provided specific guidance for the postwar reconversion of engineering schools, including MIT, that had been forced to adopt an instrumental (i.e., very limited) approach to engineering training during the war (SPEE 1940; Report of the Committee 1944).

Clearly, the Lewis committee could not ignore the potential effects of sponsored research, given the wartime program at MIT. But here, the continued assumption that MIT was foremost an educational institution still reigned dominant. Rooted as well in the hard fought commitment to academic freedom, Lewis' committee proceeded to define federal research expenditures as a matter of graduate education and research, and therefore relegated the problem to a standing committee on graduate education. This separation allowed both committees to regard the possible challenges and hazards of sponsored research as something more imagined than real; they came to view sponsored research as an invaluable opportunity for creating interesting research opportunities for graduate students. But given the volume of federal funds MIT had received, it was not surprising that the committee members came to see those funds as the "lifeblood" that would strengthen the Institute. In other words, little consideration was given to whether federally-funded research might take MIT in a direction that was at odds with the broader ideals of higher education (Lewis Survey 1949, pp. 49–64).[6]

The key historical significance of the Lewis Committee's report was that it created a strong consensus among the MIT faculty about the future direction of the Institute. The administration, beginning with Compton, was a definite partner to this transformation and, in fact, to the Lewis Report itself (Akera 2012b). Yet even if the Report provided a postwar blueprint for MIT, not all departments immediately subscribed to the ideals of engineering science that came to be a necessary – and sometimes contradictory – to the Lewis committee's vision. The extent to which a science-based approach to engineering was embraced varied from department to department. Only during the early 1960s was there a concerted effort, led by Dean of Engineering Gordon S. Brown, to install engineering science as the educational doctrine for MIT's School of Engineering. Even then, only the support of the Ford

[6]The seductive power of federal research funds was best shown at Georgia Tech, which also saw significant volumes of federal research by the early 1950s. When a faculty member pointedly asked at a faculty meeting whether Georgia Tech was accepting "tainted" money, the administrator responded that the only taint that mattered was that "there t'ain't enough of it"(McMath et al. 1985, esp. pp. 212–217, 256–270; personal correspondence with August Giebelhaus).

Foundation and subtle pressure from a physics-dominated administration, managed to push all of MIT's engineering departments in this direction.[7]

While many schools looked to MIT (and later to Stanford and to the University of California) as the model institution of the Cold War era, it took several rounds of conversation within ASEE and the engineering accreditation organization, ECPD, to fully establish science based engineering as the dominant trope among U.S. research universities (Leslie 1987; Seely 1993; Reynolds and Seely 1993). For those familiar with the broad outlines of the history of engineering education in the U.S. context, the principal efforts of the ASEE during this period will be quite familiar. The focal points were the Grinter Report (1955), often cited for encouraging the introduction of engineering science into U.S. engineering schools, and the ASEE Goals Report (1968), which recommended that a master's degree (indeed, an undesignated master's degree) be the first professional degree in engineering, a conclusion that many dismissed as ahead of its time.

A more nuanced analysis of these two reports is needed to appreciate the historical significance of these two studies. An important entry point to this understanding is the relationship between ASEE and the Engineers' Council for Professional Development (ECPD). ECPD, and more specifically, its Committee on Education served as the principal accrediting body for U.S. engineering schools. Established in 1937, ECPD began as a relatively weak organization that adopted a qualitative, peer-based system of accreditation designed to uphold the regional variation in the quality and emphasis of U.S. engineering schools. It also accredited individual degree programs, as opposed to entire schools, in a move initially designed to protect disciplinary interests. However the decision to accredit degree programs resulted in an unintended consequence, namely the proliferation of new engineering degree programs such as engineering management, engineering physics, and even sales engineering. These developments, as well as the appearance of the engineering technology degree in the years after 1945 were sometimes uncomfortable for an organization dedicated to the "professional development" of engineers (Akera n.d.; Reynolds and Seely 1993).

The postwar shift toward engineering science in engineering curricula provided ECPD with an opportunity to promote a tighter professional identity of engineers as "scientific." Led by Cornell's Dean of Engineering Solomon Cady Hollister, who struggled with these issues at his own institution, ECPD proposed the establishment of new standards for engineering degree programs. However, in 1951, when this process was put in motion, ECPD still recognized ASEE to be the primary body responsible for defining engineering curricula. As a consequence ECPD handed off the task of defining a new curricular standard to ASEE, and engineering educators set out upon yet another major investigation of engineering education. The major recommendations of the Committee for the Evaluation of Engineering Education, chaired by the University of Florida's Graduate School Dean, Linton E. Grinter,

[7] The history of MIT's engineering degree programs has not received full treatment but some of the struggle over engineering science is described in Wildes and Lindgren (1985, pp. 310–319). The relevant primary sources may be found primarily in AC 12 College of Engineering. MIT Institute Archives and Special Collection, MIT Libraries, Cambridge, Mass.

squarely addressed the rise of the engineering science ideology. Its initial proposal was to bifurcate U.S. engineering degree programs into "professional-general" and "professional-scientific" accreditation, with the former being more common; in addition the committee proposed to provide special recognition to academic departments with a strong research faculty (ASEE 1953). These recommendations were withdrawn amidst a storm of protest from those schools apparently separated from the research emphasis of the professional-scientific category. Still, in the end an underlying commitment to a more science-based curriculum held sway within the engineering education community (Hollister 1979, pp. 194–196; Reynolds and Seely 1993).

The greatest historical significance of the Grinter Report lay not in the report itself, but in the subsequent changes that were made in ECPD accreditation policy. In the second of the two wartime reports produced by SPEE, a study committee recommended applying a quantitative standard to the humanistic and social scientific portion of the curriculum (Report of the Committee 1944). They proposed that that no less than 20 % of the curriculum be devoted to the "humanistic-social" stem, finding no other way to ensure that academic departments would uphold the society's vision for more robust liberal-professional training in the face of pressure to devote additional classes to specialized technical and scientific subjects. During an early conversation with Grinter's committee, ECPD officials indicated they were willing to consider counting courses in all areas of the curricula as a strategy for enforcing standards for engineering education. This was indeed the direction Grinter's committee took. The ECPD then discussed the recommendations they had helped produce and approved curricular requirements for one full-year of the basic sciences, one full-year in engineering sciences, one full-year in humanities and social sciences, and a half-year spent on synthesis and design. From 1956 forward, no U.S. engineering degree program could present a curriculum with more than a year of training within a specific engineering discipline without risking a challenge during the accreditation process (Akera n.d.; 2008).

As might be expected, engineering schools began to challenge the new ECPD standards in short order, complaining that the rule would guarantee that students learned everything *except* engineering. This concern prompted ASEE to renew discussions of the issue as early as 1961. These debates led to another large-scale investigation, directed this time by the former Dean of Engineering and President of Penn State University, Eric A. Walker and the committee's report in 1968 – *The Goals of Engineering Education.* Without the complete support of members of his own investigative teams, Walker came to the controversial conclusion (which he considered logical and unassailable) that it was no longer possible to provide the scientific and analytical foundation in science and in engineering, broad liberal training, specialized disciplinary and sub-disciplinary knowledge, a feel for engineering practice, and a capacity for engineering design within the confines of a 4-year undergraduate program. His conclusions were also backed by linear projections that showed that the need for a graduate education at the master's level would replace an undergraduate credential within a matter of decades (ASEE Goals Report 1968; Seely 1993).

Walker attempted to soften the blow by casting his report as a future projection for the profession (hence the name, the "goals" report), but his recommendations produced strong resistance and created fissures within ASEE that the organization's position as the voice of American engineering education was threatened and diminished (Reynolds and Seely 1993). In the wake of the Goals Report, ECPD (renamed the Accreditation Board for Engineering and Technology–ABET in 1980) and later the National Science Foundation and the National Academic of Engineering stepped up to help define U.S. engineering curricular standards. This was to complete a transition that really had already been put into motion through S.C. Hollister's maneuvers at the time of the Grinter Report.

Another perspective on the development in engineering education during the Cold War era can be gained by reviewing the engineering origins of the 1960 *Master Plan for Higher Education in California*. As described by historian of education John Aubrey Douglass (2000), the California Master Plan was a seminal document that served to affirm and extend California's tri-partite system of higher education. It firmly established a tiered system of junior colleges, state colleges, and the University of California system. The main impetus for the Master Plan was California's postwar commitment to "democratize" higher education, namely to provide much broader access to higher education; the impending wave of Baby Boomers, augmented by internal migration into California, which produced quite astounding projections for higher education enrollments in the state by the 1970s; and the need for the State of California to retain fiscal control over the costs of higher education. The resulting system, politically orchestrated by the UC President Clark Kerr, concentrated state allocations for research in a way that remade the University of California into the most powerful and envied university system in the world.

But for reasons that can only be briefly outlined here, this political dialogue also was driven by the growing need for an engineering workforce, especially in California. At the national level, the number of undergraduate engineering degrees (bachelor's) conferred annually in the United States rose from 11,358 in 1940 to 37,808 in 1960.[8] The underdevelopment of engineering degree programs in California during the prewar years, paired with the booming defense industries that transformed the state's economic base during the postwar period ensured that enrollment growth in engineering was even more pronounced in California. An important agent for this postwar expansion was the novel, unified engineering program established at UCLA by its inaugural Dean of Engineering, Lewellyn M.K. Boelter. Boelter had embraced the vision of a more fundamental, unified undergraduate engineering curriculum when he created UCLA's first College – and Department – of Engineering in 1944. However, the postwar explosion of Southern California's

[8] Enrollment data from Story and Armsby (1951), 4; and Landis (1981), 784. There is some inconsistency in the two data sets, but the difference suggests simply that the growth may have been even more pronounced than reported here. While comparable figures for California are not available for the entire time period, undergraduate engineering enrollments in the University of California system rose from 2,606 in 1950 to 3,183 in 1960, which underrepresents the total change because of the growth of engineering degree programs in the California State Colleges, as described below. Story and Armsby (1951), p. 6 and Tolliver and Armsby (1961), p. 472.

aviation industry, and the emphasis the industry place on new research and development capabilities, produced some incredible twists in Boelter's instructional program. Much of the enrollment growth in UCLA Engineering occurred within a new continuing education program serving the region's expanding military economy by offering master's degrees. As an indication of the opportunistic dynamic that drove this expansion, a majority of these evening extension courses took place off-campus at venues provided by defense contractors and taught by corporation employees hired as adjunct instructors to the university. These courses were open to employees from other firms, demonstrating the unusual spirit of cooperation, networking, and labor mobility that already characterized this region prior to the rise of Silicon Valley to the north.[9]

However, Boelter's efforts are merely illustrative of the opportunism that ran rampant through California's higher education institutions during the early postwar era. Thus, of equal note was the emergence of nearly a dozen new engineering degree programs not only at other UC campuses, but within the California State Colleges, all responsive to distinct regional rationales for expanded engineering workforces. The growth of the state college system drove the political conversations leading to the 1960 Master Plan. While engineering was only one of the new baccalaureate programs launched by the state colleges based on a key 1947 state legislative decision, the rapid and unregulated expansion of engineering programs produced a series of crises and prompted policy articulations that eventually led to the Master Plan. For example, the 1948 Strayer Report pointed to the difficulties related to reserving pre-professional, "occupational" training in engineering to the state colleges while still permitting four-year "professional" training in engineering for the UC system. And yet a 1953 Engineering Agreement which was forged during the furor surrounding the Grinter Report, affirmed a dual system of accreditation in which state colleges offered practical and occupationally-oriented engineering degree programs while the University of California was given responsibility for training more scientifically-oriented graduates. A 1958 revision of the 1953 agreement then stemmed from the state colleges' desire to emulate Boelter's program of continuing engineering education at the master's level. This push came from San Jose State College, which was connected to the early development of Silicon Valley (Douglass 2000, pp. 170–235, 252–255).[10]

The California Master Plan addressed these and other difficulties by restricting admissions to the UC system and the California State College system (top 12.5 % and 33 % of California high school graduates, respectively); essentially limiting

---

[9] Some aspects of Boelter's program are described in Wisnioski (2009). See also UCLA (1995). Other relevant sources are in the records of the College of Engineering (RS 38 and RS 52) located at the University Archives, UCLA Library, Los Angeles, CA; and the records of the College of Engineering (CU39) and UC President's Office (CU5) located at Bancroft Library, University of California, Berkeley, CA.

[10] Further discussions about broadened access in engineering education, and the subsequent phenomenon of "academic drift," may be found in the following chapter in this volume, Steen Hyldgaard Christensen and Byron Newberry, "The role of research in academic drift processes in European and American professional engineering education outside the university sector."

the CSCs to a teaching mission; and broadening access to junior colleges through state support for these institutions, while developing articulation agreements that facilitated the transfer of academically-capable students into the CSC or UC systems (Douglass 2000, pp. 265–297). Once again, this compromise was the product of broad fiscal, political, and demographic issues. Yet, the solution bore the substantial imprint of discussions about engineering education. For example, the state's engineering workforce needs, and the perennial national discussions of "engineering manpower crises" (such as that surrounding Sputnik), created a compelling rationale for expanding the state college mission. Indeed, the engineering directors and deans first worked out articulation agreements between the junior colleges and the CSC and UC system in response to the same crisis. Moreover, longstanding discussions about the inverted nature of the engineering profession – the proliferation of land-grant institutions and undergraduate professional B.S. degrees had produced more professional engineers than technicians due to the public attention given to the former. But the need for technicians (soon renamed technologists) remained, a need mapped onto the plan by vastly expanding the state's system of junior colleges. The California Master Plan also had a direct impact upon national conversations about engineering education: the demographic projections of the California Master Plan, and the methods used to produce them led Eric Walker to his major recommendations in the 1968 Goals Report concerning the ascendant role of the master's degree in engineering (Engineering Advisory Committee 1965).

These three historical developments – MIT, ASEE/ECPD, and developments in California – were all influential, but they did not represent all of the new patterns for engineering education during the Cold War. In fact, the situation looks rather different when we bring into view a broader institutional ecology that includes less prominent state universities, state college systems, private engineering colleges, as well as liberal arts colleges with engineering degree programs and a smattering of public technical institutions centered on its engineering and technical degree programs. The historical study of this larger group of institutions after 1945 is significantly underdeveloped. Yet, in terms of the number of engineering graduates, these institutions are far from insignificant. Here we can only take brief note of the experience of two institutions from this diverse cohort.

One institution from this outer ring worth examining for comparative purposes is the University of Texas at Austin. Despite the language of its charter in the state's constitution labeling the University of Texas a "university of the first class," the Texas state legislature regularly restricted state resources to the university's teaching mission until the 1980s. While UT Austin successfully created strong research programs in selected areas such as nuclear engineering, petroleum engineering and water resources (all topics of critical interest to the state), the legislature invested little in the university's research infrastructure. Ironically, one reflection of this policy was the recognition UT Austin's College of Engineering received during the 1960s and 1970s for its innovative work in undergraduate education. Most notable was its early experiments in personalized, student-centered approaches to instruction as carried out by its Bureau of Engineering Teaching (UT Austin 1967; Stice 1971).

The downturn in defense spending during the late 1970s, declining oil revenues, and stagnation within the agricultural sector prompted the Texas legislature to reverse traditional policies. Pressed by local boosters, the legislature embraced the new high-tech economy represented by Stanford and Silicon Valley. Austin and UT successfully competed for the Microelectronics and Computer Technology Corporation (MCC) – then said to be the U.S. response to the Japanese "Fifth Generation Project" in artificial intelligence – with a variety of state and private commitments and appropriations. MCC itself fizzled, but the state legislature's willingness to join in on the game of leveraging state funds to secure private investments and federal research contracts proved more enduring. The strict emphasis upon undergraduate education faded into the background as the University of Texas ascended into the ranks of the top engineering schools and universities (Wilson et al. 1981; UT Austin 1983).

The University of Maine's main campus in Orono affords another point of comparison. The University of Maine is located in another state dependent upon natural resource extraction, but one facing long-term economic and demographic stagnation. The heyday of Maine's economy lay in a golden past of timbering, shipbuilding, and maritime commerce, although coastal tourism, fishing, and a strong pulp and paper industry continued to provide an economic base up until the present that generated a steady regional demand for engineers.

Probably the most interesting aspect of the University of Maine's encounter with the postwar ascent of engineering science has to do with the school's struggles with accreditation. Maine's faculty followed national trends in gradually replacing shop courses with more fundamental subjects during the early postwar years. A research-based approach to engineering gained a significant foothold during the 1970s as tenure standards, faculty workload, hiring policies, and salaries shifted more towards the model adopted by other research universities. Yet the College of Engineering and Technology (College of Engineering and Science after 1973) began running up against accreditation problems following the Grinter Report. The difficulties came to head during the 1970s, not over curricula, but because of facilities, faculty-student ratios, and faculty salaries. Successive ECPD visiting accreditation evaluation teams concluded that the University of Maine program could not be competitive without additional resources and support for recruiting and retaining competent faculty.[11]

In other states, including Texas, negative accreditation reviews usually provided university presidents with a compelling argument that convinced state legislatures to provide additional funds for engineering degree programs. However, in Maine the stagnant economy limited the legislature's options; more intriguingly the uni-

[11] See especially the correspondence related to accreditation in the folder, Engineers' Council for Professional Development (ECPD), 1935–1967. Maine, University at Orono, president's office files. Department of Special Collections, Raymond H. Folger Library, University of Maine. Orono, ME.

versity president sided with the legislature. The president warned the dean of engineering not to use accreditation to pressure the administration into making investments it could ill afford to make, and even labeled his request to follow national trends in engineering faculty salaries as extravagant. At the University of Maine, the pace of life for engineering faculty remained somewhat different, as pre-tenure publication standards settled at perhaps half that of more prominent research universities, while the faculty remained more committed to and directly engaged with their undergraduate students. An emphasis upon teaching remains an important part of the University of Maine's faculty identity, even as many pursue world-class research, especially in areas connected to the state's key industries (Sandford interview 2012).

Late and Post-Cold War Changes in Engineering Education

From the Cold War to the present, two other significant changes took place in U.S. engineering education. The first, emphasized by historian of engineering Matthew Wisnioski (2009, 2012) was the diffusion of radical ideals from the 1960s into engineering education. Wisnioski describes the "long decade of the 1970s," during which these values became normalized within engineering disciplinary identities and practices. This shift was then followed by a conservative turn that began during the dual oil shocks of the 1970s. By the 1980s, engineering and national cultures alike accepted a neoliberal doctrine, manifested in attention to "national competitiveness," and more recently, to economic globalization. The first topic is covered extensively in Wisnioski's recent book, *Engineers for Change*, while the latter is a subject of our current research. Both topics can receive only summary treatment in this chapter, but our intent here is to provide some context for other chapters in this volume that explore some of these developments in greater detail.

A quick review of engineering education and professional journals from the 1970s leaves no doubt concerning the lasting impact on the engineering profession of the rhetoric of social responsibility and environmentalism, as articulated during the 1960s.[12] Topics related to the nation's energy resources, the environment, and to a lesser extent, social and urban problems became an integral part of national conversations about engineering education and engineering professional training. For the most part, the emphasis lay with developing technological solutions to social problems, the trope most familiar to engineers and engineering educators alike from the point of view of their disciplinary practice. Of equal significance, most of the changes occurred within individual courses, curricula, and degree programs. Yet apart from the occasional wholesale evolution of a degree program, such as the transformation of many sanitary engineering departments into environmental engineering departments, the radicalism of the 1960s had little lasting effect upon the

[12] These range from the *Journal of Engineering Education* to undergraduate magazines such as *The Spectrum* published by the undergraduate engineering students at Pennsylvania State University.

deeper structure of U.S. engineering education. Research remained the dominant focus, although student protests, and a course evaluation movement did bring some attention back to the quality of undergraduate instruction and teaching methods.[13]

The other imprint of the 1960s was an increased emphasis upon gender, race, and ethnic inclusivity within the engineering profession. This is a subject addressed explicitly and with a broader time horizon in Chap. 8 of this volume ("Meritocracy, technocracy, democracy: understandings of racial and gender equity in American engineering education"; See also Slaton 2010). Slaton documents the persistent pattern of exclusion that occurred despite well-intentioned programs that sought to bring more women and minorities into engineering. For the purposes of this overview, it is probably sufficient to note that the inclusionary initiatives that originated during the 1960s had a dual origin. They drew, on the one hand, on progressive social ideals that grew out of the Civil Rights and Women's movements of this period. Of equal importance, however, were the cyclical crises related to the size of the U.S. engineering workforce – the "engineering manpower crisis" that began with the Korean War, and persists in present-day rhetoric. The 1960s and early 1970s, in particular, was a period when elevated interest in the sciences led to concerns about under-enrollment in engineering, producing a strong interest in augmenting engineering enrollments just at the historic moment when women and minorities were demanding greater access to white, male-dominated occupations. The pressure to produce more engineers ebbed during the late 1970s but has resurfaced at various intervals in response to national economic challenges, inviting renewed efforts at inclusion. Yet, despite over 40 years of commitment to greater gender and racial diversity in engineering, the changes have been slow to come, as noted by Slaton in her chapter.

We also invite readers to consider the large body of literature on women in engineering, including sociological, ethnographic, and other forms of social scientific studies of the specific mechanisms that both facilitate and propagate women's inclusion and exclusion from engineering programs. Some of the most interesting, recent studies include works by Amy Bix, Wendy Faulkner and others – but the relevant literature is much wider when questions are asked about underrepresentation within STEM (Science, technology, engineering and math). The National Science Foundation also devotes significant attention to diversity in its regular reports on the state of science and engineering (Bix 2004; Faulkner 2000a, b, 2007; Frehill 1997; Gill et al. 2008; Kohlstedt and Longino 1997; National Science Board 1987; See also Committee on Maximizing…et al. 2007). There is as well a growing literature on Asian-American and foreign student (and faculty) experiences, and associated discussions about the "brain drain," overrepresentation, and discrimination, and

[13] See CDL 3/A12, UT College of Engineering Records for a collection of reports on course evaluations. University of Texas Archives, Briscoe Center for American History, University of Texas at Austin, Austin, TX. Related discussions, especially as they relate to sustainable development, may be found in Chap. 10 of this volume (Lucena, "Bridging Sustainable Community Development and Social Justice"), as well as in Jamison (2012).

now a reverse brain drain. This and the associated literature on affirmative action and reverse discrimination suggest how contentious diversity issues have become in the United States in recent decades.

Some engineering educators also emerged from the 1960s suggesting that broader, socially informed strategies were required to solve vexing societal problems. This included the authors of an early study by ASEE's Liberal Studies Division, *Liberal Learning for the Engineer* (ASEE 1968). However, it is notable that by 1975, the members of the division had themselves rejected the major findings of the report, calling for more diverse approaches that, in practical terms, recognized the limited influence of the humanities and social sciences faculty (ASEE 1975). While the 1970s did produce a significant number of new programs in "science, technology, and society," technology policy, and other related fields, few proceeded to operate in an integrated way that sought to directly influence engineering professional identities. Most such programs evolved primarily to function as external critics or observers of engineering and its disciplinary practice. More importantly, all of the changes described in this section produced few basic changes in the structure or content of the engineering curriculum.

More substantial changes in engineering education occurred during the 1980s, as concerns about economic competitiveness replaced Cold War emphases. The dual oil shocks and U.S. economic stagnation during the 1970s contributed to fading interest in social issues, as U.S. engineering schools began to focus upon bolstering U.S. industrial capacity. These concerns were heightened during the 1980s by the rise of the Japanese economy, the U.S. trade imbalance, and the emergence of neoliberal economic doctrine (Harvey 1989, 2007). While contemporary writers such as Thomas Friedmann (2005) present the latest economic trends as if they comprised a radical new era of globalization, from a historical standpoint we consider the emphasis on "national competitiveness" in the 1980s and the new rhetoric of economic globalization to be part of the same historic moment.

This being said, U.S. institutions responded differently to the different phases of this global economic transition. Within engineering education, many efforts during the 1970s were in fact directed towards manufacturing, extending the push to reintroduce "real" engineering subjects back into the engineering. While this focus upon hands-on activities, usually defined as "design", continued into the 1980s, Orientalized fears, initially about "Japan Inc.," and then the "Asian Tigers," prompted a few calls for a national industrial policy and more centralized planning. The activities in Austin, Texas surrounding the MCC and numerous state programs to invest in R&D high-tech commercialization exemplified these responses. A definite interest in Japanese management techniques emerged with significant curricular impact in industrial engineering and associated engineering and business disciplines (Vogel 1980; Fallows 1989, 1994; Grayson 1983, 1984a, b).

By the 1990s, interest in industrial policy began to wane as engineering educators, policymakers, and industrialists alike came to accept the global spread of industrial capitalism as inevitable. The more recent focus has therefore been on defining a distinctive role for U.S. educated engineers. Astute observers worry about the limits of this kind of "up-skilling" strategy rooted in continued nationalistic assumptions and sentiments – especially as corporate supporters underwriting this approach themselves operate in global rather than national environments. Nevertheless, the primary focus of recent curricular initiatives has been to emphasize entrepreneurship and innovation, as well as broad professional competence as a means of retaining a distinctive advantage for U.S. engineers in the global economic arena (Bordogna et al. 1993; National Academy of Engineering 2004, 2005; Committee on Underrepresented Groups 2011).

It is also worth noting that the *method* used to carry out these three, recent phases of educational reform map onto the political economic regime within which they unfolded. Thus, during the 1970s, the emphasis on manufacturing resulted from a return to university partnerships with regional industries. The iconic efforts of the 1980s was based more on centralized policies and planning, as represented by the National Science Foundation's Engineering Education Coalitions, a federal attempt to produce and disseminate new approaches to engineering education. Each of the NSF's EECs were supported through multi-million dollar, multi-year grants issued to groups of engineering schools willing to undertake major engineering education reform efforts as administered under a well-defined management plan. While the details are beyond this account, continued academic commitments to research, including commitments by the engineering educators themselves to cast their own work as research output, undermined the effort to produce transformative programs that supported NSF's new "national vision for engineering education" (Bordogna 1989; Meade 1991; Coleman 1996).

The EECs were by no means a total failure, but their limited accomplishments prompted a somewhat wider group of leading institutions and engineering educators to push through ABET's Engineering Criteria 2000 (ABET Inc., 1997). A similar set of concerns prompted the National Academy of Engineering to produce a pair of reports that offered a modified vision for engineering and engineering education (National Academy of Engineering 2004, 2005). Both initiatives, along with further changes in the funding priorities of NSF's Engineering Directorate, map onto the continued evolution of U.S. neoliberal doctrine. That is, they are not the product of federal efforts to formulate a more autonomous federal industrial policy, but rather reflect the coalescence of public and private interests cooperating more subtly to articulate a new national vision (Harvey 2007). In concrete terms, the most significant change associated with EC 2000 has been ABET's decision to abandon accreditation based on courses taken in favor of a regime based upon student outcome assessment based upon institutionally-defined learning outcomes connected to wider criteria. The limitations of, and the resistance that has surfaced to, this accreditation regime requires separate treatment, and is hopefully addressed by other chapters in this volume. But not only is this approach the main policy direction for U.S. engineering degree programs today, but it is also being adopted widely outside the U.S. (Lattuca et al. 2006; Seely 2012).

Conclusions

It has been our desire in this chapter to provide historical background knowledge for the other chapters in this volume, so we did not set out to offer broad conclusions about American engineering education. Nevertheless, we can offer the following comments about the historical developments described above. The first is to reemphasize the highly diverse institutional ecology for engineering education in the United States, a situation that still prevails today. Beyond being a hybrid of British apprenticeship and continental formal education traditions, the U.S. system of engineering education was developed in a manner responsive to diverse geographic, industrial, and political interests. From a peer based system of accreditation that acknowledged regional differences, to the political and economic differences among the fifty states and their interests in higher education, quite different approaches to engineering education emerged and persisted, despite countervailing national efforts to standardize engineering education under several different visions.

The middle section of this article focuses on the extent to which engineering science became the dominant ideology for engineering education during the Cold War. But it does so in a way that continues to highlight divergent experiences, and perhaps more importantly, the process by which new ideas about engineering education took hold within the United States. Our most important historical observation here is that engineering educators possessed a distinct if also evolving body of practice for adapting engineering knowledge to "changing times and needs." While described only briefly in the above account, this first involved a well-defined body of investigative practice that was firmly rooted in the Progressive Era's reform traditions, before evolving to a more centralized, planned approach that drew upon the kind of public-private partnerships represented by the State of Texas' competitive bid for MCC, the NSF Engineering Education Coalitions, and more recently, ABET's new accreditation standards. Each of these later developments grew from the rise of neoliberal economic doctrine; they also reflect the different ways that neoliberalism was manifested within the American economy. To put this focus on reform practice in somewhat more formal terms, we could say that engineers possess an ethnomethodologically accountable, which is to say describable, body of practice for reexamining the epistemological foundations of their discipline in response to a change in social and historical context. Our account suggests that it may be possible to identify where these practices come from, and perhaps also how they evolve. This is a point of some significance to the sociology of knowledge, and merits closer scrutiny.

Speaking more practically in terms of the actual structural changes in the U.S. system of engineering education, the broad outlines of the historical account suggests that engineering education witnessed an early period of significant variation, partial uniformity set by the curricular standards from the land-grant institution, to a more unified vision of science-based engineering following World War II. The most recent period, marked by ABET's EC 2000 accreditation regime, places us in an era that once again seems to favor greater institutional variation. This approach is consistent with the underlying ideal of educating distinctive engineering graduates with a strong capacity for entrepreneurship and innovation. This transition, unlike the previous shift toward engineering science, was based on relaxing the previously-introduced quantitative standards for accreditation, deeming those rules to be too rigid and likely to stifle innovation not only in the work produced by our graduates, but in the engineering education programs themselves. Whether ABET evaluators can emerge from the shadow of earlier accreditation practices remains to be seen; existing evidence suggests that the transition has not been easy. Still, it is too early to reach final conclusions; indeed the intention of these volumes is to cultivate dialogues about alternative possibilities. This being said, it may be that the full institutional apparatus for achieving such changes still remain to be forged in the United States.

References

ABET, Inc. (1997). Engineering criteria 2000: Criteria for accrediting programs in engineering in the United States. *ASEE Prism March*, 39–40.

Akera, A. (2008). *Understanding the practice of engineering education reform: The investigative traditions of the American Society for Engineering Education, 1907–1968.* Conference paper, American Society for Engineering Education Annual Conference and Exposition, Pittsburgh, PA.

Akera, A. (2010). Exploring the legacy of a liberal-professional vision for engineering education. Unpublished manuscript. Talk presented at the 3rd Union College Symposium on Engineering and Liberal Education, May 2010, Schenectady.

Akera, A. (2012a). Liberal learning revisited: A historical examination of the reasons, frustrations, and continued prospects for Engineering and Liberal Arts Integration. *Proceedings of the 2011 ASEE Annual Conference and Exposition*, Vancouver.

Akera, A. (2012b). The MIT Lewis survey: Creating a cold war blueprint for a technological university, 1947–1949. *Proceedings of the 2012 ASEE Annual Conference and Exposition,* San Antonio.

Akera, A. (n.d.). Rhetoric and engineering science ideology in American engineering education reform, 1952–1955. Unpublished manuscript.

Alder, K. (1997). *Engineering the revolution: Arms and enlightenment in France, 1763–1815.* Princeton: Princeton University Press.

ASEE Committee on Evaluation of Engineering Education. (1953). *Preliminary report of the committee on evaluation of engineering education* (AC 12, box 2/ASEE). Cambridge: MIT Institute Archives and Special Collections.

ASEE Goals Report. (1968). See Final report: Goals of engineering education.

ASEE, Humanistic-Social Research Project. (1968). Liberal learning for the engineer. *Journal of Engineering Education, 59*, 303–342.

ASEE, Humanities-Social Sciences Project. (1975). Liberal learning for the engineer: An evaluation five years later. *Engineering Education, 65*, 301–334.

Bix, A. S. (2004). From 'engineeresses' to 'girl engineers' to 'good engineers': A history of women's U.S. engineering education. *Feminist Formations, 16*(1), 27–49.

Bordogna, J. (1989). Entering the '90s: A national vision for engineering education. *Engineering Education, 80*, 646–649.

Bordogna, J., Fromm, E., & Ernst, E. W. (1993). Engineering education: Innovation through integration. *Journal of Engineering Education, 83*, 3–8.

Calvert, M. A. (1967). *The mechanical engineer in America, 1830–1910: Professional cultures in conflict*. Baltimore: The Johns Hopkins University Press.

Carlson, W. B. (1988). Academic entrepreneurship and engineering education: Dugald C. Jackson and the MIT-GE cooperative engineering course, 1907–1932. *Technology and Culture, 29*, 536–567.

Channell, D. F. (1989). *The history of engineering science: An annotated bibliography*. New York: Garland.

Coleman, R. J. (1996). The engineering education coalitions: A progress report. *ASEE Prism, September*, 24–31.

Committee on Maximizing the Potential of Women in Academic Science and Engineering, Committee on Science, Engineering, and Public Policy, National Academy of Sciences, National Academy of Engineering, Policy and Global Affairs, & Institute of Medicine. (2007). *Beyond bias and barriers: Fulfilling the potential of women in academic science and engineering*. Washington, DC: National Academies Press.

Committee on Underrepresented Groups and the Expansion of the Science and Engineering Workforce Pipeline, Committee on Science, Engineering, and Public Policy, Policy and Global Affairs, National Academy of Sciences, National Academy of Engineering, & Institute of Medicine. (2011). *Expanding underrepresented minority participation: America's science and technology talent at the crossroads*. Washington, DC: National Academies Press.

Douglass, J. A. (2000). *The California idea and American higher education, 1850 to the 1960 master plan*. Palo Alto: Stanford University Press.

Eddy, E. D., Jr. (1956). *Colleges for our land and time; the land-grant idea in American education*. New York: Harper and Bros.

Engineering Advisory Committee. (1965). See [University of California,] Engineering Advisory Committee.

Fallows, J. M. (1989). *More like us: Making America great again*. Boston: Houghton Mifflin.

Fallows, J. M. (1994). *Looking at the sun: The rise of the new East Asian economic and political system*. New York: Pantheon Books.

Faulkner, W. (2000a). The power and the pleasure? A research agenda for "making gender stick" to engineers. *Science, Technology & Human Values, 25*, 87–119.

Faulkner, W. (2000b). Dualisms, hierarchies and gender in engineering. *Social Studies of Science, 30*, 759–792.

Faulkner, W. (2007). "Nuts and bolts and people": Gender-troubled engineering identities. *Social Studies of Science, 37*, 331–356.

Ferguson, E. S. (1992). *Engineering and the mind's eye*. Cambridge: MIT Press.

Final Report: Goals of Engineering Education. (1968). *Journal of Engineering Education, 57*, 367–446.

Flexner, A. (1910). *Medical education in the United States and Canada*. New York: Carnegie Foundation.

Frehill, L. M. (1997). Education and occupational sex segregation: The decision to major in engineering. *The Sociological Quarterly, 38*, 225–249.

Friedmann, T. L. (2005). *The world is flat: A brief history of the twenty-first century*. New York: Farrar Straus and Giroux.

Geiger, R. L. (2000). Research, graduate education, and the ecology of American universities: An interpretive history. In R. L. Geiger (Ed.), *The American College in the nineteenth century* (pp. 153–168). Nashville: Vanderbilt University Press.

Gill, J., Sharpo, R., Mills, J., & Franzway, S. (2008). I still wanna be an engineer! Women, education, and the engineering profession. *European Journal of Engineering Education, 33*(4), 391–402.

Grayson, L. P. (1983). Japan's intellectual challenge: The strategy. *Engineering Education, 74*, 138–146.

Grayson, L. P. (1984a). Japan's intellectual challenge: The future. *Engineering Education, 74*, 296–303.

Grayson, L. P. (1984b). Japan's intellectual challenge: The system. *Engineering Education, 74*, 210–220.
Grayson, L. P. (1993). *The making of an engineer: An illustrated history of engineering education in the United States and Canada*. New York: Wiley.
Grinter Report. (1955). See *Report of the committee on evaluation of engineering education*.
Harvey, D. (1989). *The condition of postmodernity: An enquiry into the origins of cultural change*. Oxford: Blackwell.
Harvey, D. (2007). *A brief history of neoliberalism*. Oxford: Oxford University Press.
Hollister, S. C. (1979). *S.C. Hollister oral history interview (29 November 1979), Transcript 1726. Item 13/6/1858*. Ithaca: Division of Rare and Manuscript Collections, Cornell University Libraries.
Jamison, A. (2012). Turning engineers green: Sustainable development and engineering education. In S. H. Christensen, C. Mitcham, Li Bocong, & Yanming An (Eds.), *Engineering, development and philosophy: American, Chinese, and European perspectives* (pp. 7–22). Dordrecht: Springer Science+Business Media B.V.
Kevles, D. J. (1977). *The physicists: The history of a scientific community in modern America*. New York: Vintage Books.
Kline, R. (1995). Construing 'technology' as 'applied science': Public rhetoric of scientists and engineers in the United States, 1880–1945. *ISIS, 86*, 194–221.
Kohlstedt, S. G., & Longino, H. E. (1997). *Women, gender, and science: New directions*. Chicago: University of Chicago Press.
Kranakis, E. (1997). *Constructing a bridge: An exploration of engineering culture, design, and research in nineteenth-century France and America*. Cambridge: MIT Press.
Landis, F. (1981). How many engineers will graduate during the eighties? *Engineering Education, 71*, 784–788.
Lattuca, L. R., Terenzini, P. T., & Volkwein, J. F. (2006). *Engineering change: A study of the impact of EC2000*. Baltimore: ABET Inc.
Layton, E. T., Jr. (1971). Mirror-image twins: The communities of science and technology in 19th-century America. *Technology and Culture, 12*, 562–580.
Layton, E. T., Jr. (1976). American ideologies of science and engineering. *Technology and Culture, 17*, 688–701.
Lecuyer, C. (1993). The making of a science based technological university: Karl Compton, James Killian, and the Teform of MIT, 1930–1957. *Historical Studies in the Physical Sciences, 23*, 153–180.
Lerman, N. E. (1997). "Preparing for the duties and practical business of life": Technological knowledge and social structure in mid-19th-century Philadelphia. *Technology and Culture, 38*, 31–60.
Leslie, S. W. (1987). Playing the education game to win: The military and interdisciplinary research at Stanford. *Historical Studies in the Physical Sciences, 18*, 56–88.
Leslie, S. W. (1993). *The cold war and American science: The military-industrial-academic complex at MIT and Stanford*. New York: Columbia University Press.
Lewis Survey. (1949). See *Report of the Committee on Educational Survey.*
Mann, C. R. (1918). *A study of engineering education, Prepared for the Joint Committee on Engineering Education of the National Engineering Societies*. New York: Carnegie Foundation.
Master Plan for Higher Education in California, 1960–1975. (1960). California State Department of Education, Sacramento.
McMath, R. C., Jr., Baylor, R. H., Brittain, J. E., Foster, L., Giebelhaus, A. W., & Reed, G. M. (1985). *Engineering the New South: Georgia Tech, 1885–1985*. Athens: University of Georgia Press.
Meade, J. (1991). Engineering coalitions find strength in unity. *ASEE Prism, September*, 24–36.
Mindell, D. A. (2002). *Between human and machine: Feedback, control, and computing before cybernetics*. Baltimore: Johns Hopkins University Press.

National Academy of Science. (2005). *Educating the engineer of 2020: Adapting engineering education to the new century*. Washington, DC: National Academy Press.
National Academy of Sciences. (2004). *The engineer of 2020: Visions of engineering in the new century*. Washington, DC: National Academy Press.
National Science Board. (1987). *Science & engineering indicators*. Washington, DC: National Science Board.
Phelan, T., Ross, D. M., & Westerdahl, C. A. (1995). *Rensselaer, where imagination achieves the impossible*. Troy: Rensselaer Polytechnic Institute.
Report of the Committee on Educational Survey. (1949). The Technology Press, Cambridge.
Report of the committee on engineering education after the war. (1944). *Journal of Engineering Education, 34*, 589–614.
Report of the committee on evaluation of engineering education. (1955). *Journal of Engineering Education, 46*, 26–60.
Reynolds, T. S. (1991). The engineer in 19th-century America. In T. S. Reynolds (Ed.), *The engineer in America: A historical anthology from technology and culture*. Chicago: University of Chicago Press.
Reynolds, T., & Seely, B. (1993). Striving for balance: A hundred years of the American society for engineering education. *Journal of Engineering Education, 82*, 136–151.
Sandford, T. (2012) Thomas K. Sandford oral history interview, Conducted by Author on 24 Jul 2012, Orono.
Seely, B. E. (1993). Research, engineering, and science in American engineering colleges, 1900–1960. *Technology and Culture, 34*, 344–386.
Seely, B. E. (1995). SHOT, the history of technology, and engineering education. *Technology and Culture, 36*, 739–772.
Seely, B. E. (1999a). The other re-engineering of engineering education, 1900–1965. *Journal of Engineering Education, 88*, 285–294.
Seely, B. E. (1999b). European contributions to American engineering education: Blending old and new. *Quadrens d'historia de l'enginyeria, 3*, 25–50.
Seely, B. E. (2012). Engineering education. In W. S. Bainbridge (Ed.), *Leadership in science and technology: A reference handbook* (Vol. II, pp. 833–841). Thousand Oaks: Sage Publications.
Sinclair, B. (1974). *Philadelphia's philosopher mechanics; a history of the Franklin Institute, 1824–1865*. Baltimore: The Johns Hopkins University Press.
Slaton, A. E. (2010). *Race, rigor, and selectivity in U.S.* (Engineering: The history of an occupational color line). Cambridge: Harvard University Press.
Society for the Promotion of Engineering Education (SPEE). (1940). Report of the committee on aims and scope of engineering curricula. *Journal of Engineering Education, 30*, 555–566.
Stice, J. E. (Ed.). (1971). *The personalized system of instruction (PSI): The Keller plan as applied in engineering education* (Bureau of Engineering Teaching Bulletin No. 4). Austin: University of Texas at Austin.
Story, R. C., & Armsby, H. H. (1951). 1950 Enrollment in engineering colleges. *Journal of Engineering Education, 41*(6), 1–15.
Tolliver, W. E., & Armsby, H. H. (1961). Engineering enrollments and degrees in institutions with ECPD-accredited curriculums: 1960. *Journal of Engineering Education, 51*, 466–476.
UCLA School of Engineering and Applied Science. (1995). *50th Anniversary historical review 1945 to 1995*. Los Angeles: UCLA School of Engineering and Applied Science.
University of California Engineering Advisory Committee. (1965). *The engineering master plan study for the University of California* (Copy in University Archives). Berkeley: Bancroft Library, University of California.
University of Texas at Austin, College of Engineering. (1967). *Engineering Teaching Effectiveness Colloquia, 1966–1967*. CDL 3/A12, College of Engineering Records, University of Texas Archives.
University of Texas at Austin, College of Engineering. (1983). *Planning document 1983*: College of Engineering, The University of Texas at Austin centennial issue. CDL 3/A13, UT College of

Engineering records. University of Texas Archives, Briscoe Center for American History, University of Texas at Austin, Austin.
UT Austin. See University of Texas at Austin.
Vogel, E. F. (1980). *Japan as number one: Lessons for America.* New York: Harper Colophon Books.
Wildes, K. L., & Lindgren, N. A. (1985). *A century of electrical engineering and computer science at MIT, 1882–1982.* Cambridge: MIT Press.
Williams, R. L. (1991). *The origins of federal support for higher education: George W. Atherton and the Land-Grant College Movement.* University Park: Pennsylvania State University Press.
Wilson, M., McMains, H., & Smilor, R. (1981). A survey of research and development activities in the state of Texas. In *Texas 2000 Project, Texas past and future: A survey: Economic development issues* (pp. 150–167). Austin: Office of the Governor.
Wisnioski, M. (2009). 'Liberal education has failed': Reading like an engineer in 1960s America. *Technology and Culture, 50*, 753–782.
Wisnioski, M. (2012). *Engineers for change: Competing visions of technology in 1960s America.* Cambridge: MIT Press.

Atsushi Akera M.A. and Ph.D. in the History and Sociology of Science, University of Pennsylvania. Associate Professor, Department of Science and Technology Studies, and Director, First Year Studies at Rensselaer Polytechnic Institute, Troy, New York. As Director of First Year Studies he has been active in the development of new pedagogic strategies for engineering education. Associate Editor of *Engineering Studies,* officer for the Liberal Education/Engineering & Society Division, American Society for Engineering Education; and elected member of the Executive Council to the Society for the History of Technology. His current research, with co-author Bruce Seely, is on the history of engineering education reform in the United States (1945-present). Publications include *Calculating a Natural World: Scientists, Engineers and Computers during the Rise of U.S. Cold War Research* (MIT Press, 2006).

Bruce Seely Ph.D. in History of Technology from the University of Delaware. Dean, College of Sciences and Arts and Professor of History and of Science, Technology and Society, Michigan Technological University. He is the president of the Society for the History of technology (2013–2014), and served as program director for Science and Technology Studies at the National Science Foundation (2000–2002). He is the author of *Building the America Highway System: Engineers as Policy Makers* (Temple University Press, 1987); and (with Mark Rose and Paul Barrett) *The Best Transportation System in the World: Railroads, Trucks, Airlines, and American Public Policy in the Twentieth Century* (Ohio State University Press, 2006). He also authored numerous book chapters and articles on the history of engineering education, including "The Other Re-engineering of Engineering Education, 1900–1965," *Journal of Engineering Education* 88, no. 3 (July 1999): 285–294, which received the William Elgin Wickenden Award of the American Society for Engineering Education, for the best article published in the Society's journal in 1999. He received the 2004 Olmsted Award of the Liberal Education Division of the American Society for Engineering Education for contributions to the liberal arts within engineering education. Other scholarly interests include the history of transportation and transportation policy and the societal implications of nano-scale science and engineering.

Chapter 2
The Role of Research in Academic Drift Processes in European and American Professional Engineering Education Outside the Universities

Steen Hyldgaard Christensen and Byron Newberry

Abstract 'Academic drift' refers to a long term process induced by educational systems' dynamics whereby vocationally and professionally oriented post-secondary education institutions with a focus on professional training, teaching, and learning strive to become like universities by incorporating university structures and emulating their values, norms, symbols and practices. In this process they increasingly aspire to research and scholarship. However, the role of research in academic drift processes in professional non-university engineering education has attracted relatively little attention in the literature on academic drift as the focus has up till recently largely been on the introduction of more theory in the curriculum at the expense of practice, on the vertical extension of study programs, and on the introduction of university courses in the engineering college sector. In this chapter we will examine three examples of research drift that have taken place in professional non-university engineering education institutions in Ireland, The Netherlands, and the United States, respectively, from the massive expansion of higher education in the 1960s to the present. More precisely we will examine and compare research drift in Irish Institutes of Technology, Dutch Hogescholen, and three American institutions – a public technical institute, a state teacher's college, and a sectarian liberal arts university, and with an eye to recent developments in Denmark. In reviewing the literature, we have the following questions in mind: What are the driving forces behind academic drift in non-university engineering education in Europe and the

*Equal co-authors on this chapter.

S.H. Christensen (✉)
Department of Development & Planning, Aalborg University,
Vestre Havnepromenade 5, Aalborg 9000, Denmark
e-mail: steenhc@plan.aau.dk

B. Newberry
Department of Mechanical Engineering, Baylor University,
One Bear Place 97356, Waco, TX 76798, USA
e-mail: Byron_Newberry@Baylor.edu

S.H. Christensen et al. (eds.), *International Perspectives on Engineering Education*,
Philosophy of Engineering and Technology 20, DOI 10.1007/978-3-319-16169-3_2

United States? Are these driving forces of a similar nature or do they differ? Is academic drift desirable for vocationally and professionally oriented programs, and if not, can it be avoided? What research mission are former designated non-university engineering education institutions in Europe and the United States aspiring to fulfill? What kinds of tension and dilemma does this new mission create in the abovementioned kinds of institution?

Keywords Vocational non-university engineering education • Academic drift • Research drift • Mergers • Driving forces • Structural dynamics

Introduction

> There is no doubt that colleges and universities in this country model themselves upon each other....All one has to do is read catalogues to realize the extent of this isomorphism. (Riesman 1956, p. 25)

David Riesman, in his 1956 book *Constraint and Variety in American Education,* was among the first to introduce the notion of "institutional isomorphism" as a characteristic mechanism of mimetic behavior among higher education institutions in the United States. His conceptualization was later given wide traction in organizational theory by the work of Paul J. DiMaggio and Walter W. Powel on institutional isomorphism and collective rationality in organizational fields (DiMaggio and Powel 1983). For Riesman the higher education system could be aptly comprehended as an "academic procession". Portraying the American higher education system, Riesman pictured what later came to be called "academic drift" as a kind of reptilian procession – referring to the movements of these animals. In this reptilian procession the most prestigious institutions in the hierarchy are positioned at the head of the reptile, followed by a middle group constituting the body of the reptile, and at the tail the less prestigious schools. The most prestigious institutions constantly watch over each other to keep abreast, while those at the middle emulate those at the head, and those at the tail emulate those in the middle. As a result each group starts to look more and more like those they emulate and the institutional forms within them become blurred and less distinctive. As the reptile moves forward the mimetic process would be repeated as prior positions have been changed. Over time a less diversified system would tend to emerge. (Riesman 1956, p. 25; O'Meara 2007, p. 124; Morphew and Huisman 2002, p. 492).

As an illustration of Riesman's observation, the American higher education researcher Christopher C. Morphew in 2002 noted that in the United States since 1990 more than 120 public and private 4-year colleges representing nearly 5 % of all 4-year post-secondary institutions had changed their names and become universities (Morphew 2002, p. 207). He further claimed that the conversion of these institutions from colleges to universities represents a significant trend in higher education in the United States and deplored that "higher education researchers have not studied this trend, with the result that we know little about the institutions undergoing this transformation and what characteristics they might share" (Morphew 2002, p. 207).

Our point of departure for this chapter is that we have been wondering what is actually going on here, how can this significant and seemingly ubiquitous trend among post-secondary education institutions worldwide be explained, and how does it affect engineering education institutions – our main concern here – outside the university which are engaged as vital actors in the process. An initial hypothesis that we will further explore and qualify in the following is the assumption that under conditions of uncertainty, organizational decision-makers will tend to mimic the behavior of other organizations within their environments. Here organizational decision-makers would likely mimic the behavior of successful top tier organizations to which they have some network connections via boundary spanning personnel (see Morphew and Huisman 2002, p. 497). Conditions of uncertainty might be created among other things by globalization, internationalization, marketization of higher education, strong competition for funding and students among institutions, greater access and widening participation, efficiency pressures related to resource constraints created by state withdrawal of funding, students as consumers, and credentialing pressures on students to be able to compete on the labor market and more (Molesworth et al. 2011; Tuchman 2009).

In examining academic drift we have the following questions in mind: What are the driving forces behind academic drift in non-university engineering education in Europe and the United States? Are these driving forces of a similar nature or do they differ? Is academic drift desirable for vocationally and professionally oriented programs, and if not, can it be avoided? What research mission are former designated non-university engineering education institutions in Europe and the United States aspiring to fulfill? What kinds of tension and dilemma does this new mission create in the above-mentioned kinds of institution?

The chapter is structured as follows. In the second section we start by zooming in on major differences between, and dynamics of change in European and American higher education. Our aim is to introduce an ideal type with respect to structural transformations of non-university engineering institutions, and to present a theoretical framework for academic drift with a particular focus on research drift. In the third section we examine two European examples of academic drift – Irish *Institutes of Technology* (IoTs) and Dutch *Hogescholen* (HBOs) – with a particular focus on the transition from a teaching culture to a research culture taking place in these institutions. In section "Structural Dynamics in the American System of Higher Education" we offer a more detailed examination of the American system of higher education and dynamics of change in the system. Finally in the fifth section research drift in American engineering education is examined.

Mass Higher Education and Ensuing Structural Dynamics impacting on Vocationally Oriented Professional Schools Outside the University Sector

In 1996 Martin Trow (Trow 1996, p. 25), discussing continuities and change in American higher education, noted an important contrast between educational systems in Europe and the United States. In the United States, he claimed, a diversified system of higher education able to cope with a considerable expansion of student enrolments was put in place already by the turn of the twentieth century, though the numbers that characterize mass higher education were still to come. According to Bruce Doern (2008, p. 9; see also Henderson and Kane 1991; Eckel 2008), the architecture of the American higher education system is among, if not the most complex and diverse multi-level system in the world. It is composed of more than 4,000 public and private colleges ranging from elite research universities to 2-year technical and community colleges and 4-year colleges. More specifically, the system consists of:

- Comprehensive colleges and universities that provide undergraduate and graduate level education
- Research universities that provide undergraduate and graduate level education and support the granting of PhDs through their research mission
- Community and junior colleges that offer associate degrees, baccalaureate-track courses, and vocational education and training

Martin Trow noted, that in Europe by contrast the transformation of systems of élite higher education into systems of mass higher education took place from the 1960s and early 1970s onward. Prior to the 1960s post-secondary education in Western Europe can be described as *university-dominated.* Higher education was the exclusive province of the university and university-level specialized colleges, including university-level engineering colleges. Vocational training in engineering, teacher training and nursing were not regarded as higher education and were offered by separate professional schools either to prepare for a specific occupation or to prepare for a profession (Kyvik 2009, p. 3).

In the 1960s and early 1970s a transition from *university-dominated systems* to *binary systems* of higher education including engineering took place in many European countries. It became increasingly clear that a broadly educated population could no longer be formed within and by the universities alone. Hence the new types of institutions were created *ab initio* to deal with increasing numbers, a more diversified student body and a rapidly growing need for manpower in advanced industrial societies (Slantcheva-Durst 2010). These new institutions were of a comprehensive nature and have been variably called "universities of applied science", "university colleges", "institutes of technology" or "polytechnics" (Kyvik and Lepori 2010, p. 4). In the United States by contrast there is not a distinct, comprehensive system of polytechnic institutions. There are, however, institutions of technology ranging from elite institutions such as MIT and Caltech, to numerous local/regional technical and technological institutions serving local industries and trade (Doern 2008, p. 9).

Guy Neave (1979, pp. 156–157), the grand old man in comparative studies of higher education, has pointed to a set of objectives for institutions belonging to this new higher education sector that was created as an alternative to the autonomous university tradition in Europe. The objectives mentioned by Guy Neave are:

- Meeting the demands for vocational, professional and industrially based courses
- The creation of a separate sector of higher education outside the universities
- Greater public control to ensure continued responsiveness to social and economic demands of the locality.
- Increased standing of vocational and professional education.

In professional engineering education the objectives mentioned above would thus apply, by the time of their implementation, to British Polytechnics, French Instituts Universitaires de Technologie, so-called IUTs, German Fachhochschulen, Dutch Hogescholen, Belgian Hogescholen in the Flemish part of Belgium, Hautes Écoles in the French part of Belgium, Hautes Écoles Spécialisées in the French part of Switzerland, Ammatikorkeakoulu in Finland, Irish Institutes of Technology, so-called IoT's etc. The objectives are characterized by their work orientation and orientation towards the needs of the local community and industry for a skilled workforce to boost growth and competitiveness in the regional economy. They would also apply to a broad range of study programs in higher professional education in Europe at the bachelor's level such as nursing and other health education programs, social work, teacher training, economics and business administration, information technology, and other non-academic vocationally oriented degree programs (Kyvik 2009, p. 4). See also (Neave 1978, 1979; Labaree 1997, 2006; Morphew 2002; Huisman and Morphew 1998; Morphew and Huisman 2002; Kyvik and Skodvin 2003; Kyvik 2007, 2009; Apesoa-Varano 2007; Jónasson and Jóhannsdóttir 2010; Leiho 2010).

What makes the binary policy in general and the British binary policy in particular so fascinating is that one of its purposes was to prevent academic drift. In 1965 British Secretary of State Anthony Crosland, who was the architect behind the British binary policy, warned against academic drift saying: "For more than a century, colleges founded in the technical college tradition have gradually exchanged it for that of universities. They have aspired to an increasing level of work, to a narrowing of student intake, to a rationalization of course structure, and to a more academic course content" (Pratt 1997, p. 12). With the objectives of the new type of institution it was intended to put an end to the academic drift tradition.

For better or worse, for the non-university, professional higher education institutions this tendency to mimic universities among aspiring "wannabe" universities represents a shift from the above-mentioned vocational objectives towards more theoretically oriented academic values, norms, and attitudes. Martin Trow, in 2003 gave an illuminating description of the strive, both institutionally for upward movement and vertical extension of study programs, and among cosmopolitan and academically oriented faculty for vertical distinctions through research and scholarship.

> The older research universities, with their international reputations and prestige, their high academic standards, their great libraries and laboratories, their relatively favorable funding and provision for research, their links to government, and the high status of their staff and students everywhere exert a powerful pull on all other kinds of colleges and universities. Other newer institutions that have grown up in every modern society tend to look toward these elite institutions as models, and in many cases, hope to emulate them as closely as possible, and over time perhaps to be accepted into the charmed circle of those elite universities, to gain the prestige, the levels of funding required to gain and hold prestigious and distinguished academic staff. Of course, many new institutions understand that achievement of those ambitions cannot be expected quickly, but for many, they remain the models for what a college ought to strive to be. Success as they conceive it is movement toward higher standards, more selective admissions, an academic staff who holds degrees from research universities and want to do research. These institutions feel they ought to be called universities, they ought to be empowered to give degrees, even higher degrees; they ought to be able to initiate and support research (Trow 2003, p. 3).

The notion of academic drift as we use it here is meant to refer to the set of phenomena described by Martin Trow as cited above. Academic drift may be seen as corresponding to what the Australian scholar Malcolm Skillbeck has alternatively called *academic creep* (Skillbeck 2003, p. 5) and to a certain extent to what the Dutch scholar Aant Elzinga has called *epistemic drift* (Elzinga 1985). In this chapter, however, we prefer to stick to the notion of academic drift as this is the standard use in the literature that we are reviewing. In the remainder of this chapter academic drift in professional non-university engineering education is therefore understood as follows.

First, academic drift at a curricular level encompasses a *cognitive dimension*. From this perspective, academic drift refers to a tension between practice-oriented and science-oriented curricula. Related to engineering it thus refers to the process whereby knowledge derived from practical engineering work experience and intended to be useful for industrial practice gradually loses its close ties to practice. Instead, engineering knowledge becomes increasingly theoretical and oriented toward engineering disciplines, including mathematics and natural science (Harwood 2010).

Second, academic drift in recent years, as already noted, has come to encompass a *research dimension*. The notion of *research drift* was originally coined by Svein Kyvik and Benedetto Lepori in 2010 (Kyvik and Lepori 2010, p. 9) as a descriptor of the transition in aspiring "wannabe" universities from vocational training and teaching primarily at the bachelor's level to research-led and research-informed professional education taking place at the master's level and beyond. From this perspective research drift refers to a cultural tension between the basic beliefs, attitudes, norms and values of a teaching culture as opposed to the basic beliefs, attitudes, norms and values of a research culture.

Third, academic drift also encompasses an *institutional dimension*. From this perspective, it refers to: (1) the question concerning the appropriate locus for educating professional engineering students for engineering practice, and (2) the question concerning the relative market value of credentials earned by students in diverse educational settings. More precisely, academic drift here refers to a tension between what are considered "noble" and "less noble" institutions (Furth 1982; Teichler 2008), and accordingly to a tension between narrow vocational training taking place in less prestigious, less selective, and less intellectually demanding institutions and broad professional research-oriented and research-informed academic education taking place in prestigious, selective, and intellectually demanding and stimulating institutions (Burgess 1978).

Finally, academic drift also refers to a *structural dimension*. In this dimension academic drift operates across the entire non-university higher education sector to transform educational systems.

According to Svein Kyvik (2009), European non-university institutions of higher education seem to a great extent to have gone through three different, though overlapping, phases of transformations since the 1960s. The binary policy mentioned above was the product of phase two below. Formulated in an ideal typical fashion the three phases of transformations are:

1. *Fragmented expansion.* This phase is a reflection of the elite origins of higher education. The fragmented nature of educational expansion was aimed at differentiation and diversification by means of geographical and institutional decentralization. As a result, dual systems consisting of short-cycle vocational education and post-secondary higher education were established by the early 1960s, with a clear division between universities and the college sector. In this model, the college sector is fragmented into many small and specialized professional schools that offer short-cycle 2- or 3-year vocational courses. These small schools, based on vocational training, are not regarded as higher education institutions. Each of the schools has distinct vocational cultures and is subject to different public regulations.
2. *Horizontal integration.* This process aimed at field contraction, authority unification, institutional de-differentiation, program coordination, and regionalization. The outcome of this process, which took place from the mid-1960s and early 1970s until the early 1990s, may be characterized as a gradual transition to a binary model where the college sector came to be organized in comprehensive vocational, multi-profession colleges, sometimes termed polytechnics, alongside the university sector. Massification and expansion differentiated this second stage in higher education development from the elite origins of the university sector. To be able to cope with the massive expansion of higher education and an increasingly diversified student body, the new kinds of non-university institution were made more equal to the traditional universities than before and gradually appeared to be the less expensive part of the higher education system. The college sector now became subject to a common system of regulations.
3. *Vertical integration.* This phase, aiming at academization, field coupling, student mobility, structural convergence, network building and organizational integration, largely took place from the early 1990s onward. The outcome of this phase is characterized by a blurring of boundaries of the binary divide and a gradual

transition to a unified system of tertiary education. In unified systems, both traditional academic studies as well as vocational programs are offered within universities. New demands also suggested that professionals should be trained more in research activities to apply and update innovative knowledge in their work (Griffioen and De Jong 2012, p. 2). Unified systems have been created in three different ways: by upgrading polytechnics, by merging traditional universities and other higher education institutions, and by incorporating professional schools into universities (Kyvik 2009).

British polytechnics were upgraded to university status in 1992. In Germany the gap between universities and Fachhochschulen narrowed down or simply eroded from 2001 to 2004 as the outcome of the Bologna process. Briefly put, the Bologna process refers to the attempt by the European ministers of higher education to create a European higher education area aiming at "greater compatibility and comparability of the systems of higher education" in order to "promote citizens' mobility and employability". The ultimate goal is to increase the international competiveness of European higher education on a global scale (Bologna Declaration 1999). At the core of the Bologna process was "the adoption of a system essentially based on two main cycles, undergraduate and graduate" – bachelor and master – (Bologna Declaration 1999), as these were seen as generally accepted exit points for professional practice. The importance of the Bologna Process in the context of our present purpose was that it came to serve as an opportunity for policymakers and other constituencies to reconsider institutional identities and the distribution of roles and status between the institutional types in the system (Witte et al. 2008, p. 218).

In contrast to German Fachhochschulen, French IUTs were nested into universities already from the start (Christensen 2012). Yet a different attempt to seek parity with the university would be merging engineering colleges with universities. This attempt has been the dominant trend in Denmark since the mid-1990s. In Denmark, professional engineering colleges – former so-called Teknika – were created in the early twentieth century and were from the start nested into technical vocational schools for the crafts. In 1962 Danish Teknika gained independence from the supervision by the technical vocational school leadership and became part of a binary system of engineering education (Frandsen and Harnow 2011). The following figure illustrates the development of Danish professional engineering education institutions from the early 1970s until 2011 (See Christensen and Ernø-Kjølhede 2011, p. 290).

In the table a lack of year indicates that only an informal and loosely defined association with a university has taken place presently. However it also indicates that a future merger is likely to take place with the respective university. The end of structural reforms in professional engineering education in Denmark is destined by an act of the Danish Parliament to be completed no later than by 2015 (Table 2.1).

In the table it is noteworthy that the only institution – the Engineering College of Horsens, that merged horizontally into a polytechnic type of institution – presently seems to have regretted its decision and now wants to merge vertically with a university instead. Moreover, in 2012 the Engineering College of Aarhus finally merged

Table 2.1 Merging Danish professional engineering education institutions from the mid-1970s to 2011

University colleges		Engineering colleges		Universities		Engineering academies
VIA	←	Horsens (2008)				
		Helsingør (1995)	→	Technical University of Denmark	←	The Danish Academy of Engineering in Copenhagen (1995)
		Haslev (1997)				
		Aalborg (1974)	→	Aalborg Unviersity	←	The Danish Academy of Engineering in Aalborg (1974)
		Esbjerg (1995)				
		Copenhagen				
		Sønderborg (1997)	→	University of Sothern Denmark		
		Odense (2006)				
		Herning (2006)	→	Aarhus University		
		Aarhus				

Based on Frandsen et al. (2011, pp. 149–152)

with Aarhus University. The Engineering College of Copenhagen, contrary to what was the original intention as presented in the figure, decided in 2012 not to merge with Aalborg University but instead with Technical University of Denmark, a process which has now been completed.

Two Examples of Institutional Dynamics and Research Drift in Non-university Engineering Education Institutions in Europe

Irish Institutes of Technology (IoTs)[1]

The decision by the Irish Government in the late 1960s to establish a number of Regional Technical Colleges (RTCs) distributed throughout the country represented a significant step toward the transition to a *binary system* of higher education in Ireland. RTCs were renamed Institutes of Technology (IoTs) in 2000. Prior to the late 1960s and early 1970s higher education in Ireland took place within *a university-dominated system* consisting of five universities (see e.g. Universities Act 1997). These are: (1) University College Dublin: the university originated in a body

[1] The authors take full responsibility for this section. However we have incurred a considerable amount of debt as we have made an effort to validate the findings of the section trough personal communication with Mike Murphy and William Grimson, Dublin Institute of Technology. Dr. Mike Murphy is Director and Dean at College of Engineering and Built Environment. William Grimson is a chartered engineer, former Head of the Department of Electrical Engineering, and Registrar with overall responsibility for academic quality enhancement.

founded in 1854 as the Catholic University of Ireland with John Henry Newman as the first rector, receiving a Royal Charter and incorporated into the National University of Ireland in 1908. (2) University of Dublin/Trinity College: founded by Royal Charter of Queen Elizabeth in 1592. (3) The National University of Ireland, Maynooth: the college had previously been St. Patrick's College, which was admitted to the Pontifical University in 1896, and subsequently became a recognized college of the National University of Ireland in 1910. (4) The National University of Ireland, Galway: established in 1845 as one of three Queen's Colleges (Galway, Belfast and Cork), and becoming a founding member of the National University of Ireland in 1908. (5) University College Cork: like Galway, originally one of the Queen's Colleges and one of the founding members in 1908 of the National University of Ireland (Duff et al. 2000).

The system reflected a cultural divide between catholic and Anglican protestant institutions, Trinity College being of Anglican protestant origin. Today this cultural divide is of minor significance. In 1989 two new universities were established, Dublin City University and The University of Limerick. Both were formally established by statute in 1989 from existing, but relatively new, national institutes for higher education thereby increasing the number of universities in the Republic of Ireland to seven (Hazelkorn and Moynihan 2010b, p. 176).

Irish Regional Technical Colleges (RTCs) were established gradually from the 1970s onward to provide short-cycle – mainly sub-degree level – courses of 2-year duration in engineering, construction studies, business studies, applied science, and art and design on a regional basis. Five Regional Technical Colleges were established in 1970. Four more were added during the 1970s, along with another three in the 1990s, and a final one in the year 2000, for a total of 13 RTCs. There is also a 14th RTC – Dublin Institute of Technology (DIT) – which is an exception that enjoys special privileges. DIT is by far the largest Irish higher education institution, with some 12,000 full time students and a further 7,000 part-time students (Skillbeck 2003, p. 19). It has its own legislation and has the right to confer academic awards from apprenticeship to Ph.D. under the Qualifications (Education and Training) Act 1992, whereas the other RTCs only have delegated authority from the Higher Education and Training Awards Council (HETAC) as specified by the Regional Technical Colleges Act 1992 (see Dublin Institute of Technology Act 1992; Regional Technical Colleges Act 1992). As part of the Bologna process, Ireland agreed to and implemented the standard grid of level indicators for short-cycle, first cycle, and second cycle qualifications awarded to students within the transnational *Framework for Qualification of the European Higher Education area* (Adelman 2009).

DIT was formally established in 1978 by the City of Dublin Vocational Committee, and was later given special status by the Dublin Institute of Technology Act of 1992. Its history dates back to 1885 when an artisans' exhibition was organized in Dublin, supported by carpenters, joiners, coopers, bookbinders, bricklayers, etc. As a result of the success of this exhibition, Kevin Street Technical School was opened in 1887 and provided evening courses for the working class. A College of Music opened in 1890, a College of Commerce in 1901, a College of Marketing and Design in 1905, a College of Technology in 1911, and a College of Catering in 1941. These were all supported through public funding and governed via Dublin Vocational Education committee. In 1978 Dublin Institute of Technology (DIT) was established on an *ad hoc* basis as a coordinating body for the work of these six colleges. DIT was formally established by statute on 1 January 1993 with full awarding powers (Duff et al. 2000, pp. 3–12; Coate and Mac Labhrainn 2008, p. 200; Hazelkorn and Moynihan 2010b, p. 176; Clancy 2008, p. 126).

When most of the RTCs were put in place in 1992, their principal function was defined by the Regional Technical Colleges Act 1992 in the following way:

> The principal function of a college shall, subject to the provisions of this Act, be to provide vocational and technical education and training for the economic, technological, scientific, commercial, industrial, social and cultural development of the state with particular reference to the region served by the college.....A college shall have the following functions....
>
> (c) ...Engage in research, development and consultancy work...
> (d) ... Enter into arrangements with other institutions in or outside the State for the purpose of offering joint courses of study and of engaging jointly in programmes of research, consultancy and development work....
> (e) ...Exploit any research, consultancy or development work...

In 2000 the RTCs were renamed "institutes of technology" (IoTs). This could be seen as a critical moment for the binary system as the new designation was an official recognition that university-level teaching and research should take place in the IoTs, most notably at DIT "subject to such conditions as the Minister may determine" as stated in the DIT Act of 1992. Unofficially the name "Institute of Technology" was perceived to convey a sense of higher rank and status similar to the name and rank of the renowned American elite institutions "Massachusetts Institute of Technology" or simply MIT, and "Stevens Institute of Technology". Rank and status became increasingly important as the influence of global league tables and performance indicators for research came to be an issue of some import in IoTs (Coate and Mac Labhrainn 2008). The functions (c), (d), and (e) mentioned in the above citation seem to suggest that already by 1992 a drift of policy by the Irish Government had taken place regarding the non-university sector with respect to academic drift and research drift.

However, the 1997 University Act reconfirmed that the function of the universities is to carry out all kinds of research ranging from basic to applied. By implication this interpretation follows from the lack of qualification of research in the 1997 University Act. The 1997 University Act only pronounced that a "university shall promote and facilitate research". This pronouncement delayed the development of research in IoTs considerably (Hazelkorn and Moynihan 2010b, p. 177).

Corresponding to Kyvik's phase model, horizontal integration took place from the late 1960s onward where a reoriented educational system was seen as a key facilitator of the pursuit of economic growth. During the strong growth period of the Irish economy from 1993 to 2008 – the so-called "Celtic tiger" era, corresponding roughly to Kyvik's phase three from 1990s onward – there was a policy shift in the rhetoric of the global knowledge economy. Policy emphasis shifted from access and widening participation in higher education to world-class excellence in research and innovation. The Department of Enterprise, Trade and Employment (DETE) in 2006 formulated the new policy emphasis this way:

> Ireland by 2013 will be internationally renowned for the excellence of its research, and will be to the forefront in generating and using new knowledge for economic and social progress, within an innovation driven culture (cited in Hazelkorn and Moynihan 2010b, p. 179).

The new rhetoric was promoted in white papers, position papers and national and transnational reports on the future needs of higher education in Ireland (see e.g. OECD 2004; Forfás 2007). More recently the requirement for world-class excellence in teaching and research was recognized through the *National Strategy for Higher Education to 2030* as laid down in the so-called Hunt Report (The Hunt Report 2011). These papers and reports are all signifiers of a neo-liberal agenda that includes new public management approaches endorsed by the EU and further stimulated by transnational agencies such as OECD and the World Bank. According to Daniel B. Saunders much of this shift is:

> …due to the infiltration of economic rationality within higher education, which has resulted in the prioritization of revenue generation and efficiency, corporate governance replacing shared and collegial models of decision making, faculty acting like entrepreneurs, and students being treated and identifying themselves as customers while simultaneously changing their goals and motivations in ways that correspond to the central ideas of neoliberalism (Saunders 2012, p. 66).

In this policy context the Irish non-university sector is structurally subordinated to the policies and expectations of state authorities, supranational organizations, societal stakeholders and academia, but not only. The non-university sector as a whole, as well as individual institutions acting separately, have also contributed in shaping their own trajectory of development (Kyvik and Lepori 2010, pp. 8–9). The neo-liberal agenda has been embraced and strongly supported by the Irish Government (see e.g. Donally 2004).

However, as Deirdre Lillis noted in 2007, between the *espoused theory* promoted through the rhetoric of excellence by the Irish Government and as reflected in institutional mission statements, and the *theory-in-use* in day-to-day practices in the Regional Technical College sector, there seems to be a chasm. The increase of student numbers is still serving as the main measure of performance as opposed to the measures of excellence in teaching, research and innovation promoted in official policy thinking (Lillis 2007).

The Irish system of higher education seems to have traversed Kyvik's three phases in a much less clear cut manner than was the case in the British experience, which in some sense might be seen as an ideal typical case. As a result, the Irish system of higher education may still to some extent be characterized as a diversified system. In Ireland, Kyvik's second phase was extended considerably, and only as recently as 2012 has a proposal for a transition to Kyvik's third phase been proposed in a policy paper by Simon Marginson (2012). Marginson, an Australian professor of higher education from the University of Melbourne, was commissioned by the Irish Government in the extension of the "National Strategy for Higher Education to

2030" report (The Hunt Report 2011) to consider a possible Technical University (TU) designation of Irish Institutes of Technology (IoTs). Marginson argued that a Technical University sector:

> ...would have the potential to eventually absorb most or all of the existing Institutes of Technology.......or to sustain a small number of TUs next to a small number of IoTs (Marginson 2012, p. 4)

As it appears from the quote, Marginson's argument is ambiguous. If only 3 or 4 of the largest IoTs are supposed to be designated as Technical Universities (TUs), a binary system will still prevail. In this case it is likely, or at least there is the risk that academic drift processes will prevail in the remaining "small number of IoTs", which may perceive themselves as having been "left behind". However here we can only speculate.

From its establishment, the Irish non-university RTC sector gave rise to a number of tensions. First among them was a tension between universities and IoTs. One of the main reasons for vertical integration in Irish IoTs was that graduates of 2 and 3 year certificate and diploma programs in IoTs were denied entry in the universities. As a result, many Irish students travelled to the UK with Certificates and Diplomas and were admitted to the latter stages of programs. This situation was untenable for the IoTs, and they came to act as main driving forces of vertical integration. However, due to its special status, vertical extension of professional programs at DIT took place earlier. This was the case in engineering, for example, where DIT offered 4-year programs since its establishment in 1978. Thus, vertical integration has been a strong driving force within DIT over a considerable timespan.

The tension between universities and IoTs also contributed to the rejection of a request by DIT in the late 1970s to be upgraded to university status, as the upgrading was strongly resisted by the universities who showed territorial attitudes. In Ireland the universities by tradition have a strong voice. Moreover the Irish Government and the Higher Education Authority (HEA) were afraid that they would also come under pressure from other IoTs aspiring to "wannabe" universities. By accepting such pressures from below on the part of some larger IoTs, the Irish government feared that this would result in policy drift.

A second tension was of a social nature. In RTCs, entry rates for children of lower socio-economic strata consisting of manual skilled, semi-skilled, and unskilled workers has been significantly higher than entry rates of these socio-economic groups in universities (Clancy 1997; O'Flaherty 2012, p. 3). However as observed by Kyvik (2009):

> In many ways, the binary model should be seen as a metaphor for the old class society, where the class a person was born into was decisive for his or her social status, cultural taste, and income. In the same way, the binary divide between universities and colleges would preserve a socially constructed and socially institutionalized division between noble and less noble higher education institutions (Kyvik 2009, p. 204).

A third tension was of a cognitive nature and was related to the research drift taking place in IoTs. A crucial issue here has been whether teaching programs can retain their standard and relevance if they are not research-informed by academics.

It has been argued (Kyvik and Lepori 2010) that a good reason for developing research is to strengthen the scientific basis of professional practice. However, shifting emphasis in engineering degree programs from teaching based on practical experience derived from engineering practice, to research-informed and research-led education creates crisis for many faculty members whose values and identities embody the core of a teaching culture. According to Ellen Hazelkorn and Amanda Moynihan most faculty:

> …held an undergraduate qualification with professional experience, but few had research credentials or practice. As the focus of attention has shifted towards more active engagement in the research enterprise, these academic staff have been asked to build up a sustainable research profile, participate in "national and international scientific networks" and develop a presence in international publications. The sheer magnitude of this transformation – on a personal and collective level – cannot be underestimated (Hazelkorn and Moynihan 2010a, p. 78)

Hence a tension and a number of ambiguities among faculty members across the teaching/research divide have created a number of impediments to developing research in IoTs. The key barriers that have been identified are: (i) Restrictive employment contracts (teaching and time constraints), (ii) Restrictive human resource practices (there has been a preference of employment of staff in the past who were inexperienced in conducting research), (iii) Lack of physical infrastructure, (iv) Poor management structure, (v) Inadequate funding models, (vi) Resource issues concerning annual leave, and (vii) Institutional ambivalence toward a research culture (Donavan et al. 2012, p. 330).

A fourth and final tension that needs mentioning here is a tension regarding the research mission of IoTs. This tension is related to: (i) what would be considered justifiable research, and (ii) what would be considered unjustifiable research in IoTs. Discussing the possible research mission of IoTs, it has been argued by John Donavan, Anthony Betts, and Stephen Jerrams, all from DIT, that academic research and development in a knowledge economy must, among other things:

> (v) inform undergraduate education and ensure that degree courses keep pace with the science, technology and methodology at the leading edge of industrial and commercial activity.
>
> ……Conversely, where research becomes distant from mainstream faculty activity and researchers pursue goals that are inconsistent with a university's mission and ethos, in all but a few elite organizations, the research is unjustifiable. To ensure that research informs teaching and learning and to avoid the proliferation of specialist research teams divorced from the academic process, R&D centers and groups must be grounded in and across faculties (Donavan et al. 2012, p. 335).

In closing this section we might therefore conclude that attempts to bridge the gap between IoTs and the universities have only been partially successful, despite the fact that a significant increase in the quality and quantity of research conducted in IoTs has occurred. However, in addition we would also argue that in many ways IoTs are facing problems of a general nature related to Kyvik's third phase of transformations of systems of higher education. Presently Christensen's former Danish home institution, the former *Institute of Business and Technology, Herning* that merged with Aarhus University in 2006, is struggling to respond to a set of challenges that to a great extent is of a similar nature.

Dutch Hogescholen (HBOs)

The Dutch system of higher education can be characterized as a *binary system* with a dividing line between research universities and institutions for higher professional education – in Dutch language the latter type of institutions is termed *Hogescholen voor Hoger Beroepsonderwijs* (HBOs). As in Ireland each sector has different historical roots and is based on different rationales. The history of the Dutch university sector dates back to 1575, when the University of Leiden was founded as a reward for the bravery of Leiden's citizens when fighting against the Spaniards during the 80-year war (1568–1648). As the outcome of the war the Netherlands gained independence from Spain. Soon after the founding of the university in Leiden, the universities in Groningen (1614), Amsterdam (1632) and Utrecht (1636) were established. Over the centuries additional universities were established, with some of the more recent ones being part of an explicit government economic policy aimed at boosting economic growth in some disadvantaged regions. Examples include the University of Twente (1961) and the University of Maastricht (1976). Some institutions belonging to higher education were founded by protestant and catholic clergymen, such as the protestant Free University in Amsterdam (1880) and the research-intensive Roman Catholic Radboud University in Nijmegen (1923). All of the above mentioned institutions are funded by the Dutch government and are part of the national accreditation system (de Wert and Boezerooy 2007, pp. 11–12; National Institution for Academic Degrees and University Evaluation 2011, p. 6).

Presently there are 13 Dutch universities, 9 of which provide teaching and conduct research within a wide range of academic disciplines, namely Erasmus University Rotterdam, Leiden University, Maastricht University, Radboud University Nijmegen, Tilburg University, University of Amsterdam, State University of Groningen, Utrecht University, Free University Amsterdam. Three others, Delft University of Technology, Eindhoven University of Technology, and University of Twente, have a main focus on engineering and technology. The final one – Wageningen University and Research Center – has an exclusive focus on agriculture. In addition, the Dutch university sector also includes the open university and a number of small institutes with university status (de Wert and Leijnse 2010, p. 199; de Boer and Goedegebuure 2007, p. 46).

By 2010 the HBO sector consisted of 45 publicly-funded *hogescholen.* Many of these existed long before the sector as a whole gradually became part of the Dutch binary system of higher education from the 1960s onward. The idea of the existence of two sectors can be traced back to the early 1900s when the Domestic Science and Technical Education Act was passed by The Dutch Parliament in 1919. However, when looking exclusively at legislation, as argued by Jeroen Huisman and Frans Kaiser

(Huisman and Kaiser 2001, p. 27), it could be claimed that the Dutch binary system of higher education has only been officially known since 1986. In 1986 a redefinition of the status of the higher professional education sector took place. The 1968 Act of Secondary Education stipulated that *hogescholen* belonged to secondary education and was therefore not part of a binary system of higher education. In 1986 the HBO sector was taken out of its secondary education status and jurisdiction and given separate legislation (HBO Act 1986), thus formally becoming part of a binary system of higher education. Moreover in the 1993 Higher Education and Research Act a further step in the upgrading of *hogescholen* took place as the two higher education sectors were now defined as "equal but different" in status. The 1993 Act is still serving as the legal foundation of the Dutch higher education system (de Boer and Goedegebuure 2007, pp. 47–48). As a signifier of their new status, Dutch *hogescholen* have adopted the name *Universities of Applied Science* in an international context.

According to Harry de Boer and Leo Goedegebuure (2007, p. 48) numerous merger processes in the HBO sector aiming at horizontal integration started in 1983 after the publication of a government White Paper titled *Scale-enlargement, Task–Allocation and Concentration.* The title of the White Paper is mimicking the rationale of Kyvik's second phase, aimed at field contraction, authority unification, institutional de-differentiation, program coordination, and regionalization. As a result, the number of *hogescholen* was drastically reduced from well above 400 in 1983 to a little more than 40 in 2005. Despite the aim of establishing fewer, larger and more multi-sectoral teaching institutions for professional practice, a characteristic feature of the HBO sector is that *hogescholen* differ considerably in scale in terms of size and range of fields of study. As observed by Egbert de Wert and Frans Leijnse (2010, p. 201), 15 out of the present 45 *hogescholen* are large multi-sectoral institutions encompassing a broad range of fields of study with student enrolments ranging from 12,000 to 35,000 students. In the middle category, 15 *hogescholen* cover more than one subject area with student enrolments amounting to less than 10,000 students. In the category of smaller institutions, the remaining 15 *hogescholen* focus mainly on one or two areas such as teacher training, fine arts, agriculture or hotel management. Their student enrolments range from several hundred to a few thousand students. In 2007, student enrolments in Dutch higher education totaled 587,500 students. The HBO sector took the lion's share of higher education students amounting to 65 % (374,000) against 35 % in universities. In 2007 the number of engineering student enrolments at the bachelor's level in the HBO sector totaled 60,000 students representing 16 % of all HBO students at the bachelor's level.

Although the term "research" in relation to *hogescholen* appeared already in the Higher Education and Research Act from 1993 – indicating a transition to Kyvik's third phase of vertical integration – it was not given clear definition. The 1993 Act only pronounced that:

> Hogescholen have as a task to provide higher professional education. They can carry out research to the extent that this is connected with the education at the institution (Cited in de Wert and Leijnse 2010, p. 202).

The European policy drift taking place in the context of the Bologna Process and the Bologna Declaration in 1999 resulted in a treaty between the Dutch Minister of Education and the *hogescholen* (2001) which defined a novel set of tasks for all *hogescholen* in addition to their traditional teaching tasks. These are:

1. To raise the quality of the educational programs and the quality of the teaching staff.
2. To add to the theoretical body of knowledge of the different professions.
3. To help the professional field to innovate (cited in Griffioen and De Jong 2012, p. 4)

This definition indicates that the HBO sector is now committed to fulfill a tripartite goal: (1) an educational goal, (2) a theoretical goal, and (3) an external goal. Moreover all of these goals would have the potential of implying a research dimension. In principle this might involve the risk that the three goals might stimulate research following entirely different research trajectories thus blurring the distinctive role of research in the HBO sector. However, according to Egbert de Wert and Frans Leijnse (2010, p. 207) from analyses of institutional annual reports the following common components have been identified in institutional research strategies and priority setting:

1. Initiatives for research emanate from the needs of professional practice
2. Research should be relevant for the quality and innovation of education and the professionalization of the teaching faculty
3. Research should be practice-driven in that it is oriented to solve practical problems and to intensify collaboration with external constituencies

The creation of a special position at a professorial level termed *the lectorate* – the position of *lector* should not be conflated with the traditional position of *lecturer* in the Anglo-Saxon tradition – was seen as a means to enhance the quality of professional education and the qualities of the teaching faculty as stated above. The purpose of the new position as a *lector* is to respond to the knowledge needs of SMEs and professional organizations and to stimulate the development of research skills among the teaching faculty by recruiting these to research projects in which they conduct research on part-time basis. Moreover *lectores* are expected to create so-called "knowledge circles" each group consisting of 10–15 staff members. Finally they are expected to provide external funding and contracts and to develop professional network in their domain (OECD 2008, p. 75; de Wert and Leijnse 2010, p. 208). Despite a favorable review of the *lectorate,* OECD in 2008 expressed the following reservation regarding its current impact on developing research capabilities among teaching faculty in the HBO sector:

> It is important to note that the resources required to properly develop the research capacities of the HBOs are very substantial given that these efforts almost start from scratch. In this context the *lectoren* system is scarcely more than a drop in the ocean and may simply constitute the equal distribution of scarcity. This suggests that it would be advisable to target resources and expand the system very selectively through a bid-based system oriented to the development of research in the HBOs. The aim would be to create competitive research environments and promote strategic thinking in HBOs. In this process, HBOs should obtain resources and abilities to attract young research-trained people to start their research programs. These programs should have a regional orientation and should be able to attract external money (OECD 2008, p. 81).

It is noteworthy that, for HBOs starting from scratch in becoming practice-oriented research institutions, OECD recommends as an effective strategy that they should hire young research-trained people from the outside instead of a developing research competence among an aging teaching faculty. There is the risk that such a strategy might backlash. However here we can only speculate.

As in Ireland the history of the binary system in the Netherlands has been and still is characterized by a number of tensions. First among them is a tension between the university sector and the HBO sector. In 2001 Huisman and Kaiser noted that on the one hand tensions have grown with the upgrading of the HBO sector, and on the other hand have been further aggravated with the vocational drift taking place in universities. As a result of the latter, universities move more into the area of professional education instead of solely stressing academic education (Huisman and Kaiser 2001, p. 32). Regarding the changing distribution of roles between the two sectors there is the risk that instead of co-existence based on a clear division of labor it may instead lead to fierce competition among them. Hence a major issue is the fear by universities that the trend toward the extension of the research function and vertical extension of degree programs in the HBO sector will constitute a threat to their privileged status in terms of access to research resources and funding (de Wert and Leijnse 2010, p. 200).

At the faculty level a tension related to educational attainment may be observed. Until recently HBOs refrained from serious efforts to raise the qualification levels of their faculty. By contrast Dutch universities since the 1980s have put much effort into upgrading their faculty by increasing the number of Ph.D. courses and setting the Ph.D. as a minimum requirement to enter the academic faculty. As a result of this situation HBOs

> …have soldiered on with a teaching faculty which for 47% has a bachelor degree only (most of them at the UAS level) and (thus) no research qualifications whatsoever (de Wert and Leijnse 2010, p. 210).

In closing this section we may therefore say that the tensions and impediments found in Dutch *hogescholen* to a great extent resemble those found in Irish IoTs. As a result, *hogescholen* still struggle to overcome the old teacher's culture and to view research and knowledge production as part of their core competence.

Structural Dynamics in the American System of Higher Education

The United States system of higher education is diverse and complex, comprising a multitude of both private and public institutions that span a broad range of sizes, types, and missions. At the national level, the government neither administers institutions nor mandates any particular institutional forms, though it does have influence through various policies, such as for student loans and research funding (Tuchman 2009). Publicly funded institutions are administered at or below the state level, and the systems that have been put in place vary considerably across the 50 states. And while state governments play a critical role in directing and funding public higher education, they also permit a great deal of institutional autonomy. Michael Dobbins, Christoph Knill, and Eva Maria Vögtle classify the U.S. system of higher education as ideal-typical *market-oriented*, which means that, "Instead of shaping and designing the system, the state tends to promote competition, while ensuring quality and transparency" (Dobbins et al. 2011, p. 672). With the flexibility to adapt to the market, institutions in the U.S. system, both public and private, are capable of relatively rapid growth and change. Thus, not only is academic drift possible in the U.S. system, it might be perceived as necessary for institutional survival in the marketplace (Jaquette 2013).

David Labaree (2006) has posited four broad tiers of higher educational institutions that have arisen in the course of U.S. history. The oldest and most prestigious institutions, established prior to the mid-1800s, are the *colonial colleges*, which comprise many of today's elite private universities, along with some early state colleges that subsequently evolved into top tier state flagship universities. Labaree's second tier consists of state *land grant colleges*, arising out of the Morrill Acts of 1862 and 1890, which facilitated the creation of institutions in each state to promote practical education, particularly in the agricultural and mechanical arts. As will be discussed in more detail below, many of these institutions later evolved more cosmopolitan missions and have now earned state flagship status, rivaling those in the first group for positioning among the top tier of U.S. research universities. The third tier, developing from the mid-1800s, were the *normal schools* – state vocational institutions for the purpose of training school teachers, catering primarily to women. Through the early to mid-twentieth century these normal schools evolved into *state teachers colleges*, then to *state colleges,* and finally into *state universities* (Labaree 2006; Graham 1978). These institutions now typically represent the second tier of state institutions, behind the flagships. Labaree's final tier comprises *the junior/community college system*, a sub-baccalaureate system arising from the early twentieth century onward.

While Labaree's tiers aptly describe major segments of the U.S. higher education landscape, they do not completely capture its historical scope and complexity. For example, Claudia Goldin and Lawrence Katz (1999) report that the greatest expansion of the number of institutions of higher education in U.S. history occurred in the latter half of the nineteenth century, with over 400 new institutions coming into existence between 1860 and 1900 alone. But the vast majority of these – 5 out of every 6 – were either relatively small private colleges (often sectarian) or independent professional schools (Goldin and Katz 1998, 1999; Grubb and Lazerson 2005). Baylor University, the home institution of Newberry, fell into the first of these categories, but also overlapped the second. Founded by the Baptist denomination in Texas in 1845, in what were still relatively frontier-like conditions, the institution offered classical-style collegiate studies, but within a few years had also started offering a professional course in law.

Toward the end of the nineteenth century, therefore, there was a profusion of vocational and professional educational programs across a wide range of institutions.

> Quantum leaps in the number of professional schools of law, engineering, business, pharmacy, and medicine in the late nineteenth century testified to the perceived value of formal vocational training and sent signals that ricocheted through American education (Kett 1994, p. 244).

But in the period during and after the turn of the twentieth century there was a significant horizontal integration of such professional programs, sending the number of independent (non-college/university-affiliated) institutions into decline. Between 1900 and 1934 the percentage of professional students being trained at independent institutions dropped from 48 % to 19 % (Goldin and Katz 1999). One reason was that the time period leading up to the turn of the twentieth century witnessed the rise of scientific approaches in the professions, accompanied by the growth of societies of learned professionals and of professional licensing requirements (Goldin and Katz 1998, 1999; Grubb and Lazerson 2005). Professionals were increasingly expected to have a broader, more liberal, and more rigorous training, and "the schools providing the credentials were increasingly required to have state approval, research facilities, and unassailable reputation" (Goldin and Katz 1998, p. 15).

Despite their practical and vocational educational objectives, land grant schools had already embarked upon a trajectory to provide a broader and more comprehensive education (Grubb and Lazerson 2005), so they were well positioned to meet this need for more refined professionals. But many independent professional schools, with their small size and narrow focus, were less able to adjust to the broader educational requirements and were thus often absorbed into, or supplanted by, colleges and universities. One mechanism for this was through the rise of the comprehensive state universities that occupy Labaree's third tier. These institutions often developed by way of the expansion of a normal school – or state teachers college – through additions and/or mergers (Grubb and Lazerson 2005; Goldin and Katz 1999). Wayne State University is an example of this. It was created in 1933 by the merger of several previously independent institutions, including the Detroit Medical College, the City College of Detroit, and the Detroit Teachers College (previously the Detroit Normal Training School for Teachers) (Britannica 2013). Finally, many liberal arts colleges and denominational institutions absorbed or otherwise initiated professional programs. For example, between 1903 and 1923 Baylor University assimilated previously independent professional schools of medicine, nursing, pharmacy, and dentistry, along with launching new programs in business, music, and education (Russell and Murray 2013; Baylor 2012).

Following the analysis of Labaree (2006), we might view these developments as a two-way drift, or convergence. Liberal arts colleges, ranging from the elite Ivy League colleges to small regional colleges, trended toward offering vocational/professional programs, while professionally/vocationally-oriented institutions, ranging from public land grant schools to independent professional institutes, trended toward offering a more well-rounded liberal education. Several authors have analyzed the drift of liberal arts institutions towards professional education. These studies indicate that many of the institutions that self-identify as liberal arts schools must continue do so for reasons of either tradition or marketing, because such a label does not correlate with the current realities of their curricular offerings (Delucchi 1997; Grubb and Lazerson 2005; Brint et al. 2005; Baker et al. 2012). "At many liberal arts colleges, more than 60 % of degrees are awarded in professional fields" (Neely 1999, p. 36).

With respect to engineering education specifically, by the early twentieth century it was already embedded in a variety of diverse institutional settings. The Engineers' Council for Professional Development (now ABET) issued its first accreditations of engineering programs in 1936. By 1937 there were 107 institutions with accredited engineering programs (ECPD 1971). Not surprisingly, 51 of these, or just under half of the total, were state flagship universities and state land grant institutions (14 and 37, respectively). Perhaps a little more surprisingly, the next largest group – 30, or just under a third of the total – consisted mostly of institutions that could be classified, at least historically, as private liberal arts colleges. But many of these – though not all – had evolved, or were evolving, into major research universities (Trow 2000). This group included institutions such as Harvard, Yale, Princeton, Swarthmore, Tulane, Syracuse, the University of Southern California, and Santa Clara University. The remaining group – 26 in total – comprised technological institutes, both public and private, along with a handful of military institutes. This group ranged from elite private institutions such as MIT and Caltech, to a variety of state schools of mining and technology, such as the Colorado School of Mines. One conclusion that might be drawn from this cross-section of early engineering programs is that the precedent had already been set that there were no limits, at least in principle, on the types of institutions that could offer engineering degrees in the United States – public or private, small or large, liberal or technical. This is consistent with the autonomy of U.S. institutions to respond to the educational marketplace.

Paralleling the diffusion and integration of professional and practical education within the U.S. higher education system during the late nineteenth and early twentieth centuries was the rise of the research university model. According to Gumport (1993, p. 227), "Graduate education achieved a stable American presence during the last two decades of the nineteenth century, when awarding a Ph.D. became a laudable academic goal." Marked increase in enrollments, the rise of academic science and learned societies, the trend toward professional disciplinarity and specialization (and thus departmental structure), the expansion of curricular offerings within institutions, and the infusion of research funding, first from private philanthropic sources and later from government, all helped accelerate the development of the research university structure (Goldin and Katz 1998, 1999; Gumport 1993). The development of graduate education and research helped stratify U.S. higher education, and early entrants into those arenas went a long way toward solidifying their status and prestige.

Engineering tended to go hand in hand with the development of research universities. Seven out of ten of the top eighty U.S. research universities in 2009, as ranked by the Center for Measuring University Performance, already had accredited engineering programs prior to 1950 (Lombardi et al. 2011; ECPD 1971). Industrial research laboratories proliferated in the first half of the twentieth century, and companies were eager to forge ties with universities that could supply research workers in engineering and other technical fields (Levine 1988). These ties were strengthened by new cooperative education programs that allowed engineering students to alternate between classroom studies and work in industry, starting with the University of Cincinnati in 1907 (Levine 1988; Seely 1993). Around the same time, land grant schools, led by the University of Illinois and then Iowa State, began to set up engineering experiment stations, modeled on earlier agricultural experiment stations, for the purpose of conducting practical research to benefit their local regions (Seely 1993). As university research in engineering grew through the first half of the 1900s, so did the scientific content of the engineering curriculum. This shift in balance from the practical towards the theoretical was helped by the immigration to the U.S. of scientifically trained European engineers like Theodore von Kármán, Stephen Timoshenko, and Richard von Mises (Seely 1993, 1999).

Research Drift in American Engineering Education

After World War II, the theoretical and scientific trend in engineering education accelerated, as did the emphasis on research. With the creation of the National Science Foundation and NASA in the 1950s, federal funding for scientific research expanded, including for research in the *engineering sciences*. To help garner such research funding, from the mid-1950s through the mid-1970s many schools went so far as to create new academic programs in *engineering science*, which helped to push engineering further into the theoretical realm (Newberry and Farison 2003). The seminal Grinter Report of 1955, which made recommendations for the future of engineering education, also emphasized the need for more scientifically trained engineers. This trend created a divergence between engineering academia and engineering practice, with the result that, "academic and industrial engineers developed different conceptions of engineering, and they almost stopped talking to each other" (Seely 1993, p. 383).

This led to the perception in some engineering circles of a need for a new type of academic program to fill "the 'vacuum' created by engineering colleges as they tend to shift toward engineering science" (NRC 1985, p. 5). This new type of program was meant to be a practical and hands-on educational program more closely resembling the laboratory- and shop-style engineering education that existed in the earlier part of the twentieth century (Feisel and Rosa 2005; Weese and Wolf 1994; NRC 1985).

> Among the compelling forces for creation of the baccalaureate degree in engineering technology was the growing group of graduates from the two-year community colleges desirous of a four-year degree but effectively foreclosed from the traditional engineering program. A final cause was an economy that simply could not find enough technically

> trained manpower from existing engineering schools. Those companies with engineering needs, particularly those who's very existence was tied to manufacturing, were happy to hire technical talent wherever it was to be found (Kenyon 1993, p. 363).

The ECPD had already been accrediting 2-year sub-baccalaureate engineering technology programs since 1946, which addressed a need for skilled technicians. In 1967, the ECPD began accrediting 4-year engineering technology programs, which were aimed at filling the niche between "the craftsman and the engineer" (NRC 1985, p. 7). The number of baccalaureate engineering technology programs climbed rapidly through the 1970s and 1980s, but there has been slow but steady decline in the number of programs from the early 1990s onward (NRC 1985; ABET 2003, 2010). The *engineering technology* program title originated in the context of 2-year programs based largely at community colleges and technical institutes in the mid-1900s (NRC 1985). The growth of four-year engineering technology programs, on the other hand, occurred in technical institutes as well as in public and private universities. By the early 2000s, four out of five engineering technology baccalaureate degree-granting institutions identified themselves as a university (Fox et al. 2004). While the technicians produced by 2-year programs have a distinct identity and job market, the bachelor's level engineering technology programs have consistently struggled with both identity and image relative to engineering (Kelnhofer et al. 2010; Weinsier and Brown 2008; Kenyon 1993). "For example, the lesser emphasis on theory and mathematical rigor causes engineering technology to be viewed as inferior to engineering, that is, *engineering-light*. This is perhaps the most damaging stereotype" (Kelnhofer et al. 2010, p. 3). Despite the perceived need for less theoretical engineering-type professionals, the scale of engineering technology education has remained much smaller than for engineering. The annual number of engineering technology bachelor's graduates lags that of engineers by an order of magnitude (Ford and Ball 2011).

Engineering technology programs arose in the past half-century as an extra-engineering alternative to the increasingly theoretical and research-oriented engineering education paradigm. During roughly the same time frame there was also a parallel development within engineering education that provided what could be thought of as a second alternative – a wave of new engineering programs at smaller, less prestigious institutions. Of the approximately 400 institutions currently offering ABET-accredited engineering programs in the U.S., over half are relatively new, receiving their first accreditation within the past 40 years (see Table 2.2 at the end of this chapter – the subsequent discussion will draw heavily from this table). Two types of institutions dominate this large group of new entrants into the engineering education market: smaller state institutions, or satellite campuses of larger ones, and small regional private institutions, many of which are religiously affiliated. But despite the large number of institutions, this group represents only a relatively small share of the engineering student market, and a much smaller share of the research activity.

To put this in perspective, by the start of the 1950s there were about 130 accredited engineering programs in operation. While this group only accounted for about

Table 2.2 2010–2011 Enrollment and engineering degrees for ABET accredited engineering

Date of first accredited	Institution	Institution	Median institution	Average number of engineering degrees awarded by institutional group			Average total inst.
Engineering program	Count	Fraction	Founding date	Bachelors	Masters	Doctorate	Enrollment
Private							
up to 1949	51	0.35	1861	224	218	53	10,732
1950–1969	19	0.13	1891	105	64	15	9,337
1970–1989	32	0.22	1889	96	40	16	6,261
1990–2010	43	0.30	1901	23	14	0	5,615
Total	145						
Public							
up to 1,949	81	0.33	1870	454	195	59	25,154
1950–1969	31	0.13	1898	250	156	32	23,405
1970–1989	70	0.29	1933	150	82	19	17,907
1990–2010	63	0.26	1922	48	35	15	11,360
Total	245						
All institutions							
up to 1949	132	0.34		365	195	50	
1950–1969	50	0.13		195	108	15	
1970–1989	102	0.26		133	56	7	
1990–2010	106	0.27		36	9	1	
Total	390						

Notes

Accreditation data are taken from ABET database at http://main.abet.org/aps/Accreditedprogramsearch. tutions currently offering engineering, institutions are not may have discontinued offering engineer- *technology, engineering physics,* or *computer science* programs. It also does not include any non tion of when an engineering program was first initiated. Particularly for the earliest institutions to advent of accreditation

Degree data are for 2010–2011, taken from IPEDS Data Center, National Center for Education In addtion to the institutions accounted for in the table, IPEDS data indicate an additional 11 public institutions do not appear m the ABET engineering program database. Some of these appear to be appear to be young programs which likely plan to, or are m the process of seeking accreditation are from 2010

programs by time period

Median number of engineering degrees awarded by institutional group			Median total inst.	Bachelors degrees		Masters degrees		Doctorate degrees	
Bachelors	Masters	Doctorate	Enrollment	Total	Fraction	Total	Fraction	Total	Fraction
208	143	37	8,705	11,431	0.65	10,263	0.85	2,170	0.93
61	39	15	7,180	1,988	0.11	889	0.07	92	0.04
63	22	3	4,111	3,066	0.18	836	0.07	81	0.03
18	8	0	3,663	967	0.06	111	0.01	0	0.00
				17,452		12,099		2,343	
319	128	35.5	25,498	36,780	0.64	15,420	0.60	4,490	0.76
218	*122*	24	23,085	7,741	0.13	4,524	0.18	649	0.11
111	60	10	15,077	10,473	0.18	4,834	0.19	674	0.11
27	12	11	9,500	2,880	0.05	838	0.03	88	0.01
				57,874		25,616		5,901	
				48,211	0.64	25,683	0.68	6,660	0.81
				9,729	0.13	5,413	0.14	741	0.09
				13,539	0.18	5,670	0.15	755	0.09
				3,847	0.05	949	0.03	88	0.01
				75,326		37,715		8,244	

aspx, accessed 1/29/2013. Historical information is based on initial program accreditation dates for insti-
ing are not accounted for, but are assumed to be relatively few. This does not include *engineering*
accredited or not-yet-accredited programs. The year of first accreditation may or may not be a reflec-
receive ABET accreditation, engineering programs were generally in place for decades prior to the

Statistics. http://nces.ed.gov/ipeds/datacenter/. Accessed 2/24/201)
and 17 private institutions awarding a total of 395 engineering bachelors degrees in 2011. These
established niche programs that have for some reason chosen to forego accreditation. The majority
Some may already be accredited since the latest accreditation dates reported in the ABET database

one third of the total number of institutions with accredited engineering programs by 2010, it accounted for approximately two thirds of all engineering bachelors and masters degrees awarded in 2011, and for four fifths of all engineering doctorates. So while in one sense the landscape of engineering higher education in the United States has changed dramatically in recent decades with the influx of many new types of institutions offering engineering degrees, in another sense little has changed since the traditional engineering educational institutions still account for the bulk of engineering degrees and research. To some extent, it should be no surprise that this is the case. Programs, infrastructure, and reputation take time to build, particularly for resource intensive programs like engineering, so early entry into the market confers a significant advantage. And, like compounding interest, established programs and infrastructure can attract more resources at an accelerating rate. Also, as is evident from Table 2.2, the institutions that were the earliest to offer engineering, whether public or private, tend to be the largest of their type in terms of student enrollments, and hence in terms of faculty numbers and physical facilities as well.

During the 20-year period from 1950 to 1969, the addition of engineering programs at new institutions was steady, but at a slower pace than would be true in subsequent decades. Only 13 % of institutions currently offering engineering had their programs first accredited during these years. On the private side, many of the institutions adding engineering during this time were some of the remaining larger, comprehensive private universities (e.g., Tulane University, Brigham Young University, University of Dayton). The median enrollment at these institutions is 7,180 compared to 8,705 for those accredited prior to 1950. On the public side, schools joining the engineering ranks were some additional major state universities (e.g., Arizona State, University of Georgia, University of Hawaii), as well as many of the larger second tier state universities (e.g., University of Houston, University of Texas at Arlington, University of Wisconsin-Milwaukee, Ohio University, San Diego State). In fact, the majority of the public additions at this time were large, comprehensive universities from populous states – 19 of the 31 were from California, Texas, New York, and Ohio alone. The median enrollment size for public institutions which added engineering during these 20 years is 23,085 which is a relatively modest drop from the 25,498 for public institutions having engineering prior to 1950. But despite the fact that the overall size of the 1950–1969 institutions, both public and private, is relatively close behind the pre-1950 institutions, those in the latter group currently produce on average about twice as many engineering bachelors and masters degrees as those in the former group, and over three times as many doctorates. So the institutions that started engineering prior to the mid-twentieth century still dominate, particularly in research.

From 1970 onward the pace at which new institutions adopted engineering doubled over the preceding 20 years, with just over one quarter of all institutions which currently offer engineering receiving their first accreditation between 1970 and 1989, and just more than another quarter, between 1990 and 2010. Yet despite the fact that over half of all the institutions that offer engineering began doing so during this most recent 40-year period, this group only produces less than a quarter of bachelor's degrees, less than a fifth of masters, and only a tenth of doctorates. The overall size of the institutions drops steadily too, with those in the most recent 20-year period having median enrollments of less than 4,000 for privates and less than 10,000 for publics.

With the decrease in size comes a change in the character of the institutions. Before recent decades, engineering was largely the purview of the heavyweights in the state university systems, along with larger or more elite private institutions with national reputations. Now, small regional private universities, along with many second tier state universities, and satellite campuses of larger institutions, have entered the engineering education market. For the 63 public institutions for which engineering was first accredited from 1990 to 2010, the median number of engineering bachelor's degrees awarded by the group in 2011 was only 27. Almost all of these public institutions – 59 out of 63 – are small, comprehensive *universities*, while the remaining four are public technical institutes. For the 43 private schools for which engineering was first accredited from 1990 to 2010, the median number of engineering bachelor's degrees awarded by the group in 2011 was just 18. Twenty-seven of the forty-three private institutions label themselves *universities*, with the remainder being *colleges* (with the exception of one *institute*). While the engineering programs at these institutions are young, the schools themselves are generally not. The median founding date of this group is 1901 for the private institutions and 1922 for the publics. So the decision to adopt engineering represents a departure from historical precedent for many these institutions. Of the 43 privates, 27 are religiously-affiliated institutions with liberal arts backgrounds. Three are institutions currently or historically devoted to the education of women or minorities. Nine are rooted in technical or professional education. The remaining four are small, regional universities.

Jaquette offers some clues as to why these types of smaller public and private institutions, traditionally with limited missions, have trended towards becoming comprehensive universities offering a wide range of programs, including engineering. A large factor is the *enrollment economy*, in which tuition-driven schools, both private liberal arts institutions as well as professionally-oriented public colleges faced with an era of declining state support, feel the need to compete for students. Diversification, often into new professional and graduate programs, broadens an institution's appeal, creates economies of scale, and buffers the institution against fluctuating demand for particular programs (Jaquette 2012). The only colleges that are likely secure enough in their applicant pools to be able to resist the pressures to expand and diversify are selective ones that possess a highly prestigious reputation with respect to a tightly focused educational mission. For the rest, "by becoming a university, non-selective colleges make a horizontal leap into a fundamentally different prestige market, one that does not view them from a deficit perspective" (Jaquette 2012, p. 26). But the economically-driven move to comprehensive university status comes with a potential side effect. It can lay the foundation for a subsequent research drift. Referring to second tier comprehensive universities, Grubb and Lazerson write,

> For substantial numbers of faculty hoping more prestigious universities will recruit them, and for countless numbers of administrators and trustees hoping to emulate the major research universities, the pot of gold at the end of the rainbow is to become like the selective

> colleges and major research universities by ratcheting up admissions standards, creating honors programs, dropping remedial programs, adding doctoral degrees, and expanding research (Grubb and Lazerson 2005, p. 19).

It might be instructive now to consider three examples of institutions that are in the pool of latecomers to engineering, and that evolved into comprehensive universities subject to research drift. These three comprise what were historically a public technical institute, a state teacher's college, and a private, sectarian liberal arts university.

Southern Polytechnic State University in Georgia is a public institution founded in the late 1940s as Southern Technical Institute, with the original goal of providing sub-baccalaureate technical education. In the early years it catered heavily to WWII veterans. In the early 1970s, it added 4-year engineering technology degree programs. As of the mid-2000s it has launched several bachelor's degree programs in engineering (while retaining engineering technology bachelor's programs; sub-baccalaureate technology programs were eliminated in the 1990s). The name of the institution transitioned to Southern College of Technology in the 1980s, and then to its current university designation in the 1990s (SPSU 2013). Befitting a university, curricular offerings have expanded to include business, education, sciences, social sciences, and humanities. In addition, the institution also now offers several master's level graduate programs, including a new one in civil engineering. Although not historically a research institution, SPSU appears to be striving in that direction. The institution's 2010–2013 strategic plan includes the following objective: "Increase revenue from grants and contracts to $5,000,000/year.... Pursue NSF and similar research funding ($500,000/year) for faculty course buyout" (SPSU 2010, p. 25). This recent emphasis on research is consistent with the findings of Kenneth Rennels, who studied *mission creep* in engineering technology programs.

> The data would seem to indicate that engineering technology faculty expectations are rising in terms of basic faculty credentials and research expectations. Furthermore, the rising expectations appear to be driven by internal factors [such as criteria and expectations for promotion and tenure] (Rennels 2003).

Following the story line described in the previous section of this chapter, Western Kentucky University's path to becoming a comprehensive state university took it from its origins as a *normal school* through intermediate stages as a *state teachers college* and then a *state college*, with several mergers with other institutions along the way (WKU 2013). Western Kentucky currently offers bachelor's degrees in three engineering disciplines: mechanical, electrical, and civil. These programs are relatively recent, first receiving accreditation in the early 2000s. The university had engineering technology bachelor's programs starting in the early 1970s, but in contrast to SPSU, WKU dropped engineering technology with the advent of engineering. Both SPSU and WKU are examples of a general trend for many institutions that initially have only engineering technology programs to drift in the direction of engineering, with some, like WKU, eliminating technology in the process. Western Kentucky currently offers numerous master's programs, along with a few clinical doctoral programs. With respect to research aspirations, WKU's 2010–2012 strategic plan states,

> The volume of research and scholarly activities across campus must grow, not so much to achieve specific numbers, but rather to engage as many faculty and students in research activity as possible … it is incumbent on those involved to aggressively seek external support where possible. Collectively, these grants, contracts, fellowships and other kinds of support facilitate intellectual endeavors and significantly enrich institutional quality (WKU 2010, pp. 8–9).

Baylor University (Newberry's home institution) was founded as a liberal arts college in 1845 in what was at the time the Republic of Texas. It began diversifying into professional programs in the early 1900s and has since grown into a comprehensive university. It is now classified according to the Carnegie classification scheme as a Research University/High Research Activity. It is a medium-sized private university affiliated with the Baptist denomination. While it has a national reputation, the majority of students are from Texas. With respect to engineering Baylor is a relative newcomer with the first engineering program accreditation dating to 1987. The first engineering faculty member was hired in the early 1980s. A nascent general engineering program grew slowly through the mid- to late-1990s, stabilizing for a time at approximately 10 engineering faculty members and about 25 engineering graduates per year. For the first decade and a half of its existence, the mission of the program was explicitly focused on providing a very personalized and hands-on undergraduate engineering education. Engineering faculty members were hired for their interest and skills in teaching, and there were no research expectations. The motivation for developing an engineering program had both outward- and inward-looking components. The outward component was based on the assumption that there was an important service to be provided to students who are interested in engineering and who also are attracted to a regional, religiously-affiliated university that offers a smaller, more intimate alternative to the very large public research universities. This reasoning parallels that described by Morphew (2000) in which faculty at comprehensive universities, which Baylor was at the time, justify new programs based on meeting the needs of the community. An article in the school newspaper from around the time of the program's founding supports this view:

> More and more college graduates are finding the market for their skills flooded, and therefore highly competitive. With this in mind, there is one major that almost guarantees a highly paid job upon graduation – engineering. To satisfy the need for qualified engineers, Baylor has established a new engineering sciences program… (Ledbetter 1979).

The inward component of motivation was based on the recognition that engineering typically attracts high-achieving students, which was desirable for the institution. There was also another motivation peculiar to Baylor's demographics. For several decades the university's enrollment has skewed more heavily toward women. Since engineering generally attracts a preponderance of men, the administration viewed engineering as a means of counter-balance.

In the late 1990s, the Baylor administration made a somewhat informal decision to place more emphasis on academic research, which led to new engineering faculty hires with increased, but still modest, research expectations. However, in 2002 the university launched a new strategic plan, Vision 2012, which more forcefully articulated the institution's commitment to enhance its reputation as a research university. The administration, viewing engineering and the sciences as the lynchpin for research prominence and external funding, began rapidly expanding the engineering faculty, and concomitantly the scholarly requirements upon them. The program grew from a single general engineering major with 10 faculty and 25 graduates per year in the late 1990s, to its current state of 2 departments with 3 majors (mechanical, electrical and computer, and general), over 30 faculty, and approximately 90 bachelors graduates per year. The character of the faculty and the culture of the engineering programs have transformed.

As research expectations have grown in the past 15 years, the newer faculty perceived themselves to be at a disadvantage with respect to their peers at more established research institutions, primarily with respect to availability of graduate students. Consequently the research-oriented faculty began campaigning to create a masters-level graduate program. This effort was successful, and the engineering master's program came online in 2005. But satisfaction with a master's program was short-lived. The consensus became that master's students were too transient. By the time they have learned enough and have been trained well enough to become effective researchers, they graduate. In addition, as further faculty hires were being contemplated, the now predominantly research-oriented faculty viewed a Ph.D. program as a vital selling point for attracting the highest quality faculty candidates. Thus, a second campaign was launched, this time for an engineering Ph.D. program. In 2010, the university administration approved the creation of a Ph.D. degree in electrical and computer engineering. A proposal for a mechanical engineering Ph.D. is currently pending and will likely be approved in the near term. The mechanical engineering Ph.D. proposal states:

> Advanced research programs are a critical component of prestigious universities. A doctoral program significantly increases the opportunities for high-quality research. As previously noted, most nationally-ranked, research-intensive universities have doctoral programs in engineering. A doctoral program in engineering will attract creative and talented research-oriented scholars, will increase production of high-quality research publications, and will increase competitiveness for external funding. Baylor's participation in technically-relevant, funded research will naturally yield enhanced exposure and recognition on the national and international stage (Baylor ME 2013, p. 42).

These motivations once again parallel those described by Morphew (2000) in his analysis of academic drift. The increasingly research-oriented faculty views such new programs as vital to prestige, to competitiveness with peer institutions, and to attracting highly qualified new faculty.

The three institutions just considered had quite different origins as a public, sub-baccalaureate technical training institute, a state teachers college, and a private, sectarian liberal arts college, respectively. And while there are still many differences in their institutional characteristics and constituencies, they have also been subject to a convergent evolution that has led to significant institutional isomorphism. They are all now comprehensive universities that offer a range of diverse undergraduate programs, both liberal and professional, that offer graduate programs and aspire to offer more, and that aspire to grow their research activity. The type of convergence seen in our three example institutions echoes that found by Douglas Toma in his study of how different types of institutions strive for prestige:

> I conclude that universities and colleges that are vastly different in orientation, markets served, and available resources are using roughly parallel strategies in positioning for prestige, having framed their aspirations in a similar manner (Toma 2008, p. 1).

To the extent that these institutions may be representative of other similar institutions, it would seem that there is a strong, market-driven research drift process in action. But, as argued by Frank Newman, Laura Couturier, and Jamie Scurry, the market-drive is mainly for reputation, and research is a means to that end. Four-year institutions strive "to be seen as research universities – not because of a public need but because of an internal drive for prestige" (Newman et al. 2004, p. 52). The adoption and growth of engineering programs in institutions striving to increase prestige through research likely is tied to the fact that engineering is capable of garnering significant research funding.

Conclusion

At the beginning of this chapter we posed the following questions: (1) What are the driving forces behind academic drift in non-university engineering education in Europe and the United States? (2) Are these driving forces of a similar nature or do they differ? (3) Is academic drift desirable for vocationally and professionally oriented programs, and if not, can it be avoided? (4) What research mission are former designated non-university engineering education institutions in Europe and the United States aspiring to fulfill? (5) What kinds of tension and dilemma does this new mission create in the above-mentioned kinds of institution?

Starting with question 4 and 5 we have shown that research drift and the creation of a research infrastructure in former vocationally oriented designated teaching institutions both in Europe and the United States are likely to create the following tensions and dilemmas (see also Kyvik and Skodvin 2003, p. 205).

- Allocation of resources – R&D versus teaching
- Distribution of R&D resources – quality criteria versus need for developing research skills
- Distribution of R&D resources – institutional versus individual rights and obligations
- Research-based teaching versus dissemination of advanced knowledge
- Recruitment of staff – research abilities versus professional experience
- Distribution of R&D resources – specialization versus breadth
- Vocational and regionally oriented research versus discipline oriented research
- Institutional control of R&D versus the staff's own preference

Moreover at the institutional level, vertical extension of degree programs and research drift has been a strong force in institutional survival strategies in an increasingly competitive market for higher education driven by a hegemonic neoliberal orthodoxy. The ensuing mission drift has created a tension between institutional effectiveness and prestige that must be resolved. At a policy level the prime objectives of structural transformations until the early 1990s was: (1) to ensure that higher education contributes to the economy, (2) to accommodate increasing numbers of an increasingly diversified student body in more cost-efficient ways, and (3) to take enrolment pressures away from the university. Furthermore structural dynamics have become increasingly complex as they have moved beyond the nation state to a trans-national level.

In a study of governance structures in higher education across U.S. states, Joseph Calhoun and David Kamerschen found that states with more centralized control of institutions had greater horizontal and vertical differentiation between institutions relative to curricular offerings. States with more decentralized control had greater institutional isomorphism. Calhoun and Kamerschen argue, "'Upward drift' is more likely to occur as [institutions of higher education] become more similar in an attempt to attract students, gain prestige, and increase their rankings" (Calhoun and Kamerschen 2010, p. 330). Thus decentralization and drift are linked. While European institutional forms and structures are more closely regulated by governmental policy than in the United States, the aforementioned trend toward transnational, or globalized, educational markets will likely put pressures on governmental policymakers to give institutions more autonomy, thus accelerating institutional change. An example of this was seen in the earlier section on Irish IoTs, where Irish students pursuing more favorable educational opportunities in the UK forced policy changes at Irish institutions.

In an *enrollment economy*, where autonomous or semi-autonomous institutions are striving to position themselves for prestige, there is likely to be a *Red Queen effect* (Lohmann 2004), which refers metaphorically to a passage in Lewis Carroll's *Through the Looking-Glass*, in which the Red Queen says to Alice, "Now, here, you see, it takes all the running you can do, to keep in the same place." For institutions concerned with relative rankings, ratings, prestige, and resources, they must strive as quickly as possible just to keep from loosing ground. As was illustrated in the previous section on U.S. institutions, this can lead institutions, which in many respects may have very different characteristics, to pursue highly parallel strategies for growth and change.

Acknowledgements For Steen Hyldgaard Christensen the writing of this chapter was made possible by a grant from the The Danish Council for Strategic Research (DSF) to the Program of Research on Opportunities and Challenges in Engineering Education in Denmark (PROCEED).

References

ABET. (2003). *2003 Annual report for fiscal year 2002–2003.*

ABET. (2010). *2010 Annual report for fiscal year 2009–2010.*

Adelman, C. (2009). *The Bologna process for U.S. eyes: Re-learning higher education in the age of convergence.* Washington, DC: Institute for Higher Education Policy. Available at: www.ihep.org/Research/GlobalPerformance.cfm

Apesoa-Varano, E. C. (2007). Educated caring: The emergence of professional identity among nurses. *Qualitative Sociology, 30*, 249–274.

Baker, V. L., Baldwin, R. G., & Makker, S. (2012). Where are they now? *Liberal Education, 98*(3), 48–53. *Professional Development Collection*, EBSCOhost, accessed April 21, 2013.

Baylor. (2012). *2012–2013 Undergraduate catalog*. Baylor University.

Baylor, M. E. (2013). *Mechanical engineering PhD proposal*. Baylor University Department of Mechanical Engineering, internal document.

Bologna Declaration. (1999). *Towards the European Higher European Area*. Conference of Ministers responsible for Higher Education in 29 European countries (June), Bologna, Italy.

Brint, S., Riddle, M., Turk-Bicakci, L., & Levy, C. S. (2005). From the liberal to the practical arts in American colleges and universities: Organizational analysis and curricular change. *The Journal of Higher Education, 76*(2), 151–180.

Britannica. (2013). Wayne State University. In *Encyclopaedia Britannica. Encyclopaedia Britannica Online Academic Edition*. Encyclopaedia Britannica Inc.. Available at http://www.britannica.com/EBchecked/topic/638127/Wayne-State-University. Accessed 21 Apr 2013.

Burgess, T. (1978). The officials' revolution: The British polytechnics: Ten years after. *Paedagogica Europaea, 13*(2), 45–58.

Calhoun, J., & Kamerschen, D. R. (2010, February). The impact of governing structure on the pricing behavior and market structure of public institutions of higher education in the U.S. *International Review of Economics, 57*, 317–333, Springer. Published online. doi:10.1007/s12232-010-0089-2

Christensen, S. H. (2012). Academic drift in European professional engineering education: The end of alternatives to the university? In S. H. Christensen, C. Mitcham, Li Bocong, & Yanming An (Eds.), *Engineering, development and philosophy: American, Chinese, and European perspectives*. Dordrecht: Springer Science+Business Media B.V.

Christensen, S. H., & Ernø-Kjølhede, E. (2011). Academic drift in Danish professional engineering education Myth or reality Opportunity or threat? *European Journal of Engineering Education, 36*(3), 285–299.

Clancy, P. (1997). Higher education in the Republic of Ireland. Participation and performance. *Higher Education Quarterly, 51*(1), 86–106.

Clancy, P. (2008). The non-university sector in Irish higher education. In J. S. Taylor et al. (Eds.), *Non university higher education in Europe*. Dordrecht: Springer Science+Business Media B.V.

Coate, K., & Mac Labhrainn, I. (2008). Irish higher education and the knowledge economy. In J. Huisman (Ed.), *International perspectives on the governance of higher education: Alternative frameworks for coordination* (pp. 198–215). London: Routledge.

De Boer, H., & Goedegebuure, L. (2007). 'Modern' governance and codes of conduct in Dutch higher education. *Higher Education Research & Development, 26*(1), 45–55.

Delucchi, M. (1997). "Liberal arts" colleges and the myth of uniqueness. *The Journal of Higher Education, 68*(4), 414–426.

De Wert, E., & Boezerooy, P. (2007). *Higher education in the Netherlands. Country report*. Center for Higher Education Policy Studies (CHEPS), Universiteit Twente, Enschede.

De Wert, E., & Leijnse, F. (2010). Practice-oriented research: The extended function of Dutch universities of applied sciences. In S. Kyvik, & B. Lepori (Eds.), *The research mission of higher education institutions outside the university sector: Striving for differentiation*. Dordrecht/London: Springer Science+Business Media B.V.

DiMaggio, P. J., & Powel, W. W. (1983). The iron cage revisited: Institutional isomorphism and collective rationality in organizational fields. *American Sociological Review, 48*(2), 147–160.
Dobbins, M., Knill, C., & Vögtle, E. M. (2011). An analytical framework for the cross-country comparison of higher education governance. *Higher Education, 62*(5), 665–683.
Doern, B. (2008). *Polytechnics in higher education systems: A comparative review and policy implications for Ontario*. Toronto: Higher Education Quality Council of Ontario.
Donally, R. (2004). Critical evaluation of the impact of global educational reform: An Irish perspective. *The Internal Journal of Educational Management, 18*(6), 351–359.
Donavan, J., Betts, A., & Jerrams, S. (2012). Removing barriers to conducting research in Ireland's Institutes of Technology. http://www.intechopen.com
Dublin Institute of Technology Act, 1992.
Duff, T., Hegarty, J., & Hussey, M. (2000). *The story of the Dublin Institute of Technology*. Dublin: Blackhall Publishing.
Eckel, P. D. (2008). Mission diversity and the tension between prestige and effectiveness: An overview of US higher education. *Higher Education Policy, 21*, 175–192.
ECPD. (1971). *Accreditation record volume I, 1936–1970*. New York: Engineers' Council for Professional Development.
Elzinga, A. (1985). Research, bureaucracy and the drift of epistemic criteria. In B. Wittrock & A. Elzinga (Eds.), *The university research system. The public policies of the home of scientists*. Stockholm: Almquist & Wiksell.
Feisel, L. D., & Rosa, A. J. (2005). The role of the laboratory in undergraduate engineering education. *Journal of Engineering Education, 94*(1), 121–130.
Ford, G. D., & Ball, A. K. (2011). The evolution of engineering and engineering technology educational programs in the United States. In *Proceedings of the 2011 American Society for Engineering Education Annual Conference & Exposition*. American Society for Engineering Education.
Forfás. (2007). The role of the institutes of technology in enterprise development: Profiles and emerging findings. http://www.hea.ie
Fox, P. L., Hundley, S. P., & Rennels, K. (2004). What can the past tell us about our future? Trends and developments in engineering technology. In *Proceedings of the 2004 American Society for Engineering Education Annual Conference & Exposition*. American Society for Engineering Education.
Frandsen, P., & Harnow, H. (2011). Teknikumingeniørernes uddannelse og rolle i erhvervslivet. (The education of professional engineers and their role in industry). Unpublished working paper.
Furth, D. (1982). New hierarchies in higher education. *European Journal of Education, 17*(2), 145–151.
Goldin, C., & Katz, L. F. (1998). The shaping of higher education: The formative years in the United States, 1890 to 1940. National Bureau of Economic Research, Working Paper no. 6537.
Goldin, C., & Katz, L. F. (1999). The shaping of higher education: The formative years in the United States, 1890 to 1940. *The Journal of Economic Perspectives, 13*(1), 37–62.
Graham, P. (1978). Expansion and exclusion: A history of women in American higher education. *Signs, 3*(4), 759–773.
Griffioen, D., & de Jong, U. (2012). Academic drift in Dutch non-university higher education evaluated: A staff perspective. *Higher Education Policy*, 1–9.
Grubb, W. N., & Lazerson, M. (2005). Vocationalism in higher education: The triumph of the education gospel. *The Journal of Higher Education, 76*(1), 1–25.
Gumport, P. J. (1993). Graduate education and organized research in the United States. In B. R. Clark (Ed.), *The research foundations of graduate education*. University of California Press.
Harwood, J. (2010). Understanding academic drift: On the institutional dynamics of higher technical and professional education. *Minerva, 48*, 413–427.
Hazelkorn, E., & Moynihan, A. (2010a). Transforming academic practice: Human resource challenges. In S. Kyvik, & B. Lepori (Eds.), *The research mission of higher education institutions*

outside the university sector: Striving for differentiation. Springer Science+Business Media B.V.

Hazelkorn, E., & Moynihan, A. (2010b). Ireland: The challenge of building research in a binary higher education culture. In S. Kyvik, & B. Lepori (Eds.), *The research mission of higher education institutions outside the university sector: Striving for differentiation*. Springer Science+Business Media B.V.

Henderson, B. B., & Kane, W. D. (1991). Caught in the middle: Faculty and institutional status and quality in state comprehensive universities. *Higher Education, 22*, 339–350.

Huisman, J., & Kaiser, F. (2001). *Fixed and fuzzy boundaries in higher education: A comparative study of (binary) structures in nine countries*. Den Haag: Adviesraad voor het Wetenschaps- en Technologiebeleid.

Huisman, J., & Morphew, C. C. (1998). Centralization and diversity: Evaluating the effect of government policies in U.S.A. and Dutch higher education. *Higher Education Policy, 11*, 3–13.

Hunt, C. (2011). *Report of The Strategy Group. National strategy for higher education to 2030*. Government Publications, 51 St. Stephen's Green, Dublin 2.

Jaquette, O. (2013). Why do colleges become universities? Mission drift and the enrollment economy. *Research in Higher Education, 54*, 514–543.

Jónasson, J. T., & Jóhannsdottir, G. (2010). *Defining and determining quality in HE: Potential conflicts and their effects*. Available at: http://www3.hi.is/~jtj/greinar/Working%20paper. Accessed 10 Jan 2010.

Kelnhofer, R., et al. (2010). Future of engineering technology. In *Proceedings of the 2010 American Society for Engineering Education Annual Conference & Exposition*. American Society for Engineering Education.

Kenyon, R. A. (1993). The coming revolution in engineering and engineering technology education: A new paradigm for the 21st century. *Education, 113*(3), 361–371.

Kett, J. F. (1994). *The pursuit of knowledge under difficulties: From self-improvement to adult education in America, 1750–1990*. Stanford: Stanford University Press.

Kyvik, S. (2007). Academic drift – A reinterpretation. In J. Enders, & F. van Vught (Eds.), *Towards a cartography of higher education policy change: A Festschrift in honour of Guy Neave* (pp. 333–338). Enschede: Center for Higher Education Policy Studies (CHEPS).

Kyvik, S. (2009). *The dynamics of change in higher education: Expansion and contraction in an organisational field*. Springer Science+Business Media B.V.

Kyvik, S., & Lepori, B. (Eds.). (2010). *The research mission of higher education institutions outside the university sector: Striving for differentiation*. Springer Science+Business Media B.V.

Kyvik, S., & Skodvin, O.-J. (2003). Research in the non-university higher education sector – Tensions and dilemmas. *Higher Education, 45*, 203–222.

Labaree, D. F. (1997). *How to succeed in school without really learning*. London: Yale University Press.

Labaree, D. F. (2006). Mutual subversion: A short history of the liberal and the professional in American higher education. *History of Education Quarterly, 46*(1), 1–15.

Ledbetter, R. (1979, August 31). Engineering program debuts this gall. *The Baylor Lariat*, p. 5.

Leiho, A. (2010). Academisation of nursing education in the Nordic countries. *Higher Education, 60*, 641–656.

Levine, D. O. (1988). *The American college and the culture of aspiration, 1915–1940*. Ithaca: Cornell University Press.

Lillis, D. (2007). Reconciling organisational realities with the research mission of the Irish Institutes of Technology. *Consortium of Higher Education Researchers 20th Annual Conference,* Dublin, September 2007.

Lohmann, S. (2004). Can't the university be more like business? *Economics of Governance, 5*(1), 9–27.

Lombardi, J. V., Phillips, E. D., Abbey, C. W., & Craig, D. D. (2011). *The top American research universities: 2011 annual report*. The Center for Measuring University Performance at Arizona State University.

Marginson, S. (2012). *Criteria for technological university designation*. Unpublished working paper.
Molesworth, M., Scullion, R., & Nixon, E. (Eds.). (2011). *The marketization of higher education and the student as consumer*. London: Routledge.
Morphew, C. C. (2000). Institutional diversity, program acquisition and faculty members: Examining academic drift at a new level. *Higher Education Policy, 13*, 55–77.
Morphew, C. C. (2002). 'A rose by any other name': Which colleges became universities. *The Review of Higher Education, 25*(2), 207–223.
Morphew, C. C., & Huisman, J. (2002). Using institutional theory to reframe research on academic drift. *Higher Education in Europe, 27*(4), 491–506.
National Institution for Academic Degrees and University Evaluation. (2011). *Overview: Quality assurance system in higher education*. The Netherlands.
Neave, G. (1978). Polytechnics: A policy drift. *Studies in Higher Education, 3*, 1.
Neave, G. (1979). Academic drift: Some views from Europe. *Studies in Higher Education, 4*, 2.
Neely, P. (1999). The threats to liberal arts colleges. *Daedalus, 128*(1), 27–45.
Newberry, B., & Farison, J. (2003). A look at the past and present of general engineering and engineering science programs. *Journal of Engineering Education, 92*(3), 217–226.
Newman, F., Couturier, L., & Scurry, J. (2004). *The future of higher education: Rhetoric, reality, and the risks of the market*. San Francisco: Jossey-Bass.
NRC. (1985). *Engineering education and practice in the United States: Engineering technology education*. Washington, DC: The National Academies Press.
OECD. (2004). *Review of national policies for education: Review of higher education in Ireland*. Examiners report. EDU/EC(2004)14.
OECD. (2008). *OECD reviews of tertiary education: Netherlands*.
O'Flaherty, N. (2012). *Mission impossible? Evaluating the binary divide in Irish higher education*. Unpublished working paper.
O'Meara, K. A. (2007). Striving for what? Exploring the pursuit of prestige. In: J. C. Smart (Ed.), *Higher education: Handbook of theory and research* (Vol. XXII, pp. 121–179). Springer.
Pratt, J. (1997). *The polytechnic experiment 1965–1992*. Buckingham: Open University Press.
Regional Technical College Act, 1992.
Rennels, K. (2003). Mission creep in engineering technology education? In *Proceedings of the 2003 American Society for Engineering Education Annual Conference & Exposition*, American Society for Engineering Education.
Riesman, D. (1956). *Constraint and variety in American education*. Published by the University of Nebraska at Lincoln.
Russell, L. M., & Murray, L. S. (2013). Baylor University. In *Handbook of Texas Online*. http://www.tshaonline.org/handbook/online/articles/kbb05. Accessed 20 Apr 2013. Published by the Texas State Historical Association.
Saunders, D. B. (2012). *Neoliberal ideology and public higher education in the United States*. Unpublished working paper.
Seely, B. (1993). Research, engineering, and science in American engineering colleges: 1900–1960. *Technology and Culture, 34*(2), 344–386.
Seely, B. (1999). The other re-engineering of engineering education, 1900–1965. *Journal of Engineering Education, 88*(3), 285–294.
Skillbeck, M. (2003). *Towards an integrated system of tertiary education. A discussion paper*. Dublin: Dublin Institute of Technology.
Slantcheva-Durst, S. (2010). Redefining short-cycle higher education across Europe: The challenges of Bologna. *Community College Review, 38*(2), 111–132.
SPSU. (2010). *Strategic plan 2010–2013*. Southern Polytechnic State University. Available at: http://www.spsu.edu/planningassessment/unitassessment/str09.pdf. Accessed 8 May 2013.
SPSU. (2013). *History*. Southern Polytechnic State University. http://www.spsu.edu/aboutus/history.htm. Accessed 8 May 2013.

Teichler, U. (2008). The end of alternatives to universities or new opportunities? In J. M. Taylor (Ed.), *Non-university higher education in Europe*. Springer Science+Business Media B.V.

Toma, J. D. (2008). Positioning for prestige in American higher education: Case studies of strategies at four public institutions toward "getting to the next level". In *2008 Conference of the Association for the Study of Higher Education*. Jacksonville, November 2008.

Trow, M. (1996). Continuities and change in American higher education. In A. Burgen (Ed.), *Goals and purposes of higher education in the 21st century*. London: Jessica Kingsley Publishers.

Trow, M. (2000). From mass higher education to universal access: The American advantage. *Minerva, 37*(4), 303–328.

Trow, M. (2003). *On mass higher education and institutional diversity*. Samuel Neaman Institute for Advanced Studies in Science and Technology, Technion-Israel Institute of Technology, May 2003.

Tuchman, G. (2009). *Wannabe U: Inside the corporate university*. Chicago: The University of Chicago Press.

Universities Act, 1997.

Weese, J., & Wolf, L. J. (1994). The advance toward distinction in engineering technology. *Journal of Engineering Education, 83*(1), 41–46.

Weinsier, P., & Brown, D. J. (2008). Engineering technology education: A national picture. In *Proceedings of the 2008 IAJC-IJME International Conference*.

Witte, J., van der Wende, M., & Huisman, J. (2008). Blurring boundaries: How the Bologna process changes the relationship between university and non-university higher education in Germany, the Netherlands and France. *Studies in Higher Education, 33*(3), 217–231.

WKU. (2010). *Strategic guide for 2010–2012*. Western Kentucky University. Available at: http://www.wku.edu/admin/documents/strategicguide2010-12.pdf. Accessed 8 May 2013.

WKU. (2013). *History of WKU*. Western Kentucky University. Available at: http://www.wku.edu/wkuhistory/index.php. Accessed 8 May 2013.

Steen Hyldgaard Christensen M.A. in Scandinavian Language and Literature and the History of Ideas, Aarhus University. Ph.D. in Educational Studies, Aalborg University. Senior lecturer at Aarhus University, School of Business and Social Sciences, Herning, Denmark until 2014. From 2014 adjunct associate professor at Aalborg University, Denmark. He has initiated six big international, inter- and meta-disciplinary research projects on engineering including PROCEED and coordinated five of them. He has acted in roles of editor-in-chief and co-author of four books: *Profession, Culture, and Communication: An Interdisciplinary Challenge to Business and Engineering* (Institute of Business Administration and Technology Press 2003), *Philosophy in Engineering* (Academica 2007), *Engineering in Context* (Academica 2009), and *Engineering, Development and Philosophy: American, Chinese, and European Perspectives* (Springer 2012). Besides he has co-authored *A Hybrid Imagination – Science and Technology in Cultural Perspective* (Morgan & Claypool Publishers 2011) together with Andrew Jamison and Lars Botin. In addition he has published a number of articles on engineering epistemology, culture and education. Current research interest: academic drift in engineering education and structural dynamics in higher education.

Byron Newberry B.S. in Aerospace Engineering, University of Alabama. M.S. in Aerospace Engineering and Ph.D. in Engineering Mechanics, both from Iowa State University. Professor of Mechanical Engineering, Baylor University. Baylor Fellow for teaching. Professional Engineer (PE) Texas, USA. Research interests: engineering design, engineering ethics, philosophy of engineering and technology. Aircraft structural engineering consultant. Executive board member, National Institute for Engineering Ethics. Editor of the Springer *Philosophy of Engineering and Technology* book series.

Chapter 3
Structural Transformations in Higher Engineering Education in Europe

Bernard Delahousse and Wilhelm Bomke

Abstract This chapter aims to analyze the driving forces at work which have resulted in structural transformations that have taken place in higher engineering education throughout Europe over the last four decades. After a brief discussion of the theoretical concept of academic drift, a comparative study of two higher engineering institutions in Europe is presented: the IUTs in France, nested in the French universities, and the Fachhochschulen in Germany, which were non university institutions. The study dwells on the initial missions and status of these institutions, and the academic drift processes they have been through in this globalizing time span, with regard to their autonomy, their curricula, their pedagogical methods, the recruitment of their students and staff, and their research opportunities Finally, the dynamics of these transformations will be analyzed in the light of national and international standards and requirements.

Keywords Higher engineering education • Structural transformations • Academic drift • Vocational drift • Research • Driving forces

Introduction

The evolution of higher education in Europe over the last 40 years has been marked by a double and opposite trend: on the one hand, practice-oriented institutions have turned to more science-oriented curricula; on the other hand, universities whose traditional mission is to deliver research-based knowledge have developed profession-oriented curricula. In some European countries, like Denmark, Germany,

B. Delahousse (✉)
IUT "A" Dept Mesures Physiques – Université Lille-1 Sciences et Technologies – Cité scientifique, Bâtiment P7, Villeneuve d'Ascq Cedex 59655, France
e-mail: bdelahousse@free.fr

W. Bomke
Ostbayerische Technische Hochschule Regensburg, International Office, Prüfeninger Str. 58, Regensburg 93049, Germany
e-mail: wilhelm.bomke@oth-regensburg.de

S.H. Christensen et al. (eds.), *International Perspectives on Engineering Education*, Philosophy of Engineering and Technology 20, DOI 10.1007/978-3-319-16169-3_3

the Netherlands or Belgium, this has led to a number of institutional mergers either within the framework of universities or by the creation of larger non-university entities. This phenomenon is part of "an international trend that the difference between the university and the college sector has become blurred" according to Jens-Christian Smeby (2006, p. 6). Smeby also points out that in the field of professional education the "curriculum has moved from a craft model towards an academic model" (Smeby 2006, p. 4). Similarly, Raymond Bourdoncle (2007, p. 135) first observes "the multiplication of professional university degrees, from the creation of IUTs in 1968 to the professional Masters in 2004" in France. He then sets out to show the complexity of the links between professional activities and academic ones by putting forward a distinction between *professionalization*, which derives from the creation of professional degrees by the university itself, and *universitarization* which he defines as a process of "absorption" of professional institutions, knowledge and teaching staff by the university.

This chapter is the outcome of personal reflections on the structural transformations that have taken place over the last four decades in our respective higher engineering institutions, an IUT in France and a Fachhochschule in Germany, while we have been working there as academics in the humanities. We first explore the way the concept of academic drift is delimited in the relevant literature with regard to general patterns of structural transformations in higher engineering education.

Then we present a comparative study of these two profession-oriented institutions in Europe, the French *Instituts Universitaires de Technologie* (IUTs) and the German *Fachhochschulen* (FHs), as regards their historical evolution in terms of degree of autonomy, creation or adaptation of curricula, pedagogical methods, student standing, personnel status and research opportunities. The choice of these two institutions seems to us relevant in that they have a number of traits in common: a strong focus on teaching rather than research, fixed curricula oriented toward practice including internships, close links with companies, academic staff recruitment, a particular stand with regard to universities, insistence on graduate operational skills, etc. Finally, this chapter deals with the driving forces – economic, political, social, professional, societal, technological – behind the academic drift processes, i.e. the harsh competition between the world's economies, the "knowledge-intensive" society, the internationalization of higher education as illustrated in the Bologna statement, the massification of secondary and post-secondary education, student expectations for higher qualifications and credentials to get better and more secure jobs, the demand of the teaching staff to obtain better academic recognition via research, the tremendous development of ICTs, etc.

The Conceptual Framework of Academic Drift

To delimit the concept of academic drift for our purpose, we will hereafter refer to the theoretical framework put forward by a number of authors such as Tyrrell Burgess (1972), Michael Gibbons et al. (2006), Jens-Christian Smeby (2006),

Raymond Bourdoncle (2007), Svein Kyvik (2009) and Steen Hyldgaard Christensen (2012). *Academic drift* as originally defined by Burgess refers to the dynamics of change in higher education since the massive expansion of student enrolments in the 1960s (1972); this working definition is supported by Gibbons et al. in that it closely relates transformations in higher education to the notion of massification (2006, pp. 70–76) and by Bourdoncle's concept of *universitarization,* defined as the process of integrating scientific knowledge, academic knowledge production and university teaching staff into higher professional education (2007, p. 138). In Christensen's classification (2012, pp. 147–148), academic drift encompasses four main dimensions, i.e. structural, institutional, cognitive and organizational (e.g. status and funding).

Concerning the *structural dimension*, Christensen notes that "academic drift operates across the entire non-university higher education sector to transform educational systems… from university-dominated systems, over dual and binary systems to unified systems of higher education, with stratified systems like the French as an exception" (Christensen 2012, p. 148). Like Bourdoncle, he identifies three general patterns of transformations: (1) *Expansion* designates the transformation of an institution into a higher level institution or a university, e.g. training schools in the USA turned into colleges in the nineteenth century, then into state universities of their own (Bourdoncle 2007, p. 138). But Christensen (2012, p. 149) introduces a slightly different perspective with the term "fragmented expansion" in that he refers explicitly to the initial phases of the process aiming at differentiation and diversification of institutions and curricula. (2) *Vertical integration* is characterized by the above-mentioned "absorption" of existing, rather autonomous institutions or faculties, both vocational and academic, by universities and the subsequent alignment of their policies and procedures with those of the university. Although IUTs have been parts of universities from the start, the concept of absorption, or vertical integration, can be applied to a number of local off-campus IUTs which were later integrated into new universities. (3) *Horizontal integration* generally applies to the integration of non-university institutions into polytechnic types of institutions that form coherent multi-sectoral entities beside universities, thus engendering a binary system of higher education. Numerous examples of such mergers in Europe can be given, e.g. in Germany (with the Fachhochschulen), Denmark, the Netherlands (with the Hogescholen) or Belgium (with the Hautes Écoles or the Hogeschools). These mergers are "characterized as a gradual transition to a binary model… beside the university sector" since they constitute "comprehensive vocational multi-professional colleges" (Kyvik 2009). Strictly speaking, pattern 3 does not relate to academic drift or universitarization, unlike patterns 1 and 2 which eventually lead to the extension or reinforcement of the university system. Yet insofar as in pattern 3 "the term implies that there is a macro-structure of higher education and that individual institutions are not self-sustaining entities" (ibid.), the notion of academic drift can be applied to these structural changes with some relevance.

The *institutional dimension* of academic drift refers to a tension between narrow vocational training and broad professional or research-oriented academic education (Burgess 1978). It highlights the tendency of non-university higher engineering education institutions, student body and faculty staff to access to a higher university-like status in terms of curricula, research and academic recognition. The distinction between "noble" and "less noble" institutions has become blurred, as alternative institutions, e.g. the former Polytechnics in the UK or the Fachhochschulen in Germany once regarded as second-tier, have come to rival the universities, while universities have become more profession-oriented, a convergence which is highlighted in the concept of "extended university" (Gibbons et al. 2006, p. 72).

The *cognitive dimension* of academic drift relates to the tension between theory-oriented and practice-oriented curricula that all vocational education institutions experience, i.e. the tension between scientific rigor required by the production and validation of new knowledge and competences acquired through problem-solving practices (Bourdoncle 2007, p. 144). From this perspective, practice is not only a way to develop professional skills but it also aims to improve students' theoretical understanding (Smeby 2006, p. 16). Thus an increased emphasis on theory in vocational programs has developed in the college sector as theoretical knowledge has become the "axial principle" of development in post-industrial societies (ibid. p. 8). As underlined by Christensen, it does not mean that in practice-oriented curricula "no use is made of theories, laws, concepts, etc. from the basic sciences… Instead they are regarded as just one resource among many for the solution of practical problems" (2012, p. 147). Research-based knowledge obviously plays an important role in this dimension.

The fourth dimension of academic drift that we designate here as *organizational drift* refers to the status and funding of higher engineering education institutions which constitute key issues in their current management. Autonomy is evidently a decisive status for these institutions so that they can conduct their own policy of curricular development, course design, academic staff recruitment, student selection, research, etc. This autonomy is of course highly dependent on the degree of regular funding the institutions can obtain from a variety of sustainable sources ranging from state and regional grants to industrial financing via incomes derived from continuing education and/or technological transfer contracts. Thus the status of autonomy and the degree of funding of these institutions are clearly interdependent.

In the following, we will give a short presentation of two European higher engineering education institutions, the IUTs in France and the Fachhochschulen in Germany, which were created nearly at the same time in the late 1960s or early 1970s, in a period of transition from elite to mass higher education, in order to cope with massive numbers of students, a more socially diversified student body with different aspirations, fast-growing technological progress and industry's subsequent needs for high-level graduates (ADIUT 2007, p. 3). Then we will focus on five dimensions of academic drift: structural drift, institutional drift, cognitive drift, student and staff drifts, and research drift.

The Case of French Instituts Universitaires de Technologie (IUTs)

Historical Overview of the Rationale and Missions of IUTs

French IUTs were created in 1966 in an attempt by the French Ministry of Education to respond to a number of crucial challenges[1] the country was then confronted with: (1) The fast economic and social development closely linked to technological innovation in most industrial fields, with the resulting need for higher quality and competitivity, (2) The subsequent indispensable replacement of in-house technicians, engineers and executives whose skills no longer matched technological progress and the new requirements of industry, (3) The massive increase in the number of students accessing to the university, as a result of both the post-World War II demographic growth and the raise of the compulsory school-leaving age from 14 to 16 years old, and 4. The high dropout rate among Faculty students in the 1960s, e.g. 65 % of science students left the university without a degree.

The aim of the French government, as explicitly mentioned in the founding decree of IUTs in January 1966, was threefold: (1) To train skilled middle-level graduates "capable of putting into practice the engineer's designs or the results of theoretical research, of circulating and interpreting the general instructions given by administrative, financial and commercial superiors" (ADIUT 2007, p. 5), (2) To enable students from low-income families, who traditionally terminated their studies at the end of the technical secondary school, to join a shorter and more practice-oriented degree course, thus enhancing social promotion,[2] (3) To fill in a gap in the public technological education stream between the *lycée*, which already provided 2-year post-*baccalauréat* study programs in the STS (*sections de techniciens supérieurs)* and the university, whose few ENSI (*écoles nationales supérieures d'ingénieurs*) offered students a 3-year engineering course after completion of the 2-year first cycle.[3]

The IUTs were clearly established by the Ministry of Education as integral parts of universities, which created serious tensions in the early stages with the traditional science-oriented faculties who opposed these vocational practice-oriented, hence "less noble", institutions. Although embedded in the university, these institutes were assigned a general operating framework that showed a clear demarcation line with the other faculties. They provided, and still do today, 2-year course programs that were equivalent to the first cycle of the French university system (2 + 1 + 1 + 1), leading to the *Diplôme Universitaire de Technologie* (the DUT) originally meant to

[1] See *Le Livre Blanc sur le système IUT après 40 ans d'existence: Histoire, Bilan et Perspectives.* Available at: http://www.iut-fr.net/publications/livre-blanc.html.

[2] It is to be noted that, at the same period, the same social goal was assigned to the emerging centers of continuing education in a number of French universities.

[3] Actually, most engineering high schools at that time were either parts of private Catholic universities or non-university institutions like the *Grandes Écoles*.

be a terminal degree. Most programs, namely those of the industrial departments, were organized on the basis of an average 30-week tuition per academic year, with a further 6–10 weeks' internship, compared to the university norm of 26 weeks. IUT students' workload amounted to at least 32 contact hours per week, as opposed to the 12–16-h format at the university. In order to implement this intensive school-like teaching, the IUTs introduced innovative pedagogical methods based on lab practice and constantly reviewed in order to cope with technological evolutions. All courses were compulsory, attendance was monitored and students could be dismissed for absenteeism. Students were evaluated each year by means of a continuous assessment system taking into equal consideration theoretical and practical subjects.

The IUTs benefited from the determination of the Ministry of Education to give them a special status within the universities via a policy of autonomy in terms of financial, administrative, scientific and pedagogical matters, a policy strongly resented by universities as a loss of their power. The institutes were governed by their own administration boards, composed of delegates from the different personnel categories, students, and representatives from industry (employers and trade unionists). These boards had a real power of decision concerning IUT budgets which were mostly co-funded by specific IUT-signposted grants from the Ministry and, to a lesser extent, by industry's mandatory contributions under a special tax scheme. They also had their say in the introduction of new courses or local course adjustments, in the creation of student groups, etc. Concerning pedagogical issues, their autonomy was guaranteed under the umbrella of their national bodies, the CPNs (*Commissions Pédagogiques Nationales*). As for the teaching staff recruitment, they had their own selection committees who examined the applications according to profiles they had themselves established, before submitting their choices to the Ministry for approval.

Specificities of the Student Body

One of the initial missions of IUTs was to offer new opportunities of access to higher education to students from the technical and vocational lycées who traditionally went straight to the labour market after passing the *baccalauréat*. The short study length of the DUT course, its practice-oriented contents, the adapted pedagogical methods and the relatively small size of student groups, compared to the university, were meant to attract the best students from low-income, less educated families; a survey from the French Ministry of Education in 2000 showed that "the likelihood of working-class children going on to higher education increased by a factor of 3.5 compared to an overall 2.2" over that period (Hanchan and Verdier 2005). Thus the IUTs have contributed to the reduction of social inequalities, e.g. 33 % of the students they recruited in 2004 were from worker/employee background, and the percentage of students being awarded a state grant has always been higher (32 %) than at the university (16.5 %).

Student recruitment was also characterized by two seemingly opposite mechanisms: *selectivity* and *diversification*. As IUTs were allowed to admit students up to a limited capacity allocated by the Ministry of Education, admission procedures were highly selective and provision for a minimum ratio of students with a technical or vocational baccalauréat was mandatory. However, in the 1980s and 1990s, due to the high number of applications every year, IUT admission officers tended to be more selective and use more academic criteria, so that in a number of departments the proportion of students from general lycées increased dramatically to the detriment of those from the vocational and technical streams. This soon became a source of tension with universities who blamed the IUTs for depriving them of top-level students while they were themselves confronted with *massification*.[4] Besides, the diversification of student cohorts was encouraged as "there were three other possibilities for admission aiming at attracting high-calibre students: (1) Acquisition of equivalent training in industry, (2) Completion of a diploma that would grant access to university studies, and (3) Obtaining validation of professional experience or previous learning." (Christensen 2012, p. 154).

The IUTs had clearly been invested from the start with a dual mission: to train middle-level graduates who would be "more narrowly specialized than an engineer but with a broader background than a technician" (Saumade 1998), and to prepare them for higher studies. Although the DUT was meant to be a terminal degree which was to respond to industry's needs of an adaptable highly skilled staff, there has always been the possibility for the "top 10 students" to continue their studies either at the university or in a *Grande École*. However, as a result of the tight selection procedures, together with rising unemployment rates, as well as students' and their parents' aspirations for higher degrees and qualifications supposedly ensuring better jobs, this possibility has gradually become an almost regular route for a majority of DUT holders since the 1990s, "a clear mission drift of IUTs and a policy drift of the Ministry of Education" (Christensen 2012, p. 155). Besides, the strategy of a growing number of students from the general stream deliberately opting for the IUT route in the first place so as to be in a better position to access to higher levels of education at a later stage is also a significant marker of this mission drift.

Tensions and Aspirations of the Teaching Staff

The success of IUTs in terms of professionalization has been largely due to the involvement of the teaching teams, originally composed of three categories: university personnel, teachers from secondary education and engineers or executives from industry. This tripartite system did not apply to the staff composition proper, but to the quota of teaching hours delivered by each category, i.e. a third of the total

[4] Unlike most of their European counterparts, French universities are, by law, bound to enrol *baccalauréat*-holders without restrictions, since the *baccalauréat* has a dual function: it is both the terminal secondary education exam and the initial diploma giving access to the university.

contact hours was to be taught by university personnel, another third by teachers from the lycées and the last third by professionals. This national policy aimed at ensuring that all the domains of professionalization – technical, theoretical and transdisciplinary – were catered for through these pluridisciplinary backgrounds. However if it actually worked in a number of departments, and still does here and there, it is to be noticed that it generated real tensions. The first tension concerned the professionals: they were often recruited on the basis of personal/professional relations and/or designation by their company, but after two decades of active participation it gradually became difficult to get them involved, partly due to lack of time and motivation, partly to the low remuneration they were entitled to. As a result, their teaching quota in many departments has been, at least partly, taken over by the other categories of staff, thus entailing another dimension of policy drift.

Another tension concerned the academic category. As opposed to most European higher engineering teaching staff, the IUT personnel had the same academic status as their university colleagues, hence carrying out research in a university laboratory, but their lecturing or tutoring workload at the IUT was much more constraining in terms of energy and time devoted to pedagogical commitments and administrative tasks. As their promotions depended mostly, if not solely, on their research productivity, they realized that they were at a serious disadvantage compared with their university colleagues. Attempts have been made by the Ministry of Education and a few universities to equally recognize as promotion criteria the three types of tasks – research, administration and teaching – the academic personnel was assigned to, but up to now research has remained paramount as a general rule.

A third source of tension appeared between the secondary education staff and the university personnel: the former's major mission was dedicated to teaching activities as they were not required to do any research, and the latter expressed a staunch opposition to their recruitment on the grounds that they did not have the proper credentials and therefore contributed to the devalorization of university degrees. Yet, within IUT departments, the relationships with their academic colleagues were generally based on mutual trust, as they actively participated in the pluridisciplinary pedagogical teams and often took their share of administrative tasks. They also shared with them the drawback of being disadvantaged in their promotions, since they were still statutorily linked to secondary school procedures and criteria which took mainly into account the staff working in the lycées. Last but not least, this category offers a good example of professional drift: even though a relatively small proportion of this personnel was concerned, there was a growing claim on their part to be given the possibility of conducting research activities, which they later obtained in the 1990s.

What Structural Transformations in the IUTs?

After this overall study of the French IUT system, let us examine what structural transformations have taken place over the 40-year span since the IUTs were created. We have chosen to focus on the following five significant issues: autonomy policy, new curricula, pedagogical innovations, diversified student body, habilitation for research. First, as a result of the university reform called "Plan Université 2000" starting in the early 2000s, the IUTs have experienced a slow but regular reduction of their financial and human resources. More recently, their autonomy has been further jeopardized as the allocation of funding from the Ministry of Education has been transferred to the universities, irrespective of the IUTs' specific needs for technology and professionalization. Similarly, the creation of new teaching jobs does not respond any longer to the actual needs of these institutes. These limitations of their autonomy tend to turn IUTs into "classic" faculties since the decision-making body is now the university council, thus displaying a typical example of structural drift.

The IUTs have a long story of (re)designing curricula in order to cope with new social needs and technological requirements. They have been pioneers in developing continuing education in higher education, opening opportunities for promotion to industry's employees and technicians, as well as offering new skills to young people without qualifications. Besides, continuing education has played a crucial role for the development of IUTs not only to counterbalance the reduced state funding with regional and industrial contracts, but also to enhance its operational network of industrial partners (Convert et al. 2011). A major outcome of this trend has been the adaptation of existing curricula to this specific public with the introduction of course modules and a cumulative credit system for the obtention of the DUT degree. Another example of curricular drift was the creation in the 1990s of *Instituts Universitaires Professionalisés* (IUPs), in which the IUTs played an active role as initiators and course organizers.[5] It was one of the attempts by IUT teaching staff to offer a further professional 2-year course to first-cycle graduates, including DUT holders, thus leading to a *maîtrise,* a more prestigious university degree. This can be regarded as part of a long-term tendency of faculty and student body "to strive for an upward movement in the direction of an institutional setting or curriculum that resembles the university as the epitome of prestige" (Christensen 2012, p. 147).

The most significant example of this curricular drift was the creation of the *licence professionnelle* (DUT+1) in 2000, when the French university system implemented the common European scheme of curricular cycles (3+2+3) under the Bologna process launched by the European Commission. The French university had to align with the new Anglo-American-type system, and thus faced a crucial problem with the 2-year-cycle DUT as there was no possible equivalence with the 3-year-cycle *licence* (bachelor's degree). Most IUTs seized the opportunity of this

[5] These institutes have had to align with the new university system derived from the Bologna process in the mid-2000s and have adopted the new master's degree.

change in degree structure to promote the *licence professionnelle* for the benefit of their institutes, their students and their teaching staff, so that 60 % of these new *licences* were IUT-supported in 2006. Even though the degree is conferred by the university, its curriculum and course organization are generally based on an IUT-type model, its student body is mostly composed of DUT graduates and a high percentage of its teaching staff comes from IUTs. The main issue concerning the *licence professionnelle* lies in the difficulty to identify its different specializations and in the relevance of some degrees which are too specialized.

As regards recent pedagogical transformations in the IUTs, academic drift in its cognitive dimension is reflected in a set of novelties that are closer to the university model. As a result of the university reform, DUT courses are now modularized and organized on a semestrial basis, as in the university, with the validation of modules at the end of each semester. On top of the core modules, each department now provides for complementary modules that students can choose. The number of student contact hours in terms of classic face-to-face teaching has been reduced in most departments while project works have considerably developed: on the one hand, tutored projects enable small teams of students to work on transdisciplinary subjects with a problem-solving approach; on the other hand, personal and professional projects (PPP) which aim at reinforcing the link between the student's aspiration and the professional world have been generalized.

These transformations are also linked to a number of changes in the IUT student body. Student drift refers to the diversification of students' backgrounds as well as their diversified expectations. As pointed out by Gibbons et al. (2006, p. 77), students "are drawn from a much broader social base; the balance between the sexes is more equal; and most graduates now go, not to positions of leadership, but to join the vast middle-range salariat of the public services and private corporations". These observations apply of course to the IUT student body, but two additional considerations are to be highlighted: first, due to the importance of continuing education within IUTs, together with the validation of professional experience and the possibility for first-cycle university students to enter directly the second year of the DUT course, the proportion of mature students in IUTs has increased; secondly, due to selectivity as well as new backgrounds, together with the regular use of ICTs which encourages self-learning, a majority of students already have a good command in subject areas like natural sciences, communication or computing when entering the IUT. All these features, together with their aspiration to get on to higher degree courses, result in the student body's higher requirements in terms of course contents and degree value.

Academic drift is also to be found in the evolution of research for the IUT teaching personnel. As mentioned earlier, the IUT staff with a university status conducted their research in university labs, a state of things which still prevails. However, today a greater number of academics carry out their research in the 160 IUT-based laboratories. This research focuses mostly on pluridisciplinary, academic and applied subjects. Even though this has not had much impact on their professional promotion, it has at least facilitated their working conditions. As for secondary education personnel, their access to research being recognized, they now enjoy the same status as their academic colleagues in terms of reduced teaching workload. Their recruitment tends to be based on their previous or current research work. As the trend seems to be in the direction of a convergence between these two staff categories, the question remains as to the risks of degraded balance and complementarity between them, all the more so as the original tripartite composition of the personnel has already been reduced by the dramatic drop in the contribution of professionals from industry. In this respect, Bourdoncle points out to a difficulty inherent to research drift, when out of the three facets necessary to professional education, i.e. teaching, research and practice, he notes: "The transformation of the practitioner into a researcher ends up in the disappearance of his practical activity to the benefit of the other two" (2007, p. 146).

The Case of German Fachhochschulen (FH)

Historical Background and Missions of Fachhochschulen

Germany has a long tradition of non-university education of experts in certain fields. For a long time engineering was the major subject area in these institutions. Following France, which has an even longer history of highly regarded engineering schools, Germany saw the development of numerous technical educational institutions from the beginning of the nineteenth century onwards (Bode et al. 1997, pp. 8–21, 144–147; Becker et al. 2003, pp. 17–30). Engineering in Germany enjoyed high social and economic respect, nationally and internationally. Polytechnics, engineering schools and similar short-cycle establishments providing an engineering qualification, benefited from this positive attitude. Already in the nineteenth century some of them developed into technical universities. They sprang from the same roots as the Fachhochschulen, but lucky coincidences and state support helped them to cross the threshold early. Many other higher engineering education institutions continued to provide shorter study programs, usually with a strong link to industry and commerce (Christensen and Erno-Kjolhede 2011, pp. 285–299). In most cases, they were even financed by these and received little or no state support. Only supervision and quality control were regularly provided by the state, guaranteeing standards and keeping education in the hands of the state, whereas in many other countries professional bodies fulfilled this role.

In 1971 the German states in the Federal Republic, which still enjoy a high degree of autonomy in educational matters, transformed many of these pre-existing institutions into Fachhochschulen; some new ones were also founded at that time (Bundesministerium 2003). As was the case for British polytechnics, this process of horizontal integration has contributed to generate a binary system of higher education in Germany. The demand for higher qualified engineering staff in industry played a vital role in this development (Joschke 1981, pp. 4–11), together with the massive increase in student numbers which, as in France and the UK, required a number of structural changes to serve the new types of students and the needs of the labor market in a more cost-efficient way (Teichler 1996). In Eastern Germany numerous similar and even more highly esteemed technical Hochschulen existed. They usually could award PhDs like universities. After German reunification, most of them were transformed into Fachhochschulen, often despite fierce opposition from within the institutions concerned. Many experienced the reorganisation as a devaluation and loss of status.

The original idea was to focus the Fachhochschulen on teaching rather fixed curricula with a practical orientation, and on benefiting from their efficiency in order to provide a highly qualified workforce (Vorstand der Fachhochschulrektorenkonferenz 1990). The aim was to preserve their close connection to industry and commerce, namely regional SMEs, and to use it to optimize their curricula and teaching, to train many graduates in a short period of time providing them with a reliable knowledge base and a high affinity toward practical job demands. Similarities to schools in curricula, teaching methods including internships, control by the state and in staff salaries were a significant factor for creating the Fachhochschulen. Financial considerations had a major influence on the concept: reducing the cost of higher engineering education compared to universities was a key issue.

From the very beginning, university legislation also applied to Fachhochschulen, but at the same time they were clearly kept apart from universities proper. Actually they were meant to complement each other in the domain of engineering education and in the professional qualification of their graduates: Fachhochschulen were supposed to be equal in status to the universities but were different in nature (Christensen 2012, p. 157). Christensen notes that the new structure "marked a transition from a dual system of higher education to a binary system via the horizontal integration of former engineering schools (Ingenieurhochschulen) and higher vocational schools (höhere Fachschulen)". This important structural change led to a number of tensions in German Fachhochschulen due to their *de facto* unequal contexts in terms of student aspirations, degree awards, staff status and research opportunities.

Student Body's Aspirations

Student admission procedures in Fachhochschulen were and still are more constraining than in French IUTs: on top of passing the *Abitur* at the end of the Gymnasium, students were required to go through half a year of practical training in industry. An alternative route for admission, as in the French continuing education system, was via a 3-year apprenticeship in a craft. The majority of these students applying for admission in Fachhochschulen not only looked for a vocational training which would enable them to enter the labour market rapidly but also sought "social, political or ecological meaning in their studies" (Rau 1993, p. 40). At the end of their studies, FH graduates had very good career prospects, as the limited number of university graduates, especially in engineering, usually guaranteed that there was no real conflict in the fight for jobs with the universities (Bundesministerium für Bildung und Forschung 2003). Their business graduates also had excellent job prospects and career possibilities, and their alumni made their way to high positions in companies. Similarly, as Fachhochschulen were practically the unique providers of academically qualified staff for the social care professions, their graduates in this field had no problem to enter the labour market (Ministerium für Innovation, Wissenschaft und Forschung 2011).

However from the start a number of tensions arose in terms of degree structure and access to higher diplomas, and also as regards the salary differences with university graduates. Originally the study length for the German Dipl. Ing. awarded by Fachhochschulen was 3 years, but from the 1980s it increased to between 4½ and 5½years (Grose 2000). As a result, the degree structure was no more in keeping with the "bachelor-master-doctorate" degree system implemented by the European Union since the advent of the Bologna process and more or less adopted worldwide. Besides, to differentiate between a university degree and one earned at the Fachhochschulen, the latter had to add the postfix (FH) to the degrees and titles they conferred, a distinction which was clearly meant to safeguard the higher status and value of university degrees but which ended up in developing into a quality brand highly appreciated by industry and commerce. Another degree-related tension was that, due to the terminal nature of their courses, FH graduates were not entitled to apply directly for a master or a doctoral degree, they had first to complete a university degree *in toto* (Teichler 1996, p. 126).

Another set of tensions related to the student body was the pay differences between university graduates and Fachhochschulen ones when entering industry, but it soon became marginal. Only public service salaries remained significantly lower for FH graduates. This led to dissatisfaction and lack of interest in state jobs in times of economic prosperity. At the same time the preference for less costly staff in the public sector led to an especially high demand for cheaper FH graduates and to a high proportion of them employed in many administrative levels.

Teaching Staff: Tensions and Aspirations

The fact that all teaching staff at the Fachhochschulen have been trained in universities is of vital importance. Swiftly doctorates practically became a prerequisite for the newly employed Fachhochschule professors. The staff of the precursor secondary sector institutions had rarely held doctorates. The lack of research staff, support personnel and the lower wages were soon blamed for the lack of qualified applicants for teaching positions and the rather limited research success. Especially young professors, having recently graduated from universities, were and are often trying to move the Fachhochschulen closer to universities. They spent their formative years in the same system as university professors and, whereas university professors in Germany usually have to complete a large scale research project (called Habilitation) following their PhDs, applicants for FH professorships are required to have at least 3–5 years of experience in industry after their doctoral degree, thus entailing close company ties. This further requirement reinforced the vocational orientation of Fachhochschulen.

As regards FH faculty members and professors, tensions soon appeared in terms of teaching workload and research opportunities, just as in the French IUTs, but with regard to salaries as well. Although they often had access to research previously, a professorship at a Fachhochschule often resulted in the end of research activities, due to the high teaching workload (18 h per week), lack of support staff and restricted laboratory equipment. The salary differences between university professors and FH teaching staff also generated important tensions: the former's remuneration was about 20 % higher than the latter's. Finally, the fact that, unlike their university colleagues, FH professors were not entitled to train their own graduates for master or doctoral degrees, and so could not participate in the training of their future faculty members, was another source of tensions (HRK 2009).

Structural Transformations in the FHs

All the tensions within German Fachhochschulen described earlier were bound to lead to a number of in-depth transformations as was the case for IUTs in France, Polytechnics in the UK or Hogescholen in the Netherlands, to name but a few European higher engineering education institutions, in the same 40-year period. In order to facilitate the comparison between the transformations in the respective German and French institutions, we have chosen to focus on the same five significant issues we discussed for the IUTs: autonomy policy, study programs, pedagogical innovations, students' prospects and research.

With regard to autonomy policy, two kinds of changes can be pointed out concerning the funding and the functioning of FHs. To begin with, their financial resources have increasingly diversified through their close links to industry, which have led to a wave of new developments and growth. Short-term contracts of staff financed by project partners, new payment levels that are going beyond traditional standards, a growing differentiation between research-active and less highly regarded teaching professors are some of the consequences these changes may bring about within Fachhochschulen. Besides, a few companies have started to establish Fachhochschulen of their own, taking staff development and research organisation in their own hands. Bigger companies may continue to cooperate more readily with universities, but they may also invest in FH research or in Fachhochschulen of their own. Private higher education institutions (HEIs) may increasingly gain a stronger role in the German university landscape, a policy drift which might in the long run prove detrimental to the state FHs. A second facet of this policy drift is to be found in the attempts by Fachhochschulen to reach equality with universities, for example to get more support staff like master and Ph.D. students. Reduction of the teaching workload so as to devote more time to research is another claim. Slowly these claims are acknowledged by ministries of higher education, thus reinforcing the binary model of higher education in Germany. Other measures including money and regulations to facilitate employing project staff, reduction of teaching workload as a reward for acquiring research money or research professorships, are gradually made available to Fachhochschulen. These changes are to be assessed in the light of the privileges technical universities have long since enjoyed.

Concerning FH study programs, new curricula have sprung to life. Accreditation boards have taken over many of the former state competencies, enforcing institutional quality assurance and new administrative tasks. Joint projects with other educational institutions or companies are encouraged or are in place already. Worldwide competition for qualified students and staff will be a challenge in the near future. Elements from university degree programs cherished by the university graduates who took up positions as Fachhochschule professors are consistently being integrated in FH courses. Conversely, in many of their new course models, universities copy the better course organization and internship elements from the Fachhochschulen. By doing so, improving employability for university bachelor graduates comes within reach. Both institutions thus experience a specific curricular drift: FHs go through an academic drift as opposed to the vocational drift of the universities.

The emphasis of FHs on teaching and their strong role in the professional self-understanding of Fachhochschule professors have gradually been lost and the university philosophy of active participation in research as the basis and source of quality for teaching takes over. University teaching methods, traditions, organisational patterns and the university concept of academic freedom shape courses and institutions to an ever greater degree. Although universities copied many positive elements from the Fachhochschulen when the Bologna process made changes necessary, this led the FHs to discard the selfsame assets. Being a university or a FH does not any more dictate the choices. For example in Bavaria a bachelor degree program at a university now usually lasts six semesters, as opposed to seven in a Fachhochschule. Thus, the initial aim to provide a faster track toward engineering education through the Fachhochschulen was discarded, a clear mission drift of FHs. Only the lack of emphasis put on elements fostering employability in university bachelor programs still safeguards advantages for FH graduates in the competition for jobs.

As far as FH students' prospects are concerned, the Bologna process proved to be an important stepping stone not only for European student mobility but also in terms of degree awards. In a few German Länder the initially distinctive and vital internship parts of the degree courses were abolished, or reduced in others. As in France, the unification of degree structures (3+2+3) took place in Germany, but in both systems the prime concern of the reforms was to harmonize the first cycles. In Germany, however, this process went further than in any other European country, so much so that the gap between the two types of institutions in the binary system eroded from 1999 to 2004. This has resulted in the new right for Fachhochschulen to offer both academic research-oriented programs and professionally- or practice-oriented programs, like universities (Christensen 2012, p. 158). With this institutional drift, in that both bachelor and master degrees could be awarded by the two institutions, Germany came very close to a transition toward a unified system of engineering education (Witte et al. 2008, p. 222; Vogel 2009). However the fact that

Fachhochschulen still today have no right to award doctoral degrees renders contribution to research much less attractive.

Research is clearly a crucial issue which has concentrated tensions between universities and Fachhochschulen. The competition between the two institutions started in the 1980s and intensified considerably after the turn of the century. Recent years saw a substantial increase in research projects undertaken at FHs. This went hand in hand with an increase in research funds and state provision for first-class laboratory equipment, and companies are often partners in the financing of staff and/or equipment. Students' final projects contribute to research activities and give the opportunity to bring together company supervisors and their professors. Research has become a new internal and external indicator of status and a new source of income to professors and institutions (HRK 2005; Aspridis et al. 2001). Formerly, research activities of Fachhochschule staff were considered as an extra without any compensation in their salaries or in the reduction of their teaching workload. In recent years, this research drift has intensified as research has developed into an important asset for this kind of higher engineering institutions, insofar as it now offers FH academic staff greater freedom regarding their teaching workload and the funding of their equipment, especially when largely financed by industry. State salaries of graduates, previously a crucial differentiation between university and Fachhochschule leavers, are more and more aligned. The German governments' desire to generate more research and funding from private or company sources coincides with the options the Fachhochschulen offer. It is still not quite clear to what extent this research drift will impact the future of Fachhochschulen. It may work as an element of disruption within the institutions, it may lead to a stronger dependency on FH partners and it may bring new pressure from outside. Conversely, the outcome may be a strengthened Fachhochschule with research, Ph.D. courses, good company ties and graduates sought after by industry, administration and commerce.

The five different facets of academic drift we have discussed above illustrate the tendency of FHs and universities to integrate elements which used to be characteristic of the other type of higher engineering institutions. The current name "Hochschule", adopted by nearly all former Fachhochschulen by now, in its original meaning defined by German law, encompasses both universities and Fachhochschulen. The first step towards a merger of these two German institutions may have been taken without the intention to do so.

What Dynamics Are at Work in These Transformations?

We have shown in the previous sections of this chapter that, despite the actual differences between French IUTs and German Fachhochschulen – i.e. different status compared to the university, student credentialing perspective, academic staff status, a particular stand to research etc. – these two higher education institutions (HEIs) have always had a number of characteristics in common: socio-economic objectives, curricular orientation toward practice, strong focus on teaching rather than research, close links with companies, not to mention their creations which took place, from scratch, at the same period. Since then, they have both been subject to a number of significant changes, which we discussed above, with regard to their institutional character and their relationship to the society at large, including economy. To put it in a nutshell, Michael Tomlinson notes:

> Over time, there has been a general convergence of the education-work relationship, which, in part, has been mediated by national governments' continued emphasis on education as both a source of national prosperity and a catalyst of social and economic opportunity. Changes in the political economy of nations have led to increasing concerns over the need for strong, fit-for-purpose and efficient forms of educational provision to meet the challenges of a globally 'knowledge economy' (Tomlinson 2013, p. 1).

According to Tomlinson, there seems to be a consensus in the literature on higher engineering education that both the academic drift of HEIs and the professional drift of universities have been generated by two major significant forces: the *globalization* of the economy and the value of *knowledge* in our society; the former encompasses economic, political and social dynamics while the latter include professional, societal and technological ones – all of which have been at work in the structural transformations these two higher engineering institutions, like many others in Europe, have been through.

In our globalized society, higher education is now assigned a dual mission: to contribute to the economic success of the nation and to run its institutions in a more cost-efficient way. In the first issue, providing a competitive edge on world markets as well as on national ones is the driving force; in the second, fund-raising is a crucial factor for the autonomy and accountability of individual institutions (Bell et al. 2009, p. 5). To keep their competitive advantage, global companies require engineering graduates with top-level technical skills in order to foster innovations and guarantee quality, but also with competencies in the humanities and communication skills so as to conquer new markets or merely sustain existing ones. As a result, the pressure of the economic world has led HEIs and universities alike to introduce new curricula, e.g. the *licence professionnelle* in France, as well as new subject areas covering a broader scope of knowledge, a trend referred to as academic drift in the case of HEIs and professional drift for universities. Likewise there have been strong pressures – economic, political and social – to provide a higher number of engineering graduates to cope with increasing market demands and "to pursue research and teaching activities that will potentially be of wider economic value" (Tomlinson 2013, p. 176).

Another facet of economy as a driving force relates to the self-governance of these institutions not only in terms of competing nationally and internationally for financial resources that will sustain their autonomy, but also of implementing more cost-efficient practices to comply with more strict regulations and monitoring, such as performance indicators and benchmarking (Tomlinson 2013, p. 185). German Fachhochschulen have a long-time experience in the competition for fund-raising, due to the close ties they have maintained with companies and industry; to a lesser extent, French IUTs get similar resources from companies, even though their main funding still comes from the state. In this respect, the institutions that are more likely to capture international funds are those conducting world-leading research and those rendering the best service as regards curricular range, top-ranking credentials and teaching methods and facilities. As for their governance, universities and HEIs have become "publicly accountable institutions that need to rationalize and justify their share of public expenditure" (Henkel and Little 1999), via auditing of their practices and outputs.

A number of critiques have been expressed over this instrumentalization of higher education toward economic ends. Tomlinson argues: "Education has been reconfigured as a commodity good that should be used towards the utilitarian ends of enhancing national competitiveness. It has been economized" (2013, p. 11). Actually, if this "commoditization" of higher education mainly refers to the new managerial trend within universities throughout the world, it does not strictly apply to HEIs whose original missions already involved economic targets. However these institutions have also been impacted by the economic dynamics in that, as we have shown, the structural transformations of IUTs and FHs include curricular changes, new routes to a higher educational level, accommodating increasing student numbers and, above all, a new focus on research.

The dynamics of globalization in its political dimension is best illustrated in the implementation of the Bologna Process through the educational policies of the different national governments. In Germany, although there was no indication in the Bologna Process aiming to put an end to the binary system, the German state reform introducing changes in the degree structure, student credentialing perspective and market-oriented self-governance status came close to narrowing the gap between universities and Fachhochschulen (Witte et al. 2008). In France, the dynamics of the Bologna Process has led to a mixture of positive and negative outcomes for the IUTs. On the positive side, the degree structure was attuned to the bachelor level, via the *licence professionnelle,* and research was made available to all the teaching staff; on the negative side, there has been an operational drift away from their autonomy status, in that their financial and human resources have been largely transferred to the universities. As argued by Gombrich concerning higher education in the UK:

> …. Higher education is now subject to two complementary forces: mercantilism and dirigisme. The former is based on the belief that free markets and economic priorities should determine policy, while the latter involves the continued increase in state intervention in the structure and funding of higher education institutions (Gombrich 2000).

Although it refers a little provocatively to the British situation, this statement can also apply to the French and German institutions. Besides Bell et al. (2009) points to potential contradictions in state policies: they posit that economic success depends on substantial investment in higher education, but at the same time national governments have to reduce the budgets allocated to HEIs and universities due to global economic constraints.

The third globalization-related dynamics lies in the social dimension. The shift from elite to mass higher education may be viewed as a response to new global economic demands but it also evidences changes in the new demands for higher education. The issue of social justice is highlighted by Tomlinson: "The expansion of education and its associated theme of lifelong learning are seen to represent an inclusive model of social justice that engenders social mobility" (2013, p. 12). As early as 1963, the Robbins Report in the UK already insisted that "all young persons qualified by ability and attainment to pursue a full-time course in higher education should have the opportunity to do so" (quoted by Bell et al. 2009, p. 4). Already from the start, French IUTs and German FHs had provided for a better access of students from lower social backgrounds to higher engineering education, thus contributing to the reduction of inequalities. This dynamics of social justice has led to the creation of diversified educational routes for a diversified student body, such as continuing education, distance learning, apprenticeship, validation of professional experience, etc. In this respect, vocational institutions like IUTs and FHs are recognized as having fulfilled their economic and democratic goals, by responding to student aspirations for streamlined curricula, top quality teaching, higher credentials so as to enter the labor market in a more rewarding position (Tomlinson 2013, p. 12). However researchers in vocational education point out that social disparities still subsist and that individuals from lower socio-economic backgrounds are still underrepresented in higher education; the academic drift of HEIs might deter a number of them from pursuing their studies.

The second major driving force that we have identified earlier in this section is the value of *knowledge* in our society, which encompasses three other dynamics, i.e. professional, societal and technological. Our global society is characterized by a strong focus on knowledge as the "axial principle" of development insofar as the sources of innovation derive directly from research and development (Smeby 2006, p. 8). According to Smeby, in our post-industrial society, knowledge is a source of value, not labor, thus leading to a new professional dynamics in higher education: the production of new knowledge through research, which has always been a core mission for universities, has entailed new aspirations among the teaching staff in HEIs (Horn et al. 1992). The research drift described in both IUTs and Fachhochschulen has been engendered by this quest for new professional status and institutional prestige (Buck-Bechler et al. 1995). In turn, such a move toward research has placed new demands on professors and students alike to enhance their work-related knowledge and competences (Tomlinson 2013, p. 42). It has resulted in added value not only for curricula, hence for professors' and students' careers, but also for the economy at large. As Tomlinson puts it: "The more human capital that people can acquire, the more their productivity and value to the labour market" (2013, p. 11).

The second knowledge-related driving force refers to societal dynamics which can be identified in higher education national policies on gender equality and lifelong learning issues. The diversification of the student body in higher engineering institutions together with the feminization of work have led to a significant increase in the number of female students in higher education. Such a move has brought "an increased value to the kinds of skills and work that they undertake" as women are regarded as "highly adept at the types of 'soft skills' – communication-based, information and interpersonal skills – that the new economy requires" (Tomlinson 2013, p. 43). The feminization of the student body has not been a conflicting issue in most HEIs, even though engineering has traditionally been a male-dominated and orientated field of education. However it is to be noticed that this dynamics has actually enabled a number of engineering areas, e.g. biology, ecology, chemistry, informatics, etc. to develop significantly in IUTs and FHs. Lifelong learning, which has been strongly supported by the E.U., can also be considered as a dynamics affecting our society: it involves a totally new approach to the construction of skills and knowledge in individuals. It has an impact not only on their lifestyles, due to the inherent constraints of continuing education, for instance, but also on the public perception of higher education and what knowledge is.

Technological advances are both the outcomes and the sources or instruments of new knowledge. In this respect, they constitute the third knowledge-related dynamics. Technological innovations and developments that are created by research engender new teaching methods and support, new information resources, new knowledge. ICTs are widely used in HEIs today, and IUTs as well as FHs have largely contributed to their development both as a curriculum of its own and as an aid to other subject areas. The attractiveness of digital technology has a dynamic effect on the recruitment of students and the motivation and practices of the teaching staff; it can also be a source of income from technological transfer contracts with industry, a trait we have already discussed in previous sections.

Conclusion

In this chapter, we have discussed the structural transformations that French IUTs and German Fachhochschulen have experienced in the light of academic drift processes that have affected similar higher engineering institutions around the world since the 1960s. This trend developed against a more general post-WorldWar II background of quantitative expansion and massification of higher education (Christensen 2012, p. 163). These structural transformations were caused by a number of tensions within the HEIs themselves and between them and the universities: these tensions are related to institutional status, funding, degree-awarding system, salary differences and research. Besides the more recent globalization expansion and the advent of the knowledge society have amplified the trend:

> The traditional insular operations of nation states and their educational and welfare systems have given rise to an increasingly globalized convergence and coordination of policy, provision and practice. Increasingly, national governments have looked to align their educational systems with fast developments in a new, globally competitive knowledge-driven economy (Rizvi and Lingard 2010, quoted by Tomlinson 2013, p. 18).

According to Jónasson (2006), structural transformations of higher education systems take place in the following three steps: (1) Students' pressure to get higher credentials leads to educational expansion, as illustrated by French IUTs' curricular drift. (2) Faculty members' push for equal status and salary drive the internal structures of institutions along academic lines, as exemplified by German FHs.

(3) The constant pressure created in points 1 and 2 is modulated by national policies and labor market considerations.

As we have shown, these transformations that have affected higher engineering education in Germany and France, as well as other countries, have been convergent in their drive to implement an overall state policy of widening access to a greater number and more diversified student body and complying with the economic imperatives of efficiency and institutional accountability. Concerning credentials, there has been a significant advance in both IUTs and Fachhochschulen: in order to respond to the increasing expectations of a larger student body as well as institutional pressure, and in an attempt to attune their degree structure with international standards, FHs have been entitled to extend their degrees to the master level while IUT students can now have easier access to the bachelor level via the *licence professionnelle*. Similarly, the two institutions have registered a number of curricular changes as more theory-orientated or humanities-based subjects are integrated into their courses. Another commonality they share relates to research drift: even though research has long been a current activity in both institutions, since parts of the staff are university academics, the habilitation to conduct a research policy of their own and to accommodate research teams in their locus constitutes a crucial change, thus facilitating staff's working conditions.

Structural transformations are also to be examined in the light of the original status and mission of these two institutions, thus displaying major differences. University-embedded IUTs, which enjoyed a specific high degree of autonomy as regards funding, course design and staff recruitment, have recently lost parts of these prerogatives, thus bringing them closer to other faculties; on the contrary, non-university FHs have maintained their autonomy in their missions due to their long-time close relationships with companies. As for their respective teaching staff, while the pressure of IUT secondary education personnel has ended up in extending their access to research, the push by FH faculty members to reduce their teaching workload to the benefit of research and to reach salary equality has had little effect so far (Bassarak and Steppuhn 2002).

In the wider context of economic globalization and knowledge-driven society, we have identified a number of socio-economic driving forces that contribute to the structural transformations of higher engineering education in France and Germany, as in most developed countries: market demand for top-level engineering workforce with a broader knowledge base to ensure innovation and competitiveness – student pressure for better credentials and for social justice which also involve societal dynamics, e.g. feminization and lifelong learning – public and institutional drive for sources of funding and accountability – staff aspirations for higher status and access to research – and, to a lesser extent, technological advances. Jónasson argues that the core driving forces are actually the student body and the academic faculty, in that the former drive educational expansion while the latter contribute to revamp institutional structures along academic lines. According to him, the other "dynamics, i.e. market demand and institutional drive – to which technological advance should be added – are just external modulating or facilitating factors" (2006).

Academic drift should be regarded as a natural and irreversible process: "natural" because it interacts with the inevitable evolutions of society in its economic, political, social, cultural and technological dimensions, and "irreversible" as it constitutes a never-ending trajectory (Bell et al. 2009). Tomlinson notes: "The shifting dynamics in the interrelationship between education and work reflect broader social and economic transformations, all of which have a substantial bearing on individuals' formal (and informal) educational and labour market experiences" (2013, p. 3). Higher engineering education, indeed education in general, is not just considered important for the economy but is itself a driving force in shaping the so-called human capital of the future.

Acknowledgement The initial idea and the outline of this contribution were inspired, and indeed encouraged by Steen Hyldgaard Christensen, whose research and publications the authors of this chapter are strongly indebted to.

References

ADIUT – Association des Directeurs d'IUT. (2007). *Livre Blanc sur le système IUT après 40 ans d'existence: Histoire, Bilan et Perspectives.* Available at: http://www.iut-fr.net/publications/livre-blanc.html

Aspridis, S., Block, H.-J., Griesbach, H., & Teichgräber, M. (2001). *Evaluation des BMBF-Programms „Anwendungsorientierte Forschung und Entwicklung an den Fachhochschulen (aFuE)"* (Hochschulplanung 152). Hannover: HIS.

Bassarak, H., & Steppuhn, U. D. (Eds.). (2002). *Angewandte Forschung und Entwicklung an Fachhochschulen in Bayern.* Düsseldorf: Hans-Böckler-Stiftung.

Becker, O., Hoffmann, H. M., & Iselborn, K.-W. (Eds.). (2003). *Fachhochschulen in Deutschland und Europa. Eine Bestandsaufnahme. Festschrift anlässlich des 60. Geburtstags von Prof. Dr. hc. Dietmar von Hoyningen-Huene.* Mannheim: Pylon Verlag.

Bell, L., Stevenson, H., & Neary, M. (Eds.). (2009). *The future of higher education: Policy, pedagogy and the student experience.* London/New York: Continuum International Publishing Group.

Bode, C., Becker, W., Habbich, C., & Klofat, R. (Eds.). (1997). *Fachhochschulen in Deutschland.* München: Prestel.

Bourdoncle, R. (2007). Universitarisation. *Recherche et Formation, 54*, 135–149.

Buck-Bechler, G., Jahn, H., Klockner, C., & Tietz, H.-D. (Eds.). (1995). *Angewandte Forschung an den Fachhochschulen der Bundesrepublik Deutschland.* Weinheim: Deutscher Studien Verlag.

Bundesministerium für Bildung und Forschung. (2003). *Die Fachhochschulen in Deutschland* (4th ed.). Bonn: Bundesministerium für Bildung und Forschung.

Burgess, T. (1972). *The shape of higher education.* London: Cornmarket Press.

Burgess, T. (1978). The officials' revolution: The British polytechnics: Ten years after. *Paedagogica Europaea, 13*(2), 45–58. European Universities: Ten years after 1968. Blackwell Publishing. http://www.jstor.org/stable/1502531. Accessed May 2011.

Christensen, S. H. (2012). Academic drift in European professional engineering education – The end of alternatives to the university? In S. H. Christensen, C. Mitcham, Li Bocong, & Yanming An (Eds.), *Engineering, development and philosophy: American, Chinese, and European perspectives.* Dordrecht: Springer Science+Business Media B.V.

Christensen, S. H., & Erno-Kjolhede, E. (2011). Academic drift in Danish professional engineering education. Myth or reality? Opportunity or threat? *European Journal of Engineering Education, 36*(3), 285–299.

Convert, B., Gugenheim, F., & Jakubowski, S. (2011). La loi Faure et son utilisation par les contestataires de 1968. La professionnalisation dans une université scientifique. In *L'Enseignement Supérieur et la Recherche en Réformes – Actes du Colloque International RESUP*, 1, 175–192.

Gibbons, M., Limoges, C., Nowotny, H., Schwarzman, S., Scott, P., & Trow, M. (2006). *The new production of knowledge – The dynamics of science and research in contemporary societies*. London: Sage.

Gombrich, R. (2000, January 7). *British higher education in the last twenty years: The murder of a profession*. Lecture given to the Graduate Institute of Policy Studies, Tokyo.

Grose, T. K. (2000, March). Re-engineering in Germany. In *Prism*. Available at http://www.prism-magazine.org/march00

Hanchane, S., & Verdier, E. (2005). Educational routes and family aspirations in France, a panel data. *The European Journal of Vocational Training, 35*, 53–64.

Henkel, M., & Little, B. (Eds.). (1999). *Changing relationship between higher education and the state*. London: Jessica Kingsley.

Horn, B., Klinkmann, N., & Salustowicz, P. (1992). *Forschung an Fachhochschulen – der Weg in eine neue Identität*. Weinheim: Deutscher Studien Verlag.

HRK (Ed.). (2005). *Forschung, Entwicklung und Technologietransfer an Fachhochschulen. Dokumentation der 34. Jahrestagung des Bad Wiesseer Kreises vom 20. Mai –23. Mai 2004* (Beiträge zur Hochschulpolitik 5/2005). Bonn: Hochschulrektorenkonferenz.

HRK (Ed.). (2009). *Quo Vadis Fachhochschule? Dokumentation der 38. Jahrestagung des Bad Wiesseer Kreises vom 01.- Mai – 04. Mai 2008* (Beiträge zur Hochschulpolitik 3/2009). Bonn: Hochschulrektorenkonferenz.

Jónasson, J. T. (2006). *Can credentialism help to predict the convergence of institutions and systems of higher education*. CHER 19th Annual Conference. Systems Convergence and Institutional Diversity, 7–9 Sept 2006. Germany: University of Kassel.

Joschke, H. (1981). *10 Jahre VHB. Denkschrift des Verbandes der Hochschullehrer an Fachhochschulen in Bayern e. V. anlässlich seines 10-jährigen Bestehens*. München: VHB.

Kyvik, S. (2009). *The dynamics of change in higher education: Expansion and contraction in an organisational field*. London: Springer.

Ministerium für Innovation, Wissenschaft und Forschung des Landes Nordrhein-Westfalen. (2011). *Eine Erfolgsgeschichte. 40 Jahre Fachhochschulen in NRW*. Berlin: Dr. Josef Raabe Verlag.

Rau, E. (1993). Inertia and resistance to change of the Humboldtian university. In C. Gellert (Ed.), *Higher education in Europe*. London: Jessica Kingsley Publishers.

Rizvi, F., & Lingard, B. (2010). *Globalizing education policy*. London: Routledge.

Saumade, G. (1998). *Les Instituts Universitaires de Technologies*. CNDP: Rapport au Premier Ministre, 50 p.

Smeby, J.-C. (2006). Professionalism in a knowledge society: The academic drift of professional education in the 'new' professions. In *4th Interim Meeting, European Sociological Association's Research Network "Sociology of Professions": "Professions, Globalization and the European Project: Shifting Spheres of Opportunity"*, 30 Mar–1 Apr 2006, University of Bremen. Available at: http://www.academia.edu/565379/Professionalism_in_a_knowledge_society_The_academic_drift_of_professional_education_in_the_new-professions

Teichler, U. (1996). Diversity in higher education in Germany. The two-type structure. In L. V. Meek, L. Goedegebuure, O. Kivinen, & R. Rinne (Eds.), *The mockers and the mocked: Comparative perspectives on differentiation, convergence and diversity in higher education*. Oxford: IAU Press/Pergamon.

Tomlinson, M. (2013). *Education, work and identity: Themes and perspectives*. London/New York: Bloomsbury Academic.

Vogel, M. P. (2009). The professionalism of professors at German Fachhochschulen. *Studies in Higher Education, 34*(8), 873–888.
Vorstand der Fachhochschulrektorenkonferenz (Ed.). (1990). *Die Fachhochschulen zu Beginn der 90er Jahre – Eine Positionsbestimmung*. Bremen/Wiesbaden: FRK.
Witte, J., van der Wende, M., & Huisman, J. (2008). Blurring boundaries: How the Bologna process changes the relationship between university and non-university higher education in Germany, the Netherlands and France. *Studies in Higher education, 33*(3), 217–231.

Bernard Delahousse M.A. in English Language and Literature. Faculté des Lettres de Lille. Ex-lecturer at Université des Sciences et Technologies de Lille (USTL), France. Head of International Office at IUT 'A' of USTL until 2004. Co-editor and co-author of three previous books resulting from projects initiated and coordinated by Steen Hyldgaard Christensen: *Profession, Culture, and Communication: An Interdisciplinary Challenge to Business and Engineering* (Institute of Business Administration and Technology Press 2003), *Philosophy in Engineering* (Academica 2007), *Engineering in Context* (Academica 2009). Chief editor of *Les Langues Modernes,* the journal of the French Association des Professeurs de Langues Vivantes (APLV), 2007–2010, and co-editor since 2011.

Wilhelm Bomke M.A. and Ph.D. in History and English, University of Regensburg. Director of the International Office, Regensburg University of Applied Sciences. Author of *Die Teufelsfiguren im mittelenglischen Drama* (Lang 1990). Research interests are Medieval History, History of Engineering Education, History of Education in Germany. Co-author of three previous books resulting from projects initiated and coordinated by Steen Hyldgaard Christensen: *Profession, Culture, and Communication: An Interdisciplinary Challenge to Business and Engineering* (Institute of Business Administration and Technology Press 2003), *Philosophy in Engineering* (Academica 2007), *Engineering in Context* (Academica 2009).

Chapter 4
Engineering Brazil: National Engineering Capability at Stake

Édison Renato Silva, Roberto Bartholo, and Domício Proença Jr

Abstract This presentation of Brazilian engineering sketches its trajectory in the nineteenth, twentieth and twenty-first centuries. Well into the nineteenth century, engineering was unwelcome in Brazil: its agricultural slaver society had little use for it. Although the oldest engineering school in the Americas was founded in Rio de Janeiro in 1792, Brazilian engineering was an unwanted novelty. It took Vargas' 1930 dictatorship to bring about Brazilian engineering. Engineering in the Brazilian context became more than buildings and machines. It emerged as the core of institutional innovations, as a tool of a national development project. It bloomed in the late 1950s, leading to almost half a century of accelerated industrialization. A peculiarity of this contextualized process was the leading role of graduate studies over undergraduate education, and its emphasis on intervention in Brazilian reality in engineering education. Since the beginning of the twenty-first century, however, a new political coalition seeks to redesign the institutionality of federal universities in Brazil, jeopardizing the future of Brazilian engineering and putting national engineering capability at stake.

Keywords Brazil • History of engineering • Engineering education • National development • National engineering capability • Engineering policy

Introduction

This chapter sketches the changing contexts of engineering activities in Brazil in the nineteenth, twentieth, and twenty-first centuries. It divides this history into three periods according to the role and place of engineering in Brazilian society.

É.R. Silva (✉)
Industrial Engineering Department, Escola Politécnica/UFRJ,
PO Box 68507, Rio de Janeiro, RJ 21941-972, Brazil
e-mail: edison@ufrj.br

R. Bartholo • D. Proença Jr
Production Engineering Program, COPPE/UFRJ,
PO Box 68507, Rio de Janeiro, RJ 21941-972, Brazil
e-mail: bartholo.roberto@gmail.com; domicio@centroin.com.br

S.H. Christensen et al. (eds.), *International Perspectives on Engineering Education*,
Philosophy of Engineering and Technology 20, DOI 10.1007/978-3-319-16169-3_4

The presentation follows four interpretative keys that illuminate different facets of engineering: practice, education, development, and national engineering capability.

In 1500, Portugal colonized Brazil, beginning "economic cycles" of exploitation: Brazil-wood from trade with natives, sugar, then gold, and then coffee with African slaves. On the wake of the exile of the Portuguese Imperial court to Rio de Janeiro from 1808 to 1821, Brazil became independent under the heir to the throne of Portugal, Peter I, Emperor of Brazil, in 1822.

Engineering *practice* explains the distinction between Engineering *in Brazil* and *Brazilian* Engineering. Engineering in Brazil describes works that serve exploitation, with no concern for local prosperity or national development. Brazilian Engineering describes the pursuit of national interests, directed, and often carried out by Brazilian engineers. These might be engineering, without qualifiers, in developed countries; but the distinction is opportune for developing ones.

Engineering *education* followed practice. All engineering education took place abroad, with occasional local training of auxiliaries, until it became expedient to facilitate replacements locally. This was the rationale, and illuminates the restrictions, of the first engineering school in the Americas, established in 1792 in Rio de Janeiro. The Army School of Artillery, Fortification, and Drawing trained military officers for State-mandated civil engineering works. Private works relied on self-styled "practical engineers".

Brazilian *development* became a burning issue by the late eighteenth century. Colonial exploitation in "cycles" lacked any concern for enduring prosperity. Foreign trade, ironworks, industry, the press, and universities were forbidden. The logic of empire was monopolistical exploitation until riches gave out, leaving regions destitute in its wake. However, Brazil had an inland economy largely insensible to imperial priorities. The combination of successive "cycles" with an independent economy made development a longstanding concern. It offered a bright line for appreciating the economic and political significance of engineering.

Hence, the presentation offers a synthetic characterization of the role of engineering and engineers in Brazilian society, appreciating practice, education, development, and national engineering capability. It addresses three periods: the first, in which Brazilian Engineering can be described as an unwanted novelty; the second, in which federal universities made it a tool of State-led national development; the third, in which a new political coalition seeks to redesign the institutionality of federal universities in Brazil.

Engineering: The Unwanted Novelty

The nineteenth century Brazilian repugnance to manual labor accorded with that of most slaver societies, holding it as the mark of the slave, favoring idleness and the alleged elevation of the law as fitting activities of gentlemen. The abundance of cheap slaves and the continuity of centuries-old solutions made engineering improvements seem superfluous. From the 1820s, the "cycle" of coffee moved slaves and prosperity from sugar to coffee producing regions. Rail allowed coffee farms to spread away from the coast. There was no gradual passage to modern transportation. Rail arrived suddenly, bringing as if by magic the full panoply of industrial production, of communications, of travel. It imposed trade, goods, news, books, free laborers, technicians, and engineers on slave plantation communities. Gilberto Freyre, one of Brazil's most insightful authors, remarked in his Preface to the Catalogue of the Center for Preservation of Railroad History "to talk of rail … is to talk of a whole socio-cultural complex. Not just material engineering, but engineering that unfolds into human and social engineering" (Freyre 1982, *apud* Telles 1993, p. 119). Rail also contributed to questioning long-standing social power relations in terms of race. Telles (2009, p. 86) points out to how scandalous it was, in the construction of Paranaguá-Curitiba Railway, that the chief engineer was a dark-skinned half-breed, Antonio Rebouças, leading German engineers such as Peter Scherer, Mauricio Schwartz and Julio Kallman.

Rail changed the *practice* of engineering and the place of engineers in society, making them necessary and, like military engineers, tolerated, even commemorated. But this did not make engineering a gentleman's profession. Enrollment in Law remained five times that of Engineering through the late nineteenth century. Engineers had to know their place. Just because they were there, this did not mean they would change things just because they could. Engineers faced hindrance and death threats when they tried. In that very substantive sense, rail, employing 75 % of all engineers in 1880, remained a case of Engineering in Brazil: a tool for exploitation of riches.

There was a single Army school for engineers, which split in 1874 into schools for military and civilians – the *Academia Militar* (Military Academy) and the *Escola Central* (Central School) respectively, both offering free-ride courses through competitive examination. Engineering *education* stood aloof from engineering practice. Contents were very narrow, for the most a single curricular track that preserved an eighteenth century logic of "mathematics, physics, and natural science": rote learning with bouts of "practice", full courses as small as 2 years' duration. Bibliography was dated and predominantly French. The very structure of engineering education was inspired by the French model. That the civilian school was to be an *Ecole Centrale* intimated a subordination, presumably to the military one. But soon the civilian school gained ascendancy, and like the capstone of the French system, became a *Polytechnique* – the *Escola Politécnica*. But the French emphasis on a measure of mathematics side by side with a measure of broad, general understanding remained. Railroad engineering specifics and mechanics seem to have been left up for on-the-job training – engineering education was behind the times. One should be cautious, however, in seeing more than inspiration in this: French influence was pervasive, but diffuse, its interpretation open to local adaptation rather than following formal precepts or implementing models (Bethell 1986).

Rail offered boundless prospects. It promised *development*: an end to exploitation. For Brazilians, development meant *industrialization*. Brazilians dreaded the fate of selling fashionable commodities, hostage to the whims of importers, exhausting successive riches with little left afterwards. But there were problems, the most important of which proved to be neither financial nor technical, but institutional. Painfully, slowly, Brazilians came to realize that no amount of Engineering in Brazil, importing engineers or turnkey factories, could hope to achieve their goal (see Freyre 1988). With development at stake, they realized the need for Brazilian Engineering, capable of more than replication. They had to pursue not one, but two goals: industrialization, yes, but through *national engineering capability.*

Brazilians had to go beyond the illuminist motto of "thinking with our own heads" to add "… and doing with our own hands". This meant deliberate, conscious social engineering, abolishing slavery and changing the laws to introduce the dignity of labor, free enterprise, and industry to a slave-holding, agricultural, oligarchical society. The Republic came about in 1889 in a military coup, and the military shared the same vision. The Army began sending a few engineers a year abroad, to learn about something which would not become a reality for half a century, despite many failed attempts: how to make steel.

Engineering: Nation-Building

The Republic led to a brief war between Army and Navy, after which the "Old Republic" emerged to try to restore oligarchical power. But something had began which could no longer be prevented, only delayed. In 1930, Getúlio Vargas seized power as dictator until 1945, returned elected president in 1951, committing suicide mid-mandate to prevent a coup in 1954. Vargas capitalized on decades of institutional innovation efforts and ideas to implement an ambitious national development project to industrialize Brazil. The coup came in 1964, inaugurating a quarter century of military rule, which kept largely true to Vargas' ambitions.

In 1930 Brazil, the beginning of the Vargas Era, engineering was an activity open to all comers, just as it had been during the Empire. Experienced foremen, "practical engineers", and charlatans were as entitled to *practice* engineering as diploma engineers. Specialized tasks might require formal qualifications, but they were few. However, electricity and ferroconcrete led to a radical change in the context of practice. Both required calculations, which foremen and practical engineers could not handle, and which scared charlatans. Federal Decree 23569 of December the 11th, 1933 made a diploma from acknowledged national universities (with provision for the validation of foreign diplomas by those universities) compulsory for the legal exercise for the practice of any engineering. This institutional framework heralded the maturity of Brazilian Engineering – engineering was a profession organized, supported and accountable to the State.

Engineering *education* took a peculiar turn: schools and colleges were formed into universities under federal administration ("public universities") with tenured faculties and substantial budgets, offering free-ride education through national competitive examinations. Universities began to participate, to anticipate, to shape. Under Vargas, all Brazilian engineering schools were called upon to prepare and undertake major projects. The Empire had but two: the civilian *Escola Central*, later the *Politécnica*, and the *Academia Militar*. The *Escola de Minas de Ouro Preto* (School of Mines of Ouro Preto) would become an engineering school in the twentieth century, but under the Empire it was a research institution, specialized on Brazilian geology. On the decades following the Republic, many others were created. By Vargas' time, there were 14 engineering schools in operation. They began to work with State and private firms. Policy decisions relied upon public engineering schools to assess which opportunities could lead to development – and to provide personnel and knowledge to pursue them. Brazil was awarded its first steel complex for joining the Allies in 1943, and there were qualified Brazilian engineers for all aspects of its operation and expansion. This was the outcome of 50 years of anticipation by the Army and the ready support of public engineering schools. Conversely, Brazil's 1950s "Fifty Years in Five" program, despite the multiplication of universities and engineering schools, revealed that engineering research capability was lacking. "Fifty Years in Five" was an overambitious modernization based on a radical import-substitution model. Being merely up to date with the present proved insufficient when possible futures were at stake – capability had to be prepared ahead of needs.

In the 1960s, Coppe, part of the Federal University of Rio de Janeiro, started graduate programs tailored to specific policy priorities and in anticipation of engineering needs. It developed frameworks to manage the finance and execution of projects. Petrobras' need for underwater robots exemplifies such a collaboration. Brazil alone had to deal with deep water off shore oil exploration. There were no solutions abroad, nor interest in developing such solutions. Brazilian Engineering alone could dare to provide it. It took many years – many projects, thesis and dissertations – to formulate, engineer, and then implement a solution. Much of this work could not be made public until there was a solution, for industrial and national security issues.

From the early 1970s, graduate engineering departments took the lead in engineering education, practice, and research, defining new undergraduate specialties, forecasting and providing for possible Brazilian Engineering needs through foundations, expedient parallel structures for managing funds that did not derive from the federal budget. From 1964 to 1988, the Military Regime pursued initiatives to bring Brazil to the same level as the developed countries. Some would blossom, such as Brazil's quest for oil autonomy, began by Vargas in the 1950s; others flounder, such as Brazil's hope of domestic autarchy in computing in the mid 1980s.

Vargas' national development project relied on State-led initiatives. He reorganized undergraduate education on federal universities as part of free-ride education in all levels; modernized labor, professional, union, and public service legislation and standards; streamlined and formalized Brazil's national policies and foreign relations; established national monopolies in steel and oil; tried the same in electricity, navigation, and railroads. But Vargas' most telling institutional innovation was the National Development Bank (BNDES), to provide capital for investment in national priorities. The same intent supported Finance for Study and Projects (FINEP), the primary source of funding for engineering projects in Brazil.

Engineering capability in the narrow sense was necessary, but insufficient: development required engineers capable of pioneering, groundbreaking projects. Most engineering research takes place in federal public universities, particularly in graduate departments, which rely on foundations to propose, contract, organize, regulate, and manage projects. Brazilian development was a century-long struggle against traditional privileges. It bred the informal motto "there is everything left to be done".

Engineering: Institutional Redesign

The Military Regime ended in 1988. A neoliberal agenda became predominant in the 1990s. National plans and policies became less directive, with expectations that the market would find the best path to prosperity. Privatization dismantled whatever coherence or rationale might have existed among the large state-owned firms inherited from Vargas and expanded through the Military Regime. This has led to substantial change in Brazil's engineering context.

Today's panorama of engineering *practice* offers contrasting realities. There are sectors in which Brazilian Engineering predominates, making opportunistic and more or less integrated use of Engineering in Brazil, such as oil & gas or agribusiness (including biofuels), and, to a lesser extent, construction, mining, and aeronautics. There are sectors in which Engineering in Brazil predominates without prospect of ever leading to Brazilian Engineering, such as capital goods, automobiles, electro-electronics. As nineteenth century Brazilians were wont to acknowledge, there is more to engineering than operation and maintenance – there is design, there is ambition. Car manufacturing has been one of the largest economic and exporting sectors in Brazil for 50 years. However, it is not, and there is no real prospect that it will ever become, more than "made in Brazil" – the specter of *maquiladoras* – assembly plants in which cheap local labor work to assemble imported components into products. Further, the example of South Korea, which leveraged its Engineering to world-class standards over the last decade after building cheaper, low-cost cars to finance its development, raises many galling issues about Brazil's current choices.

Brazilian *education* policy has changed over the last 15 years. This is an ongoing process that continued even after the assumption of power by a new political coalition. It expresses a varied set of agendas, in part justified by misinformed scientometrics, in part conditioned by political goals of social inclusion, in part obedient to privatization interests, in part expressing ideologically motivated interventions on federal institutions. They come together through three main vectors:

1. CAPES, an agency of the Ministry of Education, grades graduate activities every 3 years, assessing faculty and student performance. It adopted unified criteria for all disciplinary areas in 1998, which have made publication in indexed (ISI-JCR) journals the paramount measure of performance. CAPES grades have become the overriding input to the creation, certification, and funding of graduate activities, and for qualification in federal initiatives.
2. Starting 2003, education policy for federal undergraduate education has changed from providing cadres for development to the universalization of access to

university courses, taking increased student population as proxy for social inclusion. Increased enrollment has complete priority over educational quality, as expressed by the multiplication of courses in, and "advanced *campii*" by, existing universities, the creation of new federal universities, and the use of federal funds to support enrollment in private universities, far outstripping that in public universities. Federal public universities' courses are being biased toward supposed operational employability: training, rather than education.

3. Starting 2008, federal comptrollers, who hold powers of autonomous inquiry, have repeatedly denounced federal public university freedom of project and enterprise as improper, choosing to hold individuals responsible for allegedly illegitimate or illegal institutional decisions. This ongoing political controversy clashes with Constitutional university autonomy, enshrined in Article 207. Animated by non-academic politico-ideological convictions of their own about what the role of university *should be*, comptrollers continue to question contract, funding, fund-raising or financial management by university foundations.

To assess engineering graduate activities primarily by publication misunderstands the nature of engineering. Engineering does involve the production of knowledge, but in the final analysis it is change in reality that matters. Engineering cannot be reduced to knowledge production. Engineering knowledge, in turn, cannot be reduced to scientific knowledge. Hence, the performance of graduate engineering departments cannot be assessed by the number and trend of publication in indexed scientific journals every 3 years.

To choose quantity over quality, mass over cadres, training over education is a valid policy decision. However, such an expansion of student population in federal universities and the dilution of federal education resources makes it difficult to keep pace with the cutting edge and to provide for, or anticipate, future needs – imperiling Brazil's future engineers' project capability.

Comptrollers' conception of federal universities devoid of supporting foundations leads to isolation from economy and society through the strangulation of engineering projects as part of academic activities. Considering the last century of engineering experience in Brazil, this would entail material, human and social impoverishment.

CAPES' standards are valid to all universities in Brazil – a broad spectrum that comprises privately-owned universities, ran as businesses; "communitarian" universities, associated with a given creed or faith, ran as businesses; and public universities, which offer free-ride courses: these can be owned either by the individual federal entities within the Union (state universities or municipal universities) or by the Federal government (federal universities). However, the drive to enrollment expansion and comptrollers' charges against autonomous financial management by university foundations applies *only* to federal universities. In all states but one, federal institutions are the core of university education and research. It is only in the State of São Paulo that state universities rank with federal ones. As a result, it is impossible to avoid discussing federative implications, particularly for engineering.

Top-ten engineering departments in Brazil belong to public universities: eight are federal, two are state universities in the State of São Paulo. The combined effect of the above three restrictions discriminates strongly against federal universities. Should they persist, they would endow the State universities of São Paulo with *de facto*, even conceivably *de jure* exclusivity on the freedom to pursue quality education and manage its relations with Brazilian economy and society. This outlines a major shift in the Brazilian Engineering context. It seems impossible to avoid the conclusion that it would substantially diminish Brazil's national engineering capability, with the future of Brazilian Engineering at stake.

Conclusions

It took one and a half centuries of struggle to bring about Brazilian Engineering, a century for engineering education to catch up with engineering practice, half a century more for it to become capable of meeting policy priorities, shaping and anticipating engineering needs, and yet it may take less than a decade to cripple it. This presentation of engineering in the evolving, willful Brazilian context amounts to a story of "engineering Brazil", touching practice, education, development and national capability over two centuries. What would be the moral of this story? The past exemplifies a moderately successful tale of a peripheral country's break with its colonial legacy, seeking national development on a long term basis, highlighting the value of staying the course. The present gives evidence of how easy it is to imperil dynamics that sustained over half a century of burgeoning development. Ultimately, the future might turn out to be a terrible cautionary tale on the prerequisites, demands, potential, achievements, and frailty of a national engineering policy.

References[1]

Alder, K. (2010). *Engineering the revolution arms and enlightenment in France, 1763–1815*. Chicago: University of Chicago Press.

Bethell, L. (Ed.). (1986). *The Cambridge history of Latin America, Volume 5: c. 1870–1930*. Cambridge: Cambridge University Press.

[1] A note appears the best way to annotate the synthesis presented above. The reconstruction of Brazil's history is grounded on Gilberto Freyre (1933/2005, 1959, 1987), Celso Furtado (1959/2009) and Sergio Buarque de Hollanda (1936), with François Chevalier (1977) and Leslie Bethell (1987, 1989, 1995, 2008). The history of Brazilian engineering is far more fragmentary, with pride of place for the ongoing efforts by Pedro Carlos da Silva Telles (1993, 1994, 2009, 2010), to which Paulo Pardal (1984, 1986) adds detail, and which benefits of Schultz (2001) and Alder (2010) for the role of French inspiration. The issue of steel benefits from memories of the free-docent thesis of Maria Luiza de Carvalho Proença, Domício's mother, towards full professorship, no copy of which survives. Silva & Proença Jr. (2013) offers a more extensive and annotated presentation on CAPES and Engineering with both documentary and critical references.

Bethell, L. (Ed.). (1987). *Colonial Brazil*. Cambridge : Cambridge University Press.
Bethell, L. (Ed.). (1989). *Brazil: empire and republic, 1822–1930*. Cambridge: Cambridge University Press.
Bethell, L. (Ed.). (1995). *The Cambridge history of Latin America, Volume 6, Part 1: Latin America since 1930: Economy and society*. Cambridge: Cambridge University Press.
Bethell, L. (Ed.). (2008). *The Cambridge history of Latin America, Volume 9: Brazil since 1930*. Cambridge: Cambridge University Press.
Chevalier, F. (1977). *L'Amerique Latine de l'independance a nos jours*. Paris: Presses Universitaires De France.
Freyre, G. (1933/2005). *Casa-Grande e Senzala* (47th ed.). São Paulo: Global.
Freyre, G. (1959). *Ordem e Progresso,* two volumes. Rio de Janeiro: José Olympio Editora.
Freyre, G. (1987). *Homens, engenharias e rumos sociais*. Rio de Janeiro: Editora Record.
Freyre, G. (1988). *Ferro e civilização no Brasil*. Rio de Janeiro: Record Editora.
Furtado, C. (1959/2009). *Formação Econômica do Brasil*. São Paulo: Companhia das Letras.
Hollanda, S. B. D. (1936). *Raízes do Brasil*. Rio de Janeiro: Olympio.
Pardal, P. (1984). *Memórias da Escola Politécnica*. Rio de Janeiro: Escola Politécnica da UFRJ.
Pardal, P. (1986). *140 anos de doutorado e 75 de livre docência no Ensino de Engenharia no Brasil*. Rio de Janeiro: Escola Politécnica da UFRJ.
Schultz, K. (2001). *Tropical Versailles: empire, monarchy, and the Portuguese royal court in Rio de Janeiro, 1808–1821*. New York: Routledge.
Silva, É. R. P. D., & Proença Júnior, D. (2013). *Os indicadores CAPES na Engenharia arriscam a perda da capacidade de projeto do Brasil*. Rio de Janeiro: Édison Renato Pereira da Silva.
Telles, P. C. D. S. (1993). *História da Engenharia no Brasil, v. 2, Século XX* (2a ed.). Rio de Janeiro: Clavero.
Telles, P. C. D. S. (1994). *História da Engenharia no Brasil, v. 1, Séculos XVI a XIX* (2a ed.). Rio de Janeiro: Clavero.
Telles, P. C. D. S. (2009). *História da Engenharia Ferroviária no Brasil*. Rio de Janeiro: Notícia & Cia.
Telles, P. C. D. S. (2010). *Escola Politécnica da UFRJ, a mais antiga das Américas, 1792*. Rio de Janeiro: Synergia.

Édison Renato Silva M.Sc. and D.Sc. in Production Engineering, Coppe, Universidade Federal do Rio de Janeiro, The University of Brazil, Rio de Janeiro. Adjunct Professor, Industrial Engineering Department, Escola Politécnica, Universidade Federal do Rio de Janeiro. He was nominated 2014–2015 International Scholar of the Society for the History of Technology. His main research interests include the history and epistemology of engineering, design research and methods for systematic literature review.

Roberto Bartholo M.Sc. in Production Engineering, Coppe, *Universidade Federal do Rio de Janeiro,* The University of Brazil, Rio de Janeiro, and Dr. in Friedrich-Alexander-Universität Erlangen-Nürnberg, Germany. Professor, Production Engineering Program, Coppe, Chairman, The Laboratory for Technology and Social Development (a multi-institutional national research group), *Universidade Federal do Rio de Janeiro,* The University of Brazil, Rio de Janeiro. Senior Visiting Researcher and Chairman, Production Engineering and Local Development Network, *Universidade do Estado do Amazonas* (Amazonas State University). Coordinates the *Network Africa-Brazil Dialogues on Social Innovation,* with fellow researchers in the Cape Peninsula University of Technology, South Africa, University of Botswana, Makerere University, Uganda, University of Nairobi, Kenia and Université Mohammed V, Maroc. Coordinates the Brazilian research group in the Choices Project: *Leveraging Buying Power for Development – Ethical Consumption and Sustainable Procurement in Chile and Brazil.*

Domício Proença Jr M.Sc. in Science and Technology Policy and Politics and DSc in Strategic Studies in Production Engineering, Coppe, *Universidade Federal do Rio de Janeiro*, The University of Brazil (Rio de Janeiro). Awarded the Brazilian Order of Merit (for National Defense). Professor, Production Engineering Program, Coppe and Public Policies, Strategies and Development Program, Institute for Economics; Chairman, The Group for Strategic Studies (a multi-institution national research group), *Universidade Federal do Rio de Janeiro*, The University of Brazil (Rio de Janeiro). He has the first Brazilian Ph.D. in Strategic Studies, and has worked for 25 years in research, education, policy advice, private consulting and the media in subjects related to international security, national defense and public security; he has published on war, technology, arms industry, strategy, peace missions, drug trafficking, police organization and policy, guerilla and terrorism.

Chapter 5
Engineering Education in India: A Comprehensive Overview

Balasundaram Subramanian

Abstract This chapter is an attempt to survey the steady evolution of engineering education in India from colonial times to the present in terms of institutional landmarks and defining moments in Indian history. The survey urges that engineering education in India, especially in its contemporary reflexes, has to be understood against the backdrop of periodic reviews, reforms, recommendations of various committees and the compulsive social, economic and political matrix informing policy-making. In particular, it highlights the achievements as well as the shortcomings of the concerted attempts to consolidate teaching and research in keeping with the imperatives of national objectives and market needs. It reviews reform initiatives in recent times, and it locates in the convergence of the technical sciences and the liberal arts disciplines the potential for improving human resources in India.

Keywords Colonial history • Engineering education • Knowledge economy • Policy-making • Reforms

The Colonial Interlude

> We have no enemy now in India, except popular ignorance, and that we are doing our best to remove by the most complete system of State education that has yet been devised in any country (J. G. Medley, Principal (1863–1871), Thomason College of Engineering, Roorkee).

Engineering Education in India, to borrow an expression from Leibniz, is heavy with the past and big with the future. To understand the cumulative presence of the past, one has to go back to the colonial interlude, which marks India's inexorable march toward modernity and westernization. From 1773 onward, colonial dispensation in India comes increasingly under imperial supervision; with Crown rule replacing Company Raj after the Great Mutiny of 1857, a complex canvas of administration

B. Subramanian (✉)
School of Humanities & Social Sciences, Indian Institute of Technology Mandi (Kamand Campus), 175005 Katola P.O., Himachal Pradesh, India
e-mail: bs@iitmandi.ac.in

S.H. Christensen et al. (eds.), *International Perspectives on Engineering Education*, Philosophy of Engineering and Technology 20, DOI 10.1007/978-3-319-16169-3_5

emerges, scripting India's progress to a welfare state and ultimate self-determination with Benthamite efficiency. It is a passage assisted no doubt by the massive influx of technology in the wake of the industrial revolution, with steam navigation, railroads, roads and telegraphs decisively influencing the emergence of India as nation-state. It is indeed a passage that goes hand in hand with the establishment of a heavily centralized public administration. Centralization – the root impulse to organize in order to manage – may have tended to go against the grain of a certain native disposition of managing not to organize at all – a trait that had once prompted John K. Galbraith to brand famously India as a "functioning anarchy". Nevertheless, it cannot be gainsaid, that it was predominantly a growing class educated on western lines that propagated widely and instilled firmly the notions of nationalism and democracy in the country.

Besides the unifying effect of education in the English tongue, education fitting the needs of this vast centralized system of law and administration emerges as major instrument of social change. Also, transfer of technology makes its own set of institutional demands for the organization and advancement of knowledge altogether different from its traditional transmission within the confines of caste-based professional networking. Progressively, choice of profession is influenced by economic considerations rather than by caste affiliation, viz., by ascription to birth in a certain group or community. More importantly, exposure to the new education enabled Indian intellectuals to reflect critically on a moribund tradition, its mores and values and to revive its quintessential civilizational cementing force by a process of progressive reform. This dialectic of the foreign and the familiar leading to a virtual renaissance of Indian culture, especially in Bengal, was prompted no doubt by the reformist bid to establish educational institutions like the Hindu College in 1817 in Calcutta, modeled on western curricular structure. Notably, this preceded the famous Macaulay Minute of 1835 (Subramanian 2010, p. 165) that makes a complete departure from native educational systems and catapults India onto the trajectory of western education in the English medium through government schools and complemented by a complete range of subjects in the arts and the sciences.

An even more significant milestone was Sir Charles Wood's Education Despatch of 1854, "which sets out a blueprint for the future development of western education throughout British India. It included provision for the establishment of provincial education departments, government and voluntary schools, and universities based on the London model of affiliated colleges in Calcutta, Madras and Bombay" (Whitehead 2003, p. 5). Soon after India becomes Crown Colony, Lord Stanley's Despatch of 1859 reinforces Wood's policy and establishes firmly the requisite structure for overseeing education and related policy at the provincial level.

The "Beautiful Tree"

It has often been held against Macaulay that his educational policy resulted in the demise of native educational systems. Indeed, this is the substance of the charge made by the Mahatma in his lecture at Chatham House (London, 10.10.1931):

> ...India is more illiterate than it was fifty or a hundred years ago, and so is Burma, because the British administrators, when they came to India, instead of taking hold of things as they were, began to root them out. They scratched the soil and began to look at the root, and left the root like that, and the beautiful tree perished.

Myths have a compelling appeal, and soon the myth of the "beautiful tree" found willing adherents, eager to graft on to it engineering and engineering education in India that obtained prior to colonial rule. Scholars like Dharampal have documented Indian technology and agriculture on the ascendant till mid-eighteenth century, not to speak of an attendant comprehensive system of education embracing the bulk of its youth, however informal its structure (Dharampal 2000, p. 10). The sustained depletion of India's economy through an insidious system of taxation, the erosion of its social and cultural fabric by laying the axe at the root of its education system, in short the uprooting of indigenous wisdom by the scheming ingenuity of its colonial masters form the stuff of much colonial and post-colonial discourse, fuelled no doubt in part by professional malcontents eager to reclaim a romanticized past. It has to be conceded though that Dharampal's investigation of Indian science and technology in the eighteenth century is a special pleading for revisiting the past and its embedded value systems and to reconstruct self and society in that light in order to eschew the pitfalls of blind westernization (Dharampal 2000, pp. 13–14). Further, it cannot be denied that, for example, technical training institutions were set up largely to train middle level technicians for maintenance of public works and for military purposes. It may be argued that British intent was not to develop intrinsic human resources in India, yet by the early twentieth century through an osmotic effect, there is considerable Indian presence in the top echelons of the civil service and in academic establishments. Bhaskaran's trenchant observation merits, therefore, earnest consideration: "It is a popular superstition, impossible to eradicate, that our educational system was devised to produce and only turned out inferior employees of government and commercial offices. On the contrary it has produced in the past and is still turning out in large numbers, young men and women equipped to meet all the needs of a healthy society" (Bhaskaran 1967, p. 220).

The Industrial Imperative

In the third quarter of the eighteenth century, engineering education in India sought to address itself in the main to the technological deficit obtaining between local skills and the demands brought about by the industrial revolution. In addition, colonial rule, to acquire a firm grip on subject territory, required exhaustive information, so to speak, on the lie of the land. The Great Trigonometric Survey of 1818, for example, was prompted by this need and called for skilled experts (Arnold 2004, p. 25). Madras had established a Survey School as early as 1794. Following Wood's Despatch, the school at Madras was turned into a civil engineering school in 1858, and elevated in the following year to a civil engineering college. Likewise, civil engineering colleges came to be established in Poona (1854) and in Sibpur/Calcutta

(1856). Alongside Roorkee (1847) these locations describe geographically a quadrilateral, with each institution designed to serve, so to speak, its watershed. These institutions developed in stages; the college at Madras on shifting to the impressive redbrick building in suburban Guindy in 1920 came to be known as the Guindy College of Engineering, and forms today the influential core of the Anna University. The Poona Engineering Class and Mechanical School was renamed by 1911 as the College of Engineering, Pune, just as the college at Calcutta came to be known finally as the Bengal College of Engineering. (Renaming thus is not just an index of institutional advancement; it is equally reflective of the attempts to comprehend training in engineering skills within an academic frame.)

The construction of the Upper Ganges canal from 1842 for major irrigation works necessitated the establishment of a training school in Sahranpur, and soon it became in 1847 the College of Engineering on moving to nearby Roorkee. It was renamed in 1854 as the Thomason College of Civil Engineering in honor of its initiator James Thomason, the Lieutenant General of the North Western Provinces (Vir et al. 2011, p. 5). The transformation of the College to a university was accomplished soon after Independence (1949); laudably, it was inducted into the IIT league by an act of Parliament in the year 2001. Within a quarter century of its establishment in 1847, the institution at Roorkee had become more or less the role model for other engineering institutions in the country.

At Roorkee, programs of instruction were intended in the main for engineers (mostly military personnel), for assistant engineers (chiefly civilians for the Public Works Department) and last but not least for a large native class, who were educated as sub-overseers, sub-surveyors, estimators, and draftsmen (Medley 1873, p. 41). To go by Medley's first-hand account, Roorkee had also a "library, model room, and museums in the college, and an excellent press, whence a good many useful works have issued, chiefly relating to Indian engineering" (Medley 1873, pp. 41–42.). Indeed, the "Roorkee Treatise on Civil Engineering" bears eloquent testimony to native engineering skills. In fact, in acknowledgement of excellent mechanical skills obtaining among the native gunsmiths of Munger, the largest railway workshop in India was set up in 1862 at picturesque Jamalpur in Munger District, Bihar State. Jamalpur became home thus in 1905 to the first of many Centralized Training Institutes, namely the Indian Railways Institute of Mechanical and Electrical Engineering (IRIMEE). In sum, Roorkee and Jamalpur set the tone and tenor for engineering education, the ambitious collegiate education model and the more practical polytechnic model respectively.

The Victoria Jubilee Technical Institute may be said to be the precursor of the polytechnic model. Established in Bombay in 1887 to commemorate the Diamond Jubilee of Queen Victoria, it trained licentiates in electrical, mechanical, and textile engineering and technology (Sen 1989, p. 227). The Indian Education Commission of 1882 under the stewardship of Sir William Hunter made a series of excellent recommendations to improve technical education. Rather than yoke technical education to the colonial imperative, Sir William intended to forge an efficient industrial society in India by a concerted development of human resources. Understandably, he met with little success.

Milestones on the Path to Nationhood

All the same, with rising nationalist ardor, private initiative, and government support, the number of higher technical institutions in India too rose from 5 in 1919 to 21 by 1939; likewise the number of diploma schools increased from 8 to 23. Several leading institutions emerge during India's struggle for independence, notably: Institute of Technology of the Benaras Hindu University (1919, now an IIT), Harcourt Butler Technological Institute, Kanpur (1920), Indian School of Mines, Dhanbad (1926), Maclagan College of Engineering, Lahore (1930), University Department of Chemical Technology, Mumbai (1934), Engineering College at the Aligarh Muslim University (1935), Delhi Polytechnic (1941), Laxminarayan Institute of Technology, Nagpur (1943), Alagappa Chettiar College of Technology, Guindy (1944), etc., (Vir et al. 2011, pp. 4, 7).

Other scientific developments run more or less parallel to the foundation of these institutions, leading to the emergence of an Indian scientific community in earnest dialogue with its western counterparts from 1890 onwards. Publications like the *Imperial Gazetteer* documenting every aspect of India ranging from anthropology to zoology, professional associations and guilds, learned bodies and academic societies like the Bengal Medical Association (1885), Indian Association for the Cultivation of Science (1876), Indian Medical Congress (1894), Indian Science Congress Association (1914), Indian Institution of Engineers (1920), privately endowed research institutions like the Tata Institute of Science (1909, now: Indian Institute of Science), numerous government funded research establishments like the Indian Agricultural Research Institute (1905) at Pusa, the Central Research Institute (1906) at Kasauli etc., – all contribute significantly to the growth of a scientific community. The Imperial Civil Service and a host of similar institutions like the Indian Educational Service provide the solid administrative ballast. Research too gets a boost with the setting up in 1911 of the Indian Research Fund Association into which government funds and private donations were channeled (Arnold 2004, p. 144). Nevertheless, post-graduate education and research languished for the most part, and many Indian scholars went abroad for higher studies.

The British Nobel laureate Sir Archibald Vivian Hill was commissioned by the Viceroy's Council to report on the state of scientific and industrial research in India as part of a post-war reconstruction and reorganization plan. The report (1944) while praising India's scientific community, held that the war "had left India's scientists 'sorely cut off … from intellectual contacts with the rest of the world"; consequently, the scientific and technical resources of India had "not been utilized, or developed for war purposes to anything like the same degree as those of the other major countries" (Arnold 2004, pp. 196–197). Hill's report underscored also the need to give lead and direction to research through greater central coordination. This suggestion resulted ultimately in the strengthening of the apex body, the Council of Scientific and Industrial Research (est. 1942) under the direction of the eminent scientist Shanti Swarup Bhatnagar, especially in the post-Independence era.

India's Tryst with Destiny

It is not well known that the Hill report provided the platform for the establishment of the IITs along the lines of the MIT. It was Sir Nalini Ranjan Sarkar, Member of the Viceroy's Executive Council (Dept. of Education, Health and Lands) who took the cue from this report. He headed the 23 member committee set up in 1945 on the advice of Sir Ardeshir Dalal (Member-in-Charge of Planning and Development of the Viceroy's Executive Council) to go into the need for new institutes of technology to produce adequate technical manpower for the development of post-war India. Sir Ardeshir had in fact on his return from the USA in 1944 spoken of an "Indian MIT", spurring thus the imagination of policy-makers in India (Vir et al. 2011, p. 11). The Sarkar Committee in its interim draft of February 1946 recommended the establishment of four higher technical institutions along the pattern of the MIT, one for each of the major geographical regions.

In a notable departure from staid university fare, the Sarkar Committee emphasized the need for blending scientific training with a broad human outlook and recommended the inclusion of subjects like industrial administration, economics, mathematics, statistics, chemistry and physics. Further, it laid heavy emphasis on workshop and laboratory training. Teachers were to be allowed to have consultancy and research besides adequate leave to go back to the industry for keeping up to date with developments (Vir et al. 2011, pp. 13–14).

Soon after Independence, these recommendations found favor with the visionary ideas of India's first Prime Minister Jawaharlal Nehru. In stark contrast to Gandhi's plea for technology "cut to size" in keeping with simple needs of an India of villages, Nehru's call for the dynamic industrialization of India rested on the plank of heavy industries, public sector investment, nationalization of basic infrastructure and a planned economy. The compelling need for higher technical institutions was recognized right away, and the Eastern Higher Technical Institute was set up in May 1950, initially at Calcutta, and later renamed in November 1950 as Indian Institute of Technology Kharagpur, located strategically not far from Calcutta and its heavily industrialized hinterland. An Act of Parliament declared it in 1956 to be an Institute of National Importance (INI) and gave it autonomous status.

Other IITs followed soon in its wake: Bombay (1958), Kanpur (1959), Madras (1959); besides, the College of Engineering Delhi is converted into an IIT in 1961, the same year in which all IITs are declared by an Act of Parliament to be institutes of national importance. The purpose of such an act is to invest these institutions with a great measure of autonomy in charting their course. The academic credibility and international standing of these institutions owe much to this self-regulatory structure. A certain amount of internationalization took place in the early phase of institution building, with Government of India inviting developed countries to assist in the setting up of the IITs. American initiative came thus to IIT Kanpur, while IIT Madras became the largest recipient of German educational development aid. This helped the institutions to keep abreast of international academic developments. Besides, residential campuses almost the size of small townships, contribute by way of concentration of students and staff to the professionalization of the disciplines. Also, industry based projects and summer internships played a decisive role in the making of the young engineer.

The careful selection of higher secondary school leavers by an exacting nationwide Joint Entrance Examination (JEE) has played a significant role too in nurturing the IITs as institutions of international excellence, reflected by the very high proportion of students of the Bachelor of Technology program leaving mostly for American universities and highly paid jobs. Recent criticism that the JEE despite its stiff acceptance rate (top 2 % out of roughly 400,000 candidates) has become an avenue rather for numerates than literates has brought about significant changes from this year (2013), with a two-stage examination process in place. The JEE Main Examination subsumes now the entrance examinations held earlier severally by the numerous engineering institutions at the federal and state levels, while the JEE Advanced Examination poses the sterner challenge of entry into the IITs. Even as this promises improvement in the quality of undergraduate material, major concerns still remain about the quality of research at the IITs, not to speak of other engineering institutions.

State of Research

The need as well as scope for research was initially circumscribed by the limited objectives set out by the colonial educationists; further, the early universities were of an affiliating nature and did little to promote research. In fact, it was nationalist sentiment surrounding the establishment of the Tata Institute (1909) or the BHU (1915) that gave slight impetus to research (Saha and Ghosh 2011, p. 111). After Independence, the IITs have played a significant role in vitalizing research by meshing it with teaching; success at the Graduate Aptitude Test in Engineering (GATE) has become the standard entry ticket for public funded post-graduate education and research India wide. As part of Quality Improvement Programs (QIP) the IITs have encouraged teachers at engineering colleges to enroll for doctoral research; in addition to part-time doctoral positions for the industry, the IITs have had numerous tailor-made post-graduate courses for the public sector industries and the military establishment. Industrial consultancy and sponsored research have stayed the course at the older IITs, with much funding coming initially from government agencies, public sector undertakings and defense laboratories. Of late, the IITs have also put in place a reasonably well-funded system of post-doctoral fellowships. Nevertheless, scholars like Sen (1989, p. 247), Saha and Ghosh (2011, pp. 111–112), and Subbarao (2013, p. 64) have not hesitated to point out the inherent infirmities in research development. Natarajan has listed some of the persistent problems: research positions go often to the rejects of the job market; state universities find it hard to modernize infrastructure; much research is of the incremental variety, more theoretical than practical (cf. Saha and Ghosh 2011, p. 112).

It may also be said that tardy progress on the research front was initially the result of ill-defined policies and the failure to scale up institutions suitably for the purpose for want of funds. India's initial success with the First Five Year Plan (1951–56) led planners to accept more or less the Soviet style of planned economy. More ambitious 5-year plans came at a price though (Rothermund 1993, p. 130). The consensus among planners that industrialization alone held the key to the economy led to the establishment of capital-intensive public sector industries, low on productivity and high on employment. Curbs on private enterprise by the notorious "permit-license-quota" regime and import substitution measures, while engendering a protectionist market for the new industries, did little to encourage research other than attempts at reinventing the wheel.

Periodic performance audit at the IITs did much to address institutional shortcomings; for the first time in 1983 the Nayudamma Committee, chaired by Dr Yelavarthy Nayudamma, Director-General of the CSIR, initiated a comprehensive review of all IITs. Its findings, published 1986, are directed at reinforcing institute-industry partnership and at dedicated research for uplifting the living standards of India's vast rural hinterland. More importantly, the Committee sought to liberate research from the maze of obsolescence, both in terms of men and material, of old-fashioned bureaucracy and outmoded equipment (Vir et al. 2011, p. 121).

The Growth Story

Unwittingly, these recommendations came at a time when the collapse of the Soviet bloc was just around the corner and about to trigger off untrammeled globalization and a free market economy. Dwindling foreign exchange reserves compelled India in 1990 to depart from its populist socialist stance and to embrace willy-nilly liberal market values. The changes in its wake are phenomenal. Within a span of two decades, India has become an economic powerhouse, indeed a Prometheus unchained. India dominates today the IT and Software sector, Indian industry has entered into major joint ventures and mergers with multi-national concerns, companies like the Tatas today have a sizeable global reach, a state like Tamil Nadu, for instance, has become the major hub of the automotive manufacturing sector; besides, massive infrastructural investment in transport and communication has transformed the economy completely.

The telling impact of liberal reforms on the education sector can be hardly overlooked. Students have come to realize increasingly that good education alone can secure for them a professional future. For example, in recent years, 150,000 Indian students alone have invested over $2 billion on education overseas. [In comparison, central and state governments together invest annually roughly $3.7 billion in higher education (Panagariya 2008, p. 432).] From 157 engineering institutions at the start of the 80s, India now has over 3,500 engineering colleges and over 1,750,000 engineering students (CABE annual report,[1] 2012–2013). The booming economy has also in part marginally reversed the earlier trend of Indian students and researchers seeking their fortunes abroad, especially in the United States, not to speak of many Indian scholars settled there returning to their homeland and to jobs in many universities and industrial establishments. [From 1985 to 2000, Indian students earned more than 13,000 science and engineering doctoral degrees at U.S. universities, mainly in engineering and physical and biological sciences. They also earned by far the largest number of U.S. doctoral degrees awarded to any foreign group in computer and information sciences. Among IIT alumni alone, 25,000 are thought to be working or studying in the United States (Clark 2007, p. 9).]

The economic unshackling of India has converged somewhat fortuitously with a major revolution in telecommunications and a burgeoning IT industry with massive investments from the private sector, an industry whose potential was little recognized at first, thus eluding luckily stern bureaucratic scrutiny and likely snares. By the turn of the millennium, it had become abundantly clear that liberal reforms had to be backed by massive infrastructure and human resources development. In particular, the educational sector required urgent attention, more so because of the imperative to cash in on the so-called demographic dividend, the 70 % of population aged below 30. Besides, the cyber era and the influx of newer and newer technologies had set in train profound socio-economic, cultural and political transformations. One may indeed speak legitimately of Indian society in a state of complete ferment.

Given, however, the very nature of social dynamics, it is open to question whether academic education can ever manage to keep pace with social change, let alone anticipate it. In addition, universities have had to accommodate within the academic fold the rapid spread of emerging technologies, setting up new departments, research facilities and recruiting experts. In India, all major institutions charged with overseeing higher education have responded impressively to these new challenges. Education being a concurrent subject (cf. Schedule VII, Constitution of India), both the centre and the 29 states are equally responsible for the care of education at all levels. So, in 2002, upon the recommendation of the Mashelkar Committee, the centre made bold to convert all the Regional Engineering Colleges, established originally to satisfy regional aspirations and manpower needs between 1956 and 1960, into the National Institutes of Technology (NITs). Today there are 20 NITs, all declared as institutions of national importance, functioning with a great deal of autonomy and central funding, much on the model of the IITs, though under a different Act.

The liberating impact of this move is reflected in the vastly improved quality of education and research besides placement record. The IITs too have witnessed a decisive phase of expansion, with eight new IITs being set up to meet the growing challenges. The expansion of the Indian Institutes of Management (IIMs), the setting up of the Indian Institutes of Science Education and Research (IISER), proposals to upgrade some of the older technical institutions of repute (like the

[1] Central Advisory Board of Education, Ministry of Human Resource Development, Government of India.

Bengal College of Engineering Sibpur) into Indian Institutes of Engineering Science and Technology (IIEST) or the recent bid to expand the network of Indian Institutes of Information Technology (IIIT) through public-private partnership – all point to the complementary array of institutions needed to buttress technical education. The latest statistics set out in the appendix detail the institutions under the dispensation of the Ministry of Human Resources Development.

Private Enterprise

Yet, it is the private sector that has contributed largely to the growth of educational institutions in the last two decades, trying to match the staggering growth of the economy with a phase of near breathless expansion. Unlike the much older, benevolent private institutions, say the Birla Institute of Technology and Science Pilani or the A.C. College of Technology Madras, with undisputed academic credentials, a large number of the newer private establishments (self-financing institutions) was set up largely, often with political patronage, to bridge an ever-widening market gap (Varshney 2006, p. 3). To review the work of the prime advisory and regulatory body, the All India Council of Technical Education (est. 1945) in the light of the new situation, government set up the U.R. Rao Committee in 2002. Its report reviewed all engineering disciplines, architecture and town planning, the applied arts, hotel management, business management and pharmacy. In the main, it called for restructuring the AICTE to meet the challenges of globalization. It levels sharp criticism at the unregulated growth of private engineering colleges, their abysmal infrastructure, weak faculty resources, exorbitant fees and lack of research facilities. The unbridled growth of private colleges had led to several distortions, notably: (1) regional disparities on account of overcrowding of colleges in select regions to the neglect of other areas, and consequent oversupply in some markets and shortfalls elsewhere (2) the graduate growth rate far exceeded the economy's growth rate (3) poor standards make for unemployability, thus accentuating unemployment rather oddly in a market facing serious shortage of skilled manpower (4) manifestly, little effort had been made to understand the manpower needs of the industry and to offer tailor-made courses for the purpose.

NBA and the Washington Accord

Above all, the Rao report calls upon the National Board of Accreditation (est. 1987), an autonomous wing of the AICTE, to tighten up its lax institutional and program accreditation procedures, and to bring every institution under its scrutiny. In sum, the Rao Committee report "has pointed out, the AICTE needs to focus on ensuring that its standards are met at already existing institutions, new institutions are opened in areas that need them, substandard institutions are closed and that faculty shortages are reversed by investing in postgraduate education and encouraging talented students to remain in India to pursue careers in academia" (Clark 2007, p. 9). Uniform accreditation of over 3,000 institutions is a daunting task, and the NBA is yet to become a full signatory to the Washington Accord. The projected roadmap to full status in June 2014 is paved with good intentions, but ground realities suggest a long haul. In the short run, a two-tiered accreditation framework with Tier I status for top-rated Indian institutions like the IITs is more likely to set in train the entire process of quality assurance, continuous improvement, training of university coordinators and program administrators, and training of faculty for outcomes-based accreditation (cf. NBA website).[2]

Reform and Self-Renewal

In the last two decades, the IITs have witnessed large-scale expansion, with new IITs being opened in underserved areas of the country. The imperatives of globalization have made the IITs address the need to raise research and development to internationally competitive levels. The Rama Rao Committee set up in 2004 developed a roadmap for the future, based firmly on the maxim that "excellence is a journey and not a destination". It advocated induction of foreign nationals as faculty besides joint appointments with industry. It also encouraged giving research incentives and the induction of bright B.Tech students into challenging Ph.D. programs; it wanted to enrich the science and humanities component of the undergraduate program besides promoting design and business centric projects. It wanted the IITs to profile their research and innovation through better IP management (cf. Kakodkar Report 2011, p. 15).

Building upon the Rama Rao report and striking an even bolder approach to a brave new future is the report of the Kakodkar Committee tabled in April 2011, and predicated upon "our national development aspirations, growing economy with inclusive participation, creating opportunities for our youth and building our competitiveness in the emerging knowledge-driven global economy". It is based upon the firm conviction that the "IITs are by far the only institutions, which can lead this process on a scale commensurate with the needs of our country". At the same time, it notes that "with only 7,500 undergraduate (UG) and less than 1,000 Ph.D. students graduating every year, the output of the IITs is inadequate for the future". Kakodkar observes ruefully: "Clearly the world has passed us by. If India has to be among the three largest economies of the world, the IIT system has to grow several folds in terms of research output, the number of PhDs and student graduation." To achieve this aim, the IITs have to be more accessible to a greater number of talented Indians. And more importantly, transparent as well as greater representative governance are indispensable: "The IITs should have standards benchmarked against the best universities around the world. One of the essential ingredients for this is a good

[2]As this text goes into press, it may be noted that India has become a signatory to the Washington Accord as of June 15th, 2014.

governance system with an independent and fully empowered Board with representation from key constituents such as scientific establishment, industry, alumni, faculty and Government." In practice, the key recommendations read thus (Kakodkar 2011, pp. 162–163):

- Make IITs the Primary Research Institutes, with a focus on high quality frontier research and technology development within the Indian context.
- Scale up Ph.D. students from less than 1,000 Ph.D. graduates per year today to 10,000 Ph.D. graduates by 2020–25 from about 20 IITs (15 existing IITs plus 5 new to be set up over the next several years in states where there are no IITs).
- Scaling Ph.D. scholars' admissions to include enabling bright UGs being admitted for Ph.D. at the end of their third year, teachers from other institutes joining for Ph.D. and significant numbers from industry joining sponsored/part-time Ph.D. program. It is strongly recommended that a fellowship scheme covering all categories of PhD students is in place.
- The faculty: student ratio is 1:10; while the UG: PG ratio is close to 1:1.
- Each IIT should aim to acquire technology leadership in at least 3–4 areas.
- Research groups in one or more IITs to take up large projects together to address major national challenges
- Set up research parks at each of the IITs similar to the IIT-Madras Research Park.

While making research integral to the IITs, the Committee has also recommended an Executive M.Tech program for about 10,000 working professionals from industry through live video classes to enhance the knowledge base in the industries.

In keeping with the focus on innovation, the Ministry of Human Resources Development (MHRD) has also introduced in May 2012 the "Universities for Research and Innovation" Bill in Parliament with the aim of creating institutions "recognized universally for their quality in teaching, learning and research" (CABE annual report 2012–13, p. 13).

In a more recent development, alarmed by the fall in global rankings of premier engineering institutions, the MHRD has announced independent third-party reviews of all the 15 IITs (Indian Express 2013a, p. 6). Parameters for the review include among other things: global character of the institution (in terms of international student enrollment, visiting foreign faculty and courses with international participation), internationalization (in terms of publication and citation index), alumni engagement quotient, adequacy of facilities and teaching, contribution to national development goals (NDGs), transparence in governance structures and, last but not least, student and faculty diversity in terms of gender equity.

Much of the scaffolding for these bold measures and blueprints has come from the pivotal role of the National Knowledge Commission (est. 2005 with Sam Pitroda as chairperson) in fortifying higher education. Primarily, it has sought to develop appropriate institutional frameworks (1) to strengthen the education system, promote domestic research and innovation, facilitate knowledge application in sectors like health, agriculture, and industry (2) leverage information and communication technologies to enhance governance and improve connectivity (National Knowledge Commission, March 2008).

The advances made in information and communication technology hold great promise for distance education, enhancing in particular the reach and quality of engineering education, offsetting in a way acute shortage of faculty and at times woeful quality of instruction. The National Program on Technology Enhanced Learning (NPTEL) is an initiative by the seven IITs (IIT Bombay, Delhi, Guwahati, Kanpur, Kharagpur, Madras and Roorkee) and Indian Institute of Science (IISc) for developing curriculum-specific video and web based course contents in engineering, sciences and the humanities (Natarajan et al. 2009, pp. 71–77). Spurred by the initial success and encouraged by the MOOCO (massive open online courses) floated by Western campuses, in a most recent initiative, seven IITs, the IT industry and NASSCOM are set to revolutionize higher technical education in India by offering free online courses (Indian Express Editorial 2013b).

Polytechnics

Free online learning supplementing the curricular offerings of India's dedicated television education network "Gyan Darshan" (with telecasts in English and regional languages) is bound to make a qualitative impact on the middle-level technical education sector too. From about 50 polytechnics at the time of Independence, India now has around 1,300 training institutions. But the irresistible appeal of entering the boardroom has held out against the enticements of the shop floor. India thus produces today more engineering degree graduates than diploma holders. This unhealthy trend is in need of urgent remedy, for the industry faces a serious shortfall of skilled technicians. Over the years, clear improvements to the quality of polytechnic education (cf. G.R. Damodaran Committee 1970) have been made, the curriculum revised and fine-tuned to meet industry requirements, and sandwich programs devised for lateral entry to university courses. Today, however, the emergence of a large service sector, which hinges considerably on technology applications, has necessitated a relook at the scope of polytechnic education. The new curriculum sports a considerable IT component; besides, more diploma courses, especially in soft skills, make the polytechnics attractive for women students. More significantly, the MHRD has proposed a Scheme of Community Development through Polytechnics (CDTP) and their extension centers. This aims at "providing non-formal, short-term, employment oriented skill development programs, through AICTE approved Polytechnics, to various sections of the community, particularly the rural, unorganized & disadvantaged sections of the society, to enable them to obtain gainful self/wage employment". In a significant reinforcement of the polytechnic system, the MHRD has decided recently to set up 200 Community Colleges in states and union territories in order to redress the gross mismatch between supply and demand for skilled workers. The Community Colleges are being planned within the overall framework of National Vocational Education Qualification Frameworks (NVEQF) (CABE annual report 2012–13, p. 17).

Women in Engineering

In the last couple of decades, women have taken increasingly to undergraduate engineering programs, and the industry has paid recognition to their excellence (Parikh et al. 2004, pp. 193–201). From 124,606 women enrolled for engineering subjects in 2000–2001, the number has risen to 276,806 in 2009–2010 according to a report of the University Grants Commission of India (UGC). The most recent upgrading of the Indira Gandhi Institute of Technology into the Indira Gandhi Delhi Technical University for Women (IGDTUW) in May 2013, has been prompted by the need to recognize women's aspirations and to empower them as engineers in accord with changing social trends and perceptions. This is in addition to the 14 exclusive women's engineering institutions, mostly in the private domain, and spread across various regions of the country, complementing several women's universities set up by many state governments over the last three decades and over 4,000 women's colleges, public and private. Significantly, women students constitute today 41 % of the total student enrollment in higher education.

The Uncharted Future

Academic organization and transmission of knowledge has to come to grips today with the global dimension of the new knowledge economy. In particular, technical education in India faces the twin challenge of being internationally relevant and internally useful. While the application of technology to rural life offers much scope for research and innovation, it has become increasingly clear to scientists that the heady initial optimism of solving rural problems by technical intervention alone is hardly feasible. Problem solving has to take into account the received framework of tradition and experience. Likewise, the global reach of multi-nationals and industries today calls for training in foreign languages, and in social and cultural competence. In other words, policy makers are recognizing increasingly the indispensable role of the humanities and social sciences, disciplines that were studiously ignored earlier, if not grudgingly acknowledged and conveniently consigned to the so-called service sector. In many engineering institutions, it has now become standard practice to fling fresh undergraduates in at the deep end of the pool in a bid to expose them to the requirements of engineering design and innovation. With marginal mentoring and with little theoretic input either from the social sciences or from the engineering disciplines, raw undergraduates have been compelled to reflect on product design, technology assessment and attendant lifestyle changes. This near autodidactic course component – reminiscent of the WPI and CDIO models – has in considerable measure reinstated, if not rehabilitated, the humanities and the social sciences in the engineering curriculum. Indeed, some of the IITs have been introducing full-fledged graduate studies programs in the arts and social sciences in a move to rediscover the true integrative power of knowledge in its diverse forms. Indeed this recent, not wholly unintended development may be the unforeseen prelude to improving the quality of human resources, to internationalizing education and to enabling academic mobility, reviews and solemn recommendations of committees notwithstanding. Besides, there is an urgent need to rejig the cluttered curriculum in keeping with current multi-disciplinary requirements of the emerging technologies, discarding outdated topics in favor of a unified common core structured around a combination of solid and fluid mechanics, electro-magnetic fields, heat transfer, and digital signal processing. Such a move may also result in freeing up a little more learning time for an imaginative encounter with the humanities.

Toward the end of our survey, it may not be out of place altogether to point to Panagariya's puzzler. He asks pertinently how a "dysfunctional system such as the one India currently has can produce so many students able to compete with the best in the world. Such students come not merely out of the top, well-run institutions outside the ambit of the UGC, such as the IITs and IIMs, but also lesser universities and colleges" (Panagariya 2008, p. 443). To him, the answer to this paradox lies in the enduring intellectual tradition of India, the excellence of private schools, and the centralized testing procedures of the universities. Privileged access to schools instructing in the English medium tilts no doubt the balance in favor of a small, elite band, if the recent report of most seats in the IITs going to major metropolises is any indication. Nevertheless, the picture is one of acute contrasts: on the one hand, world-class institutions of the caliber of the IITs with their unmistakable quest for excellence, and on the other the continuing failure to take adequate funding and reform to where it all matters, when it comes to improving demographic quality, namely primary education. Even today, the myth of the "Beautiful Tree" has a haunting ring to it; like the plum tree in Brecht's eponymous poem, it cannot be recognized by the fruit, for it bears none; yet, you can tell it by the leaf.

Appendix

Select Statistics

Table 5.1 Number of colleges, universities, and students

Year	Colleges	Universities	Students (million)
1857–58	27	3	0.00025
1947–48	496	20	0.2
1950–51	578	28	0.2
1960–61	1819	45	0.6
1970–71	3,277	93	2.0
1980–81	4,577	123	2.8
1990–91	6,627	184	4.4
2001–02	11,146	272	8.8
2005–06	17,625	335	10.5

Sources: Central Advisory Board of Education (2005, table 5.1); Government of India (2005–06b, chap. 10), see Panagariya (2008, p. 441)

Table 5.2 Number of institutions and enrolment

Number of institutions/enrolment	2010–11	2011–12
Universities	523	574
Colleges	33,023	35,539
AICTE approved technical institutions	11,809	13,507
Distance teaching universities/institutions	200[a]	200[a]
Enrolment in the universities and colleges (in millions)	16.975	20.327
Enrolment in open distance learning (ODL) system (in millions)	3.745[b]	3.856[b]
Enrolment in post school diploma/PG diploma (in millions)	1.856[b]	23.02[b]
Intake in AICTE approved technical programmes (in millions)	2.615	3.014

Source: CABE annual report 2012–2013, p. 56. UGC annual report 2011-12/AICTE annual report 2011–12/Statistics of higher and technical education 2009–10 (Provisional)
[a]Repeated at the level of 2009–10 as per Prof. N.R. Madhava Menon report of committee to suggest measures to regulate the standards of education being imparted through distance mode
[b]Estimated

Table 5.3 Programs, number of institutes, and intake

S. no.	Program	No of institutes	Intake
01	Engineering	3,495	1,761,976
02	Management	2,450	385,008
03	Master of computer application	1,241	100,700
04	Pharmacy	1,145	121,652
05	Architecture	126	5,996
06	Hotel management & catering technology	105	8,401

Source: CABE annual report 2012–2013, p. 62

Table 5.4 Major centrally funded institutions

(i) Central universities	44[a]
(ii) Deemed university	130
(iii) Technical institutions	16 – Indian Institutes of Technology (IITs) 30 – National Institutes of Technology (NIT)
(iv) Management institutions	13 – Indian Institutes of Management
(v) Information technology institutions	4 – Indian Institutes of Information Technology (IIIT)
(vi) Science & research councils	5 – Indian Institutes of Science Education and Research (IISER)
	1 – Indian Institute of Science (IISc)
(vii) Planning & architecture institutions	3 – School of Planning & Architecture
(viii) Training institutions	4 – National Institutes of Technical Teachers' Training & Research (NITTTR)
(ix) Planning & consultancy institutions	1 – NUEPA & 1 – EdCIL
(x) Area/sector specific institutions	7 [1-Indian School of Mines (ISM), Dhanbad; 1-Sant Longowal Institute of Engineering and Technology; 1-North Eastern Regional Institute of Science & Technology (NERIST), Itanagar; 1-Central Institute of Technology (CIT), Kokrajhar; 2-National Institute of Industrial Engineering (NITIE), Mumbai and National Institute of Foundary & Forge Technology (NIFFT), 1-Ghani Khan Choudhury Institute of Engineering & Technology (GKCIET), Malda, West Bengal
(xi) Institutions of national importance	33[b] [7 IITs, 20 NITs, 5 Universities and 1-Hindi Institution]

Source: CABE annual report 2012–13, p. 74

[a]Of which, 39 are being given maintenance and development grant by MHRD through UGC. The IGNOU, New Delhi, the Central Agricultural University, Imphal and the Indian Maritime University, Chennai are being funded by MHRD, Ministry of Agriculture and the Ministry of Shipping and Transport respectively. The funding for South Asian and Nalanda Universities is being made by the Ministry of External Affairs

[b]These institutions are included amongst the existing IITs/NITs/Universities/Institutions

Table 5.5 Women's enrollment: rise across streams

Stream	2009–10	Change (%)	2000–01
Arts	2,772,580	62	1,711,487
Science	655,257	72	1,129,255
Commerce/management	545,712	68	915,719
Engineering and tech.	124,606	122	276,806
Medicine	107,177	89	202,803
Law	89,256	33	67,196
Education	180,771	223	55,907
Agriculture	15,253	74	8,769
Vet. sciences	4,519	29	3,511
Others	62,140	118	28,499
Total	5,649,102	70	3,325,927

Source: UGC annual report 2010–11, p. 52, table 2.4; here: from: Kasturi (2011)

References[3]

Arnold, D. (2004). *Science, technology and medicine in colonial India. The New Cambridge history of India III*. Cambridge: Cambridge University Press.

Basu, M. (2013, August 8). IIT JEE success skewed in favor of urban, high-income students. *The Indian Express*, pp. 1–2.

Bhaskaran, R. (1967). *Sociology of politics: Tradition and politics in India*. Bombay: Asia Publishing House.

Clark, N. (2007, January). Engineering education in India. A story of contrasts. *World Education News and Reviews*, pp. 1–13. Available at: http://www.wes.org/ewenr/PF/07jan/pffeature.htm. Accessed 31 Jul 2013.

Dharampal. (2000). *Indian science and technology in the eighteenth century*. Mapusa: Other India Press.

Kakodkar Report. (2011). *Taking IITs to excellence and greater relevance*. Report of Dr Anil Kakodkar Committee. Appointed By MHRD to Recommend Autonomy Measures to Facilitate IITs Scaling Greater Heights (see note above).

Kasturi, C. S. (2011, January 10). Number of women choosing engineering doubled since 2000–01. *The Hindustan Times*.

Medley, J. G. (1873). *India and Indian engineering*. Three Lectures delivered at the Royal Engineer Institute. London: E.& F. Spon.

Natarajan, R., Ananth, M. S., & Singaperumal, M. (Eds.). (2009). International engineering education. *Proceedings of the INAE-CAETS-IITM Conference, Indian Institute of Technology. Madras 1–2 March 2007, IIT-M*. Singapore: World Scientific Publishing.

National Knowledge Commission. (2008). *Report of working group on engineering education*. Available at: http://knowledgecommission.gov.in/downloads/documents/wg_engineer.pdf

Panagariya, A. (2008). *India. The emergent giant*. Oxford: University Press.

Parikh, P. P., & Sukhatme, S. (2004). Women engineers in India. *Economic and Political Weekly, 39*(2), 193–201.

Rao, U. R. (2003). *Revitalising technical education*. New Delhi: AICTE review report. AICTE.

Report (Damodaran Committee) of the special committee on reorganization and development of polytechnic education in India, 1970–71. http://www.teindia.nic.in/mhrd/50yrsedu/f/M/B/Toc.htm

Rothermund, D. (1993). *An economic history of India*. London: Routledge.

Saha, S. K., & Ghosh, S. (2011). Engineering education in India: Past present and future. *Propagation: A Journal of Science Communication, 2*(2), 111–119.

Sen, B. (1989). Development of technical education in India and state policy –A historical perspective. *Indian Journal of History of Science, 24*(4), 224–248.

Subbarao, E. C. (2013). India's higher engineering education. Opportunities and tough choices. *Current Science, 104*(1), 55–66.

Subramanian, L. (2010). *History of India, 1707–1857*. New Delhi: Orient BlackSwan.

The Indian Express. (2013a, September 2). IITs get ready for third-party performance assessment. *The Indian Express*.

The Indian Express. (2013b, July 22). Editorial "Access granted: A new initiative by IITs and IT organization promises to widen the scope of India's knowledge economy." *The Indian Express*, p. 10.

Varshney, L. R. (2006). *Private engineering education in India: Market failures and regulatory solutions* (Science, technology, and public policy paper). Cambridge: Massachusetts Institute of Technology.

[3]**Note**: *All Annual Reports of the Central Advisory Board of Education (CABE) besides most reports of committees constituted by the MHRD can be had at:* http://mhrd.gov.in.

Vir, D., Dhrubajyoti, S., Patnaik, P., & Hazra, A. K. (2011). *Sixty years in the service of nation. An illustrated history of IIT Kharagpur*. New Delhi: Orient BlackSwan.
Whitehead, C. (2003). *Colonial educators. The British Indian and Colonial Education Service, 1858–1983*. London: I.B. Tauris.

Balasundaram Subramanian B.Sc. in Chemistry followed by an M.A. and Ph.D. in German Studies from the Karnatak University Dharwad, India. Presently, he heads the School of Humanities & Social Sciences at the newly established Indian Institute of Technology Mandi in the Himalayan State of Himachal Pradesh in India and is responsible for designing the humanities curriculum of the undergraduate program in the engineering disciplines. Prior to this assignment, he was Professor of German Studies at the Indian Institute of Technology Madras (till 2007) and also Professor of German Studies at the Jawaharlal Nehru University New Delhi (from 2008 till 2011). Publications notably on Rilke, Goethe and Weimar Classicism.

Chapter 6
Engineering Education in Slavic Languages Countries

Maria Kostyszak, Jan Wadowski, and Marcin Zaród

Abstract This chapter presents the historical common core of engineering education in Central-Eastern Europe. It also discusses the consequences of the fall of the communist system with the Soviet Union as its leader (1989–1991) and European Union expansion (2004). General information is provided about the Czech and Belarus Republics with the situation in Ukraine and Russia discussed in detail. In-depth analysis of Polish engineering education is provided with post-Soviet legacies, EU Union directives and local policies presented including social context of engineering education and religion influence. This chapter also shows didactic trends in Poland and Russia, both in formal and non-formal practical engineering education as well as in engineering ethics development. Since the authors come from Poland, their local perspective determines also the range of research.

Keywords Central-Eastern Europe • Transition • Post-communism • Reform • Social pressure • STS (Science and Technology Studies) • Reports

Introduction

There are about 140 ethnic groups speaking Slavic languages, mostly located in Central-Eastern Europe. For clarity of the outline, we will concentrate our investigation on the Northern Slavic countries: Russia, Ukraine, Poland, Czech Republic, and Belarus.

M. Kostyszak (✉)
Institute of Philosophy, Wrocław University, Wrocław ul. Koszarowa 3,
Gęsiniec, ul. Tęczowa 24, 57-100 Strzelin, Poland
e-mail: syntropia7@gmail.com; maria.kostyszak@uwr.edu.pl

J. Wadowski
Department of Humanities and Social Sciences, Wroclaw University of Technology,
Wybrzeze Wyspianskiego 27, 50-370 Wroclaw, Poland
e-mail: jan.wadowski@pwr.edu.pl; jan.wadowski@gmail.com

M. Zaród
Institute of Sociology, University of Warsaw, Karowa 18, 00-927 Warszawa, Poland
e-mail: m.zarod@is.uw.edu.pl

S.H. Christensen et al. (eds.), *International Perspectives on Engineering Education*,
Philosophy of Engineering and Technology 20, DOI 10.1007/978-3-319-16169-3_6

Central-Eastern Europe was strongly influenced by political changes in 1989 (with the collapse of the communist system as well as the Soviet Union) and in 2004 (the enlargement of the European Union). The first date marks the division inside the Soviet political zone, resulting in political freedom in the countries under discussion. The second date affected Poland and the Czech Republic, pushing their engineering education into the western academic community. Belarus, Ukraine and Russia remained partially outside of this system but far from totally isolated because of their access to the Schengen Zone and Erasmus mobility program.

Economic changes in 1989 affected heavy industries in these countries, e.g., the Minsk Tractor Factory (Belarus) survived the crisis while the Warsaw Mechanical Workshop (Poland) was relocated to another city and reduced its output. The same changes affected the shipyard industry (e.g., Gdańsk Shipyard) and economic changes also transformed electronics industries (e.g., Tesla in the Czech Republic).

The collapse of the Soviet Block had an especially severe impact on the Russian science and engineering system; e.g., the Academogorodok (Russian academic semi-city in Nowosybirsk in Siberia) suffered from unemployment and brain-drain related emigration (Bird 1994). Another example was the destruction of the seed bank in the Institute of Plant Industry in Petersburg, where a unique plant collection that survived the Second World War was destroyed in order to make space for luxury housing (Rosenthal 2010).

Such disruptions have created multiple administrative crises and deprived the technical education community of solid knowledge about what is taking place within it. As a result, we have chosen to limit ourselves to a case-study approach. Any generalizations should be considered no more than hypotheses. Only recently have Poland and the Czech Republic started to gather and publish state-scale educational reports, especially within the OECD Eurydice Research network. Unfortunately those reports focused on higher education in general with only minor data points on engineering education. With the situation in Ukraine and Belarus remaining volatile, only partial data are available for these countries.

Economic upheavals since 1989 undermined as well humanities and social science education and research. This negative impact was only intensified by the 2007 economic crisis thus further limiting collaborative pedagogical engagement and research on engineering education.

Common Experiences

From 1945 until 1989, all Central-Eastern European Countries were under the strong influence of the Russian educational system. Russia, Ukraine, and Belarus were parts of the Soviet Union, while Czechoslovakia (later divided into the Czech Republic and the Slovak Republic) and Poland remained formally independent.

Bonds between the Soviet Union and smaller states were multiple. Industrial policies affected engineering education. One example is the way the development of electronics (Eastern precursor of IT technologies) was promoted (in terms of advice and political pressure) in the Czech Republic more strongly than in Poland. As an influence of the Soviet Union and previous historical conditions, technical universities played a primary role in engineering education. With some exceptions (e.g., chemistry and mathematics departments in universities have strong links to industry), separations between technical and general university education remain distinct (even in form of academic titles). Market reforms in the early 1990s and the emergence of private educational institutions did not have much effect in engineering education. Most private institutions focused on lower cost curricula that do not require student laboratories (with exception for computer sciences).

Aside from political and economic pressures, there are other lasting educational legacies. Russian science and engineering was strongly present in smaller countries in the form of engineering books and manuals. One example might be the Landau-Lifschitz theoretical physics course that was a common part of engineering curricula throughout the Soviet block (Hall 2005).

Another aspect was scientific collaboration, which did not occur on a peer-to-peer basis. The Russian Academy of Sciences was the hub, while local Academies (Polish, Czech, Ukrainian) were at the rim as satellite institutes. Scholarly work in Moscow or Petersburg was often an important element in the scientific careers of local scientific elites. For engineering researchers in Magnitogorsk (heavy industry), Baikonur (aerospace), or Academogorodok (electronics, biochemistry) an academic period in Moscow or Petersburg was of much importance for their professional development. In the 1980s, the role of these Russian scholarships diminished, while Fulbright opportunities became more available due to the Perestroika and Gorbachev reforms. Nevertheless, such exchange opportunities remain an important experience in the lives of current Slavic engineering professors. Since 1989 some of those programs have remained active; the Polish Academy of Sciences, for example, has a research branch in Moscow used by chemical engineers and physicists to develop collaboration with their Russian colleagues.

European history resulted in national minorities becoming resident in different countries. For example there are still large communities of Poles living in Belarus, Kazakhstan, and Ukraine. The Polish academic system has special scholarships for them. The 1989 crisis also resulted in the immigration of Russians and Ukrainians to western countries. Some of those immigrants stayed in Poland, some moved elsewhere.

Political changes resulted in changes in national borders with educational implications. The transformation of the University and Technical University (two distinct institutions) in Lvov is a good example. Today it is an important research institution in Ukraine. But there was an important Lvov/Polish Mathematical School (Stefan Banach, Stan Ulam, Hugo Steinhaus) that operated in those institutions between 1918 and 1939, when Lvov was part of Poland. Another example is the Technical University in Gdansk, which shares German-Polish ancestry. Such shared legacies are easily found in Eastern and Central Europe.

The role of vocational education is another example of a common feature. Czechoslovakia, Poland, and Soviet Russia adopted the German style vocational high-school system, with the presence of technical high-schools. What was unique for Central and Eastern Europe (and Germany in some parts) was that some technical high schools became more prestigious than ordinary high schools. This was an important feature for engineering education, as they became another important source of potential students. Between 1945 and 1989 the technical universities accepted mixed students from high schools and technical high schools. Students in the former group were perceived as having better backgrounds in mathematics and physics; those in the latter were acclaimed for their "practical approach".

During the 1990s all these countries suffered declines in high-school technical education (possibly connected with industrial decline). The overall number of students in higher education rose, but the rate of high schools/technical high school graduates rose as well (Birzea 1994). As a result, technical universities had more students with a general background than students having vocational training in high schools. This affected traditional branches of engineering education (electrical, mechanical, civil) more than newer counterparts (IT, electronics, biochemistry). In the newer branches, general backgrounds and/or non-formal experiences were more important than formal high-school technical training. Details on vocational education structure in Poland can be found in the "Report on the State of Education 2011" from the Institute of Educational Research (a branch of Polish Ministry of Education) (Federowicz and Wojcuk 2011).

The crisis in vocational education affected engineering education. It diminished an important source of graduates and forced technical universities into opening vocational courses to supplement formal vocational training. In general, an increase in the number of students, including students in engineering universities, seems to be another common feature. For example, between 1993–1995 and 2003, the number of students in Russia nearly doubled, reaching six million (Arapov 2006).

Differences and Divisions

The collapse of the Soviet Union significantly altered common experiences across Central and Eastern Europe. Belarus and Ukraine gained formal independence and the Russian influence in Poland and the Czech Republic rapidly decreased. State companies were privatized or went bankrupt due to inflation, high-labor costs and inefficiency. Details of sociological and economic changes can be found in the literature (Dunn 2004).

The collapse of heavy industry was more severe in industrial regions (Silesia in Poland, Magnitogorsk in Russia, Donets in Ukraine). Generally speaking, citizens of Poland and the Czech Republic suffered less from the transition than people from Belarus, Ukraine, and Russia. Current unemployment rates, life expectancies, and infant mortality continue to document the differences. The division of Czechoslovakia resulted in the formation of the Czech and Slovakian Republics as independent states, both of which became members of the European Union.

The expansion of the European Union in 2004 was a turning point in the history of engineering education. Poland, the Czech Republic, and Slovakia accepted the

Bologna process for higher education in 1999. Russia and Ukraine followed in 2003 and 2005, retaining a larger degree of independence in education. Belarus remains outside this agreement.

The Bologna process was a series of agreements between governments concerning higher education. Effectively it introduced a division between the first and the second degree of academic education. The engineer degree is the technical counterpart of the bachelor of sciences. All engineer training in Poland, Czech Republic, and Slovakia have some similar core courses in mathematics, physics, and technical subjects (e.g., CAD drawing, basics of programming). Mathematics is frequently divided into an algebra course (matrices, linear equations, and basics of topology) and a mathematical analysis course (calculus). The contents of particular courses may differ to some extent, but in principle they are interchangeable within the Bologna system.

Introduction of the Bologna system put a strain on engineering education. In theory, engineers should be able to go to work immediately upon completing their degrees (bachelor: 3 years; master: 2 more). In practice, degrees often take longer than specified (e.g., 4 years for the bachelor degree). This is connected with problems in higher education. While formal measurements show rising knowledge (both Poland and Czech Republic improved their math and science scores in the Program for International Students Assessment or PISA), high school curricula have gradually become more narrow (Federowicz and Wojcuk 2010). For example, the old curriculum in Poland (used before the 1999 educational reform) included calculus in the extended mathematics course in high school. Additionally, although Ukraine and Russia joined the Bologna process they have retained distinct educational characteristics; Belarus was not accepted to the convention. Some high-schools even offer extra curriculum or non-formal calculus training, in order to prepare graduates for first year demands.

Divisions became even more visible after the 2004 EU expansion when Poland, Slovakia, and the Czech Republic became members. It resulted in strengthening the scientific collaboration with other European countries. Access to the Schengen zone allowed people to travel through Europe without passports, which resulted in the rise of popularity of Erasmus student exchange programs. EU structural funds became important sources of money in development (347 billion euros from 2007 to 2013). Some of these funds were put into infrastructure development (European Development Fund). Some formed the basis for extra training (European Social Fund).

The framework programs connected with the Lisbon Declaration were another source of funds. Both the Czech Republic and Poland co-signed this declaration after being accepted to the EU. The increase of financial investment in research and development to 2 % of GDP was the primary goal of the Lisbon Declaration. Both states failed in this regard: Poland invested 0.9 %, the Czech Republic 1.4 %. Differences further emerged in the EU Innovation Scoreboard. Poland was qualified into the worst group (modest innovators), the Czech Republic into the middle one (moderate innovators) (Hollanders and Es-Sadki 2013).

Countries that signed the Lisbon Declaration decided to increase the competitive character of their potential on the basis of knowledge. That is why reorganization of teaching methods is considered to be one of the most significant factors for improving the education level. Distance learning is accepted as one of the instruments to offer access to education to those who otherwise would be excluded. Interdisciplinary modules were created during the project to support the study of mechanical engineering: development of courses such as *Design for X* (DFX), courses on visual prototyping, courses on rapid prototyping (RP) and rapid tooling (RT). The project initiated by the Silesian University of Technology was realized with the participation of many East European countries.

Izabela Wagner, a Polish sociologist of science, claims that Poland and Russia have

> maintained public education, especially in science, on a high level. Children in public schools (most often) start learning biology with the evolution theory, math and physics on a high level since the first years of their education. In these societies we can find values attached to science and research. (Wagner 2011)

Such statements are supported by the Relevance of Science Education (ROSE) study. This was conducted in 2009 among 15-year old students from 40 countries (including Russia, Poland, and the Czech Republic). It showed that students from Poland and Russia have a higher interest in science and engineering education than students from the Czech Republic, Germany, UK, or Finland (Sjøberg and Schreiner 2010).

According to the report prepared by the Eurydice educational institution in 2009/2010 on education in 31 EU and associated countries, all strive to adapt their systems of higher education to rapid changes taking place in society. According to the report, social policies are not always properly implemented in European academic institutions. Some political declarations – e.g., regarding more equitable access to higher education for all social groups – are neither provided sufficient financial support nor are their realization adequately monitored. There is an urgent need to decide, on the basis of empirical information and critical reflection, on the main problems connected with the social dimensions of higher education, especially in face of economic constraints. The report concerned higher education in general, but all its results apply especially to engineering education.

The Situation in Russia

Maria Dobrayakova and Izak Fourmin in their overview on "Higher Engineering Education in Russia: Incentives for Real Change" (2010) are quite skeptical about verbal commitments to improvements in engineering education that are not effectively realized. The positive changes that have taken place in Russian engineering universities are limited to buying new technical facilities, and the introduction of new regulatory mechanisms (the national unified examination along with the Bologna process) have no specific relation to engineering education and have in fact left the substance of engineering education untouched. The education process remains unchanged: lectures still make up the core form of communication between teachers and students; there are very few interdisciplinary courses (the lack of which hardly encourages innovative thinking); and students mostly cannot modify their individual curricula (about 30 % of students are given only one to two elective course options and 57 % reported that all of their courses were required).

Changes in Russian universities have to do more with survival than striving for progress; they are reactive rather than proactive. Employers are fragmented, their efforts are scattered, there is no active or central body that could direct and consolidate positive efforts. For changes in the system of engineering education to take place, at least one of the stakeholders must be actively engaged: state, employers, students, or universities themselves. The state, even though it organizes competitions for the best, does not set a framework for the development of engineering education, and in fact all grant funds are spent on purchasing new equipment. There is no doubt that technical facilities are necessary, but according to Dobraykova and Fourmin they are not enough to maintain and advance the quality of education. Employers complain about the quality of graduates but quite often have nothing to offer them (engineering salaries are on average the same or slightly lower than for graduates of business programs). Although employers sometimes try to collaborate with universities, their efforts are scattered and do not produce any noticeable effect at the national level.

Vassiliy M. Zhurakovsky, Ministry of Education of the Russian Federation, argues in an article on the history of the establishment and development of engineering education in Russia, that necessary changes have been introduced. The main effort is to form a new generation of engineers who can address the demands of the post-industrial society by advancing the scientific-technical potential of higher educational institutions in Russia. This next generation will be able to solve the problems concerning the quality of education in the area of high technologies on the basis of both Russian and foreign pedagogical experience. The important areas of humanization, fundamentalization, and professionalism in forming the content of engineering education in high technologies are also discussed (Zhurakovsky et al. 2008).

Corporations operating in the industries of natural gas, oil, uranium, and other resources have become important sources of funds for engineering education. Gazprom (a major Russian gas company) closely cooperates with nine technical universities and one vocational school. Two of them are officially called Gas and Oil Universities (Gubkin State, Tyumen State). Such universities are better funded and have access to hands-on training in the company. For example, Gazprom is a member of the Board of Trustees in the Gubkin State University of Oil and Gas. Their cooperation focuses on gas engineering, mining, and the petroleum industry. The NORRIC Report (NORRIC 2005) on Higher Education in Russia includes the following relevant description of an engineering program at one state university:

> One of the institutes offering courses within metallurgy is the Moscow State Institute of Steel and Alloys. The institution offers 11 different engineering specializations in different aspects of metallurgy. All specializations last 5 1/2 years. The degree awarded is that of the Specialist and the professional title that of "inzener-tehnolog" (инженер-технолог/

engineer-technologist), with the specialization also being mentioned. The institute also offers Bakalavr/Magistr (Bachelor/Master) degrees in the field of metallurgy. After completing a 4-year Bakalavr programme, graduates may continue their studies on a 2-year Magistr programme in order to obtain a Master of Science and Technology degree. Finally, it is possible to obtain a Specialist degree carrying the title of engineer–researcher (inzener-issledovatel - инженер-исследовател).

The time scheduled for the preparation of a dissertation at the Moscow State Institute of Steel and Alloys is 1.5 months for a Bachelor's degree and 3 months (+ practical research period) for engineers. For the Master's degree one semester is scheduled for writing a dissertation.

The Situation in Ukraine

The formation of the Ukrainian Research and Academic Network URAN, a joint initiative of the Ministry of Education and the National Science Academy, opened a new stage in the development of Internet education in the Ukraine in 1998. The initiative was supported by a number of foreign funds and organizations: the NATO Research Program and the German Research Network in particular. In 2000 the URAN infrastructure was represented by six regional centers, DonSTU being one of them. The successful joint work of Donetsk State Technical University and the Siemens company made it possible to establish a Siemens Engineering and Technical Centre-Ukraine branch. The German Engineering Faculty graduates belong to its engineering staff. The engineering service in the field of information technologies in the Ukraine and abroad is the main activity of the Center. In the process DonSTU and Siemens thus developed and implemented a new conception of interactivity at the labor market.

Ivano-Frankivsk National Technical University of Oil and Gas is the biggest educational-research institution with the highest (IV) accreditation level and is the main centre of petroleum engineering education and science in the Ukraine. The University has started three-level training programs leading to the award of the Bachelor of Engineering, Specialist of Engineering, and Master of Engineering. New forms and methods of teaching are being introduced: a rating system of student knowledge evaluation and the modular system of training; various types of business games, simulation and drama activities, team project methods, and computer and information processing technologies.

The Situation in the Czech Republic

As in other Central European countries, 1989 was an important date for higher education in the Czech Republic. Petr Mateju and Natalie Simonova (Czech sociologists from the Academy of Sciences of the Czech Republic) have commented on the 1990–2000 period as follows:

> Universities were granted almost full autonomy as early as in 1990. They have reformed their curricula, expanded programs in the humanities and social sciences, and eliminated political criteria from admission policies, both for the faculty and for the students. However, the structural changes were not as quick and profound as obtaining and mastering the freedom was. The most significant structural changes in the Czech tertiary educational system addressed in the paper are decentralization and diversification. With regard to financing, the authors argue that universities have remained dependent on the state to a high degree. Several attempts to expand multi-source financing by introducing cost sharing features (tuition fees, loans, student allowances) failed. (Mateju and Simonova 2003)

As well as in other Central-Eastern European countries, state-funded universities in the Czech Republic suffered from economic changes during the transition period. Although student enrolment rose by 60 %, this was not sufficient to meet educational demand. At the beginning, Czech Technical Universities had the liberty to set their own admission policies, without state-wide regulations.

Another important milestone was adoption of the Higher Education Act in 1998. This regulation further increased autonomy of universities (e.g. putting formal ownership and responsibility of premises to the university boards). Boards of governors supplemented university Senates in the decision-making process. It also increased university obligations to local communities (e.g., as lifelong learning centers).

Between 2009 and 2010 the state system of performance indicators changed focus from quantity toward quality. Relying on scientometrics, performance-based funding was introduced in 2009. In reference to such rankings, the Institute of Chemical Technology and Czech Technical University (both in Prague, both state funded) offer perhaps the best mixture of research and engineering training. Details of the 1998–2012 period can be found in a report by Jan Koucký (2012). The introduction of tuition fees was one significant change. Due to the budget cuts in 2007–2012, per capita student funding decreased from 50 % to 30 % GDP (considerably below the EU average). Such cuts might be especially severe in case of engineering education, because of its higher costs. Leading technical universities introduced partnership programs with industrial partners (e.g., the Institute of Chemical Technology cooperates with Unipetrol), offering research scholarships, joint research initiatives and opportunities in hands-on training in industrial chemistry, and environmental studies.

Czech engineering is also supplemented by science fairs and non-formal structures (discussed in more detail below). The source of funding is mainly the same as in the Polish case: EU development funds.

The Situation in Poland

There are 22 engineering schools in Poland, all but one state-owned. Higher education is in principle free-of-charge. Yet economic transformations following the political changes of 1989 have had an impact on the higher education system. The major technical universities are (according to rankings) Politechnika Warszawska

(Warsaw University of Technology), Akademia Górniczo-Hutnicza (AGH University of Science and Technology), and Politechnika Wrocławska (Wrocław University of Technology). These big three each have more than 20,000 students (graduate and post-graduate) (Rzeczpospolita 2013).

To adapt to historical changes, many academic institutions have significantly altered their organizations and curricula. How engineering education at the university level is organized in Poland is described by Roman Morawski, Brian Manhire and Janusz Starzyk (Warsaw University of Technology, Ohio University) in an article on "Engineering Education in Poland." They compare the Polish system, at first presenting it in detail, with the American one (Morawski et al. 1998). In Poland, most technical universities offer a 5-year program leading to the master of science degree. Undergraduate engineering programs, lasting 3.5–4 years and leading to the bachelor of science or equivalent degrees are offered by some institutions, and their number is growing. Such a program is offered, for example, at the Faculty of Electronics and Information Technology, Warsaw University of Technology, and within 2–3 years will be available at other faculties of this University. Ph.D. degrees are obtained either through doctoral programs lasting 4 years that are organized by universities, or are pursued without course work, typically by teaching and research associates of academic institutions.

In 2005 the number of university students in all fields in Poland reached an all-time high of nearly 2 million enrolled (Federowicz and Wojcuk 2011). Initially, the increase was mostly connected with the creation of private educational sector. The majority (about 95 %) of private schools were concentrated on the humanities. The reforms of higher education financed in 2008 (EU Human Capital Operational Program, total budget 370€ million for 2009–2013, 85 % EU/15 % local finances) resulted also in increases in engineering enrollments. The introduction of a government program of strategic science and engineering fields proved to be the key factor. Science and engineering faculties of selected branches (applied physics, chemical technology, biochemistry, IT, environmental engineering, automatics, electronics, mechanics, civic engineering) received additional funding for each student. Students in those fields also receive government scholarships in order to encourage students to take engineering degrees. It was successful in terms of popularity, as the enrolment ratio increased. But it also led to problems with lowering completion rates (Górniak et al. 2012).

Religious Impact on Education

What is distinctive about the situation in Poland is that Catholicism forms the ethos of the majority of the Poles. Traditionally, Poland has been a strongly religious country. This has positive and negative aspects. The former influences the conscience of a believer in an idealistic way that proves effective in a social sense – one may expect more loyalty, honesty, and responsibility from believers. Despite the Enlightenment criticism of religion and processes of secularization, there are in

Poland circles and clubs of teachers and scientists strongly devoted to Catholicism. One of many examples is the Chrześcijańskie Forum Pracowników Nauki (Christian Forum of Researchers and Scientists) with its active organizer Andrzej Zabołotny, where during biennale conferences scientists, researchers, and engineers from Poland and abroad meet to discuss social challenges and risks connected with technocratic tendencies. The negative aspect is connected with a false assumption by many Catholics that true belief is a sufficient basis for correct behavior and a separate professional ethics is not necessary. There are many debates in which this issue is addressed: a lot of professional ethical codes are now constructed, certain professions try to limit their obligations and regulate their proper behavior, and very often the exponents of true belief oppose them as empty facades.

In the Czech Republic, by contrast, the situation is quite different. Society is more secularized. The Czech sense of humor and irony together create flexibility with regard to both religious dogma and rigid ethics codes. Russia, Belarus, and Ukraine have been influenced by the Orthodox Church, but the communist regime tried to eliminate faith from the sphere of social life. Some of their citizens departed from the church, but in general a rich spiritual tradition (religion, literature, art) influences culture and education in these countries.

In the area of engineering, an important social division was caused by the Smoleńsk disaster of 2010, when the Polish president and many other people from the Polish elite died in an airplane crash near the Russian city of Smoleńsk. The official aviation investigation was an object of critique from the catholic-right side of Polish politics. Smoleńsk became a kind of mystical symbol and important aviation-engineering public case. Some lecturers from technical universities criticized the official technical explanation, which emphasized pilot error. It is known (in a derogatory way) as Smoleńsk-physics. This political- and religion-based division also applies to engineering students and faculties.

Engineering in Social Context

Humanities-based critical approaches to science and technology (Science and Technology Studies) are at an infant stage in Poland (and other Slavic language countries). They are conducted in the University of Poznań (Border Questions Workshop) and Mikołaj Kopernik University in Toruń. Most technical universities do not have humanities or social science departments. There is also a lack of empirical sociological research on engineering and science courses. Most publications are based rather on a philosophical approach, often appealing to Bruno Latour's actor-network theory, as in Andrzej W. Nowak's social ontology of modernity (Nowak 2011a). There are no regular STS courses for engineering students. While Latour's works are widely discussed, discussion focuses on his sociology in general rather than having an empirical STS base. One exception might be the work of Łukasz Afeltowicz on philosophy and sociology of science, focusing on the material and sociological aspect of physics and engineering (Afeltowicz 2012). Despite their relevance, such works are not widely known inside social networks of engineers.

Ethics courses are in urgent need because of changes in the social role of technical universities. Evolution toward the triple-helix university model (Etzkowitz 2008), focusing on entrepreneurship and business efficiency, does not always promote public responsibility or social criticism. In the case of environmental problems (e.g., potential shale gas reserves in Poland, nuclear power plant development, energy-climate conflict) engineering education does not really prepare graduates for dealing with the diverse perspectives of government environmental experts, NGO activists, and commercial mining interests.

Such problems were recognized in a sociological study of engineering lecturers from the AGH University of Science and Technology in Krakow (Mucha 2009). In 60 interviews with younger and senior staff, it was revealed that engineers from leading technical universities in Poland declare their ethical obligations toward taxpayers and hold reserved positions toward some technological progress (such as genetic engineering).

The same study also revealed a self-understanding of Polish engineering education. AGH staff agreed that the theoretical side of engineering in Poland remains on a solid European level. Poor financing and brain-drain affected practical sides of teaching. Despite being one of the best funded technical universities in Poland, AGH staff complained about the outdated equipment in laboratories and lack of funding for practical training of students. Scientists were also critical about the quality of math and science education in high schools.

General science education is also discussed and improved in non-formal educational networks organized by NGOs. Such networks try to participate in science policy development. For example, Obywatele Nauki (Citizens of Science) – a non-formal social movement – tries to influence university financing reform. Nowe Otwarcie Uniwersytetu (New Opening of University) conducts regular seminars on university studies, trying to include perspectives from technical and medical universities. Among the lecturers and students of engineering one might observe a rising commitment to general education. For example, Gdańsk University of Technology conducts a social-educational programme in physics and math education development named "Za rękę z Einsteinem" (Together with Einstein) for approximately 180 rural schools.

"Do-it-yourself" has played an important role in Polish economic and industrial practice. Lack of resources and goods have resulted in official government programs of this kind beginning in the 1970s inspired by books such as Adam Słodowy's "Zrób do sam" (literally DIY or do-it-yourself). Such mechanical workshops in Youth Centers and manual publications received lesser financing after 1989 (possibly due to the industrial decline and political turn), but their legacy is still readily observed in engineering education. The 2000s renaissance of DIY (hacker culture, maker movement, fabrication laboratories) brought back such legacies. In 2013 in Poland about 20 different DIY initiatives (medialabs, hackerspaces, anarchist workshops, independent technological communes, squats with mechanical workshops, large scale guerrilla gardening etc.) were started by non-formal groups or NGOs.

Formal engineering education did not yet embrace such movements, but there is a natural flow of instructors and students between TUs and DIY groups (by the Internet, during professional meetings, in classes). As such groups have problems with regular financing, their impact remains on a local or regional scale. One exception might be found in Russia, where Fab-Lab operates near St. Petersburg State National Research Polytechnic University.

Another example of the openness of science and engineering toward society is the rise in popularity of science festivals and science centers. Faculty and staff from major TUs organize or participate in such initiatives, taking the activity as a form of science popularization, building up support for engineering student recruitment or prestige development. Most such initiatives are funded by state or EU funds. Most academic staff participates as volunteers or receive little payment. They play a similar role in children's universities, where kids (ages 10–14) have lectures and workshops in TUs. There are approximately 15 regular science festivals, five science centers, and 10 children's universities in Poland. Similar activities are also found in the Czech Republic and other EU countries. A few TUs even started their own high-schools or classes in order to attract and train potential students. Such high schools are perceived as superior to regular high schools due to their academic programs in mathematics. This heavy emphasis on mathematics in many ways resembles the French model of *Grandes Écoles* starting in selected high schools. As a result of all such activities, majoring in engineering or science is perceived as a prestigious but demanding education.

Extra courses, selected high schools, and science popularization all resulted in demands for educational reform. In 2011 Poland initiated a general education reform, reshaping mathematical and science programs. It is too early to assess the effects of this reform, but one might observe that mathematical and physics curricula for high schools were simplified and reformed. For example, only the extended mathematical curriculum has a few elements of differential calculus (with no integral calculus). The basic curriculum is even more simplified. Unfortunately reform discussions did not reach all TUs. When the authors of this chapter asked lecturers working with first year students (physics, electronics, civic engineering departments) in five different TUs, not one of them could identify curricular changes.

Didactic Changes in Polish Engineering Education

EU financial support helps to improve curricula, equipment necessary to teach on an advanced level, and more. For example, in 2011–2012 Wrocław University of Technology received support for the development of 15 new humanities elective courses. Sample courses included "Man and technology: Anthropological aspects of technological development", "Ethics of new technologies", "Humanistic aspects of sustainable development", "Technological transformations of the society", "Aesthetics of space with the elements of city sociology and of urbanization", and "Engineering ethics" (the last of which will be discussed further below). These all aimed to prepare future engineers to become responsible and well informed members of society on top of their professional knowledge and skills.

Apart from internal initiatives within technical universities to improve the level of engineering education, two other strong influences are cooperation with the United States and European Community activities. The United States accrediting board for engineering and technology known as ABET recognizes programs at a few Russian and Ukrainian universities. Additionally, U.S. and Eastern European universities collaborate to hold many mutually beneficial conferences. The European Union, to which five Slavic countries now belong, requires members to fulfil certain standards regarding the engineering profession in education as well as in practice.

Many EU programs emphasize e-learning. This is connected with a rising number of students and insufficient student workshop capacity. Some basic laboratories (circuits in electronics, construction in civil engineering) are conducted in a virtual environment, using tools such as extended virtual reality simulations (Nowak 2011b). In 2002–2004 The Transport Department of the Silesian University of Technology in Gliwice (Poland) realized a project within the "Leonardo da Vinci" program titled *Integrated Knowledge Based Interdisciplinary Study Program on the Web Site* summarized in article by Markusik and Bułkowski (2005). According to them, due to the internet it is possible to project integral systems of e-teaching and e-learning. In their research they analyze the possibilities created for interdisciplinary areas, for example transport engineering. Similar ideas are also used in electrical engineering (Wańkowicz and Orzechowski 2011) and environmental engineering (Gajewski and Jarosińska 2011).

The majority of these programs were based on Moodle (or similar e-learning tools) and focused on digitizing traditional course material. Most popular forms were slideshows (occasionally supplemented with a lecturer's voice) and quizzes. Active forms (simulations, projects) were less common due to higher costs.

Apart from e-learning, there is a lack of systematic didactic research in engineering education. Several TUs introduced didactic-centered units, but they are devoted to organizational duties rather than didactic studies. Pedagogy and humanities departments are generally not interested in this branch of education. TU faculty (with engineering backgrounds) do not always have sufficient knowledge of sociological aspects of education. The use of Massive Open Online Courses (MOOCs) in regular courses is still the exception. A partial equivalent of MOOCs might be found in the Distant Learning Course Center in Warsaw University of Technology, which offers commercial blended learning (online lectures and theoretical workshops, traditional projects and laboratories in Warsaw). It offers courses on electronics and IT and uses Spice programs for e-laboratories in circuit design.

Apart from the soaring number of students, there is no clear explanation for this state of didactics in the field of science or engineering. Low salaries fail to attract teachers with practical experience. Didactics is considered an unnecessary burden, not affecting professional careers. The Polish Ministry of Science and Higher Education only recently started work on a reform policy on this subject. Until now, the only government help toward improving didactics consisted of grants for TU didactics (about 250,000€ for selected TU departments) which were ultimately used to fund salary payments rather than to fund actual didactics research and improvements.

Formalizing educational training programs has been a major policy change. All lecturers are now expected to prepare reports on outcomes and requirements of their courses. Selected courses are evaluated by student questionnaires (in-house system). Negative evaluations from the in-house process is likely to result in salary reductions or other undesirable results. Departments and universities are also evaluated by an external government agency, the State Accreditation Board. Negative evaluations might lead to canceling courses inside a TU or a reduction of government funding.

According to the 2013 annual report of the Polish Ministry of Higher Education, about 24 % of students in this year study engineering or science. There are about 15 students per each academic teacher. Biotechnology, electrical engineering, chemical technology, material engineering, automation and robotics received the highest percentage of positive or exemplary assessments from the Accreditation Board (in comparison with all academic specializations). IT received the highest percentage of negative assessments from the same source. Most likely this is connected with the recent expansion of commercial IT schools (Ministry of Higher Education in Poland 2013).

Engineering Ethics at Wroclaw University of Technology (Poland)

As was mentioned above, the Department of Human Sciences at the Wroclaw University of Technology has implemented as part of an Operational Human Resources Program titled "Young Personnel 2015 Plus", financed by the European Union, the development of new elective humanistic courses. One of the most popular of these among full and part time students deals with engineering ethics. Here we want to consider this course, which has been developed and taught by Professor Jan Wadowski, in more detail. It is delivered as a series of lectures (usually 15 h per semester) in an e-learning format beginning in 2013.

The course program consists of such topics as general basic ethics, professional ethics, theoretical and methodological assumptions of engineering ethics and also moral dilemmas of the engineering practice in the context of philosophy of technology, globalization processes and rules of conduct in engineering. The first part includes elements of anthropological philosophy and general ethics describing the structure of human activity in the context of the Aristotelian distinction between *praxis* and *poiesis*. Presenting ethics as practical philosophy, students consider – among others – problems of moral evaluation, conscience, voluntary activities, responsibility, social relations of behavior and self-evaluation as a moral (or not) person. The meaning of ethos is explained and distinctions are made between "moral" and "ethical". The course also deals with axiological problems in the context of ethical choices.

Further lectures discuss the concept of norms, both social and technical, and the process of establishing norms through reference to values. The notions of good and evil are considered and the proper or inappropriate behavior in a given society – how they function in specific discourses. Systems such as concepts of laws of nature, theories of justice, utilitarianism, and more are presented. Patterns of behavior, their attributes and specific applications, are discussed. The course also deals with the concepts of "lesser evil" and "necessary evil".

After presenting the general ethical topics, the second part of the course shifts to the issue of professional ethics. An engineer should obey primarily the general professional ethics, which means duties and rules of conduct common to all professions – such as responsibility, loyalty, reliability, diligence. Students then discuss the public trust of professions and their functions. This includes issues such as controlling goods important for life and for its quality in the whole society.

A third part of the course relates strictly to the engineering. Beginners are taught what engineering is and what engineering activities are, as nowadays a person with technical education usually also has to be a manager dealing with technological systems. Lectures call attention to four important areas of an engineer's life: conscience, employing institution, society, and the environment engineering influences.

When we compare the programs of engineering ethics offered at technical universities in Ukraine, Belarus, and Russia, the Polish one is more theoretical. In the East courses more often start with the presentation of catastrophes and failures of engineering enterprises (dams, spacecrafts, bridges, and so on). The Chernobyl disaster in 1986 (Ukraine) has had an especially large impact on such an approach.

Conclusions

As a result of the internet, we have access to numerous resources: professional organizations offering new styles and techniques in engineering education, descriptions and solutions of practical cases, discussions of experience on engineers' blogs, and rich multimedia material for teaching and testing principles of engineering education and practice within a social context. Slavic language countries may apply some of them while confronting their own problems. There are three main factors that seem to create a sort of impediment to intense and socially required progress in the area of spreading engineering knowledge and realizing competent practice. At root they all have a communist and cold war heritage and consequences.

A first factor is the relatively low level of technical development in these countries in comparison with the most developed countries. There are some advanced centers and some more advanced technological branches, but in general these circumstances inhibit the chances to develop wider horizons and adequate and flexible competence among students of technical universities.

A second factor derives from lack of recognition of the social context in which engineering operates. The possibility of application of engineering ethics, even in cases when individual engineers' awareness has become already more informed, meets various kinds of barriers (mental, habitual etc.).

A third factor is the post-communist mentality. Many citizens do not identify with the "common good", with the principles of justice and fairness, because in the communist era they were strongly manipulated and deformed by ideological propaganda. Passivity and lack of conviction that something may be changed are typical characteristics of those who were deprived of regular opportunities to exercise citizenship.

Courses in the social sciences, the humanities, and engineering ethics are gradually being introduced. Indeed, there are even courses in such specialized areas of professional ethics as the ethics of the optometrist (Wrocław University of Technology). Admittedly, students sometimes treat these as no more than official propaganda. But gradually, as respect for the individual is regained, the issues of personal responsibility and true respect for the social issues can be achieved. As Carl Mitcham and R. Shannon Duval have written in a chapter on "Honesty in Engineering", "Honesty involves honoring reality, and being honored by reality in return" (Mitcham and Duval 2000, p. 82). Although there are numerous pragmatic exemplifications of this principle, the influence of impotence remains strong. Teachers of the technical universities declare that change is necessary but not all of them appear committed to introduce it. Some of them stick to old frameworks, often complaining that students do not read enough, while at the same time they are refusing to apply new attractive methods of teaching.

When we examine big countries such as Russia (143 million inhabitants), Poland (39), Ukraine (45), the Czech Republic (10), and Belarus (9), we find examples of great success in engineering education: innovative curricula, mobility of foreign teachers and students, exchange of experience, and the introduction of attractive new teaching methods and techniques for new generations of ("digital native") students. We also find examples of confusing underdevelopment: lack of advanced equipment in laboratories, massive exodus of well-educated engineers from post-communist countries to Western Europe or to the USA and Canada along with some indolence toward progressive changes in the field.

References[1]

Afeltowicz, Ł. (2012). *Modele, Artefakty, Kolektywy. Praktyka Badawcza W Perspektywie Współczesnych Studiów Nad Nauką*. Toruń: UMK Press.

[1] General bibliographical note: For more detailed information on Poland and the Czech Republic (as well as Latvia, Lithuania, Estonia etc.), one may consult EURYPEDIA which provides descriptions of educational systems and policies in the Eurydice network countries: http://www.eacea.ec.europa.eu/education/eurypedia.

Arapov, M. V. (2006). The higher education boom in Russia. Scale, causes, and consequences. *Russian Education and Society, 48*, 7–27.

Bird, C. (1994). Down but not out in Siberia. *New Scientist*, (1911). http://www.newscientist.com/article/mg14119112.400-down-but-not-out-in-siberia-as-russias-economic-crisis-deepens-money-for-research-is-drying-up-in-one-siberian-town-devoted-to-science-the-entire-population-is-fighting-for-survival.html

Birzea, C. (1994). *Educational policies of the countries in transition*. Strasbourg: Council of Europe.

Dobryakova, M., & Froumin, I. (2010). Higher engineering education in Russia: Incentives for real change. *International Journal of Engineering Education, 26*(5), 1032–1041.

Dunn, E. (2004). *Privatizing Poland: Baby food, big business, and the remaking of labor*. Ithaca and London: Cornell University Press.

Etzkowitz, H. (2008). *Triple Helix University: University-industry-government innovation in action*. London: Routledge.

Federowicz, M., & Wojcuk, A. (2010). *Report on the state of education 2010*. Warsaw: The Educational Research Institute.

Federowicz, M., & Wojcuk, A. (2011). *Raport o Stanie Edukacji 2011*. Warsaw: The Educational Research Institute.

Gajewski, R., & Jarosińska, E. (2011). Technology Enhanced Learning w Edukacji Inżynierów. In M. Dąbrowski & M. Zając (Eds.), *Koncepcje I Praktyka E-Edukacji*. Warsaw: Fundacja Promocji i Akredytacji Kierunków Ekonomicznych.

Górniak, J., Kocór, M., Strzebońska, A., Keller, K., Czarnik, S., Turek, K., Szczucka, A., Worek, B., Jelonek, M., & Szklarczyk, D. (2012). *Study of human capital in Poland*. Warsaw: Polish Agency for Enterprise Development.

Hall, K. (2005). Think less about the foundations: A short course on Landau-Lifshitz's course of theoretical physics. In D. Kaiser (Ed.), *Pedagogy and practice of science*. Boston: MIT Press.

Hollanders, H., & Es-Sadki, N. (2013). *Innovation union scoreboard 2013*. Office: European Union Publications.

Koucký, J. (2012). *From incremental funding to quality & performance indicators: Reforms of higher education funding in the Czech Republic*. Salzburg: EUA.

Markusik, S., & Bułkowski, A. (2005). Wykorzystanie Możliwości E-Learningu W Kształceniu I Dokształcaniu Inżynierów. *E-mentor 3*(3), 10.

Mateju, P., & Simonova, N. (2003). Czech higher education still at crossroads. *Czech Sociological Review, 39*, 393–410.

Mitcham, C., & Duval, R. S. (2000). *Engineering ethics*. Upper Saddle River: Prentice Hall.

Morawski, R., Manhire, B., & Starzyk, J. (1998). Engineering education in Poland. In *ASEE conference*, Seattle.

Mucha, J. (2009). *Uspołeczniona Racjonalność Technologiczna*. Warsaw: IFIS PAN Press.

NORRIC. (2005). *The system of education in Russia*. Copenhagen: Nordic Recognition Network.

Nowak, A. W. (2011a). *Podmiot, System, Nowoczesność*. Poznań: UAM Press.

Nowak, P. (2011b). Narzędzia Edukacyjne Dla Inżynierów. *Logistyka*, (3), 306.

Rosenthal, E. (2010). Plant repository at risk in Russia. *New York Times*, August 10, 2010. http://green.blogs.nytimes.com/2010/08/06/plant-repository-at-risk-in-russia/?_r=0

Rzeczpospolita (2013). Ranking Uczelni Akademickich, *Rzeczpospolita*, September 5, 2013.

Sjøberg, S., & Schreiner, C. (2010). *The rose project. An overview and key findings*. Oslo: University of Oslo.

Szkolnictwo Wyższe W Polsce. (2013). Warsaw: Ministry of Higher Education in Poland.

Wagner, I. (2011). *Becoming transnational professional. Kariery I Mobilność Polskich Elit Naukowych*. Warsaw: Scholar.

Wańkowicz, J., & Orzechowski, A. (2011). 43. Sesja Cigre 2010. *Energetyka*, 684.

Zhurakovsky, V. M., Pokholkov, Y. P., & Agranovich, B. L. (2008). Engineering education in Russia and the quality training of specialists in the area of high technologies, Russia, October 26, 2008. http://rbth.co.uk/articles/2008/10/26/261008_technologies.html

Maria Kostyszak M.A. in English Philology, M.A. and Ph.D. in Philosophy, all from Wrocław University, Poland. Habilitation in Philosophy. Author of four books, two of them on Martin Heidegger's philosophy, including *The Essence of Technology According to Martin Heidegger*, (Wrocław University Press, 1998) numerous articles, some translations. Researcher at Department of Ethics in the Institute of Philosophy at the Wrocław University. Honourable member of Society for Philosophy and Technology, regularly participating in conferences (Utah, Delft, Lisbon in July 2013) of that organization.

Jan Wadowski He studied at the Papal Faculty of Theology in Wroclaw, Breslau, Poland. M.A. in Philosophy 1991. Ph.D. in Philosophical Anthropology in 1996, *Józef Tischner's Philosophy of the Drama of Human Existence*. The author of three books, over 40 articles and of e-learning courses: "A Man and Technology" and "Engineering Ethics". Participant of about 30 conferences. Interested in philosophical anthropology, ethics, philosophy of the civilization and the society. He is working at the Wroclaw University of Technology in the Department of Humanistic Sciences, participates in the Research Team of Philosophical Anthropology and Ethics.

Marcin Zaród M.Sc. in Applied Physics from Gdańsk University of Technology, 2009. Polymer physics and processing specialist in Centre for Molecular and Macromolecular Physics at the Polish Academy of Sciences (2009–2013). Ph.D. candidate in University of Warsaw. Science communicator and educational activist (with Political Critique and Foundation for Modern Education SPUNK). Focuses his research on boundaries and social studies of formal and non-formal engineering, scientific and educational practices. Co-author of *Szkoła Przewodnik Krytyki Politycznej* – a critical review of the Polish educational system.

Part II
Ideologies of Engineering Education

Introduction

Brent K. Jesiek and Christelle Didier

Questions about the ideological underpinnings of engineering are not new, as evidenced by the efforts of a handful of pioneering historians and sociologists who dared tackle the topic. Edwin Layton's classic *The Revolt of the Engineers* (Layton 1971), for example, showed how the professional ideals and aspirations embraced by many American engineers during the Progressive Era stood in marked tension with business imperatives and bureaucratic loyalty – and with the latter ultimately prevailing. Covering similar historical and conceptual territory but more Marxist in outlook, David Noble's *America by Design* (1979) portrayed a growing alignment of the U.S. engineering profession with market capitalism and an almost mystical ideology of quasi-autonomous technology. Both works helped contextualize the profession's development in America from the late nineteenth to mid-twentieth century and demonstrated how prevailing engineering values and attitudes were frequently interchangeable with a business ethos that was inculcated through dominant pathways of education and career development.

Still other works have helped show how partially unique configurations of ideology and engineering have emerged in other national and cultural contexts, as reflected in Ken Alder's argument that the early modern history of the engineering profession in France was "energized by a radical ideology that justified social hierarchy by reference to national service" (Alder 1999, p. xii). A growing body of cross-national comparative research by scholars such as Gary Downey and Juan

B.K. Jesiek (✉)
School of Engineering Education, Purdue University,
701 West Stadium Avenue, West Lafayette, IN 47907, USA
e-mail: bjesiek@purdue.edu

C. Didier
Département des sciences de l'éducation UFR DECCID,
Université Charles de Gaulle-Lille3,
3 Rue du Barreau, Villeneuve-d'Ascq, Lille F-59650, France
e-mail: christelle.didier@univ-lille3.fr

Lucena has also more broadly shown how engineers respond to – while to some degree shaping – prevalent meanings, such as dominant understandings of what counts as national progress, or what it means for engineers to serve government and/or private industry (Downey and Lucena 2004; Downey et al. 2007). As these works make clear, the ideological commitments of engineers and engineering not only profoundly inflect what it means to be an engineer or practice engineering; such commitments may also vary considerably by time and place.

The chapters in this part continue and extend these traditions of scholarship. They do so by reminding us of many important, recurring questions about how the ideological foundations of engineering as a modern discipline and profession resonate (or, perhaps just as importantly – may fail to resonate) with other prevalent beliefs and values – whether economic, technological, political, social, cultural, or otherwise. Qin Zhu and Brent Jesiek's Chap. 7, for example, looks to China as an underexplored yet increasingly important context for investigating the ideology-engineering nexus. More specifically, the authors identify three relevant ideological currents that can enable a better understanding of the intellectual context of engineering in China: Confucianism, Marxism, and economic pragmatism. Starting from three questions that are traditionally raised in studies of engineering ethics and professionalism by U.S. scholars (and which often take a Parsonian-functionalist approach, as represented by the authors' reference to Michael Davis' work), they first give the most common answers. Yet pushing their analysis in directions more sensitive to the Chinese ideological context allows them to propose alternate answers to these questions, thereby revealing some of the blind spots that may occur when scholars view partially unique local cultures of engineering through Western lenses. More practically, their chapter potently suggests how successful multinational collaborations in engineering may require keen sensitivity to the relevant intellectual environments of engineering education and professional practice.

Amy Slaton's Chap. 8 returns the focus to the United States, albeit with many broader implications. She begins by describing the historical dominance of two ideological logics in engineering. The first of these is *technocratic*, which paints engineering as ultimately an apolitical enterprise that can be separated from its social foundations. The second logic she proposes is *meritocratic*, which privileges individual ability and responsibility to succeed in engineering while demonstrating technical excellence. Consistent with a neoliberal worldview, these two logics pose considerable challenges for those who identify with movements toward democratization, including by promoting a far more inclusive, participatory, and liberatory climate of technical education and professional practice. Hence, pivotally important for the Slaton are questions about how the content and aims of engineering are inextricably linked to the matter of who can be (or become) an engineer, not to mention what counts as epistemic authority in engineering. These themes are illustrated through a rich variety of literature and examples, from discussion of the trials and travails of various diversity and inclusion initiatives to explorations of how some specific student populations (e.g., those with low socioeconomic status or atypical kinds of cognitive dis/abilities) are "othered" against the backdrop of a powerfully normative status quo in engineering.

Derrick Hudson's Chap. 9 explores the continued challenges of attracting, recruiting, and educating African Americans, people of African ancestry and other underrepresented groups in engineering across the United States and globally. He begins by reminding us that the numbers of African Americans in engineering have stagnated and declined since the beginning of the twenty-first century. An important connection that Hudson highlights is that the early pioneers of scholars in African American studies thought that they could easily construct a reverse mirror image of the curricula they encountered in other academic disciplines, such as history, political science, or anthropology. A glaring omission of the early pioneers is the work that would be needed in engineering education. Many early pioneers failed to take into account that work needs to also be done directly within engineering education to foster "sociotechnical" engineering undergraduates and professionals. Hudson's work compliments Cech's exploration of the "ideologies of depoliticization and meritocracy" by emphasizing that one of the nagging disconnects for African Americans and other historically underrepresented minorities is that engineering is framed in a manner that cannot address social justice or liberatory issues in society. Hudson concludes his section with suggestions for further research, highlighting the continued pivotal role of historically black colleges and universities and the need to encourage more investments to promote research and development in African universities, which account for less than 2 % of research expenditure globally.

Finally, Erin Cech and Heidi Sherick's Chap. 10, focused on the "ideology of depoliticization", serves as a fine compliment to Slaton's work.. Their chapter nicely captures and questions a pervasive view that the technical dimensions of engineering work can and should be separated from any associated political, social, or cultural considerations. Cech and Sherick's challenge echoes one present in other recent work by Caroline Baillie, Jens Kabo, and John Raeder (2012) and by Bill Williams, José Figueiredo, and James Trevelyan (2014). Such ideological boundary work – which may be contrasted with the sort of "strategic politicization" described in Zhu and Jesiek's discussion of Marxism and engineering in the Chinese context – projects a sanitized image of engineering as ultimately divorceable from anything deemed subjective, sociocultural, or humanistic – that is, anything "non-technical." As a consequence, engineering is portrayed as not only technocratic, following Slaton, but also somehow above ideology, artfully concealing the inherently value-laden and social character of engineering work behind a veil of purported objectivity and rationality. Of particular note in this chapter is the authors' discussion of how engineering education helps perpetuate this ideology, including by protecting and preserving historically dominant – but increasingly outdated – images of the profession's epistemological, ethical, and ontological foundations. In turn, this hegemonic reproduction poses considerable impediments to reforming and transforming engineering faculty, courses, curricula, and culture to meaningfully breach the boundaries between the technical and sociocultural.

In summary, the chapters in this part offer a compelling invitation for further studies that help enhance our understanding of the ideological considerations that undergird the education of engineers and their practice as professionals. Each in its own way invites increased awareness of the importance of intellectual, cultural, and ideological contexts associated with both the objects of our research, i.e., engineers and engineering, and our own work as scholars. If such ideological contents are explicitly offered, imposed, or revealed in certain contexts, what about the implicit beliefs that fail be questioned because of their invisibility? As these chapters suggest, considerations such as free market principles, efficiency, economic growth, political commitments, and techno-optimism are often inextricably bound up with questions about what counts as engineering and who can be an engineer. This section opens up opportunities for further efforts to expand the breadth and depth of ideological considerations, including through cross-institutional and cross-national comparative studies.

References

Alder, K. (1999). *Engineering the revolution: Arms and enlightenment in France, 1763–1815.* Princeton: University Press.

Baillie, C., Kabo, J., & Raeder, J. (2012). *Heterotopia: Alternatives pathways to social justice.* Alresford, Hants. (UK): John Hunt Publishing.

Downey, G. L., & Lucena, J. C. (2004). Knowledge and professional identity in engineering: Code-switching and the metrics of progress. *History and Technology, 20*(4), 393–420.

Downey, G. L., Lucena, J. C., & Mitcham, C. (2007). Engineering ethics and identity: Emerging initiatives in comparative perspective. *Science and Engineering Ethics, 13*(4), 463–487.

Layton, E. (1971). *The revolt of the engineers: Social responsibility and the American engineering profession.* Cleveland: Press of Case Western Reserve University, 2nd edn, 1986. Johns Hopkins University Press.

Noble, D. F. (1977). *America by design: Science, technology, and the rise of corporate capitalism.* New York: Alfred A. Knopf, Inc.

Williams, B., Figueiredo, J., & Trevelyan, J. (2014). *Engineering practice in a global context: Understanding the technical and the social.* Leiden: CRC Press. Taylor & Francis Group, London, UK

Brent K. Jesiek B.S. Electrical Engineering, Michigan Tech, M.S. and Ph.D. in STS, Virginia Tech. Assistant Professor, School of Engineering Education and School of Electrical and Computer Engineering, Purdue University; Associate Director of Global Engineering Program, Purdue University. Research interests: historical and social studies of engineering, engineering education, and computing; engineering epistemologies; global engineering education; and cyber infrastructures for engineering education research.

Christelle Didier B.S. in Electrochemistry Engineering, M.A. in Education, Ph.D. in Sociology from Ecole des Hautes Etudes en Sciences Sociales (EHESS), Paris. From 1993 to 2013, Assistant Professor, Lille University, France, Ethics Department. Assistant Professor, Charles de Gaulle University of Lille, Education Department. Member of CIREL (EA 4354). Co-author of *Ethique industrielle* (DeBoeck, Brussels, 1998), author of *Penser l'éthique des ingénieurs* (PUF, Paris, 2008) and *Les ingénieurs et l'éthique. Pour un regard sociologique* (Hermes 2008). She has published many articles on ethics and social responsibility in the engineering profession and education, and on the engineering profession's values (from interviews and extensive surveys). Research areas: engineering ethics and values, including historical, cultural and gender perspective, sustainable development and corporate social responsibility, social responsibility.

Chapter 7
Confucianism, Marxism, and Pragmatism: The Intellectual Contexts of Engineering Education in China

Qin Zhu and Brent K. Jesiek

Abstract Sensitivity to cross-cultural and cross-national differences in engineering education and practice is essential for globally competent engineers. Those who fail to pay close attention to the historical-cultural contexts of engineering do so at their own peril, increasing the likelihood that their gaps in knowledge and misconceptions will lead to failed collaborations, projects, and products. This chapter aims to support this thesis by describing the historical and intellectual contexts for engineering education in contemporary China. It starts by presenting a variety of controversial issues in current global discourses on China's engineering education, e.g., distinct understandings of professionalism and accountability, and different approaches to defining core bodies of knowledge, competencies, and other learning outcomes. It argues that these controversies mainly arise from insufficient understandings of three key intellectual contexts of Chinese engineering education: Confucianism (historical), Marxism (ideological), and economic pragmatism (economic). It is then followed by analyses showing how these three intellectual contexts historically contributed to shaping China's unique developmental trajectory of engineering education. The three dimensions are not presented and judged in historical sequence, but instead framed as interwoven and coproduced, with real and present implications for the culture and character of engineering education and practice. Finally, this chapter attempts to use the three-dimensional framework as an interpretative tool to reflect on the practical issues proposed in the first part. In so doing, it highlights the relevance and implications of the intellectual contexts of global engineering education and policymaking in contemporary China. The

Q. Zhu (✉)
School of Engineering Education, Purdue University,
Seng Liang Wang Hall of Engineering, Northwestern Avenue,
West Lafayette, IN 47906, USA
e-mail: qinzhu@purdue.edu

B.K. Jesiek
School of Engineering Education, Purdue University,
701 West Stadium Avenue, West Lafayette, IN 47907, USA
e-mail: bjesiek@purdue.edu

S.H. Christensen et al. (eds.), *International Perspectives on Engineering Education*,
Philosophy of Engineering and Technology 20, DOI 10.1007/978-3-319-16169-3_7

chapter's main thesis is further advanced by revisiting an influential cross-national, comparative study of engineering education, which helps show how discourses originating outside of China frequently provide impoverished or oversimplified understandings of the Chinese context.

Keywords Confucianism • Marxism • Pragmatism • Intellectual history • Global engineering education • Engineering education • Educational policy • China

Introduction

Influenced by economic, social, political, cultural, and other internationalization trends, engineering is more than ever becoming a global profession (Johri and Jesiek 2014). This reality is reflected in numerous reports and commentaries, ranging from the U.S. National Academy of Engineering's influential volume on *The Engineer of 2020* (National Academy of Engineering 2004) to current ABET accreditation criteria which explicitly note the importance of engineering graduates understanding the impacts of their work in global context (ABET 2008). In response, growing numbers of educational institutions, programs, and initiatives in Australasia, Europe, Latin America, the United States, and more recently Asia (e.g. China, Japan, Malaysia, and South Korea) are grappling with the challenge of preparing their engineering graduates to function more effectively in the ever-changing global context. Thus how to characterize and educate the "globally competent engineer" is an increasingly relevant and compelling question for many engineering education practitioners and researchers.

Among various responses to this question, one of the more thoughtful and influential comes from Downey et al., who define global engineering practice as a highly interactive form of cultural engagement. Thus global engineering competency becomes "a problem of engaging people from different cultures" (Downey et al. 2006, p. 107), and the globally competent engineer is expected to acquire the "knowledge, ability, and predisposition to work effectively with people who define problems differently than they do" (Downey et al. 2006, p. 111). According to this formulation, a globally competent engineer must understand the historical and cultural contexts of engineering education and practice in other countries and regions. The authors also observe that "statements about the benefits of global learning for engineering students typically locate those benefits in encountering and coming to understand engineers and other potential co-workers who are raised, educated, and living in countries other than their own" (Downey et al. 2006, p. 108).

This chapter similarly takes the view that sensitivity to cross-cultural and cross-national differences in engineering education and practice is essentially important for globally competent engineers. Those who ignore or disregard the historical-cultural contexts of engineering increase the likelihood that their lack of knowledge and/or misconceptions will lead to failed collaborations, projects, and products. One recent example is that of Google leaving China, in part due to an inadequate understanding of Chinese ideological and political culture (e.g. human rights, negative liberty, and internet censorship policy). This chapter aims to further illustrate this thesis by describing the intellectual contexts for engineering education in contemporary China. We focus on engineering education due to its dominance as an educational pathway in China, its position as a key to professional practice, and its implicit and explicit roles in bringing individuals into the profession.

Three Controversies

In current global discourse on Chinese engineering education, a variety of controversial issues have surfaced. Many are related to distinct understandings of professionalism and accountability, and different approaches to defining core bodies of knowledge, competencies, and other learning outcomes. One such controversy centers on the status of engineering ethics in China. As Guo (2009) has argued, for example, *engineering ethics does not as such exist in China*. However, moral reflection and regulation have actually had considerable influence on Chinese engineering education and practice, with Confucianism, Marxism, and Deng Xiaoping's development thought playing especially prominent roles. These moral ideas have real and considerable influences on current engineering practice and education (Zhu 2010).

There is also historical evidence for a sort of code of engineering ethics emerging and evolving in modern China. Su and Cao (2008) argue that the Chinese Institute of Engineers, with a history going back to the early 1900s, did not originally have a clear ethical code specifying the social responsibilities of engineers. However, the organization did uphold a strong commitment to both the nation and public during a period of struggle to break free of imperialist exploitation. Today, a similar code of ethics might not explicitly exist in Chinese institutions of engineers due to the influence of a Marxist ideology, which emphasizes governmental administration of engineering societies over professional autonomy. Chinese engineers might lack a formal code of ethics but nonetheless "have an unwritten one" (Davis 2009, p. 334). Is a code of ethics really absent just because it does not exist in Western forms? We should be careful not to define engineering ethics without adequately attending to relevant considerations of historical and cultural context.

A second controversy concerns whether *engineering in China is a profession*. In fact, Guo's argument that Chinese engineers do not have explicit ethical codes might suggest that Chinese engineering lacks some critical features of a modern profession, at least from a Western point of view. By contrast, Davis (2009) maintains that engineering is a global profession. And to the extent that engineering has distinct elements and features that Chinese engineers share with their colleagues elsewhere, they work well with engineers from other countries and regions. Davis encourages researchers to take history and culture into account and "explain" what engineering is when they want to define it. Nevertheless, Davis himself does not explore the Chinese intellectual context. In fact, and as discussed below, Confucian and Marxist approaches to communitarian thought resist some core ideas in Western definitions of professionalism (e.g., individualism and autonomy).

A third controversy concerns *who Chinese engineers were and are*. Addressing this question first requires acknowledgment of the complex historical and cultural context of engineering practice and education in China. One view is that ancient Chinese artisans and craftsmen can be viewed as engineers. For instance, Rae and Volti (1993) argue that governmental officials and bureaucrats took on roles that in part resemble modern engineering practice. Yet they do not offer convincing arguments regarding the extent to which these officials and bureaucrats were truly comparable to contemporary engineers. With artisans also, the extent to which they were operating like modern engineers is unclear.

More recently, Wadhwa and colleagues have argued that what is considered engineering and who is considered an engineer in China is not consistent with prevailing American views. For example, some auto mechanics and technicians in China are called "engineers". Further, the majority of engineers do not take engineering jobs, but become bureaucrats or factory workers (technicians or production line managers). The average level of skill and knowledge among Chinese engineers appears to be lower than in the West (Wadhwa et al. 2007; Gereffi et al. 2008). As Wadhwa and colleagues argue, a number of "technology" programs – such as "information technology" – should not be viewed as engineering programs (Wadhwa et al. 2007; Gereffi et al. 2008). However, these authors fail to link the concept of "technology" to both its linguistic origins and the pragmatic context developed since Deng Xiaoping's reform and opening-up. Technology, as a pragmatic term, has a diversity of meanings in Chinese and can in some cases include engineering, as is also the case with Western institutes of technology such as MIT. *Gongcheng jishu* (工程技术) should not be simply and literally translated into "engineering technology," but might be better (if not best) understood as "engineering **and** technology" or even "engineering (skills)".

In order to better understand and contextualize such controversies – as well as engineering and technology education in China more generally – calls for appreciation of three key intellectual contexts: Confucianism (historical), Marxism (ideological), and pragmatism (economic). These three philosophies are among the most fundamental intellectual contexts of modern engineering education in China. It is thus appropriate to begin with a sketch of each of these intellectual traditions:

- Confucianism (historical): In comparison with other philosophies (e.g. Daoism, Buddhism, and Moism) in traditional China, Confucianism is the single most influential Chinese school of thought (Shun and Wong 2004). Even today, as a sociopolitical philosophy, it shapes people's understandings of relations among humans, nature, and society, with technology playing a mediating role. As a philosophy of education, Confucianism continues to shape the values of Chinese people and cultivation of "ideal men" in society. Hence Confucianism has fundamental implications for both engineering and education, including in relation to questions like: What is the role of engineering and technology in society? What is *good* engineering? What does an "ideal person (engineer)" look like? How should people be educated?

- Marxism (ideological): Marxism is the official *zhidao sixiang* (guiding ideology) for nearly all social activities and national strategies in the People's Republic of China. As a social enterprise, engineering cannot escape the influence of Marxism. Marxism is also embedded in Chinese (postsecondary) education. College students are taught to incorporate Marxist ideologies into their future careers. At the national level, engineering students (both undergraduate and graduate) are required to take courses on Marxism. Since the 1950s, the CPC has conducted many rounds of ideological curricular reforms in colleges and universities (Andreas 2009). And for Master's degree programs, engineering students must take one Marxist course on "dialectics of nature" which provides a kind of Marxist philosophy of engineering.
- Pragmatism (economic): In contrast to Mao Zedong's "revolutionary Romanticism" guided by a radical ideology and often largely impervious to practical concerns, Deng Xiaoping's thinking was dominated by what MacFarquhar (1997) calls "pragmatism," as evident throughout the course of his political career (Joseph 2010; Wong and Zheng 2001). Since the reform and opening-up, a pragmatic economic approach initially proposed by Deng has exerted a strong influence on economic and social policymaking. Because of the interwoven relations between economic development and engineering, pragmatism is thus deeply embedded in engineering practice and education in contemporary China, and engineering education is often proposed and promoted with explicitly pragmatic goals.

These three aspects of the contemporary Chinese context are not just historically sequential phenomena; they are interwoven and coproduced, with real and present implications for the culture and character of current engineering education and practice. Further, these are not the only relevant features of Chinese intellectual life today. Another relevant theme is the ideological concept of "good life." Proposed by Xi Jinping, the new General Secretary of the Communist Party of China, this concept has its intellectual roots in both Confucianism and pragmatism. Xi's idea of promoting the "good life" as part of his national project for building a "beautiful China", which is now being integrated into China's engineering practice and education policy. As suggested by the preceding overview of controversial questions related to "engineering ethics", "engineering", and "engineers", it is clear that multiple intellectual dimensions are frequently and deeply interwoven.

Confucianism: Sociopolitical Practicality and Communitarian Ethics

It is worth stepping back to more deeply probe each intellectual tradition, beginning with Confucianism. As the most influential school of thought in Chinese culture and philosophy, it originated as a kind of "ethical-sociopolitical teaching" during the Spring and Autumn Period (770 BCE–476 BCE). As a sociopolitical philosophy, Confucianism always examines engineering and technology through ethical and political lenses. An overarching Confucian philosophy of technical projects embraces a "sociopolitical practicality", which posits that technical projects should contribute to the social welfare of the state and its people. Late in the Ming dynasty (1368–1644), this idea was systematically developed as a national philosophy or *jingshi zhiyong*. This idea has been translated as "engaging in efforts of practical use in governing the world" (De Bary and Bloom 1999, p. 765).

A story from *Zhuangzi* (a Daoist book) helps illustrate this pursuit of practical efficacy valued in Confucianism. According to the story, on his way back to the state of Jin after his travel to the state of Chu, a disciple of Confucius named Zigong saw an old man working very hard to get water from a well, putting it in a jug to irrigate his garden. Zigong felt puzzled and asked the old man, "There is a machine now that can water a hundred gardens in one day. You would get a big reward for easy work. Would you not like one?" The old man asked the Zigong to further explain how the machine worked. Zigong told the old man the machine was called a *shadoof* (counterpoise-lift) that consisted of a lever rotating on a pole with a bucket suspended at the shorter length. Because of mechanical advantage the user saved labor. The old man hesitated before responding,

> I heard from my teacher that where there are mechanical contraptions there will be mechanical business, and where there is mechanical business there are mechanical minds. With mechanical mind, you cannot preserve your simplicity. When you cannot preserve your simplicity, your spiritual life is unsettled, and the *dao* will not support an unsettled spiritual life. I am not ignorant of your contraption but would be embarrassed to use it (Ivanhoe and Van Norden 2001, p. 243).

Zigong was impressed by the moral integrity of the old man. However, when Zigong retold this story to his master Confucius, Confucius was not so inspired. Confucius argued that "for those who merely pursue their inner life and inner truth", the old man's criticism of the shadoof may "seem reasonable". But besides their inner lives, human beings also have their outer lives and they must live and "have a relationship with the outer world". In this sense, Confucius stressed, the old man "only knows one side of the truth" (Zhu 2010, p. 91). Hence from the Confucian perspective, good application of technology is able to generate practical efficacy and social prosperity (nation's economy and people's livelihood).

Further insight can be gleaned from the well-known Confucian classic *Shangshu* (The Book of Documents), which includes three doctrines that could serve as fundamental principles for a Confucian ethics of engineering: *zhengde* (rectification of virtues), *liyong* (appropriate use of resources), and *housheng* (strong protection of life). Conversely, the historical record also reveals some technical projects that were constructed with the purpose of fulfilling the emperors' personal pleasures, leading to accusations of *laomin shangcai* (wasting labor and money) or *qiji yinqiao* (magical skills and improper cleverness).

In ancient China, Confucian principles had major influences on individuals, collectives, and society, especially in terms of social hierarchy. In general, the Confucian society consisted of four major "occupations," in decreasing order or status: *shi* (gentry scholars), *nong* (peasant farmers), *gong* (artisans), and *shang* (merchants and traders). Were any of these the early predecessors of engineers? And how might they have been educated? Technical projects (e.g. structures, metalwares, mechanical devices, and weapons) were mainly conducted in family workshops or large-scale labor activities organized by the government (Li 2006). Artisans participated in both, but in comparison with gentry scholars and peasant farmers their social status was relatively low. In Confucianism, *laoxinzhe* (those who labor with the mind) have higher status than *laolizhe* (those who engage in physical labor) (Song 2002).

Hence, histories of Chinese technology commonly see artisans as the predecessors of engineers. But this is questionable. Only high-level artisans were more literate than farmers, yet much less so than scholars. According to Barbieri-Low (2007), the literacy of ancient artisans mainly involved inscribing characters on artifacts, a practice called *wule gongming*, or "engraving artisan names on products." This could be seen as an early code of ethics among artisans who took responsibility for the quality of the artifacts they produced. With years of hard work, only a small portion of high-level artisans could move into supervisory roles. Since such supervisory artisans were involved in planning and implementing whole projects and coordinating labor relations, they could be seen as early predecessors of engineers.

Yet it is arguably even more appropriate to see some scholar-officials and technical bureaucrats as predecessors of the modern engineer, especially in light of their social roles and functions. This phenomenon represents a central idea in the Confucian history of education – "practical statesmanship" – or, to adopt a more gender neutral term, "practical leadership." Hatmaker (2012), for instance, identifies a number of different roles played by engineers in contemporary society: (a) technician; (b) administrator; (c) coordinator; (d) communicator; (e) relater; and (f) caretaker. In ancient China, large technical projects frequently involved scholar-officials who assumed roles as "administrators", "coordinators", and "communicators," covering a good part of Hatmaker's characterization. The planning and building of the Dujiang Dam serves as a relevant historical example. The project was administered by Li Bing, a Confucian scholar serving as a principle governor of Shu during the Warring States period (475 BCE–221 BCE). This early scholar-official attempted to incorporate some basic management principles in administrating and coordinating the construction of Dujiang Dam (Wang et al. 2008). Such competencies distinguish official scholars managing technical projects from common artisans, as well as other scholar-officials assuming other kind of roles.

Influenced by Confucian thought emphasizing political centralization and agriculture, particularly since the Han dynasty (206 BCE–220 CE), the government favored *dayitong gongcheng* (great unified projects) such as irrigation systems, large structures, canals, and other inland waterways which required enormous labor resources. Early technical projects were therefore large-scale and complex, necessitating strategic activities such as planning, designing, coordinating, and implementing. This expansive view of technical projects in early Confucian thought continues to influence the Chinese understanding of engineering.

Since large-scale projects required well-organized operations, government officials with either management experience or technical knowledge played leading roles. These officials mainly saw technical projects as "political projects" aligned with the Confucian idea of sociopolitical practicality (e.g., political stability and social benefit). Hence, governmental officials brought Confucian thought to planning, designing, and coordinating. Such abilities remain central to technical professionals who we call "engineers" today. And because official scholars were recruited through imperial examinations, they had to be well versed in Confucian ideas and principles. In their careers, they intentionally or unintentionally applied their knowledge – much of it originating in Confucianism – to design and carry out technical projects. Given their educational experiences and application of theoretical knowledge (Confucian thought) into technical practice, these special scholar-officials exhibit attributes closely resembling those of modern engineers.

The "engineering-management" role taken by Confucian officials further developed as a political tradition in modern China in the spirit of "practical leadership." This tradition was particularly influential in the late Ming and Qing dynasties. In contrast to traditional "moral leadership," practical leadership was based on a belief that "the inner moral cultivation and exemplary leadership were not sufficient to solve the problems China was facing and professional statecraft and institutional approaches should be added" (Liu 2012, p. 96). During the late Qing Dynasty (1840–1911), Confucian officials such as Wei Yuan (1794–1857) and Kang Youwei (1858–1927) proposed that *xixue* (Western learning) could promote sociopolitical reforms and solve social problems in China. Some Confucian scholars (so called "westernizationists") also believed that only Western science and engineering could promote sociopolitical reforms in China.

Thus, in the late nineteenth century, a great number of Western books, and particularly those covering topics in science and engineering, were translated into Chinese through collaborations between Confucian officials and Western missionaries. Engineering concepts and theories were also later imported into China. Westernizationists like Zhang Zhidong (1837–1909) established a number of modern factories, militaries equipped with Western weapons, and technical schools. These schools represented the beginning of modern engineering education in China (Carroll 2008). And while westernizationists had seen the importance of modern science and engineering, they also had a deep grounding in Confucian ideology – *zhongxue weiti, xixue weiyong* (Chinese learning as the essence, and Western learning for use). Such an idea remained highly influenced by a traditional Confucian view of technology in terms of sociopolitical practicality, while Western science and engineering mainly served practical purposes, including to help "great China" resist Western imperialism and to enlighten people's minds, but without ever allowing Western learning to displace Confucianism. Hence, the more individualistic schools of thought and institutions characteristic of the West were not imported to China along with engineering. Engineering was introduced as a modern technical *occupation* but not *profession*.

In modern China, a resistance to engineering as an individual profession was thus based in the communitarian ethical values of Confucianism. The central virtues of *ren* (benevolence) and *li* (ritual) characterize Confucian ethics as relational, in contrast to an emphasis on individual autonomy and the freedom in Western ethics. According to Wong (2008), the value of individual autonomy usually includes three dimensions: (a) prioritizing individual interests over group or collective interests when these conflict; (b) giving moral permission to the individual to choose from a significantly wide range (within certain moral boundaries) of ways to live; and (c) emphasizing the importance of living according to one's own understanding of what is right and good even if others do not see it the same way. All three dimensions are central to Western professional ethics.

These three individualistic values are foreign to Confucianism. In engineering ethics, Wong's first dimension is important because it allows engineers to think of their own professional agency over group interests (e.g., those of their companies or firms). In contrast, Confucian ethics sees the individual as dependent on the group, with individual interests as part of the group's interests and vice-versa. Wong's second dimension grants engineers free will to take ethical action from a variety of options and according to an implicit or explicit code of ethics. Confucian ethics is less favorable toward legal coercion and instead emphasizes moral exhortation and inspiration by way of example. Wong's third dimension encourages engineers to make their own choices (e.g., whistle blowing) without interference or coercion from others. Confucianism does articulate the necessity to speak up when one believes their ruler is taking a wrong course of action. Yet in Confucianism, there is no mechanism proposed to protect the critical subordinate from being punished by the ruler. Confucian communitarian ethics can thus be contrasted with an individualistic understanding of ethics. For individuals, Confucianism also posits five basic social relationships: ruler to ruled, father to son, husband to wife, elder brother to younger brother, and friend to friend. The primacy of a network of such relationships further complicates the practice of individual-based professional ethics in the Chinese cultural context.

Marxism: Productive Force and Political Redness

As noted above, Confucian approaches to technical practice emphasize the sociopolitical implications of artifacts and large-scale technical projects. This more socially-oriented approach can also be contrasted with the economic and engineering approach of Marxism, which grows out of the utilization of technology to transform nature through engineering thinking.

As a socio-economic philosophy, Marxist historical materialism considers technical activity as the production process in which technology is a productive force when it is operated, maintained, and conserved by living human labor. Because of the significant role of technology in changing society, Deng Xiaoping further emphasized that science and technology constitute the *first* productive force. The productive forces are those by which society influences nature and changes it, while nature is the universal object of labor (Lorimer 1999). In criticism of Soviet thinking, Mao Zedong modified historical materialism by stressing the importance of human labor, and he glorified human capabilities of using technology to transform nature. Mao's ideology engendered two "philosophies" that continue to influence engineering practice and national development: (1) a *philosophy of nature* (material productivity), where engineering expertise serves as the superpower in transforming nature; and (2) a *philosophy of society* (social productivity), where engineering expertise should be employed to organize and manage social issues.

During the Great Leap Forward (1958–1961) and Cultural Revolution (1966–1976), Maoist thinking about nature played a major role in shaping engineering education and national development. Mao's voluntarist philosophy – believing any task could be accomplished through sheer will – held that through concentrated human exertion and energy, material conditions could be altered and all difficulties overcome in the struggle to achieve a socialist utopia (Shapiro 2001). Scientists and engineers were educated and encouraged to pursue "giant" achievements through exploitation of nature. Engineering projects were even considered "wars against nature" by Mao and Maoist theorists. Mass labor was employed in remarkable engineering projects to build large-scale dams and canals and create new irrigated farmlands in formerly fallow areas. In this sense, Maoist philosophy of society was applied to organize and manage the huge manpower and material resources in engineering projects and other related social issues (e.g., migration problems in constructing large dams). The Maoist philosophy of society has its intellectual roots in Marxist structuralist sociology, which also sees the state as a kind of mechanism.

Thus, the state can be viewed as a large engineering system with a national economy that can be developed and engineered (planned, designed, and implemented). Maoist philosophies of nature and society co-shaped a unique understanding of engineering which still has profound impacts today, including four major aspects. First, engineering involves the utilization and transformation of nature, with the purpose to construct a kind of "artificial nature". Like Marx, Mao himself endorsed the role of science in liberating humans from the material limitations set by natural world. As he explained, "natural science is the armed force by which people strive for liberty." He further elaborated that "if people want to gain liberty from nature, they need to use natural science to understand, overcome, and transform nature so as to be free from nature" (Mao 1940). This view still prevails in nearly all ideological education textbooks for engineering graduates students. For instance, in one of the most popular ideological books, Chen Changshu (a founding father of the philosophy of technology in China) sees the objective of technology (including engineering and production) as "transforming the objective world" (Chen 2001, p. 10).

Second, engineering is understood as a process that conquers nature. This view was early illustrated by a thematic phase, "Man must conquer nature" (*ren ding sheng tian*), spoken by Mao Zedong on September 15, 1956, at the 8th National Congress of CPC. Such a view still influences the majority of senior engineers and engineering administrators, most of whom were educated in 1970s and 1980s.

Third, engineering is a universal method applied in national strategies, initiatives, and planning. Therefore, a large number of national projects involve use of the term "engineering." For instance, consider *minsheng gongcheng* (people's livelihood engineering), *cailanzi gongcheng* (vegetable basket engineering), and *makesizhuyi lilunyanjiu yu jianshe gongcheng* (Marxist theoretical research and construction engineering). These in turn belong to what Marxist scholars call "social engineering."

Fourth and finally, engineering projects are usually large-scale. As already mentioned, social engineering projects often require the coordination and management of human power and material resources at the national level. Hence social engineering projects are large-scale. In the everyday usage of Chinese language, engineering is often understood as large-scale projects. In contrast to engineering, technology has a different meaning referring to technical activity or projects at any scale. In the Chinese context this unique view distinguishes engineering from technology.

In sum, Maoist philosophy understands engineering as a large-scale process that involves transforming natural resources to fulfill the socialist state's development needs in the construction of utopian engineering projects. As socialist laborers, engineers are required to adopt socialist core values such as collectivism and particularly proletarianism. During Mao's time, engineers did not have particularly high social status. Yet because most were better educated than laborers and farmers, they were considered intellectuals. As such, engineers often were accused of being too far removed from the realities of manual labor – whether of the factory worker or farmer. Some were criticized as bourgeois individualists and/or "right deviationists." Indeed, a significant number of scientists and engineers involved with the "Two Bombs and One Satellite Program" were accused of being bourgeois intellectuals due to their Western educational backgrounds (Harvey 2004).

As socialist laborers of the working class, engineers are still encouraged to engage in practical activities at the forefront of production. Partly for this reason, recent engineering graduates have a tradition of learning from technicians and laborers by working with them on the production line. Influenced by Maoist voluntarism, engineers are encouraged to exceed production plans, perhaps even with limited resources, potentially allowing them to be recognized as *laomo* (model workers).

One purpose of the Cultural Revolution was to train intellectuals (including most engineers and engineering teachers) to be proletarian intellectuals of the working class. During the Maoist period, political "redness" was increasingly prioritized, and particularly so in engineering education and other technical fields. "Red and expert" became a guiding hallmark for engineering education. As observed by Zhidong Hao,

> in the Mao era, efforts at creating a professional stratum were developed along the lines of "red and expert", and intellectuals did not achieve much autonomy. Rather, they were deprofessionalized. Intellectuals had to conform to the Maoist ideology and serve the Party's political goals. Even in their own technical fields, intellectuals were constrained by the Party objectives. (Hao 2003, p. 228)

Since Mao's era, "red and expert" has become a paradigm ensuring the political quality of engineering education and practice. Even today in both engineering schools and large industrial companies, there remains a two-track supervising system: Party committee and administrative organization. Party secretaries, whether or not with a professional background, often oversee key issues and policies, making final decisions about what can and cannot be done. Meanwhile, normally the head of the administrative organization (e.g., university President) is a member of the Party committee. In this sense, the purpose of "red and expert" is to ensure that intellectuals hold the right political direction. Andreas' (2009) history of Tsinghua University powerfully illustrates how these trends manifested in the historical development of China's most prestigious engineering school.

Although professional societies of engineers did exist prior to 1949, engineering associations established since that time have taken the form of "expertise organizations" (e.g., China Civil Engineering Society) rather than true professional organizations in the Western sense. These organizations are governed by the Ministry of Civil Affairs and other governmental departments (e.g., the Ministry of Housing and Urban-rural Development) and organizations (e.g., the Chinese Association of Science and Technology). Under the direct leadership of the Party, engineers are expected to embrace the core values of socialist ethics.

The "red and expert" idea still has broad relevance to engineering curricula in China. In engineering schools, the study of Marxist ideology is required in both undergraduate and graduate curricula. Guan (2012), for instance, has comprehensively reviewed the historical and current development of ideological education in Chinese universities, including engineering schools. Her study describes how ongoing efforts in ideological curricular reform have helped consolidate the dominant role of Marxist ideology and promote the education of engineering students as "red experts."

Economic Pragmatism: Modernization and Engineering Citizenship

In contemporary China, economic pragmatism has become the dominant strain of thought guiding social construction activities in which engineering practice and education are indispensable components. In contrast to philosophical pragmatism, economic pragmatism was mainly advocated by Deng Xiaoping in the late 1970s and early 1980s. And while there is no clear evidence relating Deweyan philosophical pragmatism to Deng's economic pragmatism, Chang notes that "the pragmatic approach of Deng Xiaoping signaled a significant step toward 'concrete problems' and toward Deweyan experimentalism" (Chang 2002, p. 61).

In Deng's economic pragmatism, technology and engineering are viewed as tools of modernization. One major initiative representing such ideas centers on the "Four Modernizations" (modernizations of agriculture, industry, national defense, science and technology). Although the Four Modernizations were first explicitly promoted by Zhou Enlai in 1963, the concept came to be widely viewed as the "brainchild" of Deng Xiaoping (Englesberg 1995, p. 100). The Four Modernizations initiative was adopted as a means of rejuvenating China's economy in the post-Mao era and was one of the defining features of Deng's tenure as the Communist leader.

In contrast to a more classical Marxist understanding of technology as a productive force, Deng further emphasized that "science and technology constitute the *first* productive force." Hence science and technology took the leading role in the Four Modernizations since they themselves independently represented one modernization and also played decisive and influential roles in the other three strands of modernization (agriculture, industry, and national defense).

In comparison with philosophical pragmatism, Deng's economic pragmatism was more like instrumentalism or what might even be called entrepreneurial "innovationism," since any economic activity is always a market experiment. As in Deng's famous "cat theory": "It does not matter whether a cat is white or black, as long as it catches mice." Unlike Mao, Deng was not especially worried about whether an activity was capitalist or socialist so long as it improved the economy. As Deng also said, "Poverty is not socialism, to be rich is glorious." Such an instrumentalist view led to the experimental creation of the four "special economic zones" (SEZs) in order to pursue "socialism with Chinese characteristics." In the SEZs, as Deng also stated, "Without the high-speed development of science and technology, there is no high-speed development of national economy" (Deng 1983, p. 86). Thus, as a tool of modernization, technology (including engineering), played a prominent role, thereby distinguishing Deng's philosophy of national development from Mao's.

Deng's economic pragmatism and instrumentalist view of engineering and technology has been continued by PRC presidents Jiang Zemin (1993–2003), Hu Jintao (2003–2013), and Xi Jinping (2013–present). Economic pragmatism has become a ruling ideology with significant impacts on nearly every aspect of socialist construction, including education. Since Deng's era, engineers have been aligned with this ideology by promoting their ability to support economic development, including by directly serving the needs of industry and leading innovation. Take for example the 10-year national "excellent engineer education and training initiative" launched by the Ministry of Education and other government agencies. To begin, the first of three kinds of engineers this initiative seeks to educate is the so-called *xianchang gongchengshi* (field engineer). It also mandates that engineering schools work closely with industry to tailor engineering graduates who can "seamlessly" serve industry (Ministry of Education 2011a). Other initiatives, such as the Ministry of Education's 2006 "National University Student Innovation Program," aim to train innovative engineers who can contribute to the pragmatic goal of increasing the economic and technological competitiveness of Chinese firms on the world stage.

Since the 1980s, the aim of educating innovative and practical engineers has focused on the rejuvenation of China after the turmoil of the Cultural Revolution (1966–1976). Inspired by economic pragmatism and emphasizing technology as contributing to economic development, engineers have gained higher social status and greater autonomy and respect (Miller 1996). Nationalism has thus become a centrally important part of engineering education.

Yet perhaps this is not surprising. As a growing body of scholarship reveals, engineers and engineering are often tightly linked to prevailing notions of national progress (e.g., Downey and Lucena 2004). As Downey et al. summarize, global engineers must therefore recognize how "dominant ideas of national progress... have played a key role in shaping dominant patterns of engineers and engineering" in different country contexts (Downey et al. 2006, pp. 113–114). The present account thus builds on previous discussions of engineering and national development in the Chinese context, as in Jesiek and Shen's (2012) study of engineering education in China during the Nationalist period. This "national ethic" evident in multiple historical periods in China can be viewed as a kind of "engineering citizenship," in that engineers have responded to and largely upheld an obligation to orient their professional expertise and engineering thinking toward national development goals and projects. This kind of "engineering citizenship" can be viewed in three ways.

First, engineers serve as *state leaders*. Most Chinese state leaders in the post-Mao era were originally trained as engineers. In fact, during the early twentieth century, Deng spent some of his formative years studying engineering and science in France. He was later named as the "chief engineer" of the reform and opening-up and Chinese modernization. Deng's successor, Jiang Zemin, studied electrical engineering and worked as an engineer for two decades. The third generation leaders President Hu Jintao and Premier Wen Jiabao were respectively trained in hydraulic and geomechanical engineering. And from 2007 to 2012, eight out of China's nine politburo's standing committee members were trained as engineers.

Second, engineers serve as *local government officials*, taking on administrative positions at lower levels of the Chinese government. Unlike in the United States, the responsibility of provincial governors and municipal mayors in China often includes extensive responsibility for technology-based economic development in local areas. They frequently visit technological companies (particularly state-owned) and are required to be familiar with technological and economic concepts. Hence, engineers are preferable for these positions. While few have experience as practicing engineers in industry, these engineering-trained governmental officials often go on to assume roles as "chief engineers" of cities and provinces. In 2013, 13 of 31 provincial governors had engineering degrees or engineering experience, with at least three holding Ph.D. degrees in engineering.

Finally, engineers are viewed as *major contributors in the great rejuvenation*. At the university level, the concept of "engineering citizenship" is interpreted in the way engineers are portrayed as major contributors in the great rejuvenation of China. This interpretation is well embedded in the aforementioned "excellent engineer education and training initiative." In fact, the program aims to educate:

> a large number and types of high-quality engineering and technical personnel having strong innovative abilities and fitting in with the societal development. These engineering and technical personnel are indispensable to establish the solid advantage in human resources for constructing an innovative state as well as achieving industrialization and modernization. They are also indispensable to improve the core competitiveness of the Chinese nation and comprehensive national power. (Ministry of Education 2011a)

Hence, developing "excellent engineering education" appears well justified in the larger context of helping China play an increasingly influential role in world affairs.

Three Controversies Revisited

Table 7.1 summarizes the three intellectual traditions that provide a contextual framework for understanding engineering, the engineer, and engineering ethics. This framework can now be used to revisit the three controversies introduced above.

First, how might we challenge the notion that there are *no engineering ethics in China*? Confucianism, Marxism, and economic pragmatism together offer a fundamental ethical system governing current Chinese engineering practice at both the macro- and micro-ethical levels (as distinguished by Herkert (2001)). At the macro-level, according to Confucianism, a good engineering project must be socially beneficial for the state and its people. And at the micro-level, Chinese engineers fundamentally are guided by communitarian ethical values from Confucianism, including relational virtues such as *ren* (benevolence) and *li* (ritual). In the workplace, and because of the virtue of *ren*, Chinese engineers are not likely to publicly criticize their peers or even inferiors. Because of the virtue of *li*, engineers are not encouraged to criticize their superiors. Enculturation into this ethos begins in engineering education. Further, a lack of mechanisms to protect engineers from being punished by their superiors means "whistleblowing" is even less common in the Chinese than Western context. Marxist ideology also plays a role in engineering. Political redness is no less important than professional expertise – and sometimes political redness serves as an evaluative condition for engineers and their work. Finally, according to economic pragmatism, engineers are educated with a nationalist ethic of "engineering citizenship," holding that the education of engineers should serve the ends of national development.

Second, why does *engineering as a profession have a different character in China?* The three traditions all uphold some form of communitarian ethics, thereby countering the individualistic ethics at the heart of most Western views of engineering as a profession. The three traditions also require engineers to be linked with larger communities (family/relatives, society and the world in Confucianism, the Communist Party in Marxism, and the Chinese state in economic pragmatism). Thus, engineers do have certain kinds of responsibility toward other members of their society. Although engineering in China does not look like a profession in the Western sense, commitments by engineers to more expansive values than simple bottom-line profit related to the three intellectual traditions may help Chinese engineers make "professional" judgments, and hence could be viewed as constituting an important kind of professionalism in the Chinese context.

Table 7.1 Three Chinese intellectual traditions

Tradition	Engineering as	Engineer as	Engineering ethics as
Confucianism	Sociopolitical practicality	Political leader	Communitarian ethics
Marxism	Productive force	Socialist laborer	Ideological redness
Economic pragmatism	Means for modernization	Pragmatic engineer	Engineering citizenship

Third, *who are the Chinese engineers?* Historically, the predecessors of engineers were not artisans. When compared with the roles played by modern engineers, it would be more appropriate to see certain Confucian government officials or technical bureaucrats, rather than artisans, as the main predecessors of engineers. In the Marxist context, contemporary Chinese engineers can also be viewed as socialist laborers, reflecting an ideological tradition of seeing engineers as linked to the working class. Finally, influenced by economic pragmatism, engineers are encouraged to be innovative and practical in addressing issues related to national development and global competition. Indeed, Chinese engineers are to some extent comparable to the long history of "state engineers" in France, where the "best French engineering schools have traditionally sent their graduates directly into state bureaucracies" (Baumgartner and Wilsford 1994, p. 71). As in France, Chinese engineer-trained-officials may also see engineering as a way of thinking or practical instrument for administering local governments and managing issues of economic development, rather than as a technological tool for solving specific problems. However, engineering is a "technical title" in China, like university professor. Through national examinations, even technicians with enough years of practical experience and who pass examinations can be promoted as engineers.

Re-considering a Western Analysis of Engineering in China

Reacting to the large and growing influence of China in the global context, many scholars have interpreted Chinese engineering for American audiences. To further highlight the implications of our account, here our framework is used to re-examine an influential example of this type of scholarship by Wadhwa et al. (2007) which examines current trends in engineering training in the U.S., India, and China.

As these authors rightly point out, the word "engineer" has varying definitions across countries. They further argue that the engineering graduate numbers gathered from the Chinese Ministry of Education are suspect because Ministry reports include "'short-cycle' degrees typically completed in 2 or 3 years … (which are) equivalent to associate degrees in the United States" (Wadhwa et al. 2007, p. 74). However, they fail to explain that the engineering students graduating from 2- or 3-year programs might better be viewed as "potential engineers" rather than fully trained or fully qualified engineers. According to Confucian principles of egalitarianism and social mobility, anyone is able on merit to move to a higher position in society. For instance, the Chinese government allows technicians graduating from 3-year programs and accumulating more than 2 years technical experience to be promoted to the status of "assistant engineers" if they pass the governmental professional-title evaluation (National People's Congress 2000, p. 1642). This policy is also linked to continuing engineering education in China, which is considered an important part of the larger engineering education system.

Wadhwa and colleagues also argue that "the Soviet development model led Chinese administrators to attach the term 'engineering' to many institutions and programs that had science- and technology-related, but not necessarily pure engineering content" (Gereffi et al. 2008, p. 15). However the authors do not clearly indicate which institutions and programs do not teach "pure" engineering content. Conversely, they fail to note that some programs having "science and technology" in their names are actually focused on educating engineers. For instance, in the "excellent engineer education and training initiative", programs such as "electronic information science and technology", "armament science and technology", and "measuring and controlling technology" are included as engineering programs (Ministry of Education 2011b). This is due to the pragmatic and broad understanding of technology in the Chinese context which views engineering as a *particular* technology, and not because the government mistakenly treats technology programs (as they are called in the U.S.) as engineering programs. Further evidence for this can be found in the remarks of Zhang Guangdou, a former Vice President of Tsinghua University and distinguished member of the Chinese Academy of Engineering who states that "higher engineering education is a technological education" (Zhang and Wang 1995, p. 29). Zhang must clearly know the Western definition of engineering since he graduated from Harvard University. Yet in Wadhwa et al. (2007), it is also argued that the data from the Chinese Ministry of Education includes "specialized fields such as shipbuilding" as engineering programs. However, it is not clear whether there really is any problem with the translation of the program's name since "marine engineering" in Chinese literally means the art of building ships, although it is treated as a subfield of engineering.

Conclusion

In sum, this chapter argues that awareness of historical-cultural contexts is crucially important for understanding engineering in cross-national perspective. Intellectual traditions in different cultures are particularly important for engineers working in cross-cultural and cross-national settings, and awareness and sensitivity to such differences could be actively taught in engineering degree programs.

More specifically, one of the most important practical implications is that awareness of the three foundational philosophies (Confucianism, Marxism, and economic pragmatism) can improve the ability of non-Chinese engineers to work effectively with Chinese colleagues. In comparison with the other two philosophies, Confucianism is more of a historical and fundamental contextual consideration defining the everyday culture of engineering practice. It also shapes the social values and the human communications within the activities of technical coordination. Marxism serves as more of an ideological context that mainly influences the ethics, standards, and regulations, engineering practice in China. Socialist values are also embedded in many engineering and development policies. Non-Chinese engineers need to well understand these policies, especially if they want to effectively work with state-owned companies. Hence, policies and policymaking in the Chinese context cannot simply be interpreted through the "native" lenses of non-Chinese engineers. Finally,

Dengist economic pragmatism has fundamental implications for understanding Chinese engineering culture, including by suggesting that Chinese engineers may define and solve technical problems in their own unique ways. It also helps non-Chinese engineers understand some of the "pragmatic parameters" in technical problems that engineers might most care about in China's fast developing economy.

In addition to the practical implications for global engineering education, this chapter has implications for global comparative studies of professions. Most such studies are mainly focused on historical events and figures. Yet to better understand the *hermeneutical meanings* of these events and figures, this chapter argues that a more fundamental *philosophical/cultural* dimension needs to complement the *historical* dimension. More specifically, regarding the comparative studies of engineering, interested scholars need to at least understand to what extent or at what level meaningful comparisons can be made among the key concepts, beliefs, and issues related to engineering practice and education in different countries. In other words, better understanding the similarities and dissimilarities of the historical events and figures across boundaries requires that one must understand what these historical events and figures mean in the contexts where they originally emerged.

References

ABET Engineering Accreditation Commission. (2008). *2009–2010 criteria for accrediting engineering programs*. Baltimore: ABET Inc.

Andreas, J. (2009). *Rise of the red engineers: The cultural revolution and the origins of China's new class*. Palo Alto: Stanford University Press.

Barbieri-Low, A. J. (2007). *Artisans in early imperial China*. Seattle: University of Washington Press.

Baumgartner, F., & Wilsford, D. (1994). France: Science within the state. In E. Solingen (Ed.), *Scientists and the state: Domestic structures and the international context*. Ann Arbor: University of Michigan Press.

Carroll, J. (2008). Ho Kai: A Chinese reformer in colonial Hong Kong. In K. J. Hammond & K. Stapleton (Eds.), *The human tradition in modern China* (pp. 55–73). Lanham: Rowman & Littlefield Publishers.

Chang, F. (2002). The problem of the public: John Dewey's theory of communication and its influence on modern Chinese communication. In X. Lu, W. Jia, & D. R. Helsey (Eds.), *Chinese communication studies: Contexts and comparisons* (pp. 47–56). Westport: Ablex Publishing.

Chen, C. (2001). *Ziran bianzhengfa daolun xinbian* [The new introduction to the dialectics of nature]. Shenyang: Dongbei daxue chubanshe [Northeastern University Press].

Davis, M. (2009). Defining engineering from Chicago to Shantou. *The Monist, 92*(3), 325–338.

De Bary, W. T., & Bloom, I. (1999). *Source of Chinese tradition* (2nd ed.). New York: Columbia University Press.

Deng, X. (1983). *Deng Xiaoping wenxuan* [Selected works of Deng Xiaoping]. Beijing: People's Press.

Downey, G. Q., & Lucena, J. (2004). Knowledge and professional identity in engineering: Code – Switching and the metrics of progress. *History and Technology: An International Journal, 20*(4), 393–420.

Downey, G., Lucena, J., Moskal, B., Parkhurst, R., Bigley, T., Hays, C., Jesiek, B., Kelly, L., Miller, J., Ruff, S., Lehr, J., & Nichols-Belo, A. (2006). The globally competent engineer:

Working effectively with people who define problems differently. *Journal of Engineering Education, 95*(2), 107–122.
Englesberg, P. (1995). Reversing China's brain drain: The study-abroad policy, 1978–1993. In J. D. Montgomery & D. A. Rondinelli (Eds.), *Great policies: Strategic innovations in Asia and the Pacific Basin* (pp. 99–122). Westport: Praeger.
Gereffi, G., Wadhwa, V., Rissing, B., & Ong, R. (2008). Getting the numbers right: International engineering education in the United States, China, and India. *Journal of Engineering Education, 97*(1), 13–26.
Guan, L. (2012). *Zhongmei daxue sixiang zhengzhi jiaoyu bijiao yanjiu* [Comparative Studies on ideological and political education in universities between China and the U.S.] (Unpublished doctoral dissertation). Dalian: Dalian ligong daxue [Dalian University of Technology].
Guo, F. (2009). *The absence of engineering ethics in China and its solutions: An STS perspective.* Paper presented at Science and Technology in Society Conference, Washington, DC, March 28–29.
Hao, Z. (2003). *Intellectuals at a crossroads: The changing politics of China's knowledge workers.* Albany: SUNY Press.
Harvey, B. (2004). *China's space program: From conception to manned spaceflight.* New York: Springer.
Hatmaker, D. M. (2012). Practicing engineers: Professional identity construction through rolc configuration. *Engineering Studies, 4*(2), 121–144.
Herkert, J. (2001). Future directions in engineering ethics research: Microethics, macroethics and the role of professional societies. *Science and Engineering Ethics, 7*(3), 403–414.
Ivanhoe, P. J., & Van Norden, B. W. (Eds.). (2001). *Readings in classical chinese philosophy* (2nd ed.). Indianapolis: Hackett Publishing Company.
Jesiek, B. K., & Shen, Y. (2012). Educating Chinese engineers: The case of Shanghai Jiao Tong University during 1896–1949. In S. H. Christensen, C. Mitcham, B. Li, & Y. An (Eds.), *Engineering, development, and philosophy: American, Chinese and European perspectives* (pp. 123–143). Dordrecht/Heidelberg/New York/London: Springer.
Johri, A., & Jesiek, B. K. (2014). Global and international issues in engineering education. In A. Johri & B. Olds (Eds.), *Cambridge handbook of engineering education research.* Cambridge: Cambridge University Press.
Joseph, W. A. (2010). *Politics in China: An introduction.* New York: Oxford University Press.
Li, B. (2006). Guanyu gongchengshi de jige wenti [Some Issues of Engineers], *Ziran bianzhengfa tongxun* [Journal of Dialectics of Nature], *28*(2), 45–51.
Liu, L. (2012). *Red genesis: The human normal school and the creation of Chinese communism, 1903–1921.* Albany: SUNY Press.
Lorimer, D. (1999). *Fundamentals of historical materialism: The marxist view of history and politics.* Chippendale: Resistance Books.
MacFarquhar, R. (1997). *The politics of China: The eras of Mao and Deng* (2nd ed.). New York: Cambridge University Press.
Mao, Z. (1940). Zai shanganning bianqu ziran kexue yanjiuhui chengli dahui shang de jianghua [Speech at the inaugural meeting of the Society for Natural Science Research in the border area]. Xinhua ribao [The New China News], March 15.
Miller, H. L. (1996). *Science and dissent in post-Mao China.* Seattle: University of Washington Press.
Ministry of Education. (2011a). Several opinions on the implement of the excellent engineer education and training program. Retrieved from http://www.moe.gov.cn/publicfiles/business/htmlfiles/moe/s3860/201102/115066.html
Ministry of Education. (2011b). The announcement on the list of the programs in the 2011 excellent engineer education and training initiative. China Education and Research Network. Retrieved from http://www.edu.cn/zong_he_862/20110811/t20110811_663946.shtml
National Academy of Engineering. (2004). *The engineer of 2020: Visions of engineering in the new century.* Washington, DC: National Academies Press.

National People's Congress. (2000). *The overview of the educational laws and regulations of the People's Republic of China (1949–1999)* (Vol. 2). Beijing: Law Press.
Rae, J., & Volti, R. (1993). *The engineer in history*. New York: Peter Lang Publishing.
Shapiro, J. (2001). *Mao's war against nature: Politics and the environment in revolutionary China*. New York: Cambridge University Press.
Shun, K.-l., & Wong, D. B. (Eds.). (2004). *Confucian ethics: A comparative study of self, autonomy, and community*. New York: Cambridge University Press.
Song, Y.-b. (2002). Crisis of cultural identity in East Asia: On the meaning of confucian ethics in the age of globalization. *Asian Philosophy, 12*(2), 109–125.
Su, J., & Cao, N. (2008). [Studies of the evolution of ethical considerations of Chinese engineers: Based on a historical inquiry into the revisions of the ethics code of the Chinese Institute of Engineers], 1933–1996. *Ziran bianzhengfa tongxun* [Journal of Dialectics of Nature], *30*(6), 14–19.
Wadhwa, V., Gereffi, G., Rising, B., & Ong, R. (2007). Where the engineers are. *Issues in Science and Technology, 23*(2), 73–84. Retrieved from http://www.issues.org/23.3/wadhwa.html.
Wang, J., Yan, R., Hollister, K., & Zhu, D. (2008). A historic review of management science research in China. *Omega, 36*(6), 919–932.
Wong, D. (2008). *Chinese ethics*. In E. N. Zalta (Ed.), *The Stanford encyclopedia of philosophy*. Retrieved from http://plato.stanford.edu/archives/spr2013/entries/ethics-chinese/
Wong, J., & Zheng, Y. (2001). *The nanxun legacy and China's development in the post-Deng Era*. Singapore: Singapore University Press.
Zhang, G., & Wang J. (1995). *Zhongguo gaodeng gongcheng jiaoyu* [Higher Engineering Education in China]. Beijing: Qinghua daxue chubanshe [Tsinghua University Press].
Zhu, Q. (2010). Engineering ethics studies in China: dialogue between traditionalism and modernism. *Engineering Studies, 2*(2), 85–107.

Qin Zhu 朱勤. Ph.D. Student in the School of Engineering Education at Purdue University. He was a Postdoctoral Visiting Scholar in the Hennebach Program in the Humanities at Colorado School of Mines in 2012. Qin received his Ph.D. in Engineering Ethics (2011) and his B.S. degree in Materials Processing and Control Engineering (2005) both from Dalian University of Technology. His main research interests include global engineering education, engineering ethics, STS, Chinese philosophy and technology, and East Asian studies.

Brent K. Jesiek B.S. Electrical Engineering, Michigan Tech, M.S. and Ph.D. in STS, Virginia Tech. Associate Professor, School of Engineering Education and School of Electrical and Computer Engineering, Purdue University; Associate Director of Global Engineering Program, Purdue University. Research interests: historical and social studies of engineering, engineering education, and computing; engineering epistemologies; global engineering education; and cyber infrastructures for engineering education research.

Chapter 8
Meritocracy, Technocracy, Democracy: Understandings of Racial and Gender Equity in American Engineering Education

Amy E. Slaton

Abstract The idea that technological labor produces both individual security and satisfaction *and* societal benefits has shaped engineering education in the United States since its inception. Educators and employers have historically cast engineering instruction as a route towards individual and collective uplift for the nation's citizens. But ideologies of racial, gender, and other categories of difference predicated on identity underlie all such claims and explain the less-than-democratic character of STEM occupations, in which minority citizens, women, LGBT persons and persons with disabilities remain under-represented despite decades-old legal proscriptions against such discrimination. This chapter explores two linked logics that perpetuate this inequitable distribution of opportunities: the technocratic understanding of engineering as an enterprise in which power relations play no part; and the related construction of engineering education as a field based solely on meritocratic judgments about eligibility and skill. Through both of these formulations American engineering supports the ongoing exclusion of certain communities based on perceived heritage and ascriptions of potential in turn based on those identities. This chapter also frames a recent strengthening of these ideologies under emergent neoliberal understandings of market, state, and the agency of individual citizens-as-learners. Finally, given the origins of engineering knowledge and practice in discriminatory social relations, this chapter asks whether improved diversity in engineering would in fact represent a liberatory change.

Keywords Engineering education • Educational standards • STEM • Laboratory instruments • Technocracy • Merit • Neoliberalism • Race • Gender • Disability • LGBT identities

A.E. Slaton (✉)
Department of History and Politics, Drexel University, 19104 Philadelphia, PA, USA
e-mail: slatonae@drexel.edu

S.H. Christensen et al. (eds.), *International Perspectives on Engineering Education*, Philosophy of Engineering and Technology 20, DOI 10.1007/978-3-319-16169-3_8

Introduction

In the United States, engineers have historically seen themselves as civic leaders, deploying the empiricism and practicality of their occupation for what are ostensibly societal benefits. From the mid-nineteenth century onward, American engineers, increasingly identified with professional organizations and formal educational systems, routinely spoke in technocratic terms of the improved public health, industrial productivity, and material developments rendered to the nation by their work. As is still often claimed today in mission statements and textbooks, keynote speeches and lists of engineering "grand challenges," the expertise of engineers serves collective aims (Kline 1995; Slaton 2012; Pfatteicher 2005). In the vast majority of such instances of self-description, however, the discipline of engineering has foreclosed the sort of reflexivity that would lead to authentic inquiry about such aims. That is, we rarely encounter frank inquiry about who in American society it is that actually benefits from the ingenuity and labor of engineers. The institutions for which engineers work—corporate, governmental, or military—are virtually never depicted in that occupation's promotional literature, let alone in technical documents, as forwarding existing structures of economic or political privilege.

The question of "for whom" engineering is practiced raises the correlate question of "by whom" it is practiced: Historically, who has become an engineer in America? Persons of what races, genders, ages, credentials, or family and institutional connections? Who in turn has not appeared among the rolls of university engineering students or faculty, or among the technical employees of industry, military or government sectors? Again: Of what race, gender, etc. are these "absent" persons? When we pose these queries, another register of question quickly arises: Which such categories have determined participation not just in engineering, but in U.S. culture more generally? And still another: What features of body or conduct or origin have delineated a person as being in each category—as a white, black or Asian person, a male or female person, a clever or slow, young or old, able-bodied or disabled person? With these questions, a raft of highly contingent social conditions busily shaping the demographics of U.S. engineering suddenly become newly visible.

At a certain level matters of inclusion have routinely received address by employers and educators concerned with diversity in technical occupations. In the United States, discriminatory habits that had historically excluded women and members of ethnic minorities from science and engineering disciplines faced powerful legal challenges beginning in the 1950s, and a wide range of educational and hiring initiatives have increased the diversity of science, technology, engineering and mathematics, or STEM, fields in the succeeding decades. Recruiting, scholarship, mentoring, and related programs have targeted groups of Americans traditionally excluded from engineering degree programs and employment opportunities (Slaton 2010).[1]

Yet, this inclusive programming has somehow left regrettable demographic patterns largely intact. Women are not proportionately represented in most STEM educational and work settings, especially in the higher reaches of universities, government agencies, and corporations. Persons of African American, Hispanic and Native American backgrounds; those of lower socioeconomic status; and persons with physical or intellectual disabilities are still drastically underrepresented across American engineering (Bayer Corporation 2011). LGBT individuals, too, face tremendous impediments to full participation in STEM occupations (Cech and Waidzunas 2011). Thus, questions about who is and is not likely to become an engineer in America, and about whose interests may be represented in the day-to-day practices of engineering occupations, can be seen as having been only tentatively confronted by most of those concerned with STEM diversity. Certainly the idea that the constituent elements of engineering, including its educational structures, epistemologies, and patronage networks, may impede democratic reform by their very nature rarely comes under discussion (Riley et al. 2009, 2014). This chapter frames ideas about how this selective address of inequity has arisen. It suggests steps we might take either to correct these instances of under-representation in engineering fields or alternatively, to reframe entirely our notions of what may constitute a more democratic industrial culture.

Importantly, this essay by no means dismisses efforts at STEM inclusion to date, but approaches those from a critical perspective that asks why such efforts have assumed the constrained form and scale that they have. How do seemingly well-intentioned efforts nonetheless promote a conservative social landscape, generation after generation? Such a perspective, interrogating both institutional and epistemic features of engineering over the last 60 years, finds that an uncritical belief in the meritocratic nature of engineering has delimited inclusive efforts in American engineering education since the civil rights movement. In the policy and curricular reforms intended to improve STEM diversity, we find an insistence on established "standards" of engineering performance. Even as college admissions or hiring priorities in engineering fields have been scrutinized for their discriminatory impacts, and in many instances reformed, unchanging presumptions about what constitutes "good" engineering in classroom, lab or factory serve to preserve exclusionary patterns of eligibility (Riley et al. 2014). Such unexamined commitments are common

[1] This essay builds on historical narratives included in my recent book on the whiteness of engineering in the United States (Slaton 2010). Here, I add two important analytical frameworks to those narratives. First, I begin to engage with intersectionality, such as the role of engineering's intertwined epistemic engagements with race, gender, disability and LGBT identities. Second, I articulate the significance of neoliberal ideologies that center on the primacy of market forces and thus camouflage social structural sources of occupational inequity in the Unites States.

in rhetoric regarding educational uplift in the twenty-first century (see, for example, Goldberg and Traiman 2001, p. 92) and deny the social foundations of technical knowledge and practice—of engineering epistemologies. Thus arises support for only a limited intervention in the prevailing social structures of engineering. For example, such standards largely foreclose the idea that students from poorer communities might appropriately spend an extra year in college to offset weaker high school math and science provisions; or that a minority-serving university might usefully receive added time and money from a federal funding agency to bring historically under-resourced engineering research programs up to competitive levels (Slaton 2010).

It is not only opponents of affirmative action who stress the necessity of preserving established expressions of rigor and selectivity in engineering, but also many advocates of enhanced minority inclusion. This aim is sometimes made explicit, as when STEM proponents defend the rigor of inclusive programming, but more often enacted through a strategic absence of critical inquiry into the content of engineering. Even many staunch advocates of enhanced STEM diversity in the United States assume that engineering need not change the pacing and content of its educational programs; its criteria for or amounts of research or scholarship funding; or, crucially, its level of reflexivity regarding these matters. But as this paper will explore, it is in those exact locations that occupational advantage and disadvantage reside.

The social functions of merit and the received criteria for meritorious engineering are inseparable. This is a linkage that today arises from and serves much broader neoliberal conceptions about the sources of economic welfare under industrial capitalism: The individual holds responsibility for his or her standing in a free market system ostensibly guaranteed to reward effort. The citizen must learn and work in ways that accord with prevailing definitions of valuable knowledge and labor, or face the consequences with only her- or himself to blame (Brown 2006; Walkerdine 2003; Wacquant 2009; Slaughter and Rhoades 2004; Gershon 2011).

The United States is of course not alone in undertaking this particular, modernized construction of productive citizenship. Globally, industrial capitalism brings stratified occupational structures and wage labor to more and more communities with each passing month. The conditions that today enable economic participation for individuals and polities worldwide, and certainly those surrounding labor in "advanced" sectors such as engineering, conduce to this stress on individual responsibility within state and corporate institutions. No doubt comparative cases will enrich this project. But the racial and gender ideologies underlying American notions of optimized productivity and of fairness are pronounced. These notions deserve careful investigation especially in light of decades-old claims of fully enlightened, legally protected civil rights in the nation. In this climate, privilege is conserved and the conservative is privileged, as this look at U.S. engineering education in the twenty-first century hopes to make clear. With that finding comes the unavoidable, disconcerting, but crucial second order concern of this paper: Should we even continue to pursue the entry of currently under-represented groups into STEM disciplines? Or should we instead cease to see such entry as having liberatory potential?

The Instrumentality of Technical Merit: Linking Content and Exclusion

A small but insistent body of scholarship over the last few years in the fields of race, gender, LGBT and disabilities studies, and in science and engineering studies, has shown that prejudicial and exclusionary treatment in STEM fields routinely occurs on the basis of perceived identity (Cech and Waidzunas 2011; Tonso 1996; Riley et al. 2009, 2014). This treatment sometimes takes the form of direct encounters between privileged and marginalized persons, either through blunt declamations of difference and ineligibility by those in authority or through the more subtle but equally damaging maintenance of an overall "chilly climate" for women, minorities, queer persons, or persons with physical or intellectual differences. Legal reforms have not done away with such discriminatory practices (Cech and Waidzunas 2011; Siebers 2010; Slaton 2010). But discouraging or exclusionary conditions are also perpetrated in engineering classrooms and workplaces through less direct expressions of privilege. Importantly, constructions of *positive* characteristics in some engineers and engineering students have rendered other persons ineligible for participation or success in the field. Traditionally in the United States these desirable traits would be whiteness, maleness, heterosexual identity, and whatever is seen to be bodily normalcy. As has historically been the case in many modern professions, the trusted practitioner in STEM occupations is often one with a particular set of ascribed identities. The veneration of objectivity and the suppression of "extra-occupational" personal attributes in the course of cognitive labor play particularly important roles in constituting professionalism in technical occupations. Claims of empiricism notwithstanding, in science and engineering the validity of findings at the bench derives from the experimenter, not the experiment; the reliability of a building material or industrial product is determined by the tester, not the test (Shapin 1989; Schaffer 1988, 1995; Traweek 1992; Knorr-Cetina 1995; Slaton 2001; Pang 2002).

To highlight the pattern by which notions of meritorious practice follow from ascriptions of eligibility, and not the other way around, I have elsewhere described cases in which ideas about whomever was undertaking an engineering task configured ideas about the validity of the work being done. In the 1970s and 1980s, as academic programs for the correction of black, Hispanic and Native American under-representation in U.S. engineering took form, many university engineering departments recognized that minority identity often went along with attendance at under-resourced and under-performing public high schools. The resulting shortfalls in math and science readiness for minority students were well understood, but at the same time the provision of resources and coursework that might have achieved parity across educational systems remained unimaginable to most educators and policy makers. Through arbitrary restrictions on the types and amounts of minority-focused post-secondary STEM programming provided, the disadvantaged student was cast as irredeemable in certain practical ways.

For example, in the majority of university STEM departments concerned with minority inclusion from about 1970 onward, the provision of remedial coursework was largely deemed to be ill advised. Many university engineering departments felt that to undertake "compensatory education" of this kind would be to lower their standards for both program admission and completion. By contrast, individual tutoring and dedicated social support systems for under-represented student groups (such as dormitories or classes earmarked for minority or women students), and brief preparatory or "bridge" courses such as weekend or summer classes for entering minority students, evaded the stigma of remediation for a department. Similarly welcome were small-scale programs that ferreted out the few so-called talented students among the many presumably untalented that made up poorer urban and rural school systems (Slaton 2010).

All of these sorts of programs found support from corporate and philanthropic sources, as they still do today, and inclusive efforts have clearly had the imprimatur of the professional worlds in which STEM finds its applications in the United States. That American industry consistently puts some value on racial and gender diversity, a point to which I return below, is without question (Holvino and Kamp 2009; Gordon 1995). Yet, these academic interventions into inequity—small, brief, and staged outside the spaces and calendars of "normal" instruction—have since their inception been distinguished in both form and content from the main body of pedagogical activity in engineering schools, creating a cordon sanitaire that could deflect perceived threats to institutional reputation.[2] With a few exceptions, institutions have felt that altering the structure of existing engineering curricula or offering divergent paths towards graduation for students of different backgrounds or inclination could mark a school as having lower caliber students and graduates (Slaton 2010). But in no cases I have found have opponents of remedial work made clear how it was that a practicing engineer whose training included, say, a set of math classes prior to or beyond the standard curriculum, would necessarily fall short of conventional skill levels. What would be missing or flawed in the resultant practitioner is nowhere articulated, any more than American critics of black participation in higher education prior to the civil rights era supplied thoroughly argued reasons for race-based exclusion from educational opportunities (Gurin and Epps 1975).

This is instead an arbitrary ascription of low potential to certain populations that has arguably followed from ideologies of class and race difference. Educational deficits in the United States historically map onto socioeconomic status, and to ascribe some inherent lack of intellectual talent to those living in communities with weak public high schools is to make a leap of logic (Brint and Karabel 1991; Hursh 2006; Ebeling and Slaton 2010). It is also an ascription that follows familiar notions of racial difference. Since the first stirrings of emancipation there have been influential countervailing voices in America insisting that to be born of particular heritage (racial, ethnic or gender minority) is to lack intellectual capacity (Duster 2003).

[2]Academic time and space, as described by Vinao, are "never neutral," but rather expressed ideals of optimally ordered and sequenced experiences (2001, pp. 133–135).

The association of identity with cognitive or moral capacity of course extends to many cultures in many periods (Carson 2006; Gould 1996), but in thinking about engineering education we find a particularly effective recourse to ideas of rigor among those who wish to confer or limit eligibility for participation by certain populations (Slaton 2010). Currently, scholarship on disability helps us see similar social instrumentalities related to bodily difference. For example, when a blind student in a university chemical engineering class suggested in 2012 that she might use an "assistive technology" that converted the visual read-out of a probe to audio signals, she was told by her professor that the resulting data would not be equivalent. Yet, the professor could identify no feature of the audio output that contravened the meaning to be derived from conventional visual readout used for this experiment; the audio signal was different in form only, it would appear (Supalo et al. 2007; Bryan 2012).

Of course, the meanings of the visual and audio signals in this case are entirely coincident with authority structures in the teaching lab. Within the purposes defined for this laboratory exercise (the conveyance of what the professor believes "the lesson" to be), there is no significance to signals apart from the professor's notion of where meaning resides. The student is not only a blind person, but an inexpert one who by definition cannot yet understand what signals do and do not "work" in this experiment...The experimenter's regress! Her advocacy for alternative mediums is inadmissible in the class on multiple levels. Historians of science have shown that even where the phenomena ostensibly observed and recorded may not differ among practitioners, the choice of representational convention itself confers or denies status to one practitioner or another (Pang 2002; Daston and Galison 1992; Slaton 2001). At bottom, the blind student's instructor made a determination of what counted as rigorous laboratory practice based on student identity (here, an identity predicated on bodily "otherness"), not on investigative procedures themselves.

The Challenge of the Social Justice Agenda in STEM Fields

When we combine the methods of science studies, which interrogate the meanings of representational conventions in science and the construction of legitimacy and certainty, with those of identity studies to reveal such elisions, it is not hard to see that invocations of rigor perform exclusionary work in STEM fields. However, the question of precisely *how* the work of engineering instantiates racial and gender privilege is extremely complex. There is no simple formula for tracing how the material, economic and political purposes towards which engineering knowledge and labor are put actually create new inequities or further existing ones. This difficulty reflects the naturalization of two cultural conditions in the United States: discriminatory ideologies in post-segregation America and the historical reputation of the sciences writ large (including social sciences) as value-free practices. I am not writing from outside either cultural condition, obviously. But I do write with the aim of criticality as far as my training within the academy, directed towards the study of power in knowledge systems (including my own), will allow. In other words: Reflexivity about our own history or social science techniques may highlight good reasons for aggregating certain aspects of STEM activity as consequential social practices. This seems like it might offer a step towards understanding how those activities accomplish distributions of power and privilege.

Many historians and social scientists today contend that normativities pervade all technical activity, whether the seemingly isolated task of conducting a compression test on a single specimen of concrete or the construction of an entire interstate system, and the social intentionalities involved in such tasks are many and layered. The very delineation of these activities as occurring at different societal levels, as typifies much existing literature in the history and sociology of engineering, is a politically freighted gesture. After all, the design of a concrete testing machine may reproduce occupational opportunity structures that follow the same lines of majority racial or ethnic advantage reflected in highway planning (Slaton 2001). My training as a historian and STS scholar and participant in emerging Engineering Studies networks has created the possibility of my belief in these contentions, at least. That such a claim does not translate meaningfully into settings where STEM content is taught and deployed, except in extraordinary cases (Riley et al. 2009; Catalano et al. 2008), begins to shed a light on the persistence of discrimination in that sector.

It seems safe to say that few historians would still maintain that human-made artifacts do not have politics, a traditional view of technology for which Langdon Winner offered his corrective some 30 years ago (Winner 1980). Socially inflected historical understandings of industry, centered on labor, have now penetrated many more general narratives of economic development. Similarly, feminist concerns have drawn scholars' attention to reproductive and other medical technologies so that those artifacts no longer seem like the inevitable result of accreting scientific knowledge. Issues of sustainability, public health and safety, and global impacts of industrialization have encouraged still other historians to pay attention to the uncertain social and environmental impacts of engineering.[3] But engineering disciplines rarely engage with any of these analytics, for the most part still tending to firewall concerns about the social impacts and origins of technologies as matters for ethical or regulatory engagement only. But neither of those two framings encourages authentic address of social justice issues. Ethics education readily predicates reform on behavioral changes on the part individual engineers and can too easily default to liability concerns. Instruction in regulatory matters is aimed at enabling compliance, not a critique of structural factors like poverty, racism or global imperialism or the role of such factors in shaping the products and processes of engineering (Catalano 2006; Riley 2008; Little et al. 2008).

A group of engineering educators concerned with social justice have articulated the many ways in which customary engineering instruction stigmatizes that kind of critique, casting it variously as a concern of "do gooders" or simply as something

[3]At the same time, happily, fewer and fewer self-identified historians of technology are using deterministic models of technological history. Multi-causal and value-laden explanations now prevail in articles found in the journal *Technology and Culture*.

outside the purview of "real" engineering (Riley et al. 2009). They have articulated an "ethic of care" in light of which the narrowed and self-serving priorities of corporate, military or state-focused engineering become clear, but it nonetheless remains extraordinarily difficult to posit a significant role for the priorities of non-engineers inside the technical classroom or workplace. These social justice scholars propose that in engineering, "the problems that are 'solved' should be authentic in the context of domination, and ring true in communities with subjugated knowledges." But they realize, too, that such an objective verges on absurdity in the terms normally used to define important engineering learning and research in U.S. institutions (Riley et al. 2009, p. 28).

Once we become aware of the belief that legitimate engineering derives from persons thought by the profession and its patrons a priori to hold the potential to be legitimate engineers, we can begin to pay attention to the problem of essentialism. This notion pervades crude understandings of diversity within engineering that promote inclusion on the basis of the presumed characteristics of different social groups. Such understandings attribute an interest in social issues to women and technical matters to men, or a concern with problems of urban infrastructure to those who live in "degraded" inner city neighborhoods. Wendy Faulkner helpfully makes the point that classifications of behaviors or values along demographic lines are not inherently oppressive; thought or belief distinctions among social groups can be meaningful. But as she puts it, given any combination of men, women, and involvements by both groups with technical tasks, we will encounter *not* "innate differences in technical ability," but rather "some differences in some settings" (2009, p. 148). Corporate diversity strategies by contrast frequently invoke the likelihood that hiring members of under-represented groups will yield untold product innovations and competitive advantage in niche markets that follow demographic lines (Holvino and Kamp 2009; Gordon 1995). This seems to me to be a set of projections indefensible on any basis other than rank essentialism or racial or ethnic stereotypes. To paraphrase Faulkner, a Hispanic product designer and Hispanic consumers may display "some commonalities in some settings," but any more certain association of STEM engagements with heritage is highly problematic.

The problems with such ascriptions go beyond inaccuracy alone. To traffic in this kind of attribution is to reinscribe gender dualities, racial categories, and a host of other potentially oppressive taxonomies. The apparent ambiguity of Faulkner's formulation, "some differences [or commonalities] in some settings," actually leads us to make two important disaggregations here, pushing us away from certainties where they could do the most discriminatory damage. First, the ambiguity forces us to trace connections between life experiences and values held by the individuals under scrutiny, rather than conflate experience and values, or worse still, heritage and experience and values. Second, the contingency of the formulation is a deterrent to the easy sorting of people and actions. It obviously doesn't prevent the most reductive and circular forms of racial or gender classification (which are likely to proceed under any circumstance, since unassailability is their primary function), but it discourages any simple ascription of group identity on the basis of behavior and thereby interrupts attributions of capacity on that basis.

That such attributions are still flourishing in 2015 is apparent. The trope of "missing" and "untapped" technical talent pervades discussions of America's global economic competitiveness beyond corporate diversity schemes; it bolsters a much broader claim of a looming national "skills gap" (National Science Board 2010; Harvard Graduate School of Education 2011; Ebeling and Slaton 2010). That kind of phrasing has always done more to deflect critique of STEM educational structures than enable it, suggesting as it does that there are expert- and worker-shaped holes that can be filled by expert- and worker-shaped people. No need to inquire about the shapes themselves. Again, the explorations now being undertaken in disabilities studies shine a bright light on the essentialist risks here. Recent media reports about the suitability of persons diagnosed with autism for certain STEM careers reinscribe problematic ideas about what constitutes a personality or learning style in need of diagnosis (Cook 2012). Although this new appreciation for the STEM-related talents of some autistic individuals challenges a conventional belief that such a diagnosis mandates treatment or correction, and may offer much-needed economic opportunities to persons with few other job options, this remains a potentially discriminatory situation. On one level, it reflects a narrowly conceived appreciation of technical talent that reifies some technical jobs as appropriately centered on repetitious or tedious labor. Just because we decide that a "type" of person enjoys a "type" of work does not mean that this kind of employment is morally defensible. To presume a "fit" between worker and job here is to ignore many ethical questions about how we understand both (Siebers 2010).

On another level, autism spectrum disorders represent a strongly contested field and here criticality about the label is nowhere to be found; a person's unusually good memory or pronounced affinity for order, repetition, or mathematical reasoning confers an identity of deviant or disabled. Without taking on the full range of epistemic challenges involved in an analysis of disease definition, we can at the very least understand that notions of "natural ability" for STEM labor reproduce eugenic ideologies and deny the existence of structural conditions under which math and science ability are or are not cultivated in individuals (see above). Certainly this kind of claim hurries non-quantitative approaches to design or technical problem solving, and other nontraditional learning or reasoning styles in engineering, rapidly towards the status of subjugated knowledges (Faulkner 2009).

The Primacy of Neoliberal Logics

The understanding among many STEM educators, employers and policy makers in the United States that achievement derives from inborn characteristics may continue a long discriminatory tradition, but its presumptions have been bolstered in the last 20 years or so by the rising influence of market-focused neoliberal ideologies (Rodgers 2011; Hursh 2006; Brown 2006; Gershon 2011). Those ideologies stress market forces as a reliable guide to and result of effective economic planning, and project a particular role for education and training in service to those forces.

American educational policy has long manifest a "vocationalist" outlook that casts education as the answer to social problems. Poverty, unemployment and civil unrest have all been ameliorated, proponents claim, through education for work (Labaree 2008; Grubb and Lazerson 2005; Popkewitz 2006, p. 124). But the stratified nature of labor in America has historically meant that "good education" is that which reproduces unequal educational experiences for different communities or populations. Some Americans receive well-resourced, intensive, and open-ended instruction as they prepare for a wide range of careers and upward mobility; others face an educational experience that is of lower quality, shorter duration, and unlikely to produce secure, rewarding and remunerative career options. Industrial capitalism has naturalized the idea that there must be managers and workers, salaried and wage labor pools and American education reproduces that structure along with attendant, variable levels of security, intellectual reward or remuneration. These are patterns that unsurprisingly follow distributions of economic resources; poorer communities in the United States, disproportionately minority, produce fewer graduates with professional or managerial credentials. Women overall hold lower paid jobs and are paid less than men for the same jobs. Divisions among groups (identified through the arbitrary classifications of race, gender, age, ability etc.) and different life opportunities thus constantly reproduce themselves. In an era in which market forces are granted primacy in social planning, policies which might address structural inequities along lines of race and gender inherent in American education gain little traction (Hursh 2006; Apple 2001).

Along these lines, recent projections of how best to increase participation in STEM disciplines forward the notion that some people are simply innately suited to high-level instruction (through 4-year university degree work or beyond) while others should not be encouraged to attend college (Harvard Graduate School of Education 2011). As noted above, the proliferation of educational standards serves a doubled essentialist purpose: defining talent and locating the talented. The projection of a national need for a scientifically and technically adept workforce entails the construction of a boundary between talented and less-talented persons that is continually delineated but rarely questioned as a project (Popkewitz 2004). Even if that boundary was to be loosened and its social instrumentalities questioned, the nation would still have some way to go towards achieving a more democratic social system. Influential studies have naturalized the absence of African American males of lower socioeconomic standing from 4-year colleges, for example, by predicting their low likelihood of success *based on their identity*. Celebratory language regarding diverse intellects and the contributions such diversity may render to society as a whole does not disguise the discriminatory power of this vision.

A report issued in 2011 by the Graduate School of Education of Harvard University, for example, is apparently dedicated to establishing the value of sub-baccalaureate education for many Americans who might otherwise find themselves pursuing 4-year degrees. The authors seem to support a systematic disavowal of "college for all" ideologies on the basis that young Americans should be offered a "menu of possible selves." The report's title, "Pathways to Prosperity," lends a note of pluralism: The pursuit of a 2-year or vocational credential is shown to be no less admirable (and its rewards, presumably, no less desirable) than attainment of a 4-year or graduate degree. Diverse life goals and talents are to be welcomed; all levels of aspiration and ability are to be cultivated and even celebrated. But tellingly, aspiration and ability are also mapped onto identity in this report; minority background disturbingly figures here as a predictor of lowered educational potential:

> Behaving as though four-year college is the only acceptable route to success clearly still works well for many young adults, especially students fortunate enough to attend highly selective colleges and universities. It also works well for affluent students, who can often draw on family and social connections to find their way in the adult world. But it clearly does not work well for many, especially young men…Similarly, among the low-income and young people of color who will make up an increasing portion of the workforce of the future, this single route does not work well either. (Harvard Graduate School of Education 2011, p. 13)

The degree to which the authors appear to accept racially and socioeconomically determined opportunity structures, both "fortunate" and not, is hardly reassuring if we are seeking to maximize opportunity for all Americans.

The Harvard GSE's frank diagnosis of discrepant educational performances among communities of different socio-economic status might appear to be a starting point for the remedy of such discrepancies. We should note, however, that the mismatch described here between certain individuals and successful prospects in the 4-year college is not cast by the authors as something to eliminate. Rather, they seek to point "low-income and young people of color" away from the pursuit of the baccalaureate degree. That is seen to be an appropriate aspiration for the more affluent student, or one with family and social connections. The poor match posited between disadvantaged citizens and 4-year higher education fails to admit the possibility of repositioning disadvantaged citizen relative to bachelor's level educational opportunities. The problem is defined as one of fit, rather than fitness, we might say, and thus in this case, the acknowledgement of identity-derived difference does not empower democratic reform.

A second worrisome trend in planning for economic inclusion through STEM is even more difficult to tease out in light of its seemingly generous intentions. In the literature of educational inclusion, success in science and engineering is routinely associated with a student's self-confidence. Women and minority students have been found by analysts to lack a sense of self-efficacy. Researchers hold that when that trait is cultivated in members of under-represented groups, greater success in STEM programs is achieved by those groups (Marra and Bogue 2006; Slaton 2011). Because such studies seek concrete sources of improved self-efficacy (classroom teamwork, support groups, or mentoring) and pay attention to the experience of schooling (not merely its outcomes), they can be instructive rather than merely circular. But this focus on self-efficacy derives from only a first-order reflexivity within STEM fields. This kind of analysis ultimately returns responsibility for performance in school to the individual, who with support achieves efficacy or fails to do so. It makes no claim for the value of any sort of collective will or attainment and posits no larger structural impediments to inclusion such as poverty, racism, or sexism. Its corrective potential resides in helping the student conform to behavioral norms, with only a very selective critique of social relations in the university. This final assignment of behavioral responsibility to the student her- or himself arises from and supports neoliberal ideas of the individual in society as the source of achievement or cause of failure; what Brown calls a focus on "self-care" (2006).

For social problems to be seen as individual problems soluble through market forces, as Brown shows to be the case with contemporary industrial culture, the projects of the market must seem unassailable to all potential stakeholders. The nearly complete characterization of technological change and innovation as a positive societal force among Americans works beautifully to deflect attention from larger social structures, in exactly this technocratic way. The conditions in which universities and certain aspirants to STEM attainment thrive and in which marginalized populations remain disadvantaged are close to invisible within STEM curricula; the apparent value of reflexive thought is nil. The selectively oppressive functions of industrial capitalism are of course not a topic of instruction in the vast majority of STEM courses. Periods of widespread doubt about the safety of science and technology and the contributions of those realms to human welfare have occurred in the United States, but this has not been a significant trend since the late 1970s. Thus, when economic sustainability, environmental justice, or global food security do today find expression in engineering curricula, it is often as part of a brief introduction or ancillary framing that would not be likely to disrupt flows of economic and political influence in the nation (Riley et al. 2009). Instead, innovation carries a totalizing positive meaning. Since 2000, as China and India have gained global economic influence, we have seen particular popularity for the notion of "innovation" as an important means by which America can regain global competitiveness and achieve economic and (especially since 9/11) military security.

The ways in which such upbeat projections deter democratic reform in STEM fields are not confined to a foreclosure of discussions about technology's ill effects and occupational inequities. Rather, the promissory nature of rhetoric about innovation means that improved opportunities can be said to always be just over the horizon, and no one need be held responsible for their absence in the current moment (Waxman 2012; McCray 2012). In the nanotechnology sector of the current day, for example, the failure of promised industrial scale-up to occur and bring with it the projected jobs is easily attributed to the natural indeterminacy of scientific discovery. Inventive serendipity brings progress but it is, after all, serendipitous and must be allowed to remains so. The global mobility of capital so celebrated in the current climate is part of the problem: American capital bears no fixed responsibility for American labor, so high-risk research and development ventures hold no moral dangers for their backers (Head 2003; Rip 2006).

Technocratic Leanings and the Deflection of Critique

Much of what I have described here could certainly apply to education or employment in the service sectors, as well as to other public or private realms beyond those associated with STEM fields. Industrial capitalism is not merely a system of technical labor and knowledge, of course, and critical literature on schooling makes it clear that humanities and liberal arts can enact similarly discriminatory effects. The conservative social lessons carried in canon-focused humanities instruction have long been clear (Bourdieu 1990; Popkewitz 2001; Apple 2000, 2001). But there is a kind of cultural instrumentality involved in technical education and labor that renders questions about equity particularly difficult in the normal course of those enterprises.

The firewall that exists between credible engineering conduct and concerted attention to social matters is well established in U.S. higher education (Slaton 2010; Kline 1995; Pfatteicher 2005). Within American universities, schools of engineering and schools of arts and letters are almost universally distinct entities. Common definitions of rigorous practice require strengthening that distinction; tenure and promotion, granting, and accreditation processes do not promote a melding of technical and arts instruction. In engineering occupations, one may certainly take an interest in arts or politics without undermining one's reputation (although there are of course limits to what counts as seemly or palatable cultural engagement). However, any prospect of testing the claims or aims of a technical field in the terms used not by engineering but by, say, history or sociology or philosophy, let alone painting or poetry, can threaten one's credibility as an engineer (Catalano 2006; Catalano et al. 2008). What is more, even within schools of engineering individual disciplines function best when they disarticulate their specialized nature relative to one another; the curricula of different departments within an engineering school or polytechnic recapitulate expectations of the profession, with distinct research, teaching and accreditation expectations for each. If civil, mechanical and chemical engineering cannot deeply engage with the nature and function of their own boundaries, it will be nearly impossible for their participants to probe how any of these disciplines choose their problems, train their future representatives, attain their institutional influence, and justify their own existence to wider publics. It is only with those questions that social origins and impacts of an expertise can be interrogated.

This uncritical assertion of boundaries for technical expertise, which renders problematic any inquiry by engineers into the social features of engineering, of course trickles down to students (Riley et al. 2009; Cech and Waidzunas 2011; Seron and Silbey 2009). Students in engineering majors in American universities are exposed to humanistic inquiry about engineering; accreditation structures mandate some content of this kind (Riley 2012; Slaton 2012). This material can take the form of laudatory historical narratives or more incisive critiques of engineering. "Liberal education," as such instruction is labeled within the American Society for Engineering Education, is today variously provided by historians, sociologists, anthropologists, and policy or STS scholars, depending on the school, and this coursework unquestionably represents a wide range of political sensibilities on the part of instructors. But even the most critical or open-ended liberal-education pedagogies may not actually empower students profoundly to question how well or poorly engineering fulfills democratic ideologies. Critical social inquiry is not part of creditable engineering epistemology; the very definition of technical work requires lip-service to the false dualism of "people" and "technology." As Wendy

Faulkner has articulated, this binary is performed despite such obvious contrary evidence as the fact that technology is made *by people*, because the binary instantiates the authority of those (engineers) who claim that social issues (such as equity) dilute technical rigor (2009, pp. 143–144; see also Slaton 2010). Raising questions about such definitions is not likely to make one appear prepared for service in the field. That is, to propose to one's civil engineering instructor that civil engineering is at times a socially irresponsible endeavor may not go so far as to alienate that professor, but nor will it constitute a sign of mastery of engineering content.

The political disciplining accomplished by educational standards of this kind is powerful. In engineering as in all fields of education today, to turn away from standards is not merely to risk acquiring the wrong bodies of knowledge, it is also to risk regressing to the naturally lower state of the undisciplined mind (Popkewitz 2004). Goldberg and Traiman warn that, "Standards mean that students grow as they learn; without them, they learn to settle" (2001, pp. 75–76). Valuable learning in the American engineering curriculum means skirting unfamiliar questions or ones that have not been certified as having value by those who derived their authority from previous standards; a conservative system indeed!

In the formation, deployment, and enforcement of standards for STEM education, objectivity compels as a tool which scientists wield deliberately and particularly well among all professionals; all fields within STEM carry some of that cache (Seron and Silbey 2009). The supposed subjectivity of non-science (i.e., social or political) inquiry helps stigmatize that inquiry within STEM institutions but it is important, Faulkner adds, not to accept that duality as any more solid than the "people/technology" one. As Riley et al. write, this marginalization of social concerns in technical education follows the logic that any practice which is intentionally more caring or more just cannot simultaneously be more scientifically appropriate (2009, p. 24). Circular as it is (or exactly because it IS circular), this construction of "scientific-ness" (as that which lessens attention to care and justice) commands our attention as an instrument of neoliberalism. The forward motion of society through the fulfillment of market functions requires a narrowly instrumental approach to knowledge about nature and to any applications of that knowledge. According to that worldview, problem choice (as in: what should be studied about nature, and what technologies thereby developed) must proceed with issues of care and justice cast as mere distractions.

Conclusions

The objective of this paper is to break open the black box of racial, gender and other inclusive projects in American engineering education to understand why inclusive efforts have only minimally disrupted conventional social relations in that field. I have tried to highlight the complexity of the issue, pointing to the roles of multiple cultural commitments ongoing in the United States today…to technology, to ideas of merit, to the neoliberal embrace of market forces and the strategic denial of the structural conditions that impede democratic reform. The motivating question of this paper might be more simply phrased this way: To what problem is STEM diversity programming the answer? That query stands in sharp contrast to the myriad questions that make up most research on STEM diversity, which focus on student experiences, teaching and learning styles, and quantitative measures of how well or poorly inclusive interventions have functioned. The tasks of engineering, to which some Americans not previously involved should now be introduced, are not themselves subject to inquiry; add minorities and stir, as the ironic catchphrase goes.

Thomas Popkewitz builds a compelling case for the ways in which education since the late nineteenth century has set the stage for inculcating precisely this set of narrowed epistemic priorities in students. Schooling, he proposes, enacts the production of the cosmopolitan citizen, the individual operating in support of a larger social structure and its privileges, "taming" if not banishing unwelcome "subjectivities" in order to produce a citizen subject to administration. Science is not coincidentally central to this education:

Cosmopolitanism makes possible the conditions of the modern state, its citizens, and the pedagogy of the school by bringing together the scientific order of reason and the individual who reasoned through science (Popkewitz 2004, p. 190). Essential to this project, Popkewitz explains, is the construction of the student as one who must be taught to distinguish between knowledge and non-knowledge. Here, he quotes Charles Eliot summarizing this prescription in an influential study of the early 1890s:

> One is fortified against the acceptance of unreasonable propositions only by skill in determining facts through observation and experience, by practice in comparing facts or groups of facts, and by the unvarying habit of questioning and verifying allegations, and of distinguishing between facts and inferences from facts, and between a true cause and an antecedent event. One must have direct training and practice in logical speech and writing before he can be quite safe against specious rhetoric and imaginative oratory. (Eliot 1892–1893, p. 424 [quoted in Popkewitz 2004, p. 205])

We can take Popkewitz even further and understand the students' historically prescribed work of distinguishing knowledge from non-knowledge to be the work of fabricating knowledge. Constructivist understandings of science indicate that the work of scientists brings the subjects of science *into being*; that is, there is no nature or material that holds meaning apart from our efforts to give those objects of our attention meaning. Applied to engineering, this might suggest that to build a bridge, HVAC system, or artificial spine comprises engineering through social relations, but it would be more precise to say that each of those tasks is necessarily both engineering and an expression of power.

The point is that technological activity is not figure to the ground of society or culture. With that organic, integrative understanding in mind, the discriminatory habits of STEM education and labor in the United States are not easily demarcated from other epistemic commitments of American science and engineering. Many dedicated educators and policy makers have worked for decades to understand STEM inequity, but in turns out that merely defining the problem is even harder than we thought. Yet, it is the recognition and embrace of precisely that difficulty that may finally lead to change.

References

Apple, M. (2000). *Official knowledge: Democratic education in a conservative age*. New York: Routledge.

Apple, M. (2001). *Education the 'right' way: Markets, standards, god and inequality*. New York: Routledge.

Bayer Corporation. (2011). Bayer facts of science education XV: A view from the gatekeepers. http://bayerus.online-pressroom.com/bayerus/?LinkServID=FABE4A9A-1372-5B6F-0B65BCC31979EA60. Accessed 10 Dec 2011.

Board, N. S. (2010). *Preparing the next generation of STEM innovation*. Washington, DC: National Science Board/National Science Foundation.

Bourdieu, P. (1990). *Reproduction in education, society and culture*. Thousand Oaks: Sage.

Brint, S., & Karabel, J. (1991). *The diverted dream: Community college and promise of educational opportunity in America, 1900–1985*. Oxford: Oxford University Press.

Brown, W. (2006). American nightmare: Neoliberalism, neoconservatism, and de-democratization. *Political Theory, 34*, 690–714.

Bryan, W. (2012). Blind CU student inspires lab changes. *Denver Post*, 12 May 2012. http://www.denverpost.com/dontmiss/ci_20607982/blind-cu-boulder-student-inspires-lab-changes

Carson, J. (2006). *The measure of merit*. Princeton: Princeton University Press.

Catalano, G. D. (2006). *Engineering ethics: Peace, justice, and the earth*. San Rafael: Morgan and Claypool.

Catalano, G. D., Baillie, C., Byrne, C., Nieusma, D., & Riley, D. (2008). Increasing awareness of issues of poverty, environmental degradation and war within the engineering classroom: A course modules approach. In *Proceedings of the 38th ASEE/IEEE Frontiers in Education Conference*, Saratoga Springs, NY.

Cech, E. A., & Waidzunas, T. J. (2011). Navigating the heteronormativity of engineering: The experiences of lesbian, gay, and bisexual students. *Engineering Studies, 3*, 1–24.

Cook, G. (2012). The autism advantage. *New York Times*, 29 November 2012.

Daston, L., & Galison, P. (1992). The image of objectivity. *Representations, 40*, 81–128.

Duster, T. (2003). *Backdoor to eugenics* (2nd ed.). New York: Routledge.

Ebeling, M. F., & Slaton, A. E. (2010). Two-year colleges and the allure of 'nano': Understanding institutional enthusiasms. In *Proceedings of ASEE Annual Meeting*, Louisville.

Eliot, C. (1892–1893). Wherein popular education has failed. *The Forum, 14*, 411–428.

Faulkner, W. (2009). The power *and* the pleasure? A research agenda for 'making gender stick' to engineers. In M. Wyer, M. Barbercheck, D. Giesman, H. O. Ozturk, & M. Wayne (Eds.), *Women, science, and technology* (pp. 143–156). New York: Routledge.

Gershon, I. (2011). Neoliberal agency. *Current Anthropology, 52*, 537–555.

Goldberg, M., & Traiman, S. L. (2001). Why business backs educational standards. *Brookings Papers on Education Policy, 4*, 75–129.

Gordon, A. (1995). The work of corporate culture: Diversity management. *Social Text, 44*, 3–30.

Gould, S. J. (1996 [1981]). *The mismeasure of man*. New York: W.W. Norton.

Grubb, W. N., & Lazerson, M. (2005). Forum: The education gospel and the role of vocationalism in American education. *American Journal of Education, 111*, 297–319.

Gurin, P., & Epps, E. (1975). *Black consciousness, identity and achievement*. New York: Wiley.

Harvard Graduate School of Education. (2011). Pathways to prosperity: Meeting the challenge of preparing young Americans for the 21st century. http://www.gse.harvard.edu/news_events/features/2011/Pathways_to_Prosperity_Feb2011.pdf. Accessed 1 Mar 2012.

Head, S. (2003). *The new ruthless economy: Work and power in the digital age*. Oxford: Oxford University Press.

Holvino, E., & Kamp, A. (2009). Diversity management: Are we moving in the *right* direction? Reflections from both sides of the North Atlantic. *Scandinavian Journal of Management, 25*, 394–403.

Hursh, D. (2006). Carry it on: Fighting for progressive education in neoliberal times. In G. Ladson-Billings & W. F. Tate (Eds.), *Education research in the public interest: Social justice, action, and policy* (pp. 46–63). New York: Teachers College Press.
Kline, R. (1995). Construing “technology” as “applied science”: Public rhetoric of scientists and engineers in the United States, 1880–1945. *Isis, 86*, 194–221.
Knorr-Cetina, K. (1995). Laboratory studies: The cultural approach to the study of science. In S. Jasanoff, G. E. Markle, J. C. Petersen, & T. Pinch (Eds.), *Handbook of science and technology studies* (pp. 140–166). Thousand Oaks: Sage.
Labaree, D. F. (2008). The winning ways of a losing strategy: Educationalizing social problems in the United States. *Educational Theory, 58*, 447–460.
Little, P., Barney, D., & Hink, R. (2008). Living up to the code: Engineering as political judgment. *International Journal of Engineering Education, 24*(2), 314–327.
Marra, R. M., & Bogue, B. (2006). Women engineering students’ self efficacy–A longitudinal multi-institution study. *WEPAN Conference, 2006*.
McCray, P. (2012). *The visioneers*. Princeton: Princeton University Press.
Pang, A. S.-J. K. (2002). *Empires and the sun*. Stanford: Stanford University Press.
Pfatteicher, S. K. A. (2005). Anticipating engineering’s ethical challenges in 2020. *IEEE Technology and Society Magazine, 24*, 4–43.
Popkewitz, T. S. (2001). The production of reason and power: Curriculum history and intellectual traditions. In T. S. Popkewitz, B. M. Franklin, & M. A. Pereyra (Eds.), *Cultural history and education: Critical essays on knowledge and schooling* (pp. 151–183). New York/London: RoutledgeFalmer.
Popkewitz, T. S. (2004). The reason of reason: Cosmopolitanism and the governance of schooling. In B. M. Baker & K. E. Heyning (Eds.), *Dangerous coagulations? The uses of foucault in the study of education* (pp. 189–224). New York: Peter Lang.
Popkewitz, T. S. (2006). Hopes of progress and fears of the dangerous: Research, cultural theses, and planning different human kinds. In G. Ladson-Billings & W. F. Tate (Eds.), *Education research in the public interest: Social justice, action, and policy* (pp. 119–140). New York: Teachers College Press.
Riley, D. (2008). *Social justice in engineering*. San Rafael: Morgan and Claypool.
Riley, D. (2012). We’ve been framed! Ends, means and the ethics of the grand challenges. *International Journal of Engineering, Social Justice and Peace, 1*, 123–136.
Riley, D., Pawley, A. L., Tucker, J., & Catalano, G. D. (2009). Feminisms in engineering education: Transformative possibilities. *NWSA Journal, 21*, 21–40.
Riley, D., Slaton, A. E., & Pawley, A. L. (2014). Social justice and inclusion: Women and minorities in engineering. In A. Johri & B. Olds (Eds.), *Cambridge handbook of engineering education research*. Cambridge: Cambridge University Press.
Rip, A. (2006). Folk theories of nanotechnologists. *Science as Culture, 15*, 349–365.
Rodgers, D. (2011). *Age of fracture*. Cambridge, MA: Harvard University Press.
Schaffer, S. (1988). Astronomers mark time: Discipline and the personal equation. *Science in Context, 2*, 115–145.
Schaffer, S. (1995). Accurate measurement is an English science. In M. N. Wise (Ed.), *The values of precision* (pp. 135–172). Princeton: Princeton University Press.
Seron, C., & Silbey, S. S. (2009). The dialectic between expert knowledge and professional discretion: Accreditation, social control, and the limits of instrumental logic. *Engineering Studies, 1*, 101–127.
Shapin, S. (1989). The invisible technician. *American Scientist, 77*, 554–563.
Siebers, T. (2010). *Disability theory*. Ann Arbor: University of Michigan Press.
Slaton, A. E. (2001). *Reinforced concrete and the modernization of American building, 1900–1930*. Baltimore: Johns Hopkins University Press.
Slaton, A. E. (2010). *Race, rigor and selectivity in U.S. engineering: The history of an occupational color line*. Cambridge, MA: Harvard University Press.

Slaton, A. E. (2011). Metrics of marginality: How studies of minority self-efficacy hide structural inequities. In *Conference Proceedings Annual Meeting of the American Society for Engineering Education*, Vancouver, B.C.
Slaton, A. E. (2012). Engineering improvement: Social and historical perspectives on NAE's "grand challenges". *International Journal of Engineering, Social Justice and Peace, 1*, 95–108.
Slaughter, S., & Rhoades, G. (2004). *Academic capitalism and the new economy: Markets, state and higher education*. Baltimore: Johns Hopkins University Press.
Supalo, C., et al. (2007). Talking tools to assist students who are blind in laboratory courses. *Journal of Science Education for Students with Disabilities, 12*, 27–32.
Tonso, K. L. (1996). The impact of cultural norms on women. *Journal of Engineering Education, 85*, 217–225.
Traweek, S. (1992). *Beamtimes and lifetimes*. Cambridge, MA: Harvard University Press.
Vinao, A. (2001). History of education and cultural history: Possibilities, problems, questions. In T. S. Popkewitz, B. M. Franklin, & M. A. Pereyra (Eds.), *Cultural history and education: Critical essays on knowledge and schooling* (pp. 125–150). New York/London: RoutledgeFalmer.
Wacquant, L. (2009). *Punishing the poor: The neoliberal government of social insecurity*. Durham/London: Duke University Press.
Walkerdine, V. (2003). Rcclassifying upward mobility: Femininity and the neo-liberal subject. *Gender and Education, 15*, 237–248.
Waxman, S. (2012). Against innovation. *Jacobin*, 10 September 2012. http://jacobinmag.com/2012/09/against-innovation/. Accessed 3 Feb 2013.
Winner, L. (1980). Do artifacts have politics? *Daedalus, 109*, 129–136.

Amy E. Slaton Ph.D. in History and Sociology of Science from the University of Pennsylvania. Professor of History in the Department of History and Politics of Drexel University. She is the author of *Reinforced Concrete: The Modernization of American Building, 1900–1930* (Johns Hopkins 2001) and *Race, Rigor and Selectivity in U.S. Engineering: The History of an Occupational Color Line* (Harvard, 2010). She produces the website STEMequity.com, and is currently writing on high-tech education and workforce development, including the engagement of community colleges in neoliberal strategies of economic uplift for American workers. Other recent projects include critical study of U.S. understandings of bodily difference/disability and technical competence, and a volume of essays on the social construction of novelty in industrial materials research.

Chapter 9
Challenges of Overcoming Structural Barriers for African American Engineers in the United States and in the African Diaspora

Derrick Hudson

Abstract This chapter outlines the academic literature that addresses the persistent underrepresentation of African Americans in engineering education in the United States and throughout the African diaspora. While the numbers of African Americans has grown over the past few decades in other professions, the numbers of African Americans in engineering have stagnated and declined since the beginning of the twenty-first century. Early pioneers of scholars in African American studies thought that they could easily construct a reverse mirror image of the curricula they encountered in other academic disciplines, such as history, political science, or anthropology. A glaring omission of the early pioneers is the work that would be needed in engineering education. Many of the early pioneers failed to take into account that work needs to also be done *directly within* engineering education to foster "socio-technical" engineering undergraduates and professionals. The depoliticization and meritocratization of engineering education has often allowed structural barriers to remain in place that hinder the success of African Americans in engineering. After summarizing some of the major explanations that attempt to explain underrepresentation of African Americans in engineering, the article concludes with suggestions for further research, highlighting the continued pivotal role of historically black colleges and universities and the need to encourage more investments to promote research and development in African universities.

Keywords African Americans • Sub-Saharan Africa • Historically black colleges and universities (HBCUs) • Engineering education • Underrepresentation • Science • Technology • Engineering • Math (STEM)

D. Hudson (✉)
Colorado School of Mines, Liberal Arts and International Studies (LAIS),
Stratton Hall, 329, 1500 Illinois Street, 80401 Golden, CO, USA
e-mail: dkhudson@mines.edu

S.H. Christensen et al. (eds.), *International Perspectives on Engineering Education*,
Philosophy of Engineering and Technology 20, DOI 10.1007/978-3-319-16169-3_9

Introduction: The Persistent Underrepresentation of African Americans and People of African Ancestry in Engineering

The African diaspora is defined to include peoples of sub-Saharan ancestral origins[1] on the continent of Africa, the Caribbean, the Americas, and worldwide. This diaspora began to emerge with the advent of the slave trade in the 1500s. Over the centuries, people of African ancestry have been able to enter other professions such as theology, law, social services, and business. However, engineering as a profession of choice and pursuit for people of African ancestry has been underrepresented than these other professions.

According to Mark Matthews, "sixty years after the Supreme Court outlawed segregated school systems in the United States, de facto racial imbalances persist in American education" (Matthews 2014). One particularly salient example is the underrepresentation of African Americans in engineering. While there is consensus of disapproval of this reality, there is no single reason that explains it (Matthews 2014). For African American men, the situation is further compounded by their overall position in higher education, where they are regularly outperformed and outnumbered by African American women. Underrepresentation of African Americans in engineering will continue to be salient as trends point in the direction of dismantling affirmative action programs, such as when the U.S. Supreme Court's ruled in the 2003 *Gratz v. Bollinger* case that the point system used by the University of Michigan for undergraduate admissions was unconstitutional. With seven other states with similar bans and others likely to follow, creative new tools will need to be implemented to enroll and retain African Americans in engineering programs in the United States (Matthews 2014).

Some of the remedies suggested are (1) recognizing that it may take five years for undergraduates to attain their engineering degrees; (2) summer bridge programs that have strong developmental teaching and conscientious advising; (3) designing lab and classroom settings that foster a welcoming setting for African Americans and other under-represented minorities in engineering environments; and (4) understanding mentors who can make a difference and provide the support to retain African Americans in engineering programs as opposed to giving up in despair (Matthews 2014).

However, a more basic set of questions have been raised to explore the assumptions of engineering education. As Caroline Baillie notes, "engineers are often not encouraged to consider *who* they engineer *for*" (Baillie 2006). As a recent UNESCO

[1] To be more precise, the peoples of African origin that are being referred to belong to the African language families of the Nilo-Saharan, which includes examples such as the Dinka and Nuer of South Sudan; the Niger-Congo, far and away the largest sub-Saharan language family to include notable examples such as Yoruba (Nigeria), Swahili (eastern Africa), and Zulu (South Africa); and Khoisan, which includes the ethnic groups of Khoe and Sandawe in southwestern and eastern Africa. The other major African language parent family, Afro-Asiatic, while including groups north of the Sahara Desert and Asia, is meant here to only include sub-Saharan groups such as the Tuareg and the Afar and Amharic of Somalia and Ethiopia.

report points out, approximately 90 % of the world's engineers work for 10 % of the world's population, generally are the richest 10 % (UNESCO 2010). As we move deeper into the twenty-first century, the subject of engineering continues to be learned in a vacuum. Engineers study the technical and practical aspects of their professions, and at more progressive institutions engineering undergraduates are encouraged to study teamwork skills and communication. However, aside from these efforts, it is rare in the United States and in many countries to find an engineering graduate who is educated in the context, whether local or global, in which they might find themselves *doing* the engineering (Baillie 2006). This is often referred to as the sociotechnical context.

This article provides a brief overview of the literature of underrepresentation of African Americans in engineering, science, and science-related professions. The next section will assess some of the major arguments and debates that have dominated the literature. The concluding section will provide some practical and future directions for more research.

Literature Review

Brian L. Yoder, writing for the *American Society for Engineering Education* (ASEE), summarizes trends in engineering education in a 2012 article. Focusing on African Americans, the trends are less than sanguine. According to U.S. Census Data (2010), African Americans comprise 12.6 % of the U.S. population. In 2005, African Americans accounted for 5.3 % of the total engineering student population and for every year up to 2012, the percentage has dropped to 4.2 % (Yoder 2012). The trend for Hispanic Americans has been better, with figures of 5.8 and 8.5 % for 2005 and 2012, respectively. Finally, the largest ethnic minority demographic in minority engineering undergraduates is represented by Asian Americans, with figures of 14.1 % in 2005 and 12.2 % in 2012. The pinnacle of engineering achievement can be said to include induction into the prestigious National Academy of Engineering, which has a 1 % representation of African Americans, according to statistics from the *Journal of Blacks in Higher Education.*

Why the persistent underrepresentation? Many suggest that the causes start in K-12 education. For African American men, the statistics are particularly troubling. Of those who graduate, only 52 % of African American men graduate from high school in 4 years, and of that group, less than a fourth enter college. In contrast, incarceration rates for African American men are seven times that of white males (Matthews and Loftus 2014).[2] Even at well-established historically black colleges and universities (HBCUs) such as Morehouse College, an all-male institution, close to 45 % of Morehouse students will not finish in 6 years and the figure nationally approaches two-thirds (Matthews and Loftus 2014). The longer time for completion

[2] One of the most thorough explorations of this topic is Michelle Alexander's *The New Jim Crow: Mass Incarceration in an Age of Colorblindness* (New York: Perseus Press, 2012).

is normally attributed to financial and academic difficulties. A final challenge is that many of these students will opt to settle for less competitive colleges and universities, which can have significant impacts on their professional careers in engineering and STEM fields.

More specifically, from an academic point of view, many African American students arrive to undergraduate education with inadequate high school preparations (Matthews and Loftus 2014). Many of these students will play catch-up for a good portion of their undergraduate careers (Matthews and Loftus 2014). These challenges add time and cost to their education.

Another factor that is often noted as a major challenge to attract and educate African Americans in engineering and science is the profound lack of awareness of the past and present contributions of African Americans in engineering, such as the recent appointment of Norman Fortenberry as executive director of the American Society for Engineering Education (ASEE) in 2011. Many argue that the point to be made here is that many African Americans often cannot envision themselves as engineers and scientists. Some studies have tried to argue that African Americans are more apt to choose professions that have a social or social justice dimension to their careers and often, engineering and science are not seen as relevant to these issues. As Mario Azevedo notes, on the continent of Africa, what most Africans know about people of African ancestry in the United States and elsewhere in the Americas is that they were enslaved and that they continue to be discriminated against. There is a superficial knowledge of famous personalities such as Jesse Jackson, Michael Jackson, and Thurgood Marshall. When one turns to the accomplishments of African Americans in engineering and the sciences, there is almost no knowledge or awareness in these fields. The same appears to be true in the United States regarding the achievements of African Americans, especially in the hard sciences (Azevedo and Sammons 2005). The UNESCO report from 2012, cited earlier, makes a similar observation that engineering is routinely overlooked in many countries around the world and that engineering needs to become more human and humane to develop a wider appeal (UNESCO 2010). One logical place to rectify this lack of awareness could be to include the accomplishments of African Americans via instruction by faculty in African American studies. However, several challenges present themselves. One of them is that most fields require many years to evolve from areas of awareness to disciplines of practice. In many respects, African American Studies has yet to settle as a discipline. In the initial years of the development of African American Studies across the United States, as well as other ethnic studies efforts, much of the drive was a corrective drive against academic exclusion. The early pioneers thought that they could easily construct a reverse mirror image of the curricula they encountered. Few anticipated the difficulty of trying to (a) create a new discipline; (b) perform corrective functions; (c) become race relations generalists; (d) do Afro-*loco-parentis* duty; (e) work as minority ombudsmen, and (f) receive "precisely the same" treatment at tenure time as their colleagues not working in the field who most likely do not have these same sets of expectations on them (Adams 2005). The challenge here is to place yet another "corrective" function on faculty that is already tasked with responsibilities that other fields do not require. While many faculty working in African American Studies may be more than willing to take on these awareness building efforts, the work of those in engineering education need to continue to encourage how to create "socio-technical" engineering and science undergraduates.

One of the major proponents to engage and develop a space to foster "sociotechnical" engineering *directly within* engineering education is the work of Caroline Baillie. Baillie, along with many others in engineering education, have devoted and explored topics to aid the practicing engineer in reflecting upon the nature and purpose within the engineering profession and how that is related to and implicated in social, economic and political issues (Baillie 2006). According to Erin Cech, engineers will incorporate considerations of social justice issues, as one topic, into their work only to the extent that they see such issues as relevant to the practice of their profession (Cech 2013). In an increasingly competitive and hostile environment in which engineers are forced to spend their lives fighting for higher profit margins, many engineers realize they are not engineering for those in need but for those who can pay. An academic literature has emerged to aid engineers and think about engineering education as a profession and to take appropriate action related to industrial development and globalization (Baillie 2006). One major aspect of the literature argues that two prominent ideologies within the culture of engineering—depoliticization and meritocracy—frame social justice issues in such a way that they seem irrelevant to engineering practice (Cech 2013). Depoliticization is the belief that engineering is a "technical" space where "social" or "political" issues such as inequality are tangential to engineers' work. Meritocracy, the belief that inequalities are the result of a properly-functioning social system that rewards the most talented and hard-working, legitimates social injustices and undermines the motivation to rectify such inequalities (Cech 2013). These aspects in the engineering education literature are linked to African Americans in the sense that as a group, social justice issues are a central aspect of the issues at stake in this community. These issues play themselves out in many areas, to include career choices and decision making processes for educational attainment. Engineering educations tends to be tangential to issues of social justice. Thus, an academic literature has emerged to explain the underrepresentation of African Americans in engineering professions.

The academic literature on African American underrepresentation is fledging and emergent. According to Lewis (2003), research in this area is important as it could inform policy and intervention efforts. However, due to the lack of more empirically based studies that is sometimes the case in the literature, intervention efforts tend to rely more on folk insight than on empirical evidence. Lewis argues that it is not uncommon for intervention programs to address factors that are not known to contribute to underrepresentation (Lewis 2003). He cites the case of intervention programs that present students with African American scientists as role models (e.g., Barisa and Holland 1993; Berrington and DeLacy 1993) or require African American students to take greater numbers of mathematics and science courses (e.g. Ellis 1993; Thomas 1984), in hopes that these measures will encourage greater numbers of African American students to consider pursuing science and science-related careers. Lewis asserts that there is no clear evidence that either

African American role models (Thomas 1984) or the number of mathematics and science courses that students take (Connell and Lewis 2003) causes or encourages students to pursue science and science-related careers.

The studies on underrepresentation of African Americans in science and engineering tend to cluster around five explanations: academic preparation, career interests, lack of educational and career planning, role models, and career opportunities (Hall and Post-Kammer 1987).

Academic Preparation

It is sometimes argued in the literature that poor preparation in mathematics and science is a pervasive problem for African American students leading to underrepresentation (Hall and Post-Kammer 1987). Proponents of this position point to data indicating that African American students enroll in fewer mathematics and science courses (e.g. Gilleylen 1993; Reyes and Stanic 1985), have low achievement scores in science, and have fewer experiences involving science (e.g. Kahle 1982; Thomas 1986). One rationale points to the inability of educational systems to enable African American students to find "science as useful out of school in the way that white students do….and that [African American] students have less awareness of how scientists work (Lewis 2003)." According to Cynthia Atman and others, African Americans and women often self-assess themselves as not being able to succeed in engineering and science professions (Besterfield-Sacre et al. 2001). Again, the explanation for this self-assessment is that African Americans point to poor performance in math and science courses in high school or if there was initial success in early math and science courses, when more advanced courses were attempted, poor performance deterred any more forward movement and persistence in these courses. A final important factor to note is that more African American families live below the poverty line than Hispanics or Asians, so they are likely to be in historically impoverished school districts that struggle to meet more diverse and intense student needs; since schools are funded in large part by state and local property taxes, the poor districts tend to have poorer schools and the rich, richer (Kozel 2012).

Career Interests

A second explanation that has been explored in the literature for the underrepresentation of African Americans in science and engineering is career interests. Some studies have been carried out to survey high school and college students. One pioneering study that surveyed college freshmen (Hager and Elton 1971) and one surveying high school juniors and seniors (Sewell and Martin 1976), showed that African American men express a greater interest in social service fields compared with White men, who prefer scientific and technical fields. The main rationale for this trend argue that African American students gravitate towards careers with social orientations out of concern for the historical disadvantaged social position of African Americans (Hall and Post-Kammer 1987; Azevedo and Sammons 2005). Other studies have explored other explanations to account for career interests to include family socioeconomic status, cultural capital, group values, social capital and effects of significant others, and institutional factors (Simpson 2001).

Lack of Educational and Career Planning

A third set of arguments that appear in the literature points to a lack of educational and career planning. This argument is substantiated by studies indicating that African American students are unaware of various career opportunities such as engineering. Moreover, African Americans are more apt to seek career guidance from family members and peers who may not be knowledgeable about engineering careers (Lewis 2003).

Role Models

While there is literature that asserts that underrepresentation persists because there is a lack of African American role models in science and engineering positions, the research has not made a clear case showing that there is a link between role models and choices of careers among African Americans. The two main rationales to understand these arguments is that African American children will not aspire to careers in science and engineering if they do not see older African Americans functioning in these roles, and secondly, in the university setting, role models are necessary as gatekeepers and sources of moral support without which African Americans feel isolated (Lewis 2003).

Career Opportunities and Economic Incentives

The final thread in the academic literature points to income potential as a major factor in a students' career decision (Lewis 2003). Acknowledging that the income potential in science and engineering is high, it would seem that African Americans, like other students, would gravitate to these careers. However, Ogbu (1978) raised issues that attempt to explain why African Americans do not choose science and engineering as career options. One major issue is that African Americans have a belief that they have fewer job opportunities than Whites or other groups and that they perceive science and engineering careers as unattainable or off-limits to them.

Future Directions

The Pivotal Role of HBCUs

Many professionals in engineering education suggest several trends that can enhance and promote the success of African Americans in STEM professions. The first is the crucial role of HBCUs. One noteworthy study by Laura Perna et al. illustrates the role that Spelman College, one of the country's premier HBCUs, is playing in promoting the attainment of women in STEM fields (Perna et al. 2009). Concurrent with other studies, African American students who attend HBCUs as opposed to predominately white colleges and universities experience less social isolation, alienation, personal dissatisfaction, and overt racism (Perna et al. 2009) and that HBCUs seem to provide a social, cultural, and racial environment that is more supportive, caring, and nurturing for students and promotes academic achievement and success (Perna et al. 2009). Second, like other research (Bensimon 2007), the findings suggest the benefits of adopting a multi-faceted institutional approach that promotes students' academic and psychological readiness to pursue advanced degrees and careers in STEM fields. More specifically, Bensimon (2007) argues that the dominant paradigm around student success places responsibility for success on the student rather than on the institution. Bensimon argues that scholars and practitioners often "assume that institutional support systems are already in place and motivated students will take advantage of them" (Bensimon 2007). Bensimon urges one to not make this assumption. The faculty and administration at Spelman College do not seem to make these assumptions and take great ownership of their role in boosting student success.[3] The supportive, cooperative atmosphere at Spelman nurtures the academic achievement of women at Spelman. This finding can be extrapolated to African Americans as a whole, especially at HBCUs which have historically been institutions to promote the success of African Americans in higher education, even in a post-Civil Rights era. While acknowledging the decline of the percentage of African Americans attaining degrees from HBCUs, they continue to serve as a viable option for African Americans who want to pursue degrees after high school, especially in the STEM professions.[4] Finally, even with Spelman's clear commitment to promoting the attainment of STEM careers by African American women, financial challenges are a major impediment for many African American women.

[3]Another useful way to think about the philosophical differences of faculty at HBCUs and traditionally White colleges and universities is the distinction between pedagogy and andragogy. Among other characteristics, andragological approaches utilize a 'coaching' versus 'teaching' tone in instruction and engagement with students.

[4]In a recent ASEE study (2011), of the top 20 institutions that award engineering degrees to African Americans, ten are HBCUs to include North Carolina A&T State University, Howard, and Southern University, to name three notable examples.

Gaps in the Academic Literature on Underrepresentation

Bradford Lewis, writing in the *Journal of Women and Minorities in Science and Engineering*, points to several areas for more research: (1) generate more empirical research; (2) develop more precise meanings of the preponderance of factors found to correlate with students' career choices; (3) define more sharply some of the constructs used to explain teacher influences on students' career choices to pursue math and science; (4) nuance the research to not only assume that there are deficiencies in the life histories of African Americans as related to science and engineering, and (5) to sharpen the explanatory model for race.

The Need for African Universities to Move towards Research and Inclusion in the Policy-Making Process

In Africa, most institutions of higher learning have had to historically focus on teaching and educating their populations as the waves of independence began to sweep across the continent in the 1950s (Atuahene 2011). With the arrival of military coups and the "lost decade" of the 1980s in much of the global south, African intellectuals had to make decisions to pursue academic careers elsewhere, most notably Europe and the United States.

According to a major report from the United States Educational, Scientific and Cultural Organization (UNESCO), published in 2010, data shows that developed, industrialized countries have between 20 and 50 scientists and engineers per 10,000 population, compared to around five scientists and engineers on average for developing countries, and down to one or less for some poorer African countries (UNESCO 2010). Recent data suggests that Africa only accounts for 8 % of research and development expenditure and 1.4 % in publications in engineering and the sciences.[5] This trend continues and is also noted in the most recent UNESCO 2010 report.

According to Felix Atume, a major issue facing the growth of the engineering profession in many sub-Saharan African countries is the lack of involvement of engineers in policy matters as many political leaders seldom take into consideration the key role that engineers and engineering can play in development (Atume 2010).

Finally, African countries should consider developing stronger south-to-south partnerships to mitigate against partnerships that are with the global North driving the research agenda. These partnerships could foster more dialogue and synergy amongst developing countries to make decisions on how to apply and utilize emerging technologies, such as nanotechnology to provide clean drinking water.

[5] As a point of comparison, Asia accounts for 21.1 % in publications and 30.5 % in R&D expenditure. Please see the *UNESCO Bulletin on Science and Technology Statistics*, Issue No. 2, September 2005.

Conclusion

As a final note, Caroline Baillie provides an example of the socio-political context constraints that the Basuto people face given the issues of globalization and the challenges of increasing the numbers of people who have been historically underrepresented in engineering and science:

> Who benefits and who pays? Who needs what and when? How will the project survive after the planners have gone? Who contributed to its planning and execution? Who decided what was needed? Who paid for it and why? What do they stand to gain? Are proceeds distributed equitably? Does it provide fair compensation for those affected? Are people treated ethically and justly both within and as a result of the project—workers, those affected but not involved and those who are 'users'? Who gets the jobs? Who makes decisions about pay and conditions? Do workers have to relocate? What effect does this have on their lives, their family's lives and those of their community? Is the engineering project contributing in any way to the increasing gap between the rich and the poor? How do you know? How do you find out? Do you feel you are in a position to do the right thing in your current job? (Baillie 2006)

As with any well-designed engineering project, a lot of questions are asked. More questions—such as the some of the neglected ones Baillie is raising—and research needs to be done to promote the future success of African Americans and other underrepresented groups in engineering.

References

Adams, R. L. (2005). African American studies and the state of the art. In M. Azevedo (Ed.), *Africana studies: A survey of Africa and the African diaspora* (3rd ed.). Durham: Carolina Academic Press.

Atuahene, F. (2011). Rethinking the mission of higher education: An anatomy of the research challenge of african universities. *The Journal of Asian and African Studies, 46*(4), 321–341.

Atume, F. (2010). *Engineering around the world: Africa. LINESCO: Engineering: Issues, challenges and opportunities for development.* Paris: LINESCO.

Azevedo, M., & Sammons, J. (2005). Contributions in science, business, film, and sports. In M. Azevedo (Ed.), *Africana studies: A survey of Africa and the African diaspora* (3rd ed.). Durham: Carolina Academic Press.

Baillie, C. (2006). *Engineers within a local and global society: Synthesis lectures on engineering, technology and society.* San Rafael: Morgan and Claypool Publishers.

Barisa, M. T., & Holland, C. (1993). *The graduate achievement program: A description of a summer enrichment in math and science for minority undergraduate students.* New Orleans, LA: Mid-South Educational Research Association.

Bensimon, E. M. (2007). The underestimated significance of practitioner knowledge in the scholarship on student success. *The Review of Higher Education, 30*(4), 441–469.

Berrington, S., & DeLacy, A. (1993). Making a difference. *Middle School Journal, 24*(4), 34–36.

Besterfield-Sacre, M., Moreno, M., Shuman,L. J., & Atman, C. J. (2001). Gender and ethnicity differences in freshmen engineering student attitudes: A cross-institutional study. *Journal of Engineering Education, 4*, 477–489.

Cech, E. A. (2013). The (Mis)framing of social justice: Why ideologies of depoliticization and meritocracy hinder engineers' ability to think about social injustices. In J. Lucena (Ed.), *Engineering education for social justice: Critical explorations and opportunities, philosophy of engineering.* Dordrecht: Springer.

Connell, S., & Lewis, B.F. (2003). *An examination of the relationship between African American students, enrollment in advanced science courses and their career considerations*. Paper presented to the national association of research in science teaching, Philadelphia.

Ellis, R. S. (1993). Impacting the attitudes of minority high school youth. *School Science and Mathematics, 93*(8), 400–407.

Gilleylen, C. E. (1993). A comparative study of the science-related attitudes and the factors associated with persisting in science of African American college students in science majors and African American students in non-science majors. Unpublished doctoral Dissertation, Indiana University of Pennsylvania.

Hager, P. C., & Elton, C. F. (1971). The vocational interests of black males. *Journal of Vocational Behavior, 1*(2), 153–158.

Hall, E. R., & Post-Kammer, P. (1987). Black mathematics and science majors: Why so few? *The Career Development Quarterly, 35*(3), 206–219.

Kahle, J. B. (1982). Can positive attitudes lead to improvement gains in science? Analysis of the 1977 national assessment of educational progress, attitudes towards science. *Science Education, 66*, 539–546.

Kozel, J. (2012). *Savage inequalities: Children in America's schools*. New York: Random House LLC.

Lewis, B. F. (2003). A critique of literature on the underrepresentation of African Americans in science: Directions for future research. *Journal of Woman and Minorities in Engineering, 9*, 361–373.

Matthews, M. (2014). A chronic disparity. *Prism: American Society for Engineering Education, 23*(8), i.

Matthews, M., & Loftus, M. (2014). Survival course. *Prism: American Society for Engineering Education, 23*(8).

Ogbu, J. U. (1978). *Minority education and caste: The American system in cross-cultural perspective*. New York: Academic Press Inc.

Perna, L., Lundy-Wagner, V., Dezner, N. D., Gasman, M., Yoon, S., Bose, E., & Gary, S. (2009). The contribution of HBCUS to the preparation of African American women for stem careers: A case study. *Research in Higher Education, 50*(1), 1–23.

Reyes, L. H., & Stanic,G. M.A. (1985). A review of literature on blacks and mathematics. Paper presented to the American Educational Research Association, Chicago.

Sewell, T. E., & Martin, R. P. (1976). Racial differences in patterns of occupational choice in adolescents. *Psychology in the Schools, 13*(3), 326–333.

Simpson, J. C. (2001). Segregated by subject: Racial differences in the factors influencing academic major between European Americans, Asian Americans, and African, Hispanic, and native Americans. *The Journal of Higher Education, 72*(1), 63–100.

Thomas, G. E. (1984). *Black college students and factors influencing their major field choice*. Baltimore: Johns Hopkins University, Center for Social Organization of Schools.

Thomas, G. E. (1986). Cultivating the interest of women and minorities in high school mathematics and science. *Science Education, 70*(1), 31–43.

UNESCO. (2010). *Engineering: Issues, challenges and opportunities for development* (p. 16). Paris: UNESCO.

Yoder, B. L. (2012). Engineering by the numbers. American Society for Engineering Education. http://www.asee.org/papers-and-publications/publications/11-47.pdf. Accessed 6 Sept 2014.

Derrick Hudson B.S., Humanities (United States Air Force Academy), M.A. Political Science (University of Central Oklahoma), Ph.D. International Relations (University of Denver). Assistant Professor of International Relations, Liberal Arts and International Studies, and Director, Master of International Political Economy of Resources (MIPER), Colorado School of Mines. Current research and teaching areas: African politics, globalization, religion and international relations, and engineering education.

Chapter 10
Depoliticization and the Structure of Engineering Education

Erin A. Cech and Heidi M. Sherick

Abstract The need for engineering students to develop nuanced understandings of the cultural, social, and political contexts of socio-technical systems has never been more obvious to engineering leaders and decision-makers. Yet, engineers often have obtuse definitions of their responsibilities to the public and seem to engage with the socio-cultural contexts and consequences of their work only in times controversy. A central underlying factor in this disengagement from considerations of social justice and equality is the *ideology of depoliticization*, the belief that engineering is a purely "technical" space in which engineers design technological objects and systems stripped of political and cultural concerns. In this chapter, we ask, what role does the culture and structure of engineering education play in promoting depoliticization? After elaborating the ideology of depoliticization, we argue that the culture of engineering pedagogy and the traditional curricular structure of engineering education (both its accreditation process and its intra-program curricula) help support and promote an ideology of depoliticization in engineering and train students to adopt this ideology within their own understandings of their professional roles and responsibilities. We end by discussing the consequences of having depoliticization embedded in the culture and structure of engineering education, and suggest possible policy solutions to *re*-politicize engineering education.

Keywords Depoliticization • Culture of engineering • Engineering education • Professional socialization

E.A. Cech (✉)
Department of Sociology, Rice University, 6100 Main St, MS28, 77005 Houston, TX, USA
e-mail: ecech@rice.edu

H.M. Sherick
Rice University, 1638 Castle Ct., 77006 Houston, TX, USA
e-mail: hmsherick@gmail.com

S.H. Christensen et al. (eds.), *International Perspectives on Engineering Education*, Philosophy of Engineering and Technology 20, DOI 10.1007/978-3-319-16169-3_10

Introduction

Engineers design technological objects and systems in an era when those objects and systems have never been more far-reaching. Large-scale sociotechnical systems, which engineers have a unique and socially validated hand in creating, touch nearly every corner of our most powerful social institutions (Verbeek 2006; Zimmerman 1995) and can reinforce (or possibly undermine) existing social inequalities along the lines of class, race/ethnicity, gender, sexual identity, and disability (Cech and Waidzunas 2011; Nye 2006; Riley 2011; Rolston and Cox 2015; Slaton 2015). Yet, the complexity of these sociotechnical objects and systems have long exceeded the ability for most "lay" individuals to fully understand them and have become "far too complex to be governable by ordinary citizens" (Zimmerman 1995, p. 89). Engineers not only engage in designing these complex socio-technical systems, but are increasingly relied upon to play the role of "public welfare watchdogs" (Cech 2014).

Accordingly, the need for young engineers to develop nuanced understandings of the cultural, social, and political contexts of socio-technical systems has never been more pertinent. Engineers' grasp of the co-construction of technical and socio-cultural realms is important for their sensitivity to how their work contributes to power hierarchies and processes of social inequity and their ability to uphold ethical standards in times of crisis. Yet, despite formal commitments to fostering engineering students' engagement with social welfare concerns, decades of literature has critiqued engineers' often obtuse definition of their responsibilities to society (Layton 1971; Petroski 1994). Engineers seem to actively engage in discussions of the contexts and consequences of their profession only in times of controversy, such as the Event Horizon oil spill (Catalano 2011).

Recent efforts in engineering education and policy (e.g. National Academy of Engineering 2004) have strived to nurture such sensitivity among future members of the profession. Engineering education, as a central place where aspiring engineers are explicitly taught their responsibilities of their professional roles, is a social location that theoretically allows for the development of engineering students' engagement with these socio-cultural contexts. Yet, mirroring cultural patterns in engineering more broadly, a study of students in four diverse U.S. engineering programs found that students' interest in public welfare actually declined over the course of their engineering education (Cech 2014). This lack of public welfare concern included factors such as whether students were interested in helping society, promoting racial understanding, and understanding the consequences of technology.

A central underlying factor in engineers' seeming disengagement from considerations of social justice and inequality is the ideology of depoliticization. As we describe in more detail below, the ideology of depoliticization is the belief that engineering is a purely "technical" space and political and cultural concerns can—and *should*—be removed from that space. This ideology emerges out of dualistic styles of thought that characterize the professional culture of engineering more broadly (Faulkner 2000) and has important consequences for the understanding that aspiring engineers develop about their professional responsibilities and how considerations of social justice fit into these responsibilities. In this chapter, we ask, what role does the culture and structure of engineering education play in promoting depoliticization? How does engineering education help reproduce this ideology among new generations of engineers?

We argue that the process of socializing students into the culture of engineering and the curricular structure of engineering education—both its accreditation process and its intra-program curricula—helps support and promote the ideology of depoliticization in engineering and train students to adopt this ideology within their own understandings of their professional roles and responsibilities. After describing depoliticization in more detail and presenting these arguments, we discuss the consequences of having depoliticization embedded in the culture and structure of engineering education, and suggest possible policy solutions to re-politicize engineering education.

Engineering Culture and the Ideology of Depoliticization

In contrast to popular belief, professional occupations are not simply collections of people who share technical expertise on a set of topics. Around—and even embedded within—this professional expertise are intricate cultural systems of meanings, practices, and epistemologies (Abbott 1988; Knorr Cetina 1999). Like other professions, engineering has its own unique, semi-autonomous culture that encompasses the beliefs systems, values, and myths built into and around engineering knowledge, practice, and tools (Cech 2013; Trice 1993). The professional culture of engineering serves to unite engineers together into a single social group, even though they may work in vastly different industries on very different projects. The culture of engineering may vary slightly by subfield, industry, and geographic region, but it is built into virtually all corners of the engineering profession.

Within this professional culture of engineering, particular ideologies serve as orienting frameworks for how engineers understand both the relationship of their profession to society and their own roles as individual professionals. Such ideologies also inform what generally counts as "legitimate" engineering work (Cech 2013). Such ideologies not only shape how individual engineers think about and enact their day to day professional work, but also the decisions profession leaders make about the direction of engineering in the future (see, for example, the National Academy of Engineering's *Grand Challenges* report [Cech 2012]).

Depoliticization in Engineering

A prominent ideology within the culture of engineering is the ideology of depoliticization (Cech 2013). Depoliticization is deeply entrenched in the professional culture of engineering and is the belief that engineering is a purely "technical" space in which engineers design technological objects and system—a space devoid of socio-cultural complexities. Depoliticization promotes an approach to engineering that assumes that political and social contexts *can* be separated out from the technical and, more importantly, that such contexts *should* be removed from engineering work. As such, this ideology may be a central factor in engineers' seeming disengagement from considerations of the co-construction of the technical and the socio-cultural.

Depoliticization is the opposite pendulum swing from ideas of technocracy that reached prominence in the 1920s (Jordan 1994). It has its roots in expressions of disillusionment and skepticism with technology brought on by WWII and the environmental movements of the 1970s (Florman 1994; Slaton 2011). The siloing of "technical" and "social" or "political" knowledge and considerations reflects a more overarching trend toward dualistic styles of thought in engineering (Faulkner 2000). In particular, Sally Hacker (1981) introduced and Wendy Faulkner (2000) expanded the idea of a "technical/social dualism" in engineering, where technical and social forms of knowledge are differentiated and separated. Depoliticization captures the notion that the separation of technical and social issues is not just a cognitive act, but a *moral* one—depoliticization prescribes how engineering work should be conducted and how engineers should approach their work.

Of course, depoliticization is an unobtainable ideology rather than a stylized notion of reality: political and cultural contexts can never be removed from technological design (e.g. Faulkner 2000; Latour 1999). Depoliticization, nonetheless, helps frame social justice concerns—such as how technology retrenches poverty, marginalizes disabled individuals, or builds sexism, racism and heteronormativity into physical objects and systems—as irrelevant to the work of engineering (Cech 2013, 2014) and delegitimizes the very socio-cultural context that provides the necessary basis for engineers' enactment of their responsibilities to the public, such as whistle blowing. In other words, depoliticization prevents engineers from understanding their work as Science and Technology Studies (STS) scholars do: as part of socio-technical systems.

In this chapter, we are thus interested in articulating the role that engineering education can play in promoting depoliticization. In particular, we focus on the culture of engineering as it manifests within the socialization of students, and the structure of engineering education via accreditation processes and engineering program design. We end by discussing the consequences of having depoliticization embedded in the culture and structure of engineering education, and suggest possible policy solutions to challenge this ideology.

Professional Socialization in Engineering Education

Depoliticization, as a prominent ideology within the culture of engineering, likely permeates engineering education programs as well (Cech 2013). As engineering programs seek to transform neophytes into practicing engineers, they not only impart upon them the intellectual tools of the trade, they also teach students how to *be* engineers—or, in common engineering parlance, how to "think like engineers." This process, called professional socialization, has been well-documented in other professions such as law, medicine and management, and is a central mechanism through which professions reproduce themselves from generation to generation (Becker et al. 1961; Costello 2005; Schleef 2006). Through their experiences in classrooms, residence halls, laboratories, study groups, assignments and internships, engineering students learn responsibilities of the engineering profession to society and what it means to be an individual representative of that profession (Dryburgh 1999).

During professional socialization, students learn, and learn to take on as their own, the beliefs and values of the culture of the profession to which they aspire. The adoption of this professional culture is not simply the adoption of a set of abstract ideologies, however. Socialization into the professional culture of engineering means that ideologies within that culture manifest in a variety of more concrete ways in students' understandings of what it means to be an engineer.

First, cultural ideologies present in engineering education can manifest in students' epistemological understandings of engineering—their definitions of what counts as reasonable and legitimate engineering knowledge, tools, and practices. In theory, a host of factors could be considered valid inputs in engineering problem-solving and design. Engineering epistemologies serve as rules for what information and practices are considered important in engineering problem definition and problem solving (Knorr Cetina 1999; Petroski 1994) and what are considered irrelevant. Ideologies within the professional culture of engineering inform these epistemologies by providing criteria for relevant inputs and outputs. The ideology of depoliticization, for instance, promotes the bracketing of information that is not strictly technical, such as questions about access and unequal burdens and benefits, from problem definition and design practices. This bracketing is illustrated in the typical structure of assignments in engineering courses, which often provide specifications for the size, shape, and mechanical functionality of the process to be designed, but little information about who will use it or what it will be used for. As engineering students learn the epistemologies of their profession, the ideology of depoliticization is likely built into what they come to understand about what counts as "real" engineering knowledge and design work.

Second, socialization means that cultural ideologies like depoliticization inform students' overall understanding of the role of their profession in society. Learning to become a professional means learning the profession-sanctioned definition of the responsibilities of one's profession in society, particularly the jurisdiction of the profession's socially-sanctioned and monopolized expertise (Abbott 1988). Jurisdictional boundaries are constantly negotiated among different professions, and must be defended from encroachment by other interested parties (Abbott 1988). As such, neophytes learn both these jurisdictions and arguments to defend (and even expand) those jurisdictions. Here, ideologies such as depoliticization influence the definitions students form about what is inside the jurisdiction of engineering. Depoliticization emphasizes a narrowly technical jurisdictional realm for engineering: if engineers do not claim jurisdiction over social issues such as the consequences of their work for public welfare, then they may not hold themselves responsible for those consequences. This has important implications, especially if these neophytes eventually become profession leaders: the notions that aspiring engineers develop about the responsibilities and jurisdictions of their profession may inform the direction they lead the profession in the future.

Third, through the socialization process, depoliticization is likely inflected in the very identities students develop as engineers. During professional socialization, neophytes usually develop a personal identification with and commitment to their profession (Becker et al. 1961; Ibarra 1999). But, the ideologies of their profession are not just layered on top of students' existing identities, these ideologies often appear in students as *personal* traits (Costello 2005). The dominant cultural ideologies in the profession serve as touchstones for the professional identities that students develop as they go through engineering training. Depoliticization within engineering education, in other words, manifests in engineering students' budding professional identities, informing the things they are *personally* committed to in their professional careers. Specifically, depoliticization may discourage new engineers from elevating considerations of social justice and public welfare to the level of technical considerations such as size, speed, and efficiency.

In sum, through professional socialization, overarching cultural ideologies within engineering such as depoliticization shape the epistemologies engineering students develop to solve problems, their overarching understanding of the responsibilities of their profession to society, and the professional identities aspiring engineers develop. Socialization in engineering education is thus a powerful process through which depoliticization is folded into engineering students' understandings of what it means to be engineers. The responsibility that accompanies the professional socialization process thus also comes with great opportunity: engineering education is an important site where depoliticization may be interrupted. However, as the next section discusses, the structure of engineering education means that such dismantling of depoliticization would be difficult to accomplish under current curricular arrangements and priorities.

Curricular Structure of Engineering Education

In addition to the professional socialization process, the very structure of the engineering curriculum may reinforce the ideology of depoliticization. Through both accreditation processes and the day-to-day pedagogical practices of engineering faculty, the typical arrangement of engineering education in the U.S. may promote the bracketing of social and political issues and the labeling of such issues as irrelevant to "real" engineering practice. We now discuss how these processes can build depoliticization into the structure of engineering education and make *re-politicization* difficult.

Although they have little formal power to shut down non-compliant engineering programs or to facilitate change in the profession beyond engineering education, the formal accreditation processes of ABET, Inc. carry tremendous *symbolic* significance: in order for engineering programs to be recognized as legitimate purveyors of engineering training, they must be accredited. Unaccredited engineering programs are disadvantaged in competing for the top students, and students without degrees from accredited programs are disadvantaged in securing top engineering jobs and professional licensure. As such, the values and commitments built into the accreditation processes can help shape the values and commitments of engineering education.

ABET, Inc., formerly known as the "Accreditation Board for Engineering and Technology, Inc.," has served as the accreditation authority in engineering for over 80 years. As an organization, ABET is composed of a board of directors, plus representatives of professional organizations from all sub-specialties of engineering. The accreditation activities themselves (site visits to schools, reviewing of program self-studies, etc.) are conducted by teams of volunteer evaluators who are usually engineers from academia and industry (abet.org). ABET's stated mission is as follows: "ABET serves the public globally through the promotion and advancement of education in applied science, computing, engineering and engineering technology" (abet.org). Alongside accrediting educational programs and evaluating quality, ABET's core mission is to "stimulate innovation" in engineering education. In practice, however, accreditation procedures tend to serve a conservative, rather than innovative, function (Abbott 1988).

Accreditation is voluntary and engineering education programs must request to be evaluated by ABET. There are several quality standards against which engineering programs are evaluated, ranging from lab space to computer facilities, faculty adequacy and program curricula. In the late 1990s, in response to criticisms about the rigidity of prior accreditation requirements, ABET changed from "bean-counting" accreditation requirements to a new set of criteria based on student outcomes; a set of competencies that students who graduate from accredited engineering programs are supposed to display, referred to as EC2000 (EC2000 report). Responding to increasing internal and external pressure to include socio-cultural concerns as accreditation requirements, ABEt also added the criteria that students graduate from their programs being able to "understand [their] professional, ethical responsibility," have a "broad education to understand social context" and have "knowledge of contemporary issues."

Although ABET's reconfigured accreditation requirements purport to make the socio-cultural context of technology more prominent in engineering education, by demarcating these contexts as separate outcomes, this reconfiguration actually may help *reproduce*, rather than undermine, depoliticization. Because socio-cultural competencies are understood as separate accreditation outcomes from more technical competencies, teaching socio-cultural contexts is effectively siphoned off from more technical training and contained within separate courses, or separate modules within existing courses. For example, a recent study (Barry and Ohland 2012) assessed the impact of curriculum reform following these changes in the ABET criteria, seeking to determine the level of professional and ethical curriculum content in place after the implementation of EC2000. While the *content* offered on topics like ethics increased within many engineering programs, engineering students did not appear to develop additional reflexivity about their ethical and social responsibilities. Of course, a nuanced understanding of socio-cultural context of technology requires that those contexts are *not* divorced from the technical considerations in which they are actually embedded. By peeling off ethics training into a separate course and codifying the relative unimportance of professional/ethical responsibilities by requiring students to take only one course on the topic, this arrangement likely reinforces, rather than undermines, the ideology of depoliticization.

Third, as noted above, accreditation evaluators are usually practicing engineers from industry and academia; few are formally trained in the socio-cultural contexts of technology. Except for their own idiosyncratic experiences, few may have the academic background necessary to judge whether students really do have "an understanding of professional and ethical responsibility," and "the broad education necessary to understand the impact of engineering solutions in a global, economic, environmental, and societal context" (abet.com). Thus, the very definitions of credibility in engineering as culturally defined by ABET undervalues the importance of socio-cultural context simply because accreditation evaluators are often not well-trained to identify and articulate the socio-cultural contexts and consequences of technology.

Through these processes, the ideology of depoliticization is threaded throughout formal procedures of accreditation and helps reinforce this ideology within the culture of engineering education. But, accreditation is not the only avenue through which the curricular structure of engineering education reproduces depoliticization. Equally consequential are widely-shared practices of curricular arrangements within engineering programs.

First is the problem of engineering faculty's own pedagogical training: even if faculty wanted to integrate socio-cultural contexts into their courses, many may lack the pedagogical tools to do so. Few faculty have ever taken a service learning, cooperative learning, or active learning course, let alone have the training to integrate socio-cultural contexts into educational spaces that otherwise promote depoliticization (Barry and Ohland 2012). Traditional engineering textbooks are also usually not written with support for dynamic modes of instruction (see Riley (2011) for a notable exception).

Second, one of the biggest challenges to undermining depoliticization in engineering education is crowding of the engineering curriculum. Faculty are under great pressure to squeeze an ever-increasing amount of content into their courses. Feedback from industrial advisory board members encourage engineering curricula to incorporate more business concepts, more inter-disciplinary cooperation, and more technological solutions into their classes (National Academy of Engineering 2004). Parallel pressures for replacing or omitting antiquated technical content are rarely expressed. As such, content not considered directly relevant to technical content are easy targets for omission. Furthermore, required courses are rigidly sequenced and tightly packed, leaving students with little flexibility to explore professional enrichment in non-engineering courses (Culver et al. 2005). As such, the typical arrangement of engineering courses not only *reflects* the ideology of depoliticization, it also *reinforces* it.

Third, and perhaps most importantly, depoliticization often looms over faculty promotion and tenure processes in engineering departments. Even if effective teaching is emphasized in promotion and tenure considerations, attempting to integrate socio-cultural contexts into otherwise technical courses is not usually the sort of effective teaching that is meant by promotion and tenure committees. Faculty who engage in pedagogical innovations may be penalized, both because the time and effort required to integrate socio-cultural contexts into their courses takes time away from research, and also because such politicization may cast those faculty as less "serious" engineers in the eyes of their colleagues (Lattuca et al. 2006). Such penalties may be particularly consequential for junior faculty.

By divvying up engineering content by technical subspecialty, by making the outcomes of those courses predominantly about technical mastery, and by devaluing socio-cultural contexts in promotion and tenure decisions, engineering programs promote a vision of engineering where technical mastery is sufficient to earn an engineering degree and competence in the socio-cultural contexts of technology is superfluous to "real" engineering work. These pedagogical and curricular structures may undergird the professional socialization process discussed above to create an educational environment where depoliticization is reinforced at multiple levels and through both formal and informal institutional processes.

Consequences of Depoliticization in Engineering Education

What are the potential consequences of a curricular structure that deemphasizes socio-cultural contexts of technology, and of professional socialization processes that embed depoliticization into aspiring engineers' epistemologies, professional identities, and their broader understandings of the responsibility of their profession to society? First, it means that engineering students may be trained with an understanding of their future roles as engineers that belies the full extent of what those roles will actually entail: engineering education presents an overly-abstracted, simplified, and decontextualized picture of the engineering profession. Contrary to the ideology of depoliticization, practicing engineering is a messy and politicized endeavor—by being trained in a social space permeated by the ideology of depoliticization, engineering students not only leave their training unprepared to deal with socio-cultural complexities inherent in "real world" engineering, but also lack the intellectual tools and epistemological scaffolding necessary to clearly recognize such complexities. As reflected in the work of Science and Technology Studies scholars (Bereano 1976; Bijker and Law 1992; Bucciarelli 1994; Faulkner 2007), engineering work is never as decontextualized as it is portrayed in engineering classrooms and textbooks. Whether they are trained to or not, engineering students who graduate and enter engineering jobs must contend with a myriad of "political" concerns such as uncertainty, regulation, public welfare, and conflicts of interest.

Second, like other ideologies at the core of cultural belief systems, the ideology of depoliticization in engineering education is likely very difficult to undermine.

Because depoliticization is codified in multiple dimensions of the culture and structure of engineering education, it is reinforced and reproduced anew through the overlapping and interdependent processes of socialization, accreditation, and pedagogy. Re-politicizing engineering education would require not only cultural shifts but systemic changes in the structure of engineering accreditation and pedagogy.

Can Engineering Education Be Re-politicized?

We have argued in this chapter that the professional socialization of engineering students and the curricular structure of engineering education impart the ideology of depoliticization into several dimensions of engineering training. Students do not just learn to value depoliticization as an abstract ideal; depoliticization comes to be a part of what it means to them to "think like engineers" and *do* engineering work. Given the cycle of influence that passes this ideology from faculty to neophytes, how might engineering education be *re-politicized*?

Recently, several schools have sought to reconfigure their curriculum to challenge depoliticization. However, institutional isomorphism makes it difficult to create lasting changes to engineering education (DiMaggio and Powell 1983). Essentially, new programs that attempt to innovate face the challenge of convincing prospective students, peer universities, and potential employers of the graduates of those programs that they are not *too* innovative. Thus, isomorphism can marginalize innovative programs that attempt to alter their pedagogical cultures and curricular structures to promote training in the social and political contexts of engineering design.

Despite these challenges, we believe there are several changes that might help to re-politicize engineering education. First, the ideology of depoliticization must be deliberately and repeatedly deconstructed in engineering classrooms and in the planning and implementation of engineering curricula (Cech 2013). Deconstruction involves overt discussions of this ideology and clear explanations of *why* it is problematic. By openly articulating the contours of this ideology, students may learn to recognize depoliticizing forces and even attempt to re-politicize their own educational spaces.

Furthermore, re-politicizing the epistemologies of engineering would help alter how students learn to "think like engineers." Such an alteration might involve pushing students to recognize and deliberate on the socio-cultural aspects of problem definition and solution. Extracurricular activities such as "Engineers Without Boarders" are also a step in the right direction. It is important that students learn that considering the social contexts and impacts of their design work is not a separate, expendable step that happens *after* a design is complete, but rather an iterative process involved at the beginning, middle, and end of design.

We also suggest several changes to the curricular structure of engineering education. First, relating to accreditation, ABET's criteria should be more specific in its expectations for outcomes related to socio-cultural contexts. While EC2000 removed the rigidity of the previous criteria, it replaces rigidity with vagueness. ABET leaders and evaluators need to be able to clearly recognize and articulate what it means to teach engineers to be competent in the social and cultural contexts of their work.

Second, as noted above, ABET evaluators are usually individuals trained as engineers. In order to competently judge whether engineering students are indeed emerging from their programs able to conceptualize socio-cultural contexts of technology, it is necessary to include among the evaluators individuals who have expertise in those contexts. We acknowledge that adding an evaluator increases the financial commitment from institutions for accreditation procedures. If this aspect of engineering education is a priority, it should be supported and embedded in the ABET evaluation process. Furthermore, the feedback ABET evaluators provide to programs after site visits need to include constructive, concrete feedback on how to improve in the areas relating to socio-cultural context (Lattuca et al. 2006).

Third, in order to undermine depoliticization, the organization of and emphases within engineering courses must shift. While we recognize that a drastic re-organization of the way engineering training is carved into courses is unlikely, technical courses could be re-politicized by introducing socio-cultural considerations in the way that engineering problem-solving is taught. In order for such content to be taken seriously by students who are steeped in depoliticization in most other realms of their engineering education, students must be held *accountable* for that knowledge: full credit on an exam question might require, for example, not only deriving the correct numerical solution to a design problem but thoughtfully articulating socio-cultural considerations of access, power, stereotypes, and unequal burdens embedded in the definition and solutions to this problem. Of course, a simple addition of content to existing course material would only exasperate the curricular crowding problem discussed above. Put bluntly, if engineering curricula is to be re-politicized, it must cover less technical content. This is a radical suggestion. But, it is widely acknowledged that engineering students rarely use *all* the content they learn in engineering courses (cf. Barry and Ohland 2012), and a great deal of the technical knowledge engineers need to do their work is learned on the job. As Culver et al. (2005, p. 19) suggest, learning the socio-technical contexts of engineering work "may be more important than learning all the power cycles." We contend that being able to recognize and articulate the socio-cultural contexts of engineering work will serve students better in the future than learning "all the power cycles."

Finally, and perhaps most importantly, depoliticization in formal and informal promotion and tenure requirements needs to be addressed. While quality teaching is usually considered important for promotion and tenure in the abstract, efforts put toward curricular innovations that integrate socio-cultural contexts into the teaching of engineering problem definition and problem solving is often considered extra and may not count as promotion-worthy activities (Lattuca et al. 2006). More consequentially, the ubiquity of depoliticization likely means that faculty who express commitment to re-politicizing engineering classrooms may *themselves* be considered less serious scholars by their colleagues. In order for engineering education to be re-politicized, faculty must be rewarded—or at least not penalized—for articulating and integrating socio-cultural contexts of engineering design.

Depoliticization is a deeply ingrained ideology within engineering. Through its integration into engineering education, this ideology is passed on to new generations of engineers. Engineering education, as the training ground for future engineering professionals, may have the strongest role in reproducing the ideology of depoliticization. But, engineering education also provides the greatest opportunity for interrupting this cultural cycle, and re-politicizing engineering for the newest generations of engineers who will lead their profession into the future.

References

Abbott, A. (1988). *The systems of professions: An essay on the division of expert labor*. Chicago: University of Chicago Press.

Barry, B. E., & Ohland, M. W. (2012). ABET criterion 3.f: How much curriculum content is enough? *Science and Engineering Ethics, 18*, 369–392.

Becker, H., Geer, B., Hughes, E. C., & Strauss, A. L. (1961). *Boys in white: Student culture in medical school*. New Brunswick: Transactional Books.

Bereano, P. L. (1976). *Technology as a social and political phenomenon*. New York: Wiley.

Bijker, W. E., & Law, J. (1992). *Shaping technology/building society: Studies in sociotechnical change*. Cambridge, MA: MIT Press.

Bucciarelli, L. L. (1994). *Designing engineers*. Cambridge, MA: MIT Press.

Catalano, G. D. (2011). *Tragedy in the Gulf: A call for a new engineering ethic*. New York: Morgan and Claypool.

Cech, E. A. (2012). Great problems of grand challenges: Problematizing engineering's understandings of its role in society [Grand Challenges, National Academy of Engineering, participation, reflexivity, technical-social division]. *International Journal of Engineering, Social Justice, and Peace, 1*(2), 85–94.

Cech, E. A. (2013). The (Mis)framing of social justice: Why meritocracy and depoliticization hinder engineers' ability to think about social injustices. In J. Lucena (Ed.), *Engineering education for social justice: Critical explorations and opportunities* (pp. 67–84). New York: Springer.

Cech, E. A. (2014). Culture of disengagement in engineering education? *Science, Technology & Human Values, 39*(1), 34–63.

Cech, E. A., & Waidzunas, T. (2011). Navigating the heteronormativity of engineering: The experiences of lesbian, gay, and bisexual students. *Engineering Studies, 3*(1), 1–24.

Costello, C. Y. (2005). *Professional identity crisis: Race, class, gender and success at professional schools*. Nashville: Vanderbilt University Press.

Culver, R., McGrann, R., & Lehmann, G. (2005). Preparing students for ABET a-k. Paper presented at the *ASEE/IEEE Frontiers in Education Conference*, Indianapolis.

DiMaggio, P. J., & Powell, W. (1983). "The iron cage revisited" institutional isomorphism and collective rationality in organizational fields. *American Sociological Review, 48*, 147–160.

Dryburgh, H. (1999). Work hard, play hard: Women and professionalization in engineering-adapting to the culture. *Gender and Society, 13*(5), 664–682.

Faulkner, W. (2000). Dualism, hierarchies and gender in engineering. *Social Studies of Science, 30*(5), 759–792.
Faulkner, W. (2007). 'Nuts and bolts and people': Gender-troubled engineering identities. *Social Studies of Science, 37*(3).
Florman, S. C. (1994). *The existential pleasures of engineering* (Vol. 2). New York: St. Martin's Griffin.
Hacker, S. L. (1981). The culture of engineering: Woman, workplace and machine. *Women's Studies International Quarterly, 4*, 341–353.
Ibarra, H. (1999). Provisional selves: Experimenting with image and identity in professional adaptation. *Administrative Science Quarterly, 44*(4), 764–791. doi:10.2307/2667055.
Jordan, J. M. (1994). *Machine-age ideology: Social engineering and American liberalism, 1911–1939*. Chapel Hill: The University of North Carolina Press.
Knorr Cetina, K. (1999). *Epistemic cultures: How the sciences make knowledge*. Cambridge, MA: Harvard University Press.
Latour, B. (1999). *Science in action: How to follow scientists and engineers through society*. Cambridge, MA: Harvard University Press.
Lattuca, L. R., Terenzini, P. T., & Volkwein, J. F. (2006). *Engineering change: A study of the impact of EC2000*. Baltimore: ABET.
Layton, E. T., Jr. (1971). *The revolt of the engineers: Social responsibility and the American engineering profession*. Cleveland: Western Reserve Press.
National Academy of Engineering. (2004). *The engineer of 2020: Visions of engineering in the new century*. Washington, DC: National Academies Press.
Nye, D. E. (2006). *Technology matters: Questions to live with*. Cambridge, MA: MIT Press.
Petroski, H. (1994). *Design paradigms: Case histories of error and judgement in engineering*. Cambridge, UK: Cambridge University Press.
Riley, D. M. (2011). *Engineering thermodynamics and 21st century energy problems: A textbook companion for student engagement*. San Rafael: Morgan and Claypool.
Rolston, J., & Cox, E. (2015). Engineering for the real world: Diversity, innovation and hands-on learning. In S. H. Christensen, C. Didier, A. Jamison, M. Meganck, C. Mitcham, & B. Newberry (Eds.), *International perspectives on engineering education: Engineering education and practice in context* (Vol. I.). Springer Science+Business Media B.V.
Schleef, D. J. (2006). *Managing elites: Professional socialization in law and business schools*. New York: Rowan and Littlefield.
Slaton, A. E. (2011). *Note to self: Save humanity (A social and culture history of the "grand challenges")*. Vancouver: American Society for Engineering Education.
Slaton, A. E. (2015). Meritocracy, technocracy, democracy: Understandings of racial and gender equity in American engineering education. In S. H. Christensen, C. Didier, A. Jamison, M. Meganck, C. Mitcham, & B. Newberry (Eds.), *International perspectives on engineering education: Engineering education and practice in context* (Vol. I). Springer Science+Business Media B.V.
Trice, H. M. (1993). *Occupational subcultures in the workplace*. Ithica: ILR Press.
Verbeek, P.-P. (2006). Materializing morality: Design ethics and technological mediation. *Science, Technology & Human Values, 31*(3), 361–380. doi:10.1177/0162243905285847.
Zimmerman, A. D. (1995). Toward a more democratic ethic of technological governance. *Science, Technology, & Human Values, 20*(1), 86–107.

Erin A. Cech B.S. in Electrical Engineering and B.S. in Sociology from Montana State University-Bozeman; M.A. and Ph.D. in Sociology from the University of California, San Diego. Assistant Professor in the Department of Sociology at Rice University. Cech's research seeks to uncover cultural mechanisms of inequality reproduction—particularly gender, sexual identity and racial/ethnic inequality within science and engineering professions. Her research has appeared in the *American Sociological Review, Social Problems,* and *Engineering Studies.*

Heidi M. Sherick B.S. in Biological Sciences, M.Ed. in Adult and Higher Education from Montana State University-Bozeman and a Ph.D. in Educational Leadership from University of Nebraska-Lincoln. She served as the Assistant Dean for Undergraduate Programs and Diversity in the College of Engineering at Montana State University (MSU) from 2001 to 2012. As Assistant Dean she oversaw all aspects of the undergraduate programs including accreditation, academic policy, curriculum, student organizations, and scholarships. She also served as the Director of EMPower, the engineering minority program. Sherick studies higher education leadership and her dissertation investigated the processes through which higher education leadership is fostered.

Part III
Reforming Engineering Education: Experiences and Cases

Introduction

Steen Hyldgaard Christensen and Niels Mejlgaard

What binds the chapters in this part together is a critical engagement with perceived challenges in and for engineering education and for engineering educators.

The diversity of images and desired identities of engineers presented in the seven chapters varies from radical departures from dominant images to instrumental ways of improving pedagogies and epistemologies constitutive of those images. Taken together the chapters seek to either explicitly or implicitly highlight a number of engineering mindsets and contexts relevant to the intersection between engineering and social justice. There is an implicit consensus among the chapters that engineering cannot be characterized by a single mindset but rather by a number of dominant ones related to a variety of contexts and presented as blinders with respect to the desire to transform engineering education. As these mindsets and related epistemologies are of a pervasive nature there is a simultaneous recognition that macro-change in engineering education is difficult and has frequently been doomed to fail. All the more so as engineering is presently in a state of what Rosalind Williams (2002, p. 70) has called expansive disintegration pulling engineering in different directions – toward science, the market, design, technological systems and socialization. Hence behind the micro-change initiatives presented in this part a strong desire for macro-change is borne out bringing to mind as a metaphor the kind of utopian thinking that Lewis Mumford in his 1922 book *The Story of Utopias* has called a yearning for a *utopia of reconstruction.*

S.H. Christensen (✉)
Department of Development & Planning, Aalborg University,
Vester Havnepromenade 5, Aalborg 9000, Denmark
e-mail: steenhc@plan.aau.dk

N. Mejlgaard
Danish Centre for Studies in Research and Research Policy, Aarhus University
Bartholins Allé 7, Aarhus C 8000, Denmark
e-mail: nm@cfa.au.dk

Social justice in this part is related to sustainability issues, energy production and use, community development, diversity and inclusiveness, new ways of learning, modes of social intervention, and more broadly international development. Taken at face value those concerns to a certain extent parallel concerns in what has been variously characterized as the emerging discipline, field or community of engineering education research/researchers (Jesiek et al. 2009). Here the challenges in engineering education are seen as lying within the following five research areas (Special Report 2006, pp. 259–261):

1. *Engineering Epistemologies*: Research on what constitutes engineering thinking and knowledge within social contexts now and in the future.
2. *Engineering Learning Mechanisms*: Research on engineering learners' developing knowledge and competencies.
3. *Engineering Learning Systems*: Research on the instructional culture, institutional infrastructure, and epistemology of engineering educators
4. *Engineering Diversity and Inclusiveness*: Research on how diverse human talents contribute solutions to the social and global challenges and relevance of our profession.
5. *Engineering Assessment*: Research on, and the development of, assessment methods, instruments, and metrics to inform engineering education practice and learning.

The optimism among engineering education researchers regarding a new beginning for not only reforming but transforming engineering education entirely by moving it out of its current mode after many years of failed reform efforts (see e.g. Borrego et al. 2008) are seen in the various chapters in this part in the sobering light of a number of recurrent tensions in engineering education. These tensions are related to questions concerning: what, how much, how, when, where, engineering student should learn, and by whom, and for whom engineering should be taught. Therefore any particular configuration of responses to these questions will only – seen from a philosophical point of view – create a temporary solution which is open to critique.

Due to the inherent normativity of any configuration of responses educational reforms are difficult to implement as they are very complex on at least five counts: (1) They are inextricably linked to perceptions of current thinking and actions on educational concerns and reforms around the world, (2) The vision behind curriculum reform is concurrently the expression of a political and a technological agenda which is open to critique, (3) Curriculum reform is both a process and a product, which involves a wide range of institutions, stakeholders and actors, (4) The process of constructing a curriculum is unique to each national and institutional setting. It is the complex outcome of negotiations between stakeholders to meet the perceived needs and requirements of companies, students and society, (5) Quite often the strategic goals of stakeholders collide.

In Chap. 11, Juan Lucena traces how engineers came to be involved with development, sustainability and communities and the consequences of this history for present day practices and projects. Through this history, he shows how a dominant configuration of responses in engineering education and practice has evolved in the US since WWII resulting in engineering mindsets characterized by the way they have come to define where and how most engineers work, think and approach problem definition and solution, and more generally what engineers value. Building on the work of Donna Riley, Lucena analyzes how the following mindsets get in the way of engineers seeing social justice:

- Dominance of military and corporate organizations
- Positivism and the myth of objectivity
- A desire to help and the persistence to do it
- A narrow technical focus
- Uncritical acceptance of authority

Taking the position of a transformative intellectual much in line with the ideology critique by the Critical Theory of the early Frankfurt School he asks whether critical engineering education can counteract the blinders of history and ideology which have made social justice a marginal concern, at best, or totally irrelevant in engineering practice and education. In responding he questions dominant models of development, dominant definitions of sustainable development, and mainstream methods of community participation and other barriers to providing socially just and sustainable solutions to communities. Lucena concludes by providing specific strategies to teach students "to see, reflect and critically question these engineering mindsets so they do not take these for granted nor assume that this is the way the world of engineering has always been and will always be."

In contrast to Lucena the overall context with respect to social justice in Chap. 13 by Jessica Rolston and Elizabeth Cox is widening participation in higher education. Mass higher education and its commitment to equality of educational opportunity challenge traditional meritocratic criteria of access. To better accommodate students from disadvantaged backgrounds and demographic categories such as ethnic groups, new immigrants or poor whites "compensatory programs" and additional nonacademic criteria have been introduced (Trow 2007). Rolston and Cox report on a program that takes a different approach to diversifying engineering education by treating the backgrounds of nontraditional students as strengths rather than simply weaknesses to be overcome. They critique the prevailing "weed out" culture in undergraduate programs as an impediment to diversity, as it draws and graduates mainly middle class students with strong backgrounds in science and math, with adverse effects for nontraditional students who have a wealth of practical knowledge but often come from high schools without strong college preparation programs. They suggest that a learning by doing initiative called *Engineering by Doing* (EbD) at the Colorado School of Mines has the potential to foster greater inclusiveness, diversity and retention of all types of students in engineering undergraduate programs. In this initiative traditional students with strong math and science skills are brought together with non-traditional students with practical skills to collaborate in projects with a specific hands-on element. Advocating an epistemological move from *knowing that* to *knowing how* they argue that students with diverse backgrounds can learn from each other's strengths, and that non-traditional students in particular would be better served as their background knowledge and experiences would be made visible and valued.

In Chap. 12 Joseph Herkert, Rachelle Hollander, Clark Miller, Frazier Benya, Chad Monfreda and Lynette Osborne discuss an initiative on social justice in the context of global energy consumption and use. Their chapter may be put into perspective by referring to Benjamin Sovacool's recent 2013 book *Energy and Ethics: Justice and the Global Energy Challenge.* Sovacool challenges the view that energy problems will be "solved" by the market and that energy policy and security problems are therefore matters best left to economists and engineers. Sovacool puts it this way:

> Left to their own devices, global energy markets will prolong the use of fossil fuels as long as they are profitable to extract and use, down to the last remaining drops of oil and lumps of coal, even if their combustion and use permanently damages the climate and ruins local communities, or if their benefits seemingly outweigh their costs (even if all benefits accrue to one wealthy company and the costs afflict thousands of penniless villagers). Similarly, research scientists and engineers will help them to do so as long as they have vested interest in the energy sector – which hundreds of thousands do (Sovacool 2013, p. 2)

In line with Sovacool Chap. 12 approaches social justice from a macro-ethical perspective concerned with distributive and intergenerational justice. Both question the idea that the energy problem facing the world today can be solved by designing technical solutions by scientists and engineers without first addressing fundamental moral questions about justice and ethics. The chapter by Herkert et al. reports findings from an energy ethics project which took place at Arizona State University. Targeting the education of engineers who have some engagement with energy the goals of the project is of a tripartite nature:

- To develop a strong intellectual basis for understanding ethical challenges posed by large-scale transitions in energy systems, as well as criteria and approaches for evaluating the ethical desirability of future energy options;
- To provide a variety of robust opportunities for students to learn about energy ethics and how it applies to energy research and development; and
- To disseminate ideas and materials broadly for use in science and engineering education.

In dealing explicitly with context in engineering education and promoting hybrid thought and imagination Chap. 14 by Andrew Jamison, Niels Mejlgaard, and Jette Egelund Holgaard should be seen in the light of what Ernest L. Boyer in his seminal 1990 book *Scholarship Reconsidered: Priorities of the Professoriate* has called *scholarship of integration.* Boyer writes:

> The point is that even as the categories of human knowledge have become more and more discrete, the need for interdisciplinary insight has increased. Indeed, the real danger is that graduate students will become specialists without perspective, that they will have technical competence but lack larger insights. To avoid such narrowness an integrative component should be built into every program. Specifically, we urge that all doctoral students be asked to carry out their special area of study in historical perspective and that time during graduate study also be devoted to social and ethical concerns. In such a program, the scholar should

find metaphors and paradigms that give larger meaning to specialized knowledge (Boyer 1990, p. 68).

Boyer contrasts the scholarship of integration with three additional kinds of scholarship – discovery, application and teaching – all of them closely interconnected. By way of providing examples from the authors' experience in teaching contextual knowledge for engineering students at Aalborg University Jamison, Mejlgaard and Holgaard illuminate their integrative effort and the way they have strived to develop *a hybrid imagination* in their programs and students. In so doing they have worked from the following understanding of the concept. A hybrid imagination is a combination in thought and action of a scientific-technical problem-solving competence with an understanding of the problems that need to be solved. It involves a mixing of scientific education and training in technical skills with an appreciation of the broader cultural implications of science and technology in general and one's own role as an engineer. Hybrid imagining and thought challenges the dominant epistemological technical-social context dichotomy in engineering degree programs.

Picking up on engineering epistemology Chap. 15 by Ulrik Jørgensen sets out to question the ideology and codes of knowledge behind the dominant construction of the epistemological core in engineering education. He argues that despite the fact that the idea of a common core with corresponding mindsets has been challenged several times throughout the history of engineering education due to the growth in professional specialties and challenges as well as new technical disciplines the idea of an engineering curriculum dominated by math and science has remained strong. As the key component of the engineering identity this core has provided engineering students with a common set of neutral and rational methods leading them to the belief that their expertise is objective and able to exclude human values and maintain independence from politics. Jørgensen identifies technology as a cornerstone in the constitution of social order and points to four different response strategies with respect to socio-technical integration in engineering. He promotes the idea that an approach emphasizing the socio-material nature of technology, which builds on lessons from technology studies, can play a vital role towards a transformation in engineering education. This idea has been promoted in new educations focusing on engineering design, and has demonstrated its strength in providing professional competences to engineers. Having a closer relation to engineering practice socio-material design would be a suitable candidate as a new core in engineering education characterized by heterogeneity instead of being dominated by a disciplinary mono-culture.

In contrast to concrete institutional micro-change and social justice informed initiatives as presented in Chaps. 11, 12, and 13 Tony Marjoram in Chap. 16 approaches social justice in engineering education from a macro-change oriented policy perspective in line with the need for re-framing the knowledge and visions outlined in Chaps. 14 and 15. In reviewing a number of transnational policy papers and reports the main trust of Marjoram's argument is that engineering and engineering education should play a key role in addressing the following UN Millennium Development goals:

1. Eradication of extreme poverty and hunger
2. Achievement of universal primary education
3. Promotion of gender equality and empower women
4. Reduction of child mortality
5. Improvement of maternal health
6. Combating HIV/AIDS, malaria and other diseases
7. Ensuring environmental sustainability
8. Development of global partnership for development

Given these goals Marjoram moves on to discuss epistemological and educational prerequisites for achieving them and to counter declining interest, enrolment and retention in engineering education. At the level of learning he argues that student-centred, project- and problem-based learning at the juncture between humanitarian engineering and technological innovation play important roles.

Alan Brent's Chap. 17, which completes this part, is thematically linked to the section in dealing with sustainable development. However, the main concern of the chapter is related to the development of learning systems and mechanisms that are appropriate for postgraduate engineering management-oriented modules. The aim of education, training and development (ETD) in these postgraduate programs is to stimulate growth and advancement of professional engineers with management responsibilities in a technical and business environment. The challenge of ETD in the programs is then to incorporate the separate, dominant epistemologies of the engineering and management sciences in an appropriate manner for the engineering management discipline. These programs typically rely on project-based learning activities through an e-learning platform, such as WeBCT/Blackboard/Moodle, and are offered under the concept of "adult learning". The chapter presents findings from a mixed methods case study on a Sustainable Life Cycle Management (SLCM) module that is offered in the South African engineering management programs. The purpose of the case study is to determine whether the web-based platform is a constraining factor for project-based learning in these programs, and to ascertain whether the learning styles on the programs serve to enhance these learning activities. Brent's research shows that there is no evidence for claiming that the web-based platform is a constraining factor. The chapter concludes "that the learning styles of the typical educators and students on the engineering management programs are conducive to project-based learning .... Specifically, more opportunities should be provided as part of the learning activities, for reflective practices".

References

Borrego, M., Streveler, R.A., Miller, R. L., & Smith, K. A. (2008). A new paradigm for a new field: Communicating representations of engineering education research. *Journal of Engineering Education, 97*(2), 147–162.

Boyer, E. L. (1990). *Scholarship reconsidered: Priorities of the professoriate.* The Carnegie Foundation for the Advancement Teaching. San Francisco: Jossey-Bass.

Jesiek, B., Newswander, Lynita, K., & Borrego, M. (2009, January). Engineering education research: Discipline, community, or field. *Journal of Engineering Education, 98*(1), 39–52.

Mumford, L. (1922). *The story of utopias.* New York: Boni and Liveright.

Sovacool, B. (2013). *Energy & ethics: Justice and the global energy challenge.* New York: Palgrave Macmillan.

Special Report. (2006, October). The research agenda for the new discipline of engineering education. *Journal of Engineering Education, 95*(4), 259–261.

Trow, M. (2007). Reflections on the transition from elite to mass to universal access: Forms and phases of higher education in modern societies since WWII. In: J. J. F. Forest, & P. Altbach (Eds.), *International handbook of higher education.* Dordrecht: Springer.

Williams, R. H. (2002). *Retooling. A historian confronts technological change.* Cambridge: The MIT Press.

Steen Hyldgaard Christensen M.A. in Scandinavian Language and Literature and the History of Ideas, Aarhus University. Ph.D. in Educational Studies, Aalborg University. Senior lecturer at Aarhus University, School of Business and Social Sciences, Herning, Denmark until 2014. From 2014 adjunct associate professor at Aalborg University, Denmark. He has initiated six big international, inter- and meta-disciplinary research projects on engineering including PROCEED and coordinated five of them. He has acted in roles of editor-in-chief and co-author of four books: *Profession, Culture, and Communication: An Interdisciplinary Challenge to Business and Engineering* (Institute of Business Administration and Technology Press 2003), *Philosophy in Engineering* (Academica 2007), *Engineering in Context* (Academica 2009), and *Engineering, Development and Philosophy: American, Chinese, and European Perspectives* (Springer 2012). Besides he has co-authored *A Hybrid Imagination – Science and Technology in Cultural Perspective* (Morgan & Claypool Publishers 2011) together with Andrew Jamison and Lars Botin. In addition he has published a number of articles on engineering epistemology, culture and education. Current research interest: academic drift in engineering education and structural dynamics in higher education.

Niels Mejlgaard M.Sc. in Political Science, Aarhus University, and Ph.D. in Development and Planning, Aalborg University. Senior researcher and director, the Danish Centre for Studies in Research and Research Policy, Department of Political Science and Government, School of Business and Social Sciences, Aarhus University. His research interests include science and technology policy, science governance, and public engagement in science and controversial technologies. Presently involved in the strategic research alliance PROCEED, a program of research on opportunities and challenges in engineering education in Denmark.

Chapter 11
Bridging Sustainable Community Development and Social Justice

Juan Lucena

Abstract In this chapter, I briefly trace the history of engineers' involvement in development, from national to international to sustainable development, and highlight when and how "sustainability" and "community participation" became important dimensions in this history. Yet throughout this trajectory, a number of engineering mindsets have come to shape engineering practice and education and contributed to making social justice invisible to most engineers, restricting their ability to contribute to a fair distribution of rights, opportunities, and resources when working in community development and humanitarian endeavors. This chapter outlines these mindsets and proposes a number of possibilities to overcome them so engineers can effectively address social justice within their practices and projects in community development.

Keywords International development • Sustainable development • Community development • Engineering mindsets • Ideology • Critical pedagogy • Social justice

Introduction

In the last decade, there has been an amazing surge in engineering activities related to humanitarian endeavors and community development. 2011 was designated as the Year of Humanitarian Engineering in Australia by all major engineering societies in that country. In the US, two of the major engineering societies created Engineering for Change (E4C), a coalition of engineering societies interested in helping communities in need. There are now organizations similar to Engineers Without Borders (EWB) in dozens of countries around the world. This surge has been preceded by a relative recent and scarce history of engineers involvement in these kinds of activities, from engineering interventions through appropriate technology in the 1960s and 1970s to Fred Cuny's humanitarian activities that span from 1969 to 1995. While in present-day engineering education these activities

J. Lucena (✉)
LAIS Division, Colorado School of Mines, Stratton Hall, Golden, CO 80401, USA
e-mail: jlucena@mines.edu

S.H. Christensen et al. (eds.), *International Perspectives on Engineering Education*,
Philosophy of Engineering and Technology 20, DOI 10.1007/978-3-319-16169-3_11

might be very attractive for multiple reasons (e.g., recruitment/retention of students in engineering, public image and societal relevance of engineering, relevance to accreditation, hands-on student learning, and so on), they also raise important questions for all of us involved in them:

- Why has engineering as a profession been a late-comer to these activities when compared with medicine, law, and nursing, professions with a long-standing history of involvement in humanitarian endeavors?
- How has the history of engineers in national and international development shaped the contemporary assumptions and practices in humanitarian relief and/or community development?
- How have ways of thinking in engineering (engineering mindsets) and their associated practices and institutions, influenced the ways in which engineers carry out humanitarian and community development endeavors? Has this influence led engineers to emphasize certain behaviors or approaches, such as engineers' desire to help, while neglecting others, such as attention to social justice?

My thesis here is that until we fully understand the history of how engineers came to be involved with development and communities and the consequences of this history for present-day practices and projects, and appreciate the influence of the engineering mindsets on how engineers define and solve problems, it will be very difficult for engineers to achieve effective, sustainable and socially just community development.

Brazil: An Example of Engineers and National Development

Historically, I first locate engineers and their relationship to development around the creation of countries throughout the nineteenth and early twentieth century. Although engineers from different countries were involved in different projects around the world outside their homelands (e.g., surveying new lands and building canals for empires, organizing warfare), it was at this time when engineers also began to serve images of progress in their own countries (Downey 2007). Brazil presents an interesting case of a country that was first a Portuguese colony, then an Empire of its own, and finally a sovereign nation-state. Throughout this transformation, engineers were challenged with the construction of a country, first, to serve the interests of the Portuguese empire and finally guided by an image of national progress: Ordem e Progresso (Lucena 2009).

Before this image took hold in early twentieth century, engineers during the Brazilian Empire (1822–1889) mapped the country and its natural resources and organized and carried out military activities for the imperial crown. During the Republic (after 1889), supported by strong state governments, regional engineers built regional infrastructures and engineering schools to support an agriculture-based economy in need of replacing free labor after the abolition of slavery in 1888. A handful of military engineers, known as The Positivists, first proposed "Ordem e

Progresso" (Order and Progress) as a national motto written in the Brazilian flag, and built and expanded communication networks (telegraph, roads, river navigation system) to promote the idea of one Brazil (Diacon 2004). Yet this image of national progress did not take hold immediately as strong regional interests dominated Brazilian politics and economy until the 1920s. It took for the government of President Getulio Vargas in 1917, and the re-writing of the constitution creating a "new State" (Estado Nuovo) to define *Order* as that achieved by a powerful centralized state and *Progress* as national industrial development in the form of import substitution mixed with Taylorism and Fordism (Williams 2001). Brazilian engineers joined these efforts at the federal and state levels and built national oil and steel works as cornerstones of a national industry and infrastructure that would make Brazil an economic power in Latin America. Once the image of Order and Progress became dominant, engineers built Brasilia as the country's administrative and political capital and made Brazil into an auto-manufacturing giant in the 1950s (Alexander 1991). During the military regimes of the 1960s and 1970s, which elevated the image of Order and Progress to new heights, engineers built the Itaipu hydroelectric (the largest in the world until the construction of the Three Gorges Dam), nuclear energy plants, and Embraer, Latin America's first and only aircraft manufacturer (Adler 1987).

Engineers and the Making of Nations

In Brazil, as in many other countries, during the period of national development, governments tried to incorporate dispersed and culturally different communities and groups of immigrants into a larger national whole. As political scientists and sociologists have shown us, this incorporation – the making of a nation – happened mainly through educational and cultural institutions and agencies that controlled and supervised aboriginal and immigrant populations (Anderson 1991). But this project of nation-making also had significant physical and material dimensions that required many engineers to be involved either as builders of physical infrastructure or as public officials in charge of institutions. For example, right after the birth of the Brazilian Republic (1889) military engineers, like Candido Rondon da Silva, following orders to build a national telegraph network, tried to make Amazon natives into national subjects as they laid out the network (Diacon 2004). In early twentieth century Mexico, under the administration of Porfirio Diaz, influential Mexican engineers involved in the ministry of public education (Secretaria de Education Publica) tried to construct "Mexican citizens" out of indigenous populations through a centralized form of education (prepa) aimed at creating Mexicans out of the dispersed ethnic communities that composed the population (Lucena 2007). In late nineteenth century US, engineers were involved in the organization and improvement of urban and industrial infrastructure as immigrant groups from Europe went to the US to supply labor for industry (Britton 2001). Engineers involved in the organization of engineering schools or systems building were contributing not only by bringing a specific service, like electricity, to people but also by integrating them into a national whole.

In these episodes of national development, ideas of national progress, often promulgated by political elites and carried out by engineers, prevailed over any kind of community development or humanitarian endeavors. Even when engineers tried to enact notions of social justice, as when US progressive engineers cared about smoke pollution experienced by city dwellers and improved the efficiency of coal burners, they were involved in nation building. How might these involvements in *national development*, that continue to this day, influence, and perhaps shape, engineers' views of communities and ways of working with them? How can engineers participate in both national development (and its associated industries, infrastructure, institutions) and in working for social justice?

Engineers and International Development

The end of World War II, and more precisely US president Truman's Point IV of his second inauguration speech, launched the era of international development (Rist 2004). In addition to national development, engineers from the US and USSR, the two sides that defined the Cold War, were challenged with an image of progress that went beyond their national borders. The new challenge called for nation building outside one's national boundaries to be done mainly though financing, science, engineering and technology. As President Truman put it,

> we must embark on a bold new program for making the benefits of our scientific advances and industrial progress available for the improvement and growth of *underdeveloped* areas. More than half the people of the world are living in conditions approaching misery. Their food is inadequate. They are victims of disease. Their economic life is primitive and stagnant. Their poverty is a handicap and a threat both to them and to more prosperous areas. For the first time in history, humanity possesses the knowledge and skill to relieve suffering of these people. The United States is pre-eminent among nations in the development of industrial and scientific techniques. The material resources which we can afford to use for assistance of other peoples are limited. But our imponderable resources in technical knowledge are constantly growing and are inexhaustible. (President Harry Truman, Second Inauguration, Jan 20, 1949 quoted in (Rist 2004))

Out of this vision, powerful institutions of international development emerged such as World Bank, International Monetary Fund (IMF), Inter-American Development Bank (IDB), US Agency for International Development (USAID), and the Peace Corps. All of these either employed engineers or funded engineering projects in private companies contracted to carry out development projects.

The dominant economic model that informed the policies and programs of international development, at least for those countries not behind the Iron Curtain, was that of Walt Whitman Rostow, leading US economist and subsequently advisor for Presidents Kennedy and Johnson.

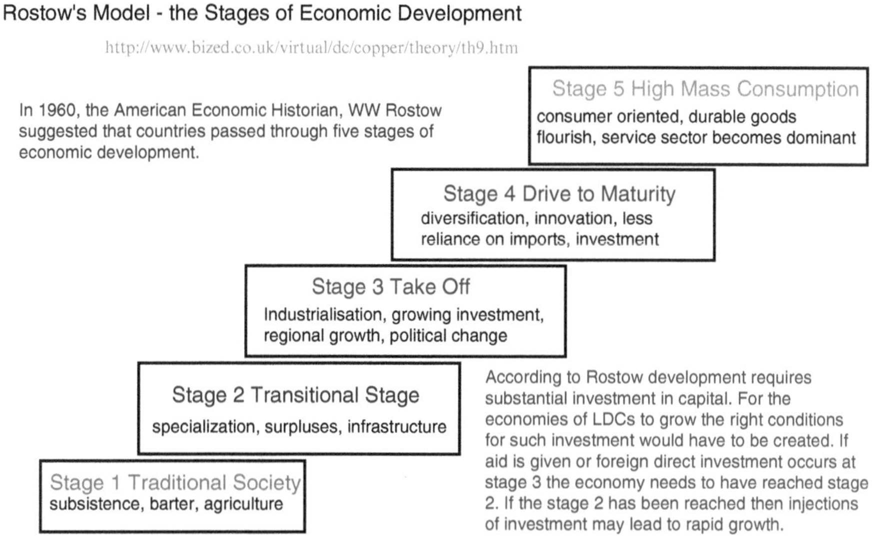

In this model, traditional societies move from transitional and take-off stages to maturity (high mass consumption) through specific economic and industrial policies and infrastructure development. Along the way traditional communities (aboriginal, ethnic, rural) and their traditional ways of life and production, are viewed as barriers to economic growth, industrialization and mass consumption (Rostow 1990). Hence, followers of this model, including those engineers who made careers in international development, were challenged to transform traditional communities and put them on a path towards modernization. In the process, communities were convinced, and often forced as when their villages stood on the path of a hydroelectric project, to abandon their means of sustenance in exchange for efficient techniques of extraction and production so they could contribute to economic growth, participate in mass consumption and be part of national development (Scott 1998).

Although there are some exceptions in the golden decades of international development (1960s–1970s),[1] most engineers involved in international development followed Rostow's model and its assumptions about communities. These engineers viewed communities as impediments to modernization and defined them in terms of what they lacked such as efficient infrastructure and manufacturing, innovation, industry, etc. (Ekbladh 2009). So in addition to the influences of working in national development, how might working in *international development* have influenced,

[1] There are a number of very engaging case studies of engineers who have challenged the ideology of development. For example, Fred Cuny questioned models of development (in the form of humanitarian aid), reconceptualizing "victims" of humanitarian crises into "partners" who needed to be employed in the solution of their own problems (Cuny 1983). Another example is that of the Volunteers in Technical Assistance (VITA) engineers who questioned international development as an instrument of Cold War politics in the 1960s and implemented an alternative model to provide technical solutions to the developing world (Wisnioski 2012).

and shaped, engineers' views of communities and ways of working with them? How might present-day desires to "help the needy" be rooted in historical commitments to modernize those who are "backward"?

Engineers and Sustainable Development

The emergence of the concept "sustainable development" has been attributed to the Brundtland report (WCED 1987) and to its subsequent acceptance and institutionalization that took place after the Rio Conference in 1992. Interestingly, Brundtland's became the dominant definition of sustainable development, accepted and adopted by most engineering organizations who wanted to promote it.[2] Perhaps this was a good faith effort to question and remedy the perils of big development and economic growth in which large numbers of engineers have been involved. Unfortunately, the adoption of this definition reinforces relationships of power and domination between countries in the global north and south and between experts and lay people. Elaborating further on why definitions of sustainable development serve the interests of experts, including those of engineers, Jeffrey Bridger and A.E. Luloff argue that

> those who depict sustainability on a macro scale portray environmental problems in such apocalyptic terms that they sometimes revert to the language of technocratic planning and administration and speak of the need for global ecological planners in international agencies who must work with national political elites and multinational corporate leaders to manage these environmental crises... The problem with this kind of solution is that relations of domination are left in place. Those who control the resources and who are responsible for many of the decisions and actions that have caused insidious environmental damage are generally charged with cleaning up their mess... The result is a crisis mentality which relies on technological solutions for much larger structural problems. (Bridger and Luloff 1999)

This reliance on technological fixes clearly appeals to engineers, especially if proposed solutions are accompanied by substantial funding from international agencies, national governments, and private corporations which have made sustainable development a key business strategy. Yet these technological solutions might not necessarily lead to sustainable *community* development since the practices that support communities reside at the local level. Bridger and Luloff propose that

> [b]y shifting the focus on sustainability to the local level, changes are seen and felt in a much more immediate manner. Besides, discussions of a 'sustainable society' or a 'sustainable world' are meaningless to most people since they require levels of abstraction that are not relevant in daily life. The locality, by contrast, is the level of social organization where the consequences of environmental degradation are most keenly felt and where successful intervention is most noticeable...sustainable community development may ultimately be

[2] A search for definitions of "sustainable development" within engineering societies will reveal a striking acceptance of Brundtland's definition without much consideration of what it means for engineering education and practice.

> the most effective means of demonstrating the possibility that sustainability can be achieved on a broader scale, precisely because it places the concept of sustainability in a context within which it may be validated as a process. By moving to the local level, the odds of generating concrete examples of sustainable development are increased. As these successes become a tangible aspect of daily life, the concept of sustainability will acquire the widespread legitimacy and acceptance that has thus far proved elusive. (Ibid)

In sum, although well intentioned, the adoption of sustainable development by organizations that employ and represent engineers has reinforced the status quo by maintaining relationships of power (e.g., North vs. South; rich vs. poor; expert vs. lay) while neglecting how sustainable practices affect local communities, and in particular how these practices might actually reinforce inequalities and social injustices.

Engineers and Communities?

Up to this point, through their involvements in national, international and sustainable development, engineers learned to view communities as groups of people to be integrated into national wholes, or as impediments to economic growth and modernization of the economy, or as lacking and in need of aid, or as entities that are invisible to technocratic definitions of sustainable development. With very few exceptions throughout this history of development like Fred Cuny (see footnote 1), up to this point engineers have engaged communities predominately through the "deficiency lens," i.e., in terms of what they lack or how they are "burdens" to higher goals like national development. For a more detailed analysis of the meanings and views of communities see (Lucena et al. 2010).

Some concerned engineers responded to this mistreatment of communities through Participatory Community Development (PCD). PCD entered development practices in late 1980s, ironically known as "the lost decade of development" with books and processes like *Listen to the People* (Salmen 1987) or *Putting People First* (Cernea 1985). These authors proclaimed more than 20 years ago certain truisms that we now take for granted in community development. For example, Cernea claimed that the role of the social analyst is to "identify, conceptualize, and deal with the social and cultural variables' that make up this missing [social] dimension [in development projects]. Even if the financial aspects of a project are apparently proceeding smoothly, these sociocultural factors 'continue to work under the surface. If the social variables remain unaddressed or mishandled, then the project will be unsustainable and fail, no matter which government or international agency promotes it." Continuing, Cernea argues that the "beneficiaries of development should have a say in implementation, and sees social scientists as playing the central role in granting this voice…putting people first is held to be 'a reversal because it proposes another starting point in the planning and design of projects than that taken by current technology-centered approaches'" (Cernea quoted in (Francis 2001).

Not many engineers involved in development practices would argue with that. Yet, although a detailed review of the critique of participatory methods is outside the scope of this chapter, it is worth noting that participatory methods have not necessarily resulted in benefits for the intended beneficiaries: communities.[3] Engineers committed to sustainable community development should be aware of these potential problems:

- *The tyranny of decision-making and control.* Participatory facilitators often override existing legitimate decision-making processes. For our purposes, we should be considering whether and how engineers filled with good intentions, the latest participatory techniques and even a strong commitment to sustainability, might be marginalizing communal decision-making processes already in place.[4]
- *The tyranny of the group.* Group dynamics put in place by participatory methods (e.g., a community meeting) might lead to participatory decisions that reinforce the interests of the already powerful (e.g., community leaders who control community resources and might end up controlling the outcome of meetings). For our purposes, we should question if engineers' interactions with others in community development projects might be reinforcing the interests of the powerful.[5]
- *The tyranny of the method.* Participatory methods like those listed above might silence or exclude others that have advantages participatory methods cannot provide. For example, participatory methods introduced in Bali, Indonesia, ignored a traditional governance system located in Buddhist temples with dire consequences for water distribution and sustainable farming (Ramaswami et al. 2007)

Being mindful of the limitations of these methods, engineers can shift decision-making power towards communities, and especially towards their more marginalized members, when working in community development. In sum, throughout the history of engineers and development a chasm between development and the interests of communities has persisted, even after the inclusion of participatory practices. Development was first about the development of nation-states, then about the geopolitics of the Cold War and economic modernization, and more recently to secure economic growth within ecological limits. Engineers have actively, and in many cases successfully, participated in each one of these stages of development. So given this history how might we put the interests of local communities at the center?

---

[3] For comprehensive analyses and critiques of participatory methods, see Cooke and Kothari (2001).

[4] See Lucena et al. (2010, Chap. 4) for the case study "The Stranger's Eyes" as an example of how this tyranny was enacted in a development project to install mills for grinding grain in various villages in Mali.

[5] See Mosse (2001) for a detailed analysis of how this happened in a participatory farming systems development project in India.

Engineering Mindsets: How Ideology Makes Social Justice Invisible to Engineers

Bridger and Lulloff's view of sustainable development challenges engineers to include the following dimensions in order to benefit communities through their practices and projects:

- Local economic diversity
- Self-reliance; local political control
- Reduction in use of energy; recycling materials
- Enhance biodiversity of local ecosystem; careful stewardship of local natural resources
- Social justice

If we take social justice to be the fair distribution of rights, opportunities, resources while minimizing risks and harms among members of a particular community, we can see that even the first four dimensions have significant elements of social justice. For example, local economic diversity challenges engineers to consider the economic relationships that exist and will be created between community and external markets prior to and after their intervention. By enhancing local economic diversity, engineering projects can serve to strengthen local market activity, generating new market opportunities and increasing revenues for community members, while disentangling local economies from external markets that might be detrimental to communities. Self-reliance/local political control challenges engineers to think about the political relationships that exist and will be created (or transformed) in a community prior to and after their technological intervention. By promoting local political control, engineering projects can enhance the political rights of community members while minimizing political control from governments or decision makers far away. Reduction in energy use, recycling materials, enhancing biodiversity and careful stewardship of natural resources challenge engineers to think about how their projects will impact the availability of valuable local resources (energy, materials, natural resources) and affect ecological relationships between community and its ecosystem.[6]

But providing an enhanced definition of sustainable development, and its constitutive dimensions grounded on social justice, is not enough if engineers, blinded by the assumptions made throughout their history in development, are not ready to see and embrace these. Besides the historical and structural constraints placed on engineers by the ways in which they have been involved in development, what else might have contributed (and continues to contribute) to engineers' difficulties in seeing and engaging in these dimensions, especially social justice?

[6] For example, engineers working with Bridges to Prosperity (B2P) build pedestrian bridges with communities that allow its members to buy and sell produce in places they could not before (local economic diversity), attend community meetings and reach voting polling places (local political control), reduce the use of fossil fuels and (re)use local materials to construct the bridges.

Engineering mindsets, as described by Donna Riley (2008), are characteristics of engineering education and practice that have evolved and come to define where and how most engineers work, think and approach problem definition and solution, and, in short, what they value. These mindsets are:

- Dominance of military and corporate organizations
- Positivism and myth of objectivity
- Desire to help and persistence to do it
- Narrow technical focus
- Uncritical acceptance of authority

According to Riley, these mindsets create significant blind spots to engineers' abilities to see social injustices and actively participate in projects and practices conducive to social justice. How might these engineering mindsets get in the way of engineers seeing social justice?

Dominance of military and corporate organizations This is very evident in most engineering schools from the job fairs, career pathways of most graduates, places of work, training and/or funding of engineering faculty, sources of funding for engineering labs and facilities, etc. Through socialization in engineering schools, which takes place via stories from professors in the classrooms, internships, company-sponsored events, etc., students learn to accept as natural the presence, dominance and hierarchies of power and decision-making within these organizations. Most students never question the power and influence that these organizations play over the organization of academic life all around them, e.g., which buildings get built, who enjoys the privileges of endowed positions, who sits at universities' board of trustees, etc. Students are also socialized into ways of decision-making and communication that might be antithetical with democratic consensus building and participatory decision-making in community endeavors. For example, after studying in depth what oil extraction has done to communities and natural environments around the world (Maass 2009), I presented students with a contrasting quote from fellow engineer and former CEO of Exxon-Mobil Lee Raymond who said: "we [oil co's] have a tremendous opportunity and a responsibility to improve the quality of life the world over. Virtually nothing is made without our energy and our products…we condemn the violation of human rights in any form, and believe our stand on human rights sets a positive example for countries where we operate." (quoted in Ibid, p. 119). Most students took for granted Raymond's condemnation and accepted the authority of his perspective –after all he is a fellow engineer in charge of the most powerful corporation in the world- in spite of the overwhelming evidence they studied before. Students were not bothered how Raymond's unsupported perspective might be silencing, at least in their head, all others that questioned the human rights record of the oil companies.

Uncritical acceptance of authority As seen in the example above, the dominance of corporations and military organizations influences how engineers accept the authority that comes from these sources. But there are other complex reasons for engineers' acceptance of authority. For example, in the US, there is a very visible political and social conservatism among engineers who uncritically accept the authority of the gospel, law, and numbers and rarely question assumptions, interpretations and power dimensions behind these.[7] More importantly perhaps is how engineering students learn to accept the authority of engineering problem solving (EPS), the core method that serves as the foundation for homeworks and exams in most engineering courses, and what this acceptance does to their ability to accept alternative perspectives and respect dissent. As Gary Downey and I reported elsewhere,

> students who complete hundreds of problem sets on graded homeworks and exams are simultaneously receiving intensive training in dividing the world of problem solvers into two parts, those who draw boundaries around problems appropriately and those who do not. The first group becomes capable of being "right," while the second, by implication, may become "wrong." One consequence is that some students emerge from engineering curricula knowing that engineering problems have either right or wrong answers, that the chief metric of ability is the frequency one is right, and that difference may be an indicator or error. In the process, such students have acquired solid grounds, seemingly mathematical grounds, not to trust the perspectives of co-workers who define problems differently. In other words, learning the five-step engineering method [EPS] can make a diversity of viewpoints suspect by definition. (Downey and Lucena 2007)

If learning EPS conditions students to reject solutions proposed by those who do not master EPS and solve problems like them, then uncritical acceptance of EPS into their lives makes them unlikely candidates to embrace social justice.

Positivism and myth of objectivity The origins and persistence of this engineering mindset are complex and varied and have different roots in different countries. For example, Ken Alder has shown how engineers of the French Revolution called for optimization of projectile trajectories over aesthetic preferences by the King to have especial decorations on cannons. Challenging royal authority, engineers tried to establish an empirically based relationship between trajectory and cannon length and thickness (Alder 1997). The history of engineering is filled with episodes where for different reasons (e.g., desire for status, access to money, boundary work vs. scientists, need for theoretical development, etc.) engineers have resorted to positive knowledge and instrumentation as main sources of knowledge (Seely 1991; Vincenti 1993; Barley and Orr 1997). This mindset has been reinforced in engineering curricula by a number of factors, including the emphasis and higher status enjoyed by math and science in academia. In her book, Riley shows the preponderance of this mindset through a series of jokes that illuminate how engineers tend to privilege positive knowledge, ("If it cannot be measured it does not exist"), and certain legitimate sources of that knowledge (e.g., instruments assumed to be void of any subjectivity).

Positive knowledge can be a powerful tool for the goals of social justice by measuring poverty, infant malnutrition, illiteracy, etc. and making them real in the minds of empiricists (Brighouse and Robeyns 2010). Yet often commitment to social justice leads us to act in spite of the absence of data, driven by principle and values.

---

[7] Interestingly with the exception of Chris Toumey who has researched the conservative and religious views of scientists (Toumey 1994), there is almost no research on conservative attitudes of engineers since the 1970s (Ladd and Lipset 1972).

In many parts of the world, there is absence of data related to the conditions of marginalized groups; their lack of political and economic power often renders them invisible to the government agencies in charge of collecting demographic data on public services or health (e.g., HIV rates among homeless). So they are not being measured but they, and their conditions, DO exist.

Also as countless STS case studies have shown, we also need to learn to accept the subjectivity in measurement tools and instrumentation (Latour 1987; Latour and Woolgar 1986). Who builds them, how they are used and calibrated, how the data is obtained, how it is interpreted orally and in writing, and how it is read by the many audiences, all of these introduce human subjectivity in every step of the acquisition of positive knowledge. So blind commitment to this mindset leads engineers not to see those injustices that cannot be empirically measured or to ignore the subjectivities involved in measuring.

Narrow technical focus Donna Riley introduces us to this mindset through a popular joke about an engineer who is about to be decapitated in a guillotine yet, instead of questioning whether justice is being served through his own execution or even showing anguish or desperation as he is about to lose his life, he is rejoicing at the opportunity to help the executioner how to figure out a technical malfunction. In *The Existential Pleasures of Engineering*, Samuel Florman provides us with a philosophical justification for this technical focus when he writes "the engineer's first instinctive feeling about the machine is likely to be a flush of pride…After the engineer's initial burst of pride has run its course, quite a different sentiment reveals itself—his love of the machine for its intrinsic beauty." (Florman 1996, p. 132). Pride and love for and aesthetic enjoyment of machines, especially if we built them, are important dimensions of our human condition as *homo fabers*. When we build these with our hands, we often come to appreciate the physical exertion required, the kinds of materials and energy involved, and how others with more dexterity and experience (often mechanics and technicians) solved problems that emerged along the way (Crawford 2009). The problem is that *making with the hands* has been almost eliminated from engineering education[8] to make room for more scientific curricula and textbook and computer mathematical idealizations of machines or physical contrivances. Graded homework, exams and labs reinforce the notion that what matters is the narrowly defined, properly bounded mathematical idealization of a physical reality void of all connections with the social world, including manual labor (see Chapter 12 by Rolston and Cox in this volume for a full analysis of the mental vs. manual divide in engineering).

At the same time, overemphasis on the technical leads engineers to ignore or undervalue the social dimensions of their work. Although ABET 2000 criteria and the *Engineer of 2020* report challenge engineering education to seriously consider the non-technical dimensions of engineering work, we are still waiting to see these

[8] Perhaps with the few exceptions of little manual work that happens in design projects and this manual work is often given to the machinist on campus. There is very little of the grade, if any, at stake for manual work.

dimensions valued in most engineering curricula. Most engineering faculty continue to significantly value mathematical idealizations of the technical over the non-technical. This valuation is reflected in curricular practices such as when the social and ethical elements of senior design projects are worth only a minimal part % of the grade in humanitarian engineering projects, clearly signaling to students not to take these seriously (Leydens and Lucena 2009). Engineering students also tend to place highest value on technical courses over non-technical ones and often wonder why they have to work hard for liberal arts classes which, according to them, do not deserve the same effort as their technical classes. So a narrow technical focus divorces students from their physical connection to making things and from the social dimensions of engineering.

Desire to help and persistence to do it As we have written elsewhere, there is a recent surge of engineering activities aimed at helping those in need around the world (Schneider et al. 2009). Historically, in the US this desire has been in tension, and often in direct conflict, with engineers' loyalty to corporate and military bottom lines (Wisnioski 2012). As mentioned above, there have been few instances when engineers have acted out of commitment to enhance the quality of life of the poor, immigrants workers, or communities in the developing world. In addition, many engineers, acting more as concerned citizens or encouraged by management in order to improve productivity, find ways to help outside of their work, in community organizations, churches, and civic projects (Geroy et al. 2000). Yet as a profession, engineering has a very recent history in dedicating and organizing educational and professional activities towards helping, especially when compared with law, nursing, and medicine (Mitcham et al. 2005). The recent emergence of organizations like Engineers Without Borders or Engineers for a Sustainable World reveal a heightened desire to help by engineers involving significant numbers of students, faculty and professionals and likely due to three historical events. First, the end of long-term loyalty between corporations and engineers has made it clear to engineers that they can no longer assume that they will have long-lasting careers with corporation. This dislocation of employer-employee loyalty has led many engineers to become freelance agents, "itinerant experts in a knowledge economy" (Barley and Kunda 2004) or individual consultants. These transformations, in addition to increasing dissatisfaction in the workplace due to budget reductions and technical work moving elsewhere, have led many engineers to seek a purpose for their work in development work or community service. But this desire to help has been motivated not by what is best for communities but by seeking a sense of purpose in one's work difficult to be found inside corporations.

Given how the history of engineers in development has shaped the way in which engineers engage communities, and the blind spots created by the engineering mindsets, social justice continues to be a missing dimension in engineering education and practice, including in many activities related to sustainable development.[9] So how might we rescue social justice and incorporate it into engineering for development so that it truly becomes engineering for *sustainable community development* (SCD)?

Can Critical Engineering Education Counteract the Blinders of History and Ideology?

Perhaps we can teach students to see, reflect and critically question these engineering mindsets so they do not take these for granted nor assume that this is the way the world of engineering has always been and will always be. Here are some strategies.

Counteracting the dominance of military and corporate organizations Teaching students different forms of organizational disobedience might challenge the dominance of military and corporate goals. For example, using the example of engineers from Volunteers in Technical Assistance (VITA), students learn how engineers *inside* the military-industrial-academic complex, who wanted to develop technologies for poor communities around the world, found a way to do so within corporations and universities. Committed to helping those who had no access or resources to the expensive lab testing and prototype development and wanting to remain distant from Cold War politics, VITA engineers found creative ways to use research labs, such as those found inside General Electric, to provide technical solutions to the questions that they were getting from poor communities (Wisnioski 2012).[10]

Students can also learn about whistle blowing, its costs and benefits, as a form organizational disobedience. For example, Roger Boisjoly, perhaps the most famous engineer-whistleblower in recent US history for disclosing the failure in decision-making prior to the Challenger disaster, visited our campus and shared with students the costs (e.g., no aerospace company will hire him again) and opportunities (e.g., he created his own firm for forensic engineering) incurred by his actions.[11]

---

[9] Note that these generalizations are drawn mainly from the history and organization of US engineering education and practice. It could be interesting to see if these apply in other national contexts, particularly in those who have emulated US educational and professional practices vs. those which are very different from the US. Also I am aware of the important exceptions from which much can be learned, e.g., US Progressive engineers in early nineteenth century, VITA, Fred Cuny, Mexican engineers of the Revolution, and present-day organizations like EWB-Australia and ISF-Colombia.

[10] Matt Wisnioski also documents other ways in which engineers have challenged the dominance of corporate and military organizations, for example, by creating *Spark*, an underground journal where they questioned and critiqued their corporate employers profit motives during Cold war weapon development.

[11] Brian Martin's *The whistleblower's handbook: how to be an effective resister* provides an excellent account of the mistakes, consequences and strategies that engineers face when speaking out against wrongdoing in a corporate setting. Also see Martin and Rifkin (2004).

We can also teach students about opportunities for humanitarian and community development engineering within the armed forces as a way to challenge the dominance of military organizations. For example, one of my engineering students and US Air Force (USAF) officer chose to revise the humanitarian operations manual of the USAF to incorporate key dimensions of community development, including social justice. Similarly, other USAF officers are researching how to use the Air Force to conduct humanitarian assistance in a hostile environment (Pavich 2004).

Students can also learn about engineers who have given up corporate/military careers in lieu of NGOs or humanitarian careers. For example, Fred Cuny, who gave up a traditional engineering career to focus on humanitarian relief efforts, serves as an exemplar to challenge conventional engineering career trajectories.[12] Or students can learn how to distinguish differences among organizations such as profit oriented, customer oriented, and engineering oriented (Harris et al. 2009, Chap. 8) and assess the companies where they want to work by asking critical questions like, does customer satisfaction go beyond prompt delivery of goods and services within budget to include public safety, accountability, transparency and relationships with the community? Is the company mainly interested in short term return to shareholders? Or does it care as well about customers' well-being, and in supporting engineers' autonomy and commitment to their professional codes of ethics?

Questioning the uncritical acceptance of authority Engineers often work and learn in organizations with rigid lines of authority so they seldom question organizational authority. In my class, students learn about the extreme consequences of engineers' acceptance of authority without critical reflection on their actions. They learn about Nazi engineers (Katz 2006; Taylor 2010) and the Engineers of Jihad (Gambetta and Hertog 2007) as extreme examples of engineers who, although very competent in their technical knowledge and skills, did not question the authority of the regimes for which they worked. Although those extreme examples are unlikely to be replicated in US settings, students also see the consequences of not questioning corporate authority as when engineers remained silent or conceded to authority in the Ford Pinto or space shuttle Challenger disasters.

Even within democratic societies, where and how might engineering students be socialized to accept authority uncritically? I often challenge students to question the authority of engineering problem solving (EPS) and its seven steps:

- Given
- Find
- Draw free-body diagram
- Identify scientific principles that apply
- Make assumptions
- Use math to solve equations, and
- Provide one solution for which they will be rewarded or punished.

[12] Other exemplars include Elena Rojas, a civil engineer who left a career in public works engineering to work with an NGO to develop community-based solutions for water supply and sanitation (Lucena et al. 2010). See also the story of William LeMessurier, who served as design and construction consultant on the innovative Citicorp headquarters tower, at onlineethics.org

After realizing the dominance of EPS in their curriculum,[13] students are invited to question, for example, who frames these problems? For what purposes? Under what kinds of assumptions? Who benefits and who doesn't when problems are pre-defined in this way and when problems are solved in this manner? After this questioning, I challenge students to redefine problems by

- *Providing their own given statements* related to a social justice issue important to them (e.g., "Given a −10 °F night temperature, a 1,500 calorie daily intake, and a 0.5 in thick coat worn by a homeless person, find the insulation material that will keep this person's body temperature to 97 °F throughout the night?");
- *Finding additional answers worth considering* (e.g., "what % of my privileged diet do I have to give up to increase the homeless person's daily calorie intake to 3,000?");
- *Drawing a relational Free Body Diagram* showing social connections to understand that this problem does not exist in isolation (e.g., network map showing homeless person in relation to shelters, food banks, police stations, available jobs, privileged neighborhoods, urban gardens, etc.);
- *Identifying alternative sources of knowledge* that might be relevant in the solution of the problem at hand (e.g., social policy, urban planning, nutrition science, distributive economics)
- *Making assumptions but critically question them* (e.g., "assuming this is a 30 year old black man… but wait, how many white males are homeless in my area? how many females? how many children?")
- Continuing to use *math* to solve the equations; and
- Providing a *number of plausible solutions* based on engineering analysis alone or engineering in combination with other sources of knowledge (e.g., "Thinsulate will keep this person's temp at 97 °F" vs. "Thinsulate + increase funding for homeless shelters + more equal distribution of food in my community…")

In sum, EPS, as the dominant method for problem solving found in most engineering science curricula, could be critically questioned and appropriated to include social justice goals.[14]

Challenging positivism and objectivity One way to teach engineering students about the myth of objectivity is to show them that engineering has always been for someone or for something. The history of engineering in different countries shows, for example, how engineers are challenged by images of progress that take different institutional, governmental, ideological, and educational forms in different places

[13] Students calculate the number of problems that they have to solve throughout their 4–5 years of engineering studies. Depending on the discipline and assumptions made during the calculations, my students have found that they solve anywhere between 1,500 and 3,000 problems using EPS.

[14] I found inspiration to appropriate EPS in the work of my colleagues in the Engineering, Social Justice and Peace (ESJP) Network such as Katy Haralampides who teaches Statistics to engineers at University of New Brunswick and Donna Riley who wrote a companion book for Thermodynamics (Riley 2011).

(Downey 2007; Lucena 2007, 2009) Engineers often respond to these challenges by building a material world (infrastructure, factories, systems, etc.) or by serving the State by rationalizing the economy or managing ministries and government agencies. Throughout these histories even those engineers deeply committed to Positivism, like the Saint-Simones in Egypt (Regnier and Abdelnour 1989) or the Positivists in Brazil (Diacon 2004), and who claimed commitment to empirical sciences as the ultimate source of knowledge, were working for someone or something and this relationship shaped the ways in which they defined and solved technical problems. Unlike EPS-type problems in engineering textbooks, engineering problems in life are always embedded in political, social, cultural and economic contexts.

In class, we study case studies showing how two groups of engineers with similar technical backgrounds and experiences can significantly disagree even when looking at the same data. This different interpretation and use of data is rooted in engineers' institutional location (from the schools where they were educated to the places where they work), way of valuing different sources of empirical knowledge (e.g., data coming from a dynamometer in a lab vs. data coming from road tests), and their ultimate goals and desires. We study the case studies of engineers' disagreement on what constituted "success" in the use of Patriot missiles in the Iraq war or what were "acceptable" launching conditions prior to Challenger explosion (Collins and Pinch 2002).

Questioning engineers' desire to help I began to question engineers' desire to help when Gustavo Esteva, a community activist from Chiapas, Mexico, came to my class and told my students: "do not go to Mexico to help. Go to listen and learn. Go to find out if the struggles of the people of Chiapas are your struggles. If so, then and only then, we can sit and talk about how we can work together." These words invited my students to question their desire to help by challenging them, first, to listen and learn and, second, to acknowledge that perhaps their desire to help is rooted in different motives far removed from the struggles of the people that they are hoping to help. Some of my students found out, for example, that seeking salvation by trying to spread the Christian faith through good works is far removed from the Zapatistas' struggle to reclaim ancestral lands or to be recognized as an autonomous community by the Mexican government.

In the US, the institutionalization of this desire to help can be traced to President Truman's Point IV about using science and technology "to relieve suffering of these people" and the emerging international development organizations and projects that employed many engineers. The origins of this desire are important for they help us understand significant discrepancies between engineers' desires and community's goals. The realization of these discrepancies, and their potential consequences for development projects, lead us to develop a list of questions for students to consider before they begin community development projects (Lucena et al. 2010):

- What are your motivations?
- What is the history and context for development in your area?

- Who benefits and who suffers from the project?
- Who is held accountable during and after your project?
- What are the possible unintended consequences of your project?
- Do you view communities as being "less than" you or your community? If so, why?

Broadening engineers' narrow technical focus The history of US engineering education is filled with attempts to broaden the education of engineers, from the early debates of the US Society for the Promotion of Engineering Education about the need for liberal education in engineering (Downey 2007) to the now regularly cited *Engineer of 2020* report (Johnston et al. 1988; Reynolds and Seely 1993; Seely 2005). Innovative programmatic developments have emerged recently, designed to counteract the narrow technical focus of engineering education in favor of more holistic and integrated approaches to engineering and its connections with domains like management, policy, STS, innovation and design such as those programs at Olin College, Lafayette College, University of Virginia, and Rensselaer. While we must applaud and continue to support these efforts, a key issue persists in most programs, including those with high percentage of courses in non-technical subjects: the pervasiveness of what Erin Cech has called *the depoliticization of engineering* (see Chapter 10 by Erin Cech and Heidi Sherick in this volume). This is the set of beliefs and practices that continue to split the world in a technical domain separate from a non-technical domain, positioning engineers as the supreme experts of the former while, in many cases, exempting them from responsibility about what happens in the latter. In engineering education, this depoliticization maintains key curricular spaces (usually the basic and engineering sciences core) well isolated from non-technical subjects, valuing the former over the latter and challenging students to be narrowly technical and serious in the core while being casual about their non-technical curriculum. As Cech writes "Engineering's status as a profession depends on its relevance to society, and depoliticization allows engineers to carry on with their socially important work (e.g. food and medicine production) without having to grapple with the messiness that comes with actually engaging with questions of the effects of engineering work on society" (Cech 2013).

While we want to respect our engineering peers' areas of expertise, we also want to constructively challenge them (and their students) to connect these technical areas to social justice, as Donna Riley has done through her companion book for Thermodynamics or as proposed above by rewriting EPS-based problems to include social justice. Through collaborative faculty workshops, we can explore ways to incorporate re-written problems into engineering science courses, engage students in problem redefinition and writing (hence enhancing their problem definition skills, underdeveloped in a curriculum that favors pre-defined problems), or make this activity for extra-credit by allowing student organizations like EWB rewrite problems based on their community development projects.[15]

---

[15] For example, EWB students in my school participate in the actual design and building of Bridges to Prosperity (B2P) for communities in the global south. I often challenge them to write

At the same time, we need to collaborate with faculty in the humanities and social sciences to open spaces in their non-technical classes for engineering students to experiment with problem re-definition. In an social science class, for example, we might allow students to bring their seemingly technical bridge project and re-write it in a way to include issues of economic exchange, migration, governance, etc. and how these affect social justice in a given locality.

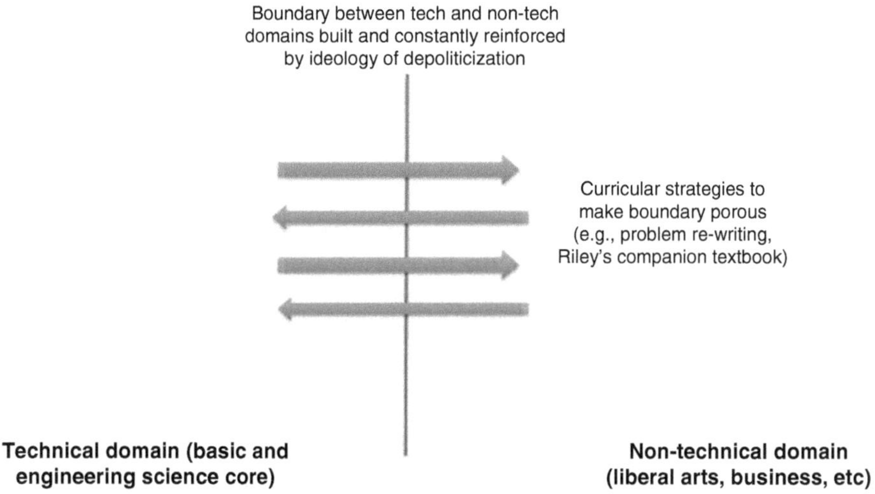

In sum, we want to create and implement strategies that challenge the boundary between the technical and non-technical domains of the engineering curriculum in order to counteract one of the most pervasive and powerful ideologies –depoliticization – that gets in the way of engineers' engaging on social justice.

Conclusion

This chapter presents a road map for engineering educators and students to question the legacy of the history of engineers in development and engineering mindsets as blinders for social justice. These blinders have made social justice a marginal concern, at best, or totally irrelevant in engineering practice and education. By removing them, my hope is that future generations of engineers will not only see the importance of social justice but will place it at the center of engineering practices in sustainable development in order to achieve *sustainable community development.*

engineering problems where they have to calculate stresses and loads on different parts of the bridge while considering how these bridges might contribute to local economic diversity, political self-determination, and social justice. Through faculty workshops I can (hopefully) establish collaborations with Statics faculty who can incorporate these problems into their courses.

While counteracting the effects of history and ideology can be daunting, especially within institutions that have been organized by these effects, my hope is that as engineering practitioners and educators we can challenge students and ourselves to

1. Appreciate the history of engineers in development, be critical of their effects and understand that they can be agents of change. Students cannot change the past but can become aware of how the past has, and continues to, shape the present and future.
2. Question models of development such as Rostow's path to modernization. But it is not enough for them to question development in the abstract. They need to see specific examples of how the ideology of development (and its associated models) operates in practices and how some engineers have successfully counteracted them.
3. Question the dominant definition of sustainable development and its hidden assumptions. As we have seen, the Brundtland definition has become accepted by most engineering organizations perhaps because it does not threaten two key premises: the need for technocratic approach and economic growth. Yet, as shown above, it is possible to critically question these premises, to refocus sustainable development on local communities, and place social justice at the center.
4. Question mainstream methods of community participation. Are these methods about extracting information in order to incorporate communities into national and global markets where they have little leverage? Or are these methods focused on enhancing people's rights, opportunities and resources, thus promoting on social justice?
5. Discern engineering mindsets and ideologies and counteract these in order to create educational and professional practices in engineering more conducive to social justice.

Then and only then, we will be taking significant steps towards an engineering education and practice with the potential to provide socially just solutions to the communities many of us want to serve.

References

Adler, E. (1987). *The power of ideology: The quest for technological autonomy in Argentina and Brazil*. Berkeley: University of California Press.

Alder, K. (1997). *Engineering the revolution: Arms and enlightenment in France, 1763–1815*. Princeton: Princeton University Press.

Alexander, R. J. (1991). *Juscelino Kubitschek and the development of Brazil*. Athens: Ohio University Press.

Anderson, B. (1991). *Imagined communities: Reflections on the origin and spread of nationalism*. New York: Verso.

Barley, S. R., & Kunda, G. (2004). *Gurus, hired guns, and warm bodies: Itinerant experts in a knowledge economy*. Princeton: Princeton University Press.

Barley, S. R., & Orr, J. E. (1997). *Between craft and science: Technical work in U.S. settings.* Ithaca/London: Cornell University Press.
Bridger, J. C., & Luloff, A. E. (1999). Toward an interactional approach to sustainable community development. *Journal of Rural Studies, 15*, 377–387.
Brighouse, H., & Robeyns, I. (2010). *Measuring justice: Primary goods and capabilities.* Cambridge: Cambridge University Press.
Britton, D. F. (2001). Smokestacks and progressives: Environmentalists, engineers and air quality in America, 1881–1951. Retrieved from http://search.ebscohost.com/login.aspx?direct=true&db=ahl&AN=A000499270.01&site=ehost-live
Cech, E. A. (2013). The (Mis)framing of social justice: Why ideologies of depoliticization and meritocracy hinder engineers' ability to think about social injustices. In J. Lucena (Ed.), *Engineering education for social justice: Critical explorations and opportunities* (pp. 67–84). New York: Springer.
Cernea, M. (1985). *Putting people first: Sociological variables in rural development.* Published for the World Bank [by] New York: Oxford University Press.
Collins, H., & Pinch, T. (2002). *The golem at large: What you should know about technology.* Cambridge: Cambridge University Press.
Cooke, B., & Kothari, U. (2001). *Participation, the new tyranny?* London/New York: Zed Books.
Crawford, M. B. (2009). *Shop class as Soulcraft: An inquiry into the value of work.* New York: Penguin.
Cuny, F. (1983). *Disasters and development.* New York/Oxford: Oxford University Press.
Diacon, T. A. (2004). *Stringing together a nation: Candido Mariano Da Silva Rondon and the construction of a modern Brazil, 1906–1930.* Durham/London: Duke University Press.
Downey, G. L. (2007). Low cost, mass use: American engineers and the metrics of progress. *History and Technology, 23*(3), 289–308.
Downey, G. L., & Lucena, J. (2007). Globalization, diversity, leadership, and problem definition in engineering education. In *1st International Conference on Engineering Education Research.* Hawaii: ASEE.
Ekbladh, D. (2009). *The great American mission: Modernization and the construction of an American world order.* Princeton: Princeton University Press.
Florman, S. (1996). *The existential pleasures of engineering.* New York: St Martin's Press.
Francis, P. (2001). Participatory development at the World Bank: The primacy of process. In B. Cooke & U. Kothari (Eds.), *Participation, the new tyranny?* (pp. 72–87). London/New York: Zed Books.
Gambetta, D., & Hertog, S. (2007). *Engineers of Jihad.* Sociology Working Papers, 2007–10. University of Oxford. http://www.sociology.ox.ac.uk/index.php/working-papers/2007.html
Geroy, G., Wright, P. C., & Jacoby, L. (2000). Toward a conceptual framework of employee volunteerism: An aid for the human resource manager. *Management Decision, 38*(4), 280–287.
Harris, C. E., Pritchard, M. S., & Rabins, M. (2009). *Engineering ethics: Concepts and cases.* Boston Cengage Learning.
Johnston, J. S., Shaman, S., & Zemsky, R. (1988). *Unfinished design: The humanities and social sciences in undergraduate engineering education.* Washington, DC: Association of American Colleges.
Katz, E. (2006). *Death by design: Science, technology, and engineering in Nazi Germany.* New York: Pearson Longman.
Ladd, E., & Lipset, S. (1972). Politics of academic natural scientists and engineers. *Science, 176*(4039), 1091–1100.
Latour, B. (1987). *Science in action: How to follow scientists and engineers through society.* Cambridge: Harvard University Press.
Latour, B., & Woolgar, S. (1986). *Laboratory life: The construction of scientific facts.* Princeton: Princeton University Press.
Leydens, J. A., & Lucena, J. C. (2009). Knowledge valuation in humanitarian engineering education. In S. H. Christensen, B. Delahousse, & M. Meganck (Eds.), *Engineering in context.* Aarhus: Academica.

Lucena, J. C. (2007). De Criollos a Mexicanos: Engineers' identity and the construction of Mexico. *History and Technology, 23*(3), 275–288.
Lucena, J. C. (2009). Imagining nation, envisioning progress: Emperor, agricultural elites, and imperial ministers in search of engineers in 19th century Brazil. *Engineering Studies, 1*(3), 24–50.
Lucena, J. C., Schneider, J., & Leydens, J. A. (2010). *Engineering and sustainable community development*. San Rafael: Morgan & Claypool (Ed: Baillie, C).
Maass, P. (2009). *Crude world*. New York: Knopf Doubleday Publishing.
Martin, B., & Rifkin, W. (2004). The dynamics of employee dissent: Whistleblowers and organizational Jiu-Jitsu. *Public Organization Review, 4*(3), 221–238.
Mitcham, C., Lucena, J., & Moon, S. (2005). Humanitarian science and engineering. In C. Mitcham (Ed.), *Encyclopedia of science, technology, and ethics*. New York: Macmillan.
Mosse, D. (2001). 'People's Knowledge', participation and patronage: Operations and representations in rural development. In B. Cooke & U. Kothari (Eds.), *Participation, the new tyranny?* (pp. 16–35). London/New York: Zed Books.
Pavich, T. (2004). Using the air force to conduct humanitarian assistance in a hostile environment. Master of Military Art and Science. Salt Lake City: University of Utah.
Ramaswami, A., Zimmerman, J. B., & Mihelcic, J. R. (2007). Integrating developed and developing world knowledge into global discussions and strategies for sustainability. 2. Economics and Governance. *Environmental Science & Technology, 41*(10), 3422–3430.
Regnier, P., & Abdelnour, A. F. (1989). *Les Saint-Simoniens en Egypte*. Cairo: Arab World Printing House.
Reynolds, T., & Seely, B. (1993). Striving for balance: A hundred years of the American Society for Engineering Education. *Journal of Engineering Education, 82*(3), 136–151.
Riley, D. (2008). *Engineering and social justice*. San Rafael: Morgan and Claypool.
Riley, D. (2011). *Engineering thermodynamics and 21st century energy problems: A textbook companion for student engagement*. San Rafael: Morgan & Claypool Publishers.
Rist, G. (2004). *The history of development from western origins to global faith*. London: Zed Books.
Rostow, W. W. (1990). *The stages of economic growth: A non-communist manifesto*. Cambridge: Cambridge University Press.
Salmen, L. F. (1987). *Listen to people: Participant-observer evaluation of development projects*. New York: Oxford University Press.
Schneider, J., Lucena, J. C., & Leydens, J. A. (2009). Engineering to help: The value of critique in engineering service. *IEEE Technology and Society Magazine, 28*(4), 42–48.
Scott, J. C. (1998). *Seeing like a state: How certain schemes to improve the human condition have failed*. New Haven: Yale University Press.
Seely, B. (1991). Scientific mystique in engineering. In T. Reynolds (Ed.), *The engineer in America: A historical anthology from technology and culture*. Chicago: University of Chicago Press.
Seely, B. (2005). Patterns in the history of engineering education reform: A brief essay. In G. Wayne Clough (Ed.), *Educating the engineer of 2020*. Washington, DC: National Academies Press.
Taylor, B. (2010). *Hitler's engineers: Fritz Todt and Albert Speer – Master builders of the third Reich*. Havertown Casemate Publishers.
Toumey, C. P. (1994). *God's own scientists: Creationists in a secular world*. New Brunswick Rutgers University Press.
Vincenti, W. G. (1993). *What engineers know and how they know it: Analytical studies from aeronautical history*. Baltimore Johns Hopkins University Press.
WCED (World Commission on Environment and Development). (1987). *Our common future*. New York: Oxford University Press.
Williams, D. (2001). *Culture wars in Brazil: The first Vargas regime, 1930–1945*. Durham/London: Duke University Press.
Wisnioski, M. (2012). *Engineers for change: Competing visions of technology in 1960s America*. Cambridge, MA: MIT Press.

Juan Lucena B.S. in Mechanical and Aeronautical Engineering from Rensselaer Polytechnic Institute. Ph.D. in Science & Technology Studies from Virginia Tech. Juan is a Professor at the Colorado School of Mines where he teaches Engineering & Sustainable Community Development and Engineering & Social Justice and directs the Humanitarian Engineering program. His books include *Defending the Nation: U.S. Policymaking to Create Scientists and Engineers from Sputnik to the 'War Against Terrorism'* (University Press of America, 2005), *Engineering and Sustainable Community Development* (Morgan and Claypool, 2010, with Jon Leydens and Jen Schneider) and *Engineering Education for Social Justice: Critical Explorations and Opportunities* (Springer, 2013).

Chapter 12
Energy Ethics in Science and Engineering Education

Joseph Herkert, Rachelle Hollander, Clark Miller, Frazier Benya, Chad Monfreda, and Lynette Osborne

Abstract Substantial global changes in energy production and use are occurring at present and will continue to occur for decades to come, with widespread ramifications for the distribution of wealth and power and humanity's social and environmental future. This raises important ethical considerations that should be addressed in the education of engineers, whose research and practice will assuredly involve energy to some degree. The Energy Ethics in Science and Engineering Education Project, funded by the U.S. National Science Foundation, sought to enhance attention to and projects in energy ethics in graduate research education concerning energy. The partners, the Consortium for Science, Policy and Outcomes (CSPO) at Arizona State University (ASU) and the Center for Engineering, Ethics, and Society (CEES) at the National Academy of Engineering (NAE), conducted a number of research, educational, and outreach activities to develop a foundational intellectual basis for understanding the ethics of energy transitions, to provide opportunities for students to learn about energy ethics, and to disseminate ideas and materials broadly. Evaluation results indicate the project has been successful in engaging students in various formats; additionally the project has illuminated a number of fundamental ideas about the interrelationships among energy, ethics, and society.

J. Herkert (✉)
College of Letters and Sciences, Consortium for Science, Policy & Outcomes, Arizona State University, 250D Santa Catalina Hall, 7271 E. Sonoran Arroyo, Mesa, AZ 85212, USA
e-mail: joseph.herkert@asu.edu

R. Hollander • F. Benya
National Academy of Engineering, 500 Fifth St. NW, 20001 Washington, DC, USA
e-mail: RHollander@nae.edu; fbenya@nae.edu

C. Miller • C. Monfreda
Consortium for Science, Policy & Outcomes, Arizona State University, P.O. Box 875603, 85287-5603 Tempe, AZ, USA
e-mail: clark.miller@asu.edu; cmonfred@asu.edu

L. Osborne
Department of Sociology, The George Washington University, 4600 W Guadalupe St B412, 78751 Austin, TX, USA
e-mail: lynetteosborne@gmail.com

S.H. Christensen et al. (eds.), *International Perspectives on Engineering Education*, Philosophy of Engineering and Technology 20, DOI 10.1007/978-3-319-16169-3_12

Keywords Energy ethics • Energy justice • Engineering education • Collective responsibility • Individual responsibility • Science education • Social justice • Socio-technical systems

Introduction

Energy production is one of largest industries in the world. Seven of the ten largest transnational corporations are energy companies. At the same time, stimulated especially by concerns with regard to global climate change, the energy sector is undergoing what is often termed an energy transition, the full dimensions of which are not yet clear. Yet in the education of engineers, all of whom have some engagement with energy, the contextual character of this ongoing engineering transition is seldom examined in depth. A collaboration of the Center for Engineering, Ethics, and Society (CEES) at the U.S. National Academy of Engineering and the Consortium for Science, Policy, and Outcomes (CSPO) at Arizona State University set out to address this lacuna with an extended effort to promote the teaching of energy ethics (National Academy of Engineering 2013). The goals of the energy ethics (EE) project have been threefold:

- To develop a strong intellectual basis for understanding ethical challenges posed by large-scale transitions in energy systems, as well as criteria and approaches for evaluating the ethical desirability of future energy options;
- To provide a variety of robust opportunities for students to learn about energy ethics and how it applies to energy research and development; and
- To disseminate ideas and materials broadly for use in science and engineering education.

The project examined the technological and socio-political plausibility of energy systems as well as issues of research ethics in energy-related disciplines, using a problem-oriented approach to ethics that required identification, assessment, and integration of diverse ethical traditions, responsiveness to real-world situations, and educational strategies in interdisciplinary settings. Ethical perspectives employed in the project ranged from traditional ethics (which considers whether actions are required, recommended, permitted, or forbidden) to issues of individual (microethical) and collective (macroethical) responsibility.

Global changes in energy production and use are occurring at present and will continue to occur for decades to come, with widespread ramifications for the distribution of wealth and power and humanity's social and environmental future. As yet unclear is the path this transition will take and the ultimate energy system that will result. One possibility is a high carbon path involving abundant new sources of fossil fuels, while another is a low carbon path involving a high proportion of renewable energy resources. In either case, the ways that energy is produced and consumed will change dramatically, based on technologies that we are beginning to see come into use. Both paths have enormous implications for how human beings will live on earth in the future (Miller et al. 2013; Kostyk and Herkert 2012).

Now is the critical time to evaluate these alternative pathways, using ethical, social and environmental as well as economic criteria. The framework for understanding energy justice must be significantly broadened (Miller 2014; Mitcham and Rolston 2013) and include attention to gender, race, class, disability, and other forms of social power in relationship to it. While traditional measures of energy justice, rooted in differential access and availability of energy among and across groups, remain important, they are inadequate. We must also consider the degree to which specific energy systems contribute to or detract from human thriving; the just and unjust distributions of benefits, costs, and risks associated with energy systems; and the role of diverse individuals, groups, and organizations in making decisions about energy futures (Bhadra 2013; Moore 2013).

We briefly summarize this project in the belief that the process it involved provides a useful model for the enhancement of engineering education. But the outcome is what is most important: six general conclusions about energy and its ethical implications. We also think the results are generally applicable; that is, that it would be beneficial for all engineering programs to introduce energy ethics into the curriculum.

Enacting the Goals

Project activities included intellectual research, pilot curriculum development, and outreach. Early in the project an interdisciplinary group of 19 scientists, engineers, social scientists and philosophers gathered in a research workshop. An engineer, a social scientist, and a philosopher made presentations about the ethical, institutional, and educational dimensions concerning energy ethics, complemented by a CSPO student's presentation on her graduate research.

Workshop findings included the following: Thinking about energy transitions from an ethical and social vantage point raises issues concerning system complexity and composition, and their effects on organization of patterns of human activity. These effects can include difficulties in how democracies engage their publics in determining energy futures. Questions of distributive and procedural justice, including social and environmental justice, arise, as do questions about professional and organizational ethics. For instance, one student participant argued that an adequate examination of siting practices requires looking outside of the NIMBY ("not in my backyard") lens and thinking about how the public relates to place, as imbued with meaning by various actors, rather than merely space. Another pointed to the ways in which aid agencies become invested in a particular "technological fix" for a particular problem which may be low in priority for communities to which they are providing assistance. In some communities, the benefits or risks of energy transitions, as well as voice, influence, or power in energy decisions, may flow disproportionately to different groups, such as men or women, different racial or ethnic groups, or groups of different abilities. Energy systems are often designed for those with specific abilities, limiting access or increasing risks to those with different ability sets (Wolbring 2011).

Historically, people have thought about questions of morality or ethics in terms of right or wrong personal action. Now, both scholarly and public talk about ethics echoes the structural differentiation characteristic of other modern discourse – for instance, biology now differentiates as ecology, genetics, etc. Energy ethics as a particular applied or practical field now raises questions about whether energy should be perceived as an unqualified good, which allows a critique that can define desirable ideals for the relationship of humanity and nature with energy, and use those ideals to direct policy and progress at critical decision moments.

Implicit institutional and ethical assumptions for energy can be identified by distinguishing between intended and non-intended purposes. Intended purposes are those at which energy aims, while non-intended are those that come along with reaching the goal. For example, coal-fired power plants have an intended goal of generating electricity with a non-intended consequence of creating smog and greenhouse gas emissions. Historically, energy policies have two intended goals: efficiency and security. But where do externalities such as environmental risks and costs fall in this equation? In energy ethics education, considerations of sustainability and development can introduce students to these issues. Class discussions can compare efficiency versus sufficiency as human development goals, and examine scores in the human development index as a function of per capita energy use. Considering the steps required to lower per capita consumption of carbon dioxide equivalents to that necessary for climate sustainability further requires students to think about the structural changes that would be needed to reach this ideal, what the costs of those changes would be, who should pay and who should say who should pay.

Overlapping with this intellectual research were various curriculum development and outreach activities: (1) a faculty and graduate student seminar on energy ethics, society and policy; (2) a set of energy ethics case studies; (3) a workshop on the social dimensions of energy transitions; (4) collaboration in two outreach and engagement events in order to incorporate ethical and social considerations into public and policy deliberations about energy futures; (5) two pilot workshops for Arizona State University (ASU) graduate students on social and ethical considerations of energy; (6) a week-long National Institute on Energy, Ethics and Society; and (7) a student-made video contest on energy ethics.

Participants in the seminar included faculty and students from science and technology policy, engineering ethics, social studies of science and technology, bioethics, applied ethics, energy, history, geography, business, chemistry, biological design, solar energy engineering and commercialization, and law. The seminar involved three major activities: (a) discussion of how humanistic and social science perspectives can be fruitfully brought to bear on discussions of energy transitions; (b) presentations of faculty and student research; and (c) presentations by outside speakers involved in Arizona's energy science, engineering, business, or policy communities. Seminar speakers included experts in electricity grid engineering and stability, state and local energy policy, the oil industry, the science and business of algae-based fuels production, utility regulation, energy consumption in the information technology sector, microgrids, and other relevant topics. The seminar also included a series of focused discussions with energy leaders about the background report (Miller and Moore 2011) and results of an Arizona Town Hall outreach exercise on "Arizona's Energy Future" which discussed key ethical, social, and policy challenges confronting the energy sector, including modules on climate change, the future of utilities, sustainability and resilience, and governance of energy systems.

The three-day Arizona Town Hall consensus conference included two days of discussions by four working groups of a series of key questions. Together the working groups involved approximately 100 individuals, representing both a geographic and demographic balance of the state's citizens, as well as key economic and policy organizations involved in the energy sector. The focus of these questions examined: the values, goals, and vision that should underlie planning for Arizona's energy future, the importance of energy for the state's economic future, the potential roles of both energy efficiency and renewable energy in creating the state's energy future, and the specific policies and strategies that the state should pursue to achieve its energy goals. The working groups developed draft reports that were synthesized by a writing team from Arizona Town Hall after each session. Finally, on the last day, the entire conference met in plenary session to negotiate on a word-by-word basis a final document of recommendations. The recommendations were subsequently circulated to state policy and business leaders and citizens through approximately 30–40 diverse events organized by Arizona Town Hall.

A second three-day workshop developed humanistic, narrative-based scenarios of Arizona's energy future, for the purposes of informing ongoing ASU energy research activities. Following traditional scenario planning methods, reconfigured to emphasize narrative story-telling among participants as the principal engagement tool, participants developed four potential future scenarios in response to the question: "How will Arizonans produce and consume energy in 2050?" The scenarios were anchored by two axes: degree of energy investment (high vs. low) and degree of centralization of energy development (centralized vs. decentralized). The resulting scenarios, and associated narratives, were intriguing, especially regarding their ethical implications, since they offered markedly different visions of Arizona society in 2050, highlighting the centrality of energy paths to future social, political, and economic organization. Key ethical considerations, such as the distribution of benefits, costs, and risks of energy production and consumption or the justness of decision-making procedures, emerged in distinct ways and required distinct forms of analysis across the four scenarios, highlighting both the significance and the path dependence of ethical analysis surrounding energy transitions (Miller et al. 2015).

CSPO also hosted one-day workshops to train graduate science and engineering students conducting energy-related research to think about the social and ethical dimensions of their own work and in energy systems more broadly. Each workshop focused on a particular energy technology (one on biofuels and another solar energy) so as to assure that the problems being considered would be relevant to the students who participated. The educational framework employed for the workshops included: (1) energy systems as complex socio-technological systems; (2) ethical theories, frameworks, principles, and codes for grappling with ethical questions and challenges; (3) energy transitions and ethical questions and challenges they raise; and (4) case studies of contemporary and historical energy transitions that illuminate key ethical challenges. The cases differed depending on the workshop audience. In the biofuels workshop, students were asked to identify ethical issues and potential solutions involving a hypothetical algal biofuels demonstration facility sited in a desert environment while the solar energy workshop held a fictionalized role-play based on controversies around the actual siting of a solar electric generating system in the California desert.

In another combination education and outreach activity, the EE project organized a National Institute on Energy, Ethics, and Society, a week-long educational seminar for ten graduate students doing energy-related research. Students were recruited by contacting over 250 faculty in energy centers and engineering departments at universities across the country and internationally. In total, there were seven participants from schools other than ASU (one international) along with three ASU students. Student topics included: ethics of cybersecurity for the energy grid, ethical issues with the development of nanoparticles for solar panels and batteries, stakeholder engagement in uncertain decisions, ethical issues with village energy development, and the implications of carbon centric discussions of climate change. All of these topics raise social justice issues for energy engineering, where the ability of different groups to influence the outcomes will be affected by status differentials that must be kept in mind for them to be overcome. In advance of the institute, students were provided with copies of recommended readings chosen to provide an orientation to the content of the workshop and to reflect the students' research interests in the broad thematic areas of energy systems and energy policy, energy ethics and social justice, fossil fuels extraction, nuclear safety, and tradeoffs involving renewable energy.

The workshop itself was organized in three phases. Phase one focused on foundational discussions of energy systems as socio-technical systems and energy ethics, with emphasis on the interrelationship of energy, ethics, and social factors. Phase two emphasized specific energy systems, in particular solar energy, fuels (both conventional and biofuels), and electric utilities. The final phase dealt with education and included presentations by the student participants of their research and take-home projects. Activities included invited talks by industry experts, scholars, and doctoral students in ASU's Human and Social Dimensions of science and technology program, a showing of the film *Gasland*, and field trips to ASU's Solar Power Lab and Biofuels Research Lab. Throughout the week students were given the opportunity to discuss their projects with the group and with individual mentors.

At the end of the week the NIEES students made presentations on their research, what they learned from NIEES that they will be applying to their research, and their follow-up plans. The students described a wide range of follow-up activities including: campus group discussions and lectures, summer high school education programs, writing articles, organizing conference sessions, building a network of advisors on ethics issues in energy development, developing ethic standards for village energy development, and writing a case study for the NAE Online Ethics Center for Engineering and Research (OEC) website (www.onlineethics.org).

Using participant observation at the student workshops and NIESS and web-based online surveys before and after the activities, an external evaluator assessed the educational activities. Overall for the biofuels workshop, close-ended questions assessing confidence in knowledge and abilities regarding issues related to energy from biofuels indicate stronger confidence after attending the workshop even for students that overwhelmingly felt confident before the workshop. After the workshop, participants also tended to be more able to provide examples of issues with energy from biofuels in most of the assessment measures. A content analysis of the workshop yielded evidence that the topics were presented in a clear, engaging manner to facilitate learning and interest. For the Solar Energy Workshop, the closed-ended questions assessing confidence in understanding issues related to solar energy and ethics indicated slightly stronger confidence after attending the workshop even for groups that felt confident before the workshop. The qualitative assessment for both workshops of ability to accurately provide examples to questions about the workshop topics indicates primarily positive, but mixed results. The evaluator recommended that examples of key workshop concepts be made more clearly and specifically so that participants could demonstrate a stronger working knowledge of the areas of concern. NIEES planning took these results into account, and the survey data indicate overwhelmingly that more participants felt confident in their understanding of a broad range of ethical concerns related to energy and ethics and in their understanding of ethics research issues after attending.

A final dual education-outreach activity consisted of a Video Challenge on the ethics of energy choices and energy research. Teams of three to four students from seven US academic institutions across the country submitted 18 videos. The videos focused on topics from fracking to wind farms, from nuclear waste disposal to smart grids, from use of public transportation to the energy costs of the meat industry.

Of the 18 videos three were determined to be gold-level quality, meaning they (a) successfully identified and depicted an ethically significant problem regarding energy, (b) clearly explained or showed the different views or sides of the issue, and (c) made a compelling argument or case for what should be done or how to handle the situation. The winning videos are available on the OEC and will be supplemented with some commentary from either judges or members of the Advisory Group from the CEES. The videos will serve as a continuing resource to faculty and students that can help spur discussion in classrooms about ethical issues in energy research and energy choices.

A final outreach activity consisted of a workshop in Washington, D.C. for a broad audience interested in energy ethics, particularly people in policy oriented positions, those involved in science and engineering education, and representatives from energy industries and professional societies. The workshop highlighted ethical, educational, and policy issues that come with various energy choices, and spurred educators and policy makers to think beyond the traditional technical aspects of energy discussions. Speakers and panelists presented alternative positions on energy ethics and policies and highlighted the project's educational activities and curriculum; discussion considered how these activities might provide a useful model for expanding energy ethics education to other universities across the country. The role of professional societies' leadership in encouraging graduate education on energy ethics was also discussed. The presenters and audience recognized that there was a strong link between the ethics and public policy activities of societies and that this linkage needs to be better addressed in professional societies' structures and policies and in their activities for members.

Summary and Conclusions

The EE project engaged substantively with numerous undergraduate and graduate students and postdoctoral fellows in engineering and other energy related fields. It involved diversified formats, from semester-long seminars to week-long short courses, from one-day research workshops to community engagement exercises. Conclusions may be summarized under six headings.

1. Energy is best understood as a complex network of socio-technological systems that integrate engineered technologies with social values, behaviors, relationships, and institutions, on the one hand, and natural resources and ecological systems, on the other. This interweaving of nature, society, and technology takes place on scales that range from the local to the global and from the individual to the organizational.
2. Energy choices involve technological and social components, embedded in a number of socio-technological systems. Energy transitions can disrupt both. Hence, current approaches to energy transition assessment, management, and policy that focus narrowly on issues of technology choice and/or energy prices are inadequate to capture either the full meaning of energy systems or the full ramifications of energy transitions for individuals and communities.
3. Energy systems are wrapped in non-obvious as well as obvious ways in modern socio-political-economic orders, and vice-versa. Thus deliberations on the ethics of energy transitions are not simply a matter of science and engineering ethics but more fully a matter of the ethics of diverse forms of individual and collective life and organization. Energy transitions are inevitably social, economic, and political transitions demanding broad assessments of ethics and justice.
4. Decisions made by scientists and engineers about designing and implementing energy research and engineered energy technologies not only incorporate notions of value, responsibility, liability, and more throughout the energy system; they also have the potential to significantly shape the human and social outcomes of energy transitions. This is also the case in public and private sector decision making that includes scientific and technical expertise.

5. Key normative and ethical questions associated with energy include: (a) the distributive justice of the costs, benefits, and risks of energy systems and of the wealth and power associated with them; (b) the procedural justice of energy governance rules, practices, and policies that determine who will have a voice in energy decisions, over what questions, and at what stage in the process; (c) the professional and organizational ethics that guide and shape resource allocations, decision-making, and standard setting by professional and organizational leaders; (d) the ethics and politics of behavior modification strategies by both private and public sector entities within the energy sector; and (e) the geopolitics and political economy of energy development, production, and consumption and their relations to patterns of energy exploitation, energy insecurity, and energy violence. These five normative issues should be incorporated into any standalone energy ethics course or other educational materials.
6. Publics are increasingly aware of and attendant to the social and ethical dimensions of energy system change and are in many parts of the world increasingly active in social mobilization around issues of energy policy. The forms of this activism are varied, as is the effectiveness of publics in asserting influence over energy policy choices.

Acknowledgements Development of the material from which this chapter was derived was supported by a grant from the U.S. National Science Foundation (Awards SES – 1032966 and 1033082; Energy Ethics in Science and Engineering Education). The views expressed in this chapter are those of the authors and do not necessarily represent the views of the National Science Foundation or the U.S. government. The authors wish to thank the many colleagues and students who have participated in this project.

References

Bhadra, M. (2013). Fighting nuclear energy. Fighting for India's democracy. *Science as Culture, 22*(2), 238–246.

Kostyk, T., & Herkert, J. (2012). Societal implications of the emerging smart grid. *Communications of the ACM, 55*(11), 34–36.

Miller, C. A. (2014). *The ethics of energy transitions*. Proceedings of the IEEE symposium on ethics in engineering, science, and technology. doi:10.1109/ETHICS.2014.6893445. IEEE: Piscataway.

Miller, C. A., & Moore, S. (Eds.). (2011). *Arizona's energy future*. Phoenix: Arizona Town Hall.

Miller, C. A., Iles, A., & Jones, C. F. (2013). The social dimensions of energy transitions. *Science as Culture, 22*(2), 135–148.

Miller, C. A., O'Leary, J., Graffy, E., Stechel, E. B., & Dirks, G. (2015). Narrative futures and the governance of energy transitions. *Futures*. doi:10.1016/j.futures.2014.12.001

Mitcham, C., & Rolston, J. S. (2013). Energy constraints. *Science and Engineering Ethics, 19*(2), 313–319.

Moore, S. (2013). Envisioning the social and political dynamics of energy transitions: Sustainable energy for the Mediterranean region. *Science as Culture, 22*(2), 181–188.

National Academy of Engineering, Center for Energy, Ethics and Society. (2013). Energy ethics. Retrieved from http://www.onlineethics.org/Topics/Enviro/Energy.aspx

Wolbring, G. (2011). Ableism and energy security and insecurity. *Studies in Ethics, Law, and Technology, 5*(1). doi:10.2202/1941-6008.1113

Joseph Herkert B.S. in Electrical Engineering from Southern Methodist University. D.Sc. in Engineering & Policy from Washington University in St. Louis. Lincoln Associate Professor of Ethics and Technology, College of Letters & Sciences and the Consortium for Science, Policy & Outcomes, Arizona State University, USA. He is Co-Editor of *The Growing Gap Between Emerging Technologies and Legal-Ethical Oversight: the Pacing Problem* (Springer, 2011), Editor of *Social, Ethical and Policy Implications of Engineering: Selected Readings* (Wiley/IEEE Press, 2000) and has published numerous articles on engineering ethics and societal implications of technology in engineering, law, social science, and applied ethics journals. He previously served as Editor of IEEE *Technology & Society* and an Associate Editor of *Engineering Studies*. He is a Distinguished Life Member of the Executive Board of the National Institute for Engineering Ethics, a former Chair of the Liberal Education/Engineering and Society Division of the American Society for Engineering Education, and a former President of the IEEE Society on Social Implications of Technology.

Rachelle Hollander Ph.D. in Philosophy from the University of Maryland, College Park. Director, Center for Engineering, Ethics, and Society (CEES) at the National Academy of Engineering (NAE), which manages the NAE Online Ethics Center (www.onlineethics.org), a widely used resource for engineering and research ethics education. Currently principal investigator on several National Science Foundation (NSF)-funded projects and subcontracts, for many years Dr. Hollander directed science and engineering ethics activities at NSF; she was instrumental in the development of the fields of research ethics and professional responsibility, engineering ethics, and ethics and risk management. She has written articles on applied ethics in numerous fields and on science policy and citizen participation. She is a Fellow of the American Association for the Advancement of Science (AAAS). In 2006, Dr. Hollander received the Olmsted Award "for innovative contributions to the liberal arts within engineering education" from the American Society of Engineering Education's Liberal Education Division.

Clark Miller B.S. in Electrical Engineering, University of Illinois, and Ph.D. in Electrical Engineering, Cornell University. Associate Director, Consortium for Science, Policy & Outcomes, and Associate Professor, School of Politics and Global Studies, Arizona State University. He has published articles and books on the governance of science and technology and the politics of knowledge systems in US and global affairs. He is the co-editor of *Changing the Atmosphere: Expert Knowledge and Environmental Governance* and *Nanotechnology, the Brain, and the Future*. He is currently pursuing research on the human and social dimensions of large-scale change in energy systems.

Frazier Benya M.A. in Bioethics and Ph.D. in History of Science, Technology, and Medicine, University of Minnesota. As Program Officer in the National Academy of Engineering's Center for Engineering, Ethics, and Society (CEES) her primary responsibilities encompass the management of CEES projects on topics ranging from professional ethics to societal impacts of energy choices and climate change. In addition she also shares responsibility for the management of the NAE Online Ethics Center for Engineering and Research (OEC). Her research concerns the development of scientific and engineering social responsibility as well as the history of bioethics during the 1960s and 1970s when the idea for the first federal bioethics commission was proposed.

Chad Monfreda Ph.D. candidate in the Human and Social Dimensions of Science and Technology at Arizona State University's Consortium for Science, Policy & Outcomes (CSPO). His research investigates knowledge and decision-making in the context of emerging markets for environmental services. He is currently a writer and thematic expert on sustainable energy for the *Earth Negotiations Bulletin* (ENB). 2007, Chad earned a master's degree in environment and resources from the University of Wisconsin, Madison, and the Center for Sustainability and the Global Environment (SAGE), where he produced a new generation of global crop maps.

Lynette Osborne B.A. in Psychology. M.A. in Sociology. Ph.D. in Sociology. Professorial Lecturer at The George Washington University and consulting Evaluation Researcher. Lynette Osborne is a sociologist who studies gender and women in science, technology, engineering, and mathematics (STEM). At GW, Dr. Osborne teaches classes on research methods, statistics, and the intersection of gender, education, and deviance. She currently conducts evaluation research on women, minorities, education, and STEM as a for the National Academy of Engineering and the Manhattan Strategy Group and has previously engaged in evaluation research for the Institute for Women's Policy Research (IWPR) and the Goodman Research Group, Inc. (GRG). Dr. Osborne was the 2006–2007 ExxonMobil Scholar in Residence at the National Academy of Engineering's (NAE) Center for the Advancement of Scholarship in Engineering Education (CASEE) where she edited a manual for writing NSF grants on gender and STEM and conducted research on challenges associated with engineering education.

Chapter 13
Engineering for the Real World: Diversity, Innovation and Hands-on Learning

Jessica Smith Rolston and Elizabeth Cox

Abstract Leaders in engineering education reform advocate integrating "real world" expertise into the undergraduate training, but these calls encompass a dizzying array of both skills and methods to develop them. The authors give form to the debate by sketching out the pedagogical value of one specific type of real world experience – practical learning with one's hands – and situating it within in the longer historical context of the mental/manual divide that characterizes U.S. engineering practice and education. Engineering by Doing (EbD), an initiative at the Colorado School of Mines, exemplifies this kind of learning by bringing together non-traditional students, with practical skills, and traditional engineering students, with solid grounding in mathematics and sciences, to work on projects with a specific practical, hands-on element. Making the background knowledge and experiences of nontraditional students visible and valued within the curriculum has the potential to improve recruitment and retention among low income and first generation students and thereby broaden participation in engineering.

Keywords Real world engineering • Diversity • Socioeconomic status • Skilled practice • Hands on learning • Contextual knowledge • Community colleges

Introduction

Leaders in U.S. engineering education point to increased "real world" skills as crucial for meeting the challenges of engineering in the future. They also identify the need to engage a greater diversity of students in engineering education in order to replace the baby boomer generation, enrich innovation and remain competitive in the global economy. This chapter proposes *Engineering by Doing* (*EbD*), an

J.S. Rolston (✉)
Liberal Arts and International Studies, Colorado School of Mines, 1500 Illinois Street, Golden, CO, USA
e-mail: jrolston@mines.edu

E. Cox
Red Rocks Community College, 13300 W. Sixth Ave., 80228 Lakewood, CO, USA
e-mail: Liz.Cox@rrcc.edu

S.H. Christensen et al. (eds.), *International Perspectives on Engineering Education*, Philosophy of Engineering and Technology 20, DOI 10.1007/978-3-319-16169-3_13

emerging concept at the Colorado School of Mines (CSM), as a mechanism for addressing both challenges and simultaneously advancing social justice in engineering education.

The crucial difference between EbD and other real world engineering initiatives is that EbD seeks to attract and retain a more diverse student body by showing students, through hands-on learning, the value of work with the hands for defining and solving engineering problems. This hands-on learning could happen both inside traditional engineering classrooms and in community and industry projects. EbD presents students with expertise in abstract science and math the opportunity to learn engineering concepts in a different way. It presents students who are not privileged by a strong background in abstract math and science coursework the opportunity to recognize their own skills and backgrounds as assets in engineering problem solving. This approach therefore has the potential to enhance learning and promote social justice, as it widens the type of skills viewed as essential to engineering and paves the way for nontraditional students to see their backgrounds and expertise reflected in engineering programs. In EbD students learn from the experiences of those with different backgrounds and skill sets, all the while developing their capabilities to engineer in the world.

This chapter begins by reviewing the recent literature on real world engineering, pointing to the vague position of practical learning within it. Next, it surveys current calls for increased diversity in engineering education, with a special focus on the shortcomings that currently exist in terms of transitioning two-year students into four-year programs. It then considers the historical legacies of the mental/manual split in engineering education to explore the institutional and cultural barriers to EbD programs. Finally, it concludes by outlining future areas of research for issues related to real world engineering, diversity and innovation.

Innovation and Practical Skills

A review of recent literature on preparing the twenty-first century engineer demonstrates that what counts as practical skills covers a wide-range of professional abilities, and that multiple approaches are being explored to developing and enhancing the practical skills of engineers. The National Academy of Engineering (NAE), a private, independent, nonprofit institution whose mission is to advance the engineering profession in the United States, conducts studies and programming that add to the growing body of knowledge on engineering and technology practice and policy. Their reports and proceedings provide engineering deans with critical insights on the future direction of the engineering profession, The NAE created the Engineer of 2020 Project to prepare for the future of engineering by asking, "What will or should engineering education be like today, or in the near future, to prepare the next generation of students for effective engagement in the engineering profession in 2020?" The Phase I report, The *Engineer of 2020* listed core attributes of future engineering professionals. One desirable attribute is "practical ingenuity," which is understood as the ability to identify problems and find practical solutions to increasingly complex global challenges impacting human welfare, such as climate change and the environment. As a means to develop these attributes, the NAE's Phase II report, *Educating the Engineer of 2020*, suggests strategies to better align engineering education with the practice of engineering. Strategies reviewed include the use of case studies to engage learners in decision-making framed in real-world environments (p. 72) and collaborations between industry and academia to produce engineers with strong theoretical backgrounds complemented with practical, hands-on experiences (p. 78).

The NAE's publication "Infusing Real World Experiences into Engineering Education" (2012) highlights the importance of grounding engineering education in real world experiences as sites where students might gain practical skills. The report reviews exemplary U.S. programs that have incorporated real world experiences into undergraduate education at four-year colleges and universities. Surveying 29 programs, the committee identified eight broad categories of real world experience programs: Capstone Programs (including senior design); Co-Op Programs (with industry); Course/Curricular Programs (includes courses other than senior or first-year design); Curricular Programs; Extracurricular Programs; First-Year Programs; Global Programs; and Service Learning Programs.

The American Society for Engineering Education also proposes skills beyond traditional technical ones that will be necessary for success in the twenty-first century workplace. Executive Director Norman Fortenberry suggests that next generation engineers will draw on a "panoply of interpersonal and management skills" – which encompass communication, teamwork and collaboration, leadership with strong moral compass, and cultural awareness – as they are solving complex problems affecting human health and welfare (2011, p. 37). He also points to the importance of practical skills, which he identifies as tool handing, such as wielding a wrench and knowing "righty-tighty, left-loosey." These skills are highlighted in a renewed focus on first-year students. First-year experience courses are restyled to not simply introduce students to engineering and design, but also to help them "develop grittiness – confidence and determination to persist through the inevitable setbacks and demanding coursework ahead" while helping them overcome "paucity of practical skills" (Lord 2011, p. 36). By creating an alternative to the "weed out" culture, hands on real world engineering, particularly at the freshman level, serves as a mechanism to welcome students into engineering with opportunities to practice engineering, learn from mistakes, and persevere in a demanding curriculum.

For these influential reformers, "practical" means applying engineering to the "real world" and engaging students in "hands-on" experiences. The emphasis on "hands-on" seems to imply that a different learning occurs when the hands, or the senses, are engaged. However, the connection of learning through the hands to deeper knowledge is not explicitly identified or theorized. More often, the need for hands-on experience is presented as overcoming the emphasis on knowing "that" versus knowing "how." From their observations of over 40 U.S. engineering programs as part of a Carnegie Foundation report on professional education, engineering education reform pioneers Sheppard and colleagues demonstrate in *Educating*

Engineers that the traditional linear engineering curriculum continues to favor knowing "that" – specific engineering principles, concepts and theories – over opportunities for knowing "how" – when, where and why those principles are applied to analyze engineering problems (2008, pp. 31–32). Hands-on experience is therefore held up as a potential strategy for students to understand the "how" of engineering and develop "engineering intuition," or the quick and immediate judgment that occurs when sizing up a problem and its possible solutions, though it appears as a black box in these calls for reform.[1]

Education scholar Mike Rose attributes the lack of research about the pedagogical value of the "intricate interplay between kinesthetics and thought" to "a tight cluster of culturally transmitted assumptions about cognition, knowledge, academic achievement, and social class that constricts our educational imagination" (2012, pp. 8, 11). He and philosopher Matthew Crawford critique the pervasive American attitude that physical labor does not require cognitive expertise because it is done by people judged as intellectually inferior.[2] As Crawford (2009, p. 21) writes, "Our testaments to physical work are so often focused on the values such work exhibits rather than on the thought it requires… It is as though in our cultural iconography we are given the muscled arm, sleeve rolled tight against biceps, but no thought bright behind the eye, no image that links hand and brain."

In contrast, Crawford and Rose identify the mental dexterity required by such work, ranging from trades like welding to service work like waitressing, as people engage and respond to the changing environment. When this is done well, it appears as "second nature" – a term that erases the mindfulness and "fusion of touch and concept" at play when "attention is focused, and all kinds of knowledge rush in on the moment, right through the fingertips" (Rose 2012, p. 9). Crawford describes how learning through the hands develops intuition in his description of "thinking as doing." Like Sheppard et al., Crawford also distinguishes knowing "that" and knowing "how," but grounds that difference in universal knowledge that can be proposed from anywhere, by any individual, as compared with practical know-how tied to a particular person, occurring in particular situations (2009, p. 161). Knowing the particular, Crawford continues, stems from regular practice and purposiveness (Ibid, p. 163). Through hands-on experience, students and apprentices begin to rely

[1] This intuition is well documented in skilled trades. Longtime miners, for example, develop "pit sense" on the job, which allows them to assess the risks of their work area. In contrast with formal safety training in classes, pit sense is acquired through direct engagement with a specific material environment. While pit sense is not always capable of being expressed verbally, it does comprise a source of knowledge upon which miners act in situations both ordinary and dangerous. The site-specific and embodied nature of this way of knowing distinguishes it from the generalized and abstract scientific understanding that informs engineering practice as well as bureaucratic accident prevention and reporting (Kamoche and Maguire 2011, pp. 726–727; Sauer 2003, pp. 81, 206–207; Rolston 2013).

[2] Rose (2012, p. 6) argues that whereas assertions of the mental handicaps of the working-class and immigrants used to be explicit, they are now more insidious, lurking behind "kinder, gentler" descriptions of different learning styles that place blame with deficient individuals, rather than unjust educational policies and socioeconomic structures.

on tacit knowledge – recognizing patterns and familiarities of problems and solutions encountered in past situations (Ibid. p. 166).

The anthropologist Tim Ingold calls this meshing of sensory perception and practical engagement *skilled practice*, finding it in craftsmen as well as hunter gatherers. "Skilled practice is not just the application of mechanical force to exterior objects, but entails qualities of care, judgment and dexterity," he writes. "Critically, this implies that whatever practitioners do to things is grounded in an attentive, perceptual involvement with them, or in other words, that they watch and feel as they work" (2001, p. 21). Learning happens as novices coordinate their movement with attention to others, meaning that "each generation contributes to the next not by handing on a corpus of representations, or information in the strict sense, but rather by introducing novices into contexts which afford selected opportunities for perception and action, and by providing the scaffolding that enables them to make use of these affordances" (2001, p. 21). Ingold argues that the cultural divide between (mental) design and (manual) implementation leads technical processes to be understood as "exercises in problem-solving rather than as forms of skilled practice" (2001, p. 28). His larger theoretical corpus positions skilled practice as an essential component of life that belies the strict western divisions between mental "problem solving" and its material execution.

Although beyond the scope of this chapter, the line of research suggested by Crawford, Rose, and Ingold could provide engineering reformers with a deeper understanding of how students, through using tools, sweating in a shop and accumulating hours on a project, can begin to accrue individual experiences from which to recognize patterns and enhance their learning. Crucially, revaluing work with the hands also provides a potential avenue for recruiting and retaining a more diverse engineering student body.

Diversifying Engineering

The 2012 President's Council of Advisors on Science and Technology report argues that for the United States to regain global competitiveness, it must produce one million more STEM professionals in the coming decade. Leaders in engineering clearly recognize the need to "draw a steady flow of bright students into engineering, prepare them well to enter into the complex job market, feed the innovation pipeline, replace the retiring baby-boomers, and all the while broaden participation to adapt to the wave of changing demographics."[3] The demographic challenges are substantial, since the profile of current engineering students is out of touch with the

[3] These challenges were posed at 2012 workshop sponsored by the National Science Foundation where stakeholders from school districts, community colleges, engineering schools, industry, and government worked to identify partnerships and pathways to help increase the supply of next generation of engineering workforce in the US who can address the National Academy of Engineering's Grand Challenges.

U.S. as a whole. Minorities are expected to represent 49 % of the U.S. population by 2039. In contrast, four-year engineering programs are dominated by economically privileged, white, heterosexual males. Of the bachelor's degrees in engineering awarded in 2011, 81.6 % went to males, 66.6 % to whites, 12.2 % to Asian-Americans, 8.5 % to Hispanics and 4.2 % to African-Americans (Yoder 2012). Statistics for the Colorado School of Mines, the home of the EbD program, are comparable, with women comprising 26.1 % and ethnic minority students comprising 15 % of the student population. Moreover, undergraduates whose family incomes were in the highest quarter also enroll in engineering programs in higher numbers (Chen 2009; Strutz et al. 2012, p. 144).

The "weed out" culture of undergraduate engineering programs draws and graduates students with strong backgrounds in abstract science and math coursework, which disadvantages nontraditional students, who are more likely to attend high schools with inadequate college prep programs (McLoughlin 2012, p. 124; Slaton 2010). Equally prepared students from low and high income families do not graduate in equal numbers (McLoughlin 2012). In STEM fields, racial disparities also exist in retention rates, as 36 % of white students complete bachelor's degrees in 5 years, compared with 21 % of African-American and 22 % of Latino undergraduates in 2011 (Boundaoui 2011). A disparity also exists in graduation rates. In 2011, while 39 % of white Americans between the ages of 25 and 29 held a bachelors degree, only 20 % of African-Americans and 13 % of Latinos were degree holders. Of the science and engineering degree holders, only 11 % were Latinos, African-American and American Indians (Microsoft 2012).

Broadening participation in engineering will need to include non-traditional students – an umbrella term encompassing minority, first-generation, low income, older students, single parents, and veterans. The non-traditional engineering student is likely to begin the engineering path in a two-year community college. Intriguingly, 42 % of students earning bachelor's degrees in engineering received community college credit over the course of their education, but "little is known about this population" (McLoughlin 2012, p. 127). Two-year community colleges play a vital role in postsecondary education. With lower tuition, open access and flexible scheduling, community colleges are the starting point for 40 % of all undergraduates as they prepare for transfer to four-year institution. Additionally, community colleges work closely with local business and industries to provide skills training for local workforce development. According to the Community College Research Center at Columbia University, approximately 50 % of Hispanics and 31 % of African Americans begin at a community college while only 28 % of whites do (http://ccrc.tc.columbia.edu/Community-College-FAQs.html).

While it is crucial to recognize diversity within the community college student population, these students tend to fall into one of two groups: those with "inadequate academic preparation" or "nonacademic issues related to life circumstances" (McLoughlin 2012, p. 137). Forty-five percent of entering community college students are the first in their families to attend college, nearly a third report incomes below 150 % of the federal poverty level, and more than 80 % work while attending school (McClenney and Arnsparger 2012). Community colleges serve students needing developmental courses, either because they have weaker academic backgrounds or did not take the required courses in math and science as high schoolers; 60 % of community college students took at least one developmental-level course, compared with 20 % at four-year schools (McLoughlin 2012, p. 129).

The experiences of community college students from vocational high school courses, fieldwork, or military service might provide the hands-on skill set to which engineering educators are now appealing. An initiative of the Obama administration seeks to help returning military service members transition into high demand fields like engineering by directing the Department of Defense to launch the Military Credentialing and Licensing Task Force to create pathways for veterans to transition into engineering (President's Council 2012). A challenge for educators of non-traditional students is to pursue a similar effort and develop mechanisms to help students recognize how their prior learning and experiences, gained either through education, work or military service is applicable to engineering.

At the same time, educators must also learn to view the practical skills that non-traditional students can contribute to engineering as an asset. According to Lisa McLoughlin, an influential scholar of engineering in community colleges, the challenge is that engineering schools have traditionally approached community college students through a deficit model instead of appreciating their strengths. Professors (perhaps unwittingly) contribute to this process when they do not attend their office hours, leaving transfer students accustomed to personal relationships and individual assistance feeling "devalued and unwanted" (McLoughlin 2012, p. 136). Furthermore, the structure of engineering programs – with few options for part-time or evening coursework – are built around students who do not have family or work obligations as many community college students do. Closer to the purposes of the EbD initiative, classes emphasizing abstract concepts and homework leave little room for students to contribute their expertise to engineering problem solving. Community college students are expected to assimilate into four-year colleges that are built around the backgrounds and experiences of elite students.

Hands-on learning offers one strategy for inspiring non-traditional students. Amy Slaton (2010) argues that the attempts to diversify engineering programs at the University of Illinois at Chicago and the Illinois Institute of Technology (ITT) failed in their progressivist missions of urban uplift part because the curriculum remained structured around abstract, scientific and math coursework rather than concrete problems related to social justice in the surrounding underserved communities. Projects such as these have additional pedagogical value for all students, since problem-based learning that makes students' expertise visible and valued improves student learning, engagement and retention (McLoughlin 2012, p. 136).

Engineering has been slow to integrate students' experiences into coursework due to the reliance on traditional science teaching methods, such as large lecture courses, rigidly defined problem assignments, and highly structured laboratory courses (Sheppard et al. 2008). The suitability of these teaching methods for student learning in the most important technical skills of engineering – the integration of knowledge, synthesis, design and innovation – increasingly comes under question, but is difficult to dislodge (Sheppard et al. 2008, p. 22). The real world initiatives summarized above are a response to these concerns, but they do not clearly address the issue of recruiting and retaining non-traditional students, especially those who originate in community colleges. This lacuna is traceable to the consolidation of engineering as a scientific, middle-class profession in the 1960s. Thus for real world engineering programs to succeed, they will have to critically address the legacies of the separation between engineering as a professional, abstract, scientific and mostly mental profession in contrast with the practical labor of technicians.

Taking Engineering Out of the Shop

The shift of U.S. engineering training from shops to schools in the late nineteenth century, and subsequently from practice-based learning to science coursework in the 1950s and 1960s, are part of the larger history of the consolidation of engineering as a professional, middle-class occupation.[4] These educational transformations both responded to and shaped reorganizations in industrial workplaces and corporations, as engineers attempted to align the field with shifting American definitions of progress revolving around low cost, mass use production (Downey 2007). Though mass consumption was imagined in the U.S. to transcend class, as it was believed to embrace the working- and middle-class alike, the history of engineering education and practice reminds us that class distinctions were integral to the production process that facilitated such consumption.

Apprenticeships dominated nineteenth century engineering training (Reynolds 1991, p. 13). At the beginning of the century, civil engineers learned technical and social skills on job sites, such as canals (Downey 2007, pp. 292–293; Reynolds 1991, pp. 13–14). Practical education was also essential as mechanical, electrical and industrial emerged as specialties by mid-century. Novices moved from positions of general labor and apprentices to journeymen and machinists, learning about economics along the way (Reynolds 1991, p. 15). Because the first engineers came from the ranks of skilled mechanics, they shared "a common occupational culture and training" with the workers below them (Zussman 1985, p. 76). The apprenticeship model, coupled with expanding industrial markets, created an entrepreneurial elite who would continue to wield influence in mechanical engineering through the first half of the next century (Calvert 1967; Downey 2007, p. 294).

The first challenge to the practice-based orientation to mechanical engineering training and work came from graduates from the land grant universities, established by the Morrill Acts of 1862 and 1890 to offer training in agriculture and the "mechanic arts," understood to include engineering.[5] The importance of academic

[4] Skilled practice remains an integral part of engineering education and practice in other countries, such as Great Britain (Morice 1988; Whalley and Barley 1997, p. 31), though it is beyond the scope of this chapter to engage in a cross-cultural comparison.

[5] Before the land grant universities were established, the military and independent polytechnics offered a classroom-based curriculum for engineering (Downey 2007, pp. 292–293; Reynolds 1991, pp. 16–23).

engineering programs grew as the expansion of industry outpaced the rate at which engineers could be trained on the job (Reynolds 1991, p. 22). As Monte Calvert (1967) relates in his still influential history of mechanical engineering, those graduates came from lower socioeconomic backgrounds and sought to establish themselves as professionals through their academic credentials rather than practical experience.

When these university-trained engineers entered industrial workplaces at the turn of the twentieth century, they appealed to their scientific training to distinguish themselves from skilled tradesmen in addition to the older generation of engineers. Their class position was not predetermined, since they "could have been treated as a type of highly skilled labor, the same as carpenters, machinists or pipe fitters, and essentially demoted to the working class" (Reynolds 1991, p. 177). Frederick Winslow Taylor's scientific management, which emerged during this period, provides perhaps the starkest example of engineers defining themselves by abstract "mental" labor in service of management directives, in opposition to the "manual" labor of blue-collar workers. Taylor attempted to break down the activities of workers into smaller and smaller motions that could be separated and reorganized to increase efficiency. Harry Braverman (1998) elucidates the class dimensions of the process, locating class conflict in the "minute-by-minute, second-by-second struggle within the labor process" (Foster 1999, p. 16). In essence, Braverman positions engineering as a class imperative based in the "specialization in those aspects of conception that were once among the most complex of craft skills" (Zussman 1985, p. 79; see also Reynolds 1991, p. 177).

At the turn of the century, pressure to emphasize scientific theories in engineering training was also percolating in university programs. In 1893, Rensselaer Polytechnic Institute president Palmer C. Ricketts lamented that engineering colleges were turning out "mechanics, rather than engineers" and that the attention paid to the machine shop resulted in "too little to head work" (Seely 1993, p. 358). Calls for more training in schools than shops were also made by Robert Thurston at Cornell (Downey 2007, p. 290). By the 1930s European engineers at MIT, Caltech and the University of Michigan were promoting a larger role for science in engineering education (Seely 1993, pp. 361–362). Such appeals for increased science and mathematics in engineering training went largely unheeded until the major transformations of research funding in the 1950s and 1960s, described below. Most schools maintained a practical component to their academic programs, and those engineers who did share this desire integrated scientific methods into engineering's focus on "problem solving and design – on doing, not knowing" (Seely 1993, p. 359; see also Layton 1971).

Though engineers were integrated into industrial workplaces as professionals rather than technicians, their work retained a practical orientation that was also reflected in engineering education through the 1930s. Experiment stations at land grant universities became the hallmark of engineering education after the University of Illinois founded the first in 1903. The practical research carried out at these stations in the name of solving problems for society reflected the Progressive Era ethos of public service (Downey 2007, pp. 295–297; Seely 1993, p. 349). This period also saw increased coordination between universities and industry in the form of coops, in which students rotated between class and work, and later corporate financial sponsorship for research to provide practical solutions for specific industrial needs (Downey 2007, pp. 297–298; Seely 1993, p. 357).

These increasingly close industry ties made corporate employment the hegemonic model of engineering careers (Downey 2007, p. 298). Engineers did not simply work for corporations, but came to identify with management and join their ranks (Downey 2007, p. 300; Reynolds 1991, p. 176). The cause and nature of this coordination has been examined with greater depth than allowed in this brief overview. Layton's original thesis explaining the alignment as a victory of the engineers favorably disposed to industry has been challenged by scholars such as Noble, who consider it instead a victory of corporate capital, and Meiksins, who characterizes it as a temporary alliance between reformist and rank-and-file engineers (Downey 2007, p. 298). While hegemonic, this corporate model came under critique from independent consultants and industrial scientists who crafted their identities as engineers based on their autonomy from big business and consideration of their "ethical obligations to clients and the public good" (Wisnioski 2012, p. 20). Even in the 1960s, when private industry and utilities employed 71 % of engineers, a "vocal minority" called for engineers to promote human welfare and social responsibility (Wisnioski 2012, p. 38).

The shift in focus of engineering training from the job site – whether a machine shop or a canal – to universities thus did not wholly remove the practical element from education, especially since the field maintained close ties with industry. The impetus to establish the field as the domain of white-collar professionals, however, meant that engineers integrated themselves into the industry as experts in abstract thinking who were loyal to management, not as tradesmen whose practical experience could invite identification with carpenters or machinists.

Science and Class Identity

If the shift from the shop to the school diminished the importance of practical training, the scientiziation of engineering in the 1950s and 1960s threatened to make it obsolete. The push to replace technology-based coursework with science and math "fundamentals" can be traced to increasingly complex technology, engineers' frustrations from being excluded from advanced engineering projects in World War II, and – arguably most importantly – the shift of funding for engineering programs from industry to the federal government, which privileged scientific rather than practical research (Downey 2007; Reynolds 1991; Seely 1993; Slaton 2010). "Only schools that embraced scientific engineering received large federal projects, and only schools with federal funding developed large research programs" (Seely 1993, p. 372). Changes in undergraduate education followed suit, requiring students to work through standardized sequences of science and math curricula (Sheppard et al. 2008 , p. 21).

The ascendance of science-based research and education widened the rift, first evident in the late nineteenth century, between engineers championing scientific theory and those favoring a more practical approach. Proposals in the 1960s to embrace the two orientations within a unified field of engineering, but with specialties at different educational institutions, were fiercely rejected because of "concerns about the stigma of practice programs" (Downey 2007, p. 303) that were viewed as "second class" (Seely 1993, p. 379). The institutional legacy was a bifurcated educational system separating engineers (and thus future managers) from technicians. "As engineering curricula nearly everywhere worked to integrate the basic and engineering sciences, technical institutes moved to fill the curricular space they left behind by expanding two-year technician programs to four-year 'engineering technology' programs" (Downey 2007, p. 304; see also Wisnioski 2012, p. 27). Engineering and vocational programs had to be distinct in reputation as well as function if engineering were to be a pathway to management (Slaton 2010, p. 34).[6]

This period's consolidation of engineering as a middle-class occupation, solidly identified with management is evident in union rates. In 1965, 95 % of engineers worked for large corporations (Reynolds 1991, p. 173). Whereas approximately 10 % of the country's 500,000 engineers belonged to unions by the mid-1950s – and sought to strengthen their position by forming a united organization called the Engineers and Scientists of America – by 1967 their membership had waned to 20,000 (Downey 2007, pp. 303–304).

Though engineering continues to be hailed by industry and the academy as a pathway of upward mobility, the increasingly science-based curriculum makes it difficult for a wide variety of students to enter and complete undergraduate engineering programs, as Slaton (2010) masterfully demonstrates in her analysis of the University of Illinois at Chicago and the Illinois Institute of Technology (ITT). Despite the progressivist missions of urban uplift present in both institutions, definitions of rigor excluded Black and socioeconomically disadvantaged students from successfully earning engineering degrees. With few exceptions, students without the benefit of graduating from elite prep academies or well-funded public schools needed extra preparation and tutoring in math and science to begin or complete the program, but found little institutional support. School leaders and evaluation standards pressured professors to prioritize research over teaching, since institutional reputations were tied up in the former rather than the latter.

The case of ITT is particularly interesting since it began as a vocational school. As its mission transformed to training professional scientists and engineers, it shed its Evening School (serving the majority of students, who also worked during the day) to focus on daytime classes that drew a more selective population (Slaton 2010, p. 93). Their Minority Engineering Program saw limited success. Refusing to adjust

[6] This unease is still evident in the pains that the editors of *Between Craft and Science* (Orr and Barley 1997) take to distinguish "technical" work from "traditional blue collar work" in the introduction.

admission standards or offer remedial courses, since leaders believed that doing so would damage the school's reputation, the pool of students from which they drew was limited to qualified, rather than qualifiable students (Slaton 2010, p. 97). Slaton succinctly identifies the "incompatibility of social reform and institutional survival" (2010, p. 107) as the crux of the problem in integrating disadvantaged students into undergraduate engineering programs.

In addition to schools, contemporary workplaces bear the traces of the division between scientific engineering and technical labor. The degradation of blue-collar work has been chronicled by generations of shopfloor ethnographers, who nonetheless staunchly draw attention to the complex "mental" expertise demanded by "manual" labor (e.g. Crawford 2009; Dudley 1994; Rose 2004). The other side of this division is the decreasing opportunity for engineers to engage in hands-on work. Some scholars and practitioners assert a continued practical element in engineering, since the field deals with the material world (Whalley and Barley 1997). At the same time, even they recognize that opportunities to engage in hands-on work are diminishing as the nature of engineering work becomes increasingly specialized and abstract. Even in workplaces that include a significant labor component, union contracts can bar supervisory personnel from engaging in physical work assigned to technicians. As sociologist Robert Zussman writes, these historical processes have culminated in engineering practice being characterized by "a near total absence of that physical, hands-on labor that is a central attribute of craftwork. Engineers manipulate symbols that refer to physical objects, mostly equipment and products, but they do not manipulate those objects themselves… physical manipulation is now the work of machinists, repairmen, mechanics, operatives, and technicians" (Zussman 1985, p. 77).

Much has changed in the world since the 1950s and 1960s when winning the space race and the military competition with the Soviet Union shifted engineering education from an applied, practical approach to one steeped in math and science. But abstract learning, institutionalized in higher education, has become even more dominant as shifts in education policy erode hands on opportunities for high school students. Today's traditional engineering students who followed college prep curriculum in K-12 may excel at advanced calculus, physics, and "pointing and clicking" but have had little opportunity to learn through the hands. Career and Technical Education (CTE), formerly known as Vocational Education, struggles to find space in public and policy discourses against the legacy of the 2002 federal regulation act, No Child Left Behind, which increased focus on academic rigor, standards and accountability. As academic course-taking has expanded, career and technical education's share of high school curriculum has decreased to 16.2 %. In 2002 from 21.8 % 20 years earlier (Silverberg et al. 2004, p. 9). Ironically, this decline in career and technical education is coming at a time when higher education, particularly engineering education, is going the other direction. Too often CTE courses are stigmatized as for the "non-college bound" or even as a deterrent to college, and in the case where CTE courses are linked to postsecondary education, they are associated with associate's or two-year degrees rather than four-year degrees (Ibid, p. 7).

Few current engineering students grew up tinkering in the garage, working on the farm, or trying their hand at shop class. The lack of these practical skills has repercussions beyond the narrow concern of limiting engineers' skill sets. It also creates distance between engineers and the artifacts and systems they design, resulting in missed opportunities to understand how they actually are made, used and disposed in the world. This understanding is furthered hindered when engineers accustomed to a narrow social universe cannot meaningfully listen to and communicate with the tradespeople – such as technicians, contractors and electricians– who actually engage their creations.[7]

Engineering by Doing

Engineering education and practice is informed by larger western cultural distinctions between mental and manual labor that value the former at the expense of the latter.[8] These distinctions are "cultural frames of great power" that "affect the way we see, think about, and value the work we do, and have important social and practical consequences for both private and public policy" (Whalley and Barley 1997, p. 27). Attempts to reintegrate practical experience into engineering education, such as in Engineering by Doing (EbD) at the Colorado School of Mines, will confront this cultural divide and its institutional legacies.

EbD at Mines is very much a work in progress. Broadly, it is imagined to reintegrate mental and manual expertise and by doing so bring value to different skill sets, experiences and perspectives of students. On engineering projects, Mines students will collaborate with students in pre-engineering courses at nearby Red Rocks Community College. The goal of these projects is to provide the scaffolding for both Mines and Red Rocks students to engage first-hand in skilled perception and action, to use Ingold's terms, while they learn from the unique experience and expertise of students from different backgrounds.

CSM and RRCC faculty with interests in innovative pedagogies are currently conducting workshops to explore how EbD can live both inside the engineering curriculum as well as through community projects outside the traditional engineering curriculum. Inside the curriculum, students can reconceptualize problems around what they are most familiar with (tools for mechanics, changing temperatures of metal for welders, beam installation for construction workers) and faculty can engage those experiences by bringing tools, machines and other tangibles that students can handle and visualize. The importance of seeing objects and using the

[7] Thank you to Juan Lucena for pushing us to make this point stronger.

[8] While powerful, the distinction between the mental and manual was never fully realized in the real world of labor, as generations of shopfloor ethnographers document the mental dexterity demanded by even the most seemingly monotonous assembly line work. Engineering education might be one of the few social arenas in which the complete separation of these two ways of knowing was accomplished.

senses can pull from research on spatial thinking and engineering problem-solving in the context of learning geometry concepts. According to undergraduate engineering educators Sharp and colleagues (2004, p. 35) "the ability to isolate vital information from a visual, representational context is an important engineering skill related to spatial thinking that engineering students need to learn." Using van Hiele K-12 Geometry Learning Theory, students are encouraged to look at an object in its entirety and make "analysis level" statements before moving on to "informal deductions." This line of research could contribute better understanding of how the senses are engaged in hands-on learning. At CSM and RRCC, faculty from both institutions are working to determine how community engineering projects will value and integrate different skills and perspectives and provide critical hands-on skills for students who did not grow up tinkering or doing manual labor.

While engaging in hands on learning provides benefits for students in general, since it actively supports integration of knowledge and skills while engaging students in the learning process, it also provides a strategy for making engineering education more inclusive. Krista Donaldson and colleagues (2008) found lower confidence in engineering-related skills among low income students. This disparity might be traceable to a lack of opportunities to see their own backgrounds and knowledge reflected in the classroom, since "students who see their own and other students' skills valued, such as occurs in a problem-based learning setting, are more engaged and learn more comprehensively" (McLoughlin 2012, p. 136; see also Smith et al. 2005). Students with the benefit of attending high schools with strong math and science coursework, therefore, are better positioned to succeed in the traditional, science-based engineering curriculum at college. On the other hand, those who lacked the cultural capital to take those classes, those who graduated from schools without the option for this coursework, and those who simply learn better with their hands struggle through engineering education because they do not see their own skills reflected or valued in the traditional curriculum. This accumulation of benefits and disadvantages is one of the ways in which privilege operates within engineering education and mitigates against a more diverse student body (see also Ohland et al. 2012). This process is evident in graduation rates, as there exists an 11 % difference between low income students and their peers even when controlling for factors such as high school grades, standardized test scores, race/ethnicity, gender and university (Strutz et al. 2012, p. 147). Hands-on learning could provide non-traditional students with an opportunity to see the value of their own experiences and thus persist through challenging curriculum.

But for EbD to be successful in reaching its goals for student learning and maintaining its own viability within the curriculum, it must be implemented with an eye to the longer historical trajectory of engineering education that informs current practice. What the history of engineering education shows is that the evacuation of hands-on learning from the four-year curriculum did not occur by accident; it was done deliberately to establish the field as a middle-class profession whose scientific orientation distinguished from mere technical labor. Thus as EbD reintegrates mental and manual learning, it also confronts institutional legacies dividing engineering and "technical" training as well as deeply engrained cultural distinctions between and judgments about the merits of the "culture of the mind" and the "culture of the hands" (Dudley 1994; see also Whalley and Barley 1997).

Of particular concern to the CSM/RRCC planning team is to insure that students with different kinds of expertise and training learn from each other's strengths. While both traditional and non-traditional students have something to gain from engaging in EbD with each other, this mutual appreciation is not guaranteed simply by having students from different backgrounds work together. Slaton's (2010, pp. 37–38) case study of engineering education in Maryland, for example, reminds us that "hands on" education takes on distinct connotations for differently privileged people. For white students at the main campus in College Park, lab and field experiences were not aimed at improving manual dexterity, but at "immersing engineering majors in a workplace culture that associated subjectivity and discretion with managerial potential." This managerial training was steeped in manly notions of physical and mental discipline. For students of color at the Eastern Shore campus, in contrast, "getting one's hands dirty held few romantic implications of heroic physicality" because these activities tracked students into lower-skilled agricultural or industrial jobs. In other words, "if engineering students at College Park gained a vital sense of their own manly capacities by shoveling gravel, at Eastern Shore shearing sheep, milking goats and making mattresses instead projected the upward limits of black vocational potential" (Slaton 2010, p. 38). EbD must, therefore, include formal mechanisms for identifying and challenging the racial, class and gender assumptions about expertise that animate the lingering mental versus manual divide in engineering education and American culture more generally.

Conclusion

In their calls for developing the "real world" skills of students, leading engineering education reformers are signaling the limitations of the regimented and abstract classroom learning that dominates the current curriculum. A consensus on what exactly those practical skills encompass does not yet exist; they range from the professional "soft" skills of communication and management to knowing how to use tools in a workshop. Nor is it exactly clear how these broadly defined skills contribute to a different kind of learning or engineering practice.

This chapter sought to clarify the pedagogical value of a very specific kind of practical learning. What distinguishes the EbD initiative at Mines from other real world engineering programs is that it is grounded in hands on projects that bring together non-traditional students, with practical skills, with traditional engineering students, with solid grounding in mathematics and sciences.

The first benefit of the program is that it provides students with an opportunity to learn with their bodies and engage a variety of senses – seeing, touching, hearing, smelling and "doing" – to understand engineering problems as they exist in the real world. This different kind of learning has the potential to enhance creativity and innovation among all students. The second benefit specifically addresses the problematic lack of diversity in engineering education. EbD aims to help recruit, retain and engage non-traditional students by providing opportunities for them to see their background experiences and expertise as not simply relevant to engineering, but vital for its continued flourishing.

The third potential benefit remains to be tested by research, but the EbD planners hope that hands on projects will help graduates grapple more effectively and holistically with the challenges of sustainability that the NAE signaled in its Grand Challenges. At a basic level, EbD makes the challenges more real by helping students to understand the life cycles of their projects by seeing and touching the material required to create and operate artifacts, as well as the byproducts and waste created in the process. But EbD does more. It situates engineered artifacts and systems within concrete social contexts, pushing students to visualize, experience and feel how the physicality of their designs and products use and waste local resources as well as impact the environment and community. Students engaged in EbD will be well positioned to address the sustainability dimensions of the NAE Grand Challenges with an eye to delivering sustainable designs, products and systems through experiential knowledge and their diverse social backgrounds.

Advocates for EbD recognize that the challenges for reforming the engineering curriculum are substantial and multilayered, encompassing the "curricular and ideological, the structural and the symbolic" (Rose 2012, p. 8). The prestige of engineering rests on its abstraction, and the cultural divide between engineers and technicians seems firmly entrenched. But if engineers are to "lead, make tough decisions and frame questions in a way that fosters creative solutions" (Fortenberry 2011, p. 37), colleges must adapt to provide real world opportunities for students to develop those skills.

Acknowledgements This chapter benefitted from generous feedback from Juan Lucena and discussions with the Engineering by Doing planning group at the Colorado School of Mines.

References

Boundaoui, A. (2011). Why would-be engineers end up as English majors. CNN. Retrieved from http://www.cnn.com/2011/US/05/17/education.stem.graduation/index.html

Braverman, H. (1998). *Labor and monopoly capital: The degradation of work in the twentieth century*. New York: Monthly Review Press.

Calvert, M. A. (1967). *The mechanical engineer in America, 1830–1910*. Baltimore: The Johns-Hopkins Press.

Chen, X. (2009). *Students who study science, technology, engineering, and mathematics (STEM) in postsecondary education*. Washington: National Center for Education Statistics.

Crawford, M. (2009). *Shop class as Soulcraft: An inquiry into the value of work*. New York: The Penguin Press.

Donaldson, K., Lichtenstein, G., & Sheppard, S. (2008). *Socioeconomic status and the undergraduate engineering experience: Preliminary findings from four American universities*. Presented at the American Society for Engineering Education Annual Conference, Pittsburgh, PA.

Downey, G. L. (2007). Low cost, mass use: American engineers and the metrics of progress. *History and Technology, 23*(3), 289–308.
Dudley, K. M. (1994). *The end of the line: Lost jobs, new lives in postindustrial America*. Chicago: University of Chicago Press.
Fortenberry, N. (2011). Teaching the practical skills. *Mechanical Engineering, 133*(12), 36–40.
Foster, J. B. (1999). A classic of our time: *Labor and monopoly capitalism* after a quarter-century. *Monthly Review, 50*(8), 12–18.
Ingold, T. (2001). Beyond art and technology: The anthropology of skill. In M. B. Schiffer (Ed.), *Anthropological perspectives on technology*. Albuquerque: University of New Mexico Press.
Kamoche, K., & Maguire, K. (2011). Pit sense: Appropriation of practice-based knowledge in a UK coalmine. *Human Relations, 64*(5), 725–744.
Layton, E. T., Jr. (1971). Mirror-image twins: The communities of science and technology in 19th-century America. *Technology and Culture, 12*(4), 562–580.
Lord, M. (2011). Seeing & doing: Revamped curricula show freshmen what it means to be an engineer. *Prism Magazine*, (September 2011), 34–39.
McClenney, K., & Arnsparger, A. (2012). *Students speak, are we listening? Starting right in the community college*. Austin: Center for Community College Student Engagement.
McLoughlin, L. A. (2012). Community colleges, engineering, and social justice. In C. Baillie, A. L. Pawley, & Donna, R. (Eds.), *Engineering and social justice: In the university and beyond* (pp. 123–142). West Lafayette: Purdue University Press.
Microsoft (2012). A national talent strategy: Ideas for securing U.S. competitiveness and economic growth (2012, September). Microsoft. Retrieved from http://www.microsoft.com/en-us/news/download/presskits/citizenship/MSNTS.pdf
Morice, P. B. (1988). Britain and European engineering education. *European Journal of Engineering Education, 13*(1), 71–75.
Ohland, M. W., Orr, M. K., Lundy-Wagner, V, Veenstra, C. P., and Long, R. A. (2012). Viewing access and persistence in engineering through a socioeconomic lens. In C. Baillie, A. L. Pawley, & D. Riley (Eds.), *Engineering and social justice: In the university and beyond* (pp. 157–180). West Lafayette: Purdue University Press.
Orr, J. E., & Barley, S. R. (Eds.). (1997). *Between craft and science: Technical work in the United States*. Ithaca: ILR Press.
President's Council of Economic Advisers and the National Economic Council. (2012). Military skills for America's future: Leveraging military service and experience to put veterans and military spouses back to work. Washington: Executive Office of the President.
Reynolds, T. S. (1991). *The engineer in America: A historical anthology from technology and culture*. Chicago: University of Chicago Press.
Rolston, J. S. (2013). The politics of pits and the materiality of mine labor: Making natural resources in the American West. *American Anthropologist, 115*(4), 582–594.
Rose, M. (2004). *The mind at work: Valuing the intelligence of the American worker*. New York: Penguin.
Rose, M. (2012). Rethinking remedial education and the Academic–Vocational divide: Complementary perspectives. *Mind, Culture, and Activity, 19*(1), 1–16.
Sauer, B. (2003). *The rhetoric of risk: Technical documentation in hazardous environments*. Mahwah: Lawrence Erlbaum.
Seely, B. (1993). Research, engineering, and science in American engineering colleges: 1900–1960. *Technology and Culture, 34*(2), 344–386.
Sharp, J., & Zachary, L. (2004). Using the van Hiele K-12 geometry learning theory to modify engineering mechanics instruction. *Journal of STEM Education: Innovations and Research, 5*(1/2), 35–41.
Sheppard, S. D., Macatangay, K., Colby, A., & Sullivan, W. M. (2008). *Educating engineers: Designing for the future of the field*. San Francisco: Jossey-Bass.
Silverberg, M., Warner, E., Fong, M., & Goodwin, D. (2004). *National assessment of vocational education final: Report to congress. Executive summary*. Washington: U.S. Department of Education.

Slaton, A. E. (2010). *Race, rigor, and selectivity in U.S. Engineering: the history of an occupational color line*. Cambridge: Harvard University Press.

Smith, K. A., Sheppard, S. D., Johnson, D. W., & Johnson, R. T. (2005). Pedagogies of engagement: Classroom-based practices. *Journal of Engineering Education, 94*(1), 87–101.

Strutz, M. L., Orr, M. K., & Ohland, M. W. (2012). Low socioeconomic status individuals: An invisible minority in engineering. In C. Baillie, P. Alice L., & R. Donna (Eds.), *Engineering and social justice: In the university and beyond* (pp. 143–156). Purdue University.

Whalley, P., & Barley, S. R. (1997). Technical work in the division of labor: Stalking the wily anomaly. In S. R. Barley & Orr, J. E. (Eds.), *Between craft and science: Technical work in U.S. settings* (pp. 23–52). Ithaca/New York: ILR Press

Wisnioski, M. (2012). *Engineers for change: Competing visions of technology in 1960s America*. Cambridge: The MIT Press.

Yoder, B. L. (2012). Engineering by the numbers. American Society for Engineering Education. Retrieved from http://www.asee.org/papers-and-publications/publications/college-profiles/2011-profile-engineering-statistics.pdf

Zussman, R. (1985). *Mechanics of the middle class: Work and politics among American engineers*. Chicago: University of California Press.

Jessica Smith Rolston Ph.D. in Anthropology and certificate in Women's Studies from the University of Michigan. Hennebach Assistant Professor, Liberal Arts and International Studies, Colorado School of Mines, Colorado, USA. She specializes in the sociocultural dynamics of mining and extractive industries, with emphases in labor, social justice, gender and corporate social responsibility, and is beginning a new research project about socioeconomic class and engineering education. She is the author of *Mining Coal and Undermining Gender: Rhythms of Work and Family in the American West* (Rutgers University Press, 2014), and her research has appeared in *American Anthropologist*, *Signs: Journal of Women in Culture and Society*, *Working USA: Journal of Labor and Society*, and *Anthropology Today*.

Elizabeth Cox M.A. in Education from Virginia Tech. Director, Red Rocks Institute for Sustainability in Education (RISE), Red Rocks Community College, Colorado, USA. Her focus is development and implementation of experiential education, focusing on its relevance and potential in engineering education. As Director of RISE, she develops active learning strategies for science, technology, engineering and mathematics education at the K-12 and two-year pre-engineering levels. She is currently coordinating the project "Engineering by Doing" between Red Rocks Community College and Colorado School of Mines as a project-centered design approach to sustainable engineering practices in community and industry.

Chapter 14
Fostering Hybridity: Teaching About Context in Engineering Education

Andrew Jamison, Niels Mejlgaard, and Jette Egelund Holgaard

Abstract The main discourses of reform in engineering education in recent years have tended to be "market-driven" and have involved adding on courses and instruction in such areas as business economics, marketing, management, and entrepreneurship, as well as various types of "on-the-job" training in companies. In response, there has been a reassertion among many educators of more traditional "science-driven" approaches to engineering education by dividing engineering fields into "subdisciplines", developing courses and even entire programs in new areas of specialization, such as sustainability science, product design, and sustainable energy planning. As a result, among different universities, as well as within many of the same ones, there is an ongoing tension or competition between market-driven approaches and science-driven approaches, often at the expense of a more balanced or comprehensive approach to education. In order to resolve this underlying tension in engineering education, this chapter proposes and exemplifies a third approach that seeks to foster a "hybrid imagination" combining a scientific-technical problem-solving competence with an understanding of the problems that need to be solved. Fostering hybridity or a hybrid imagination involves a mixing of scientific education and training in technical skills with an appreciation of the broader cultural implications of science and technology in general and one's own role as an engineer, in particular. The chapter contrasts the different ways in which matters of "context" are brought into engineering education in the different approaches. It begins with a general discussion of hybrid identities in engineering and then goes on to provide a number of examples from the authors' experiences in teaching contextual knowl-

A. Jamison (✉)
Department of Development and Planning, Aalborg University,
Erik Dahlbergsgatan 22, Malmö S-211 48, Sweden
e-mail: andy@plan.aau.dk

N. Mejlgaard
Danish Centre for Studies in Research and Research Policy, Aarhus University,
Bartholins Allé 7, Aarhus C 8000, Denmark
e-mail: nm@cfa.au.dk

J.E. Holgaard
Department of Development and Planning, Aalborg University,
Vestre Havnepromenade 5, Room: 1.211, Aalborg 9000, Denmark
e-mail: jeh@plan.aau.dk

S.H. Christensen et al. (eds.), *International Perspectives on Engineering Education*,
Philosophy of Engineering and Technology 20, DOI 10.1007/978-3-319-16169-3_14

edge for engineering students at Aalborg University in Denmark. The chapter has been written as a part of PROCEED, the Program of Research on Opportunities and Challenges in Engineering Education in Denmark.

Keywords Hybrid identities • Hybrid imagination • Contextual knowledge • Project-based learning • Turning nano-technology green • Engineering citizenship

Introduction

The point of departure for PROCEED (Program of Research on Opportunities and Challenges in Engineering Education in Denmark) is the recognition that the challenges facing engineering and engineering education in the world today have given rise to contradictory and competing response strategies, not least in the ways in which non-technical, or contextual knowledge is included in the curriculum.

The dominant response has been "market-driven" as engineering educators throughout the world have added courses and project work in entrepreneurship and innovation onto the curriculum, often sponsored by companies. A second significant response has been "science-driven" and has meant translating the challenges into new academic fields or sub-fields and adding courses in theory of science and engineering ethics to the curriculum. As such the responses have pulled engineering education into different directions and most students have therefore not been given the opportunity to understand the broader social and cultural aspects of the challenges facing engineering. That is why we have proposed the need for a more explicitly "socially-driven" strategy in which contextual understanding is integrated into the "core" engineering curriculum in order to foster what we have come to call a "hybrid imagination" (Jamison et al. 2011). In this chapter, we discuss these different meanings of context and the different ways in which contextual knowledge is included in these different approaches to engineering education, based, to a large extent, on our experiences at Aalborg University in Denmark. Is contextual knowledge primarily taught in order to provide entrepreneurial skills, professional socialization, or socio-cultural understanding? What is at issue, we suggest, are different ideas of engineering identity formation.

The Formation of Hybrid Identities

Among those who study the relations between science, technology and society, hybridity and hybridization have become popular terms. Donna Haraway (1991) has written a manifesto for what she terms cyborgs: "theorized and fabricated hybrids of machine and organism." Bruno Latour (1993) has characterized contemporary reality in terms of a "proliferation of hybrids" between humans and non-humans and calls for the overthrow of the "modern constitution" that was established in the seventeenth century separating the study of nature from the study of society.

In regard to engineering education, one can distinguish between two ideal-types of hybrid identity formation, which can be characterized as "hybridization" and "hybridity". These two types of hybrid identity formation are not mutually exclusive, and in actual educational and learning processes, they are often difficult to separate.

Hybridization corresponds to the market-oriented approaches to engineering education and to engineering in general that are characteristic of the new "mode 2" form of knowledge production, as discussed in the influential book, *The New Production of Knowledge* from 1994. As Michael Gibbons and his co-authors put it,

> Hybridisation reflects the need of different communities to speak in more than one language in order to communicate at the boundaries and in the spaces between systems and subsystems (Gibbons et al. 1994, p. 37).

Hybridity, on the other hand, corresponds to what we have come to think of as a "mode 3" form of knowledge production – "hybrid imagining" – combining the innovative and entrepreneurial spirit of mode 2 with the academic values and critical spirit of more traditional, or "mode 1" forms of science and engineering.

Where the one form of hybrid identity formation is most often generated "from the outside" and takes advantage of new commercial employment and funding opportunities, the other form is most commonly generated "from within" and represents a primarily personal decision to change. The one is primarily a form of entrepreneurship, or individual agency, while the other is a form of stewardship, or engagement in a collective process. Obviously, these are ideal types and there is no strict line of demarcation between them. The two forms of hybrid identity formation can be thought of as different poles in a continuum, with many variants in between, but for analytical purposes it can be helpful to distinguish between them since they have important implications for educational initiatives and reform processes.

Perhaps the most important aspect of this distinction for educational purposes is the degree to which the students "own" their learning process. One of the built-in drawbacks with hybridization and with externally-funded student projects and course offerings is the dependent relation that is formed between the funder and the funded, the employer and the employee. In this way the key distinction is whether students are encouraged to learn how to solve somebody else's problems (i.e. the problems facing the companies that support them) or whether they learn how to formulate problems that they themselves have come to consider important.

In any case, in relation to engineering and engineering education, it can be useful to distinguish between these two different forms of hybrid identity formation. Both are attempts to transcend the traditional disciplinary identities that were so important to scientists and engineers in the past, but which have become increasingly difficult to maintain, because of the external pressures that scientists and engineers are increasingly subjected to. But they represent different ways to respond to those pressures. While there is a fairly widespread recognition of the importance of flexibility and the mixing of skills and knowledge, there are rather different forms in which the fostering of hybrid identities has taken place.

In cognitive terms, the distinction is between interdisciplinarity, which characterized the practice of many scientists and engineers in the 1970s, who were interested in contributing actively to dealing with the social and environmental challenges and movements that emerged at the time, and transdisciplinarity, which started to be used in the 1980s as a term to characterize those working in the ever more commercialized and globalized new "mode" of knowledge production. Where the one signals a primarily internally generated process of integrating knowledge and competence from different disciplines, the other is usually an externally generated process of combining specialized "niche" competencies into specific contexts of application (cf. Krishnan 2009).

Obviously, there is a need for both types of hybrid identities, but what is perhaps most important to emphasize is that one type should not be favored at the expense of the others. In some situations, where it is primarily public participation and cultural change that need to be fostered, there need to be opportunities for interdisciplinary and cross-disciplinary approaches, and, in particular, for research in which academics and laypeople collaborate in efforts that are organized "from below" in relation to local needs and concerns. This is what we have come to identify with the notion of a hybrid imagination.

Fostering a hybrid imagination is a way to educate students in what our former colleague in Aalborg, Bent Flyvbjerg (2001), has characterized as "phronesis": the kind of common sense practical knowledge that the ancient philosopher Aristotle distinguished, on the one hand, from theoretical knowledge, or *episteme* – what we today call science – and, on the other hand, from technical knowledge, or *techné* – what we today call engineering, art and architecture. Phronesis, for Aristotle, was a kind of moral, or ethical knowledge: according to Flyvbjerg, it "concerns the analysis of values – "things that are good or bad for man" – as a point of departure for action" (Flyvbjerg 2001, p. 57).

A hybrid imagination can be defined as the combination of a scientific-technical problem-solving competence with an understanding of the problems that need to be solved. It is a mixing of scientific knowledge and technical skills with what might be termed cultural empathy, that is, an interest in reflecting on the cultural implications of science and technology in general and one's own contribution as a scientist or engineer, in particular. A hybrid imagination involves recognizing the limits to what we as a species and as individuals can do, both the physical limits and constraints imposed by "reality" as well as those stemming from our own individual limits of capabilities and knowledge. As such, a hybrid imagination is often manifested collectively, involving collaboration between two or more people, either in a project group or in relation to a broader social or cultural movement (cf. Jamison et al. 2011).

Contextual Knowledge at Aalborg University

Like other universities that were created in the 1970s, under the influence of the social movements of the times, Aalborg University has attempted to develop a more "relevant" form of education than was then being offered by the established universities. In Denmark, the new universities that were established in the 1970s – in Roskilde in 1972, Aalborg in 1974 and Odense in 1978 – were called university centres to distinguish them from the older universities. Both Roskilde and Aalborg largely did away with traditional disciplines, departments and faculties and developed a variant of problem-based learning that was organized in the form of project work carried out by groups of students (cf. Kolmos 1996; de Graaf and Kolmos 2003).

An important source of inspiration was the recently-created University of Bremen in Germany and the notion of "exemplary learning" that had been developed by the German writers Oscar Negt and Alexander Kluge (Negt and Kluge 1993). Their writings, along with those of Jürgen Habermas and other "critical theorists" had a major influence on the kind of pedagogy that was implemented at Roskilde and, somewhat less explicitly, in Aalborg, where the new university was not completely new but was, to a large extent, an expansion of an engineering school that was already there. In Aalborg, as somewhat later in Odense, where the new university was not as radical, or alternative in its pedagogical ambitions, there was a much stronger interest in contributing to regional economic development than at Roskilde.

From the outset, Aalborg University has based all of its undergraduate teaching programs on a combination of problem and project-based learning, with formalized courses playing a subsidiary or supportive role (cf. Kjærsdam and Enemark 1994; Kolmos et al. 2004). For the most part, the students are taught their subjects by carrying out semester-long projects in groups, and the task of the teacher is to advise the students, rather than instruct them. In the science and engineering fields, project work in the first year has included, since the early 1980s, a certain amount of what has come to be referred to as contextual knowledge. The particular way in which this knowledge is taught and included in the student projects varies from field to field, and has also varied from year to year, depending on who is doing the teaching, and, not least, on the relations between the main, scientific/technical advisers, who are employed by the departments responsible for the particular educational program and the contextual advisers, who, for the most part, come from outside the particular field of study.

One of the recurrent problems with teaching contextual knowledge in Aalborg, as it is for around the world is that there is relatively little incentive for academics, either engineers or non-engineers, to become "experts" in contextual knowledge. The relations between science, technology and society are marginal topics in both humanities and social sciences, as well as in engineering and natural sciences. Another, related problem is that "context" means something quite different for engineers and non-engineers (Table 14.1).

Table 14.1 Approaches to contextual knowledge in engineering education in Aalborg

Strategy	Market oriented	Academic disciplinary	Socio-cultural
Rationale	contextual knowledge is for cultivating entrepreneurship	contextual knowledge is for habituating students in a field	contextual knowledge is for fostering a hybrid imagination
Story-line	Economic innovation	Social construction	Cultural appropriation
Main contents	Innovation and business studies, economic and market analysis	Science and technology studies, philosophy of science; actor analysis	Cultural studies, history of science and technology, technology assessment

The most common approach to contextual knowledge has been to provide a kind of supplementary, or "add-on" knowledge, usually aimed at offering the students knowledge of some of the "market" conditions that affect their particular engineering or scientific field. This is similar to what is offered at many other universities throughout the world, where courses in business and marketing are commonly included in engineering programs. Typically, the lectures and advising focus on managerial issues and "entrepreneurship", and the project work often involves one or another form of market analysis of the particular technical or scientific product that the students are learning how to design and/or build in their projects.

A second approach that is used in Aalborg – particularly in the more "traditional" science and engineering fields – provides more of a complementary or extra-curricular knowledge, offering students an opportunity to reflect on the underlying values and paradigmatic assumptions of their scientific or engineering field as a way of preparing for their future professional roles. The courses usually offer an introduction to the philosophy and/or sociology of science and technology, presenting the different schools, or positions, as well as some of the methods of analysis that have been developed in science and technology studies. The social construction of technology, or SCOT, approach, as developed by Wiebe Bijker, has been especially popular (cf Bijker 1995). In the project work, the students are often encouraged to use these ideas to consider the ways in which scientific and engineering knowledge is produced, or constructed, within their fields.

A third approach, and one that has characterized our own efforts through the years is to connect, as much as possible, the technical-scientific components of the project work to broader contextual issues, and to mix something of the more instrumental ambition of the market-oriented approach with the reflective ambition of the academic approach. In the lectures we have introduced the students to the cultural history of science and technology and to some of the public debates that have taken place in relation to science and technology.[1] Students are also introduced to political

[1] *Hubris and Hybrids: A Cultural History of Technology and Science* was written to be used as a textbook in these first year courses, and has been used in a wide range of engineering education programs, including biotechnology, medialogy (or media engineering), nanotechnology, architecture and design (Hård and Jamison 2005). After it was used as a textbook for a course at Denmark's Technical University in 2010, a new book, written together with Steen Hyldgaard Christensen and Lars Botin (*A Hybrid Imagination*) was produced that has since been used in the first year in our new educational program in Techno-anthropology (Jamison et al. 2011).

and ethical issues associated with science and technology, and the contextual advising of their project work is seen as a way for them to learn how they might address, and, at best, assess the political, cultural and/or environmental implications of their particular scientific-technical project.

These different approaches to contextual knowledge reflect what we have come to think of as the different "story-lines" that have been developed by social scientists and humanists to explain the development of technology, the story-line of economic innovation, the story-line of social construction, and the story-line of cultural appropriation (cf. Jamison and Hård 2003).

The Story-Line of Economic Innovation

Since Karl Marx based his influential theory of political economy on the central role of the "means of production" in historical development, it can be suggested that the dominant approach to engineering contexts has focused on the relations between engineering and the economy. This story-line, as it has been developed during the past 150 years, has, to a large extent, been a narrative of material science-based progress, and, more specifically, emphasized the role of science and engineering in economic growth and development. It has directed attention primarily to the activities of companies and corporations, since they are generally considered to be the main sites, or contexts in which market-oriented technological development, or economic innovation takes place. The relevant contextual knowledge is almost exclusively economic and managerial.

The capitalist mode of production, as Marx characterized it, had its material base in the orientation of science and technology to the commercial marketplace. Science and engineering played a fundamental, "revolutionary" role in modern industry: "by means of machinery, chemical processes and other methods, it [modern industry] is continually transforming not only the technical basis of production but also the functions of the worker and the social combinations of the labor process" (Marx 1976/1867, p. 617).

While putting production on a scientific basis, industrialization also created, according to Marx, divisions among workers, and led to a new class of workers "whose occupation it is to look after the whole of the machinery and repair it from time to time, composed of engineers, mechanics, joiners, etc." In Marx's words, "This is a superior class of workers, in part scientifically educated, in part trained in a handicraft; they stand outside the realm of the factory workers…" (Ibid, pp. 545–546). Scientists and engineers had been given a fundamental role to play in the economy, but at the same time, they had been forced to give up their independence and apply their knowledge and skills to the requirements of the commercial marketplace, and work alongside the "ruling class" rather than the working class.

One of those who helped turn the Marxian insights into a story-line of economic innovation was the Austrian Joseph Schumpeter, who identified innovation as the main source of industrial dynamism and economic growth. For Schumpeter, innovation was a double-edged sword, however, in that it served to both create new things and destroy old ones at the same time. He called the process "creative destruction" and at its core was innovation:

> …it is by means of new combinations of existing factors of production, embodied in new plants and, typically, new firms producing either new commodities, or by a new, i.e. as yet untried, method, or for a new market, or by buying means of production in a new market. What we, unscientifically, call economic progress means essentially putting productive resources to uses *hitherto untried in practice*, and withdrawing them from he uses they have served so far. This is what we call "innovation" (Schumpeter 1928, pp. 377–378).

Drawing on the work of a Russian economist, Nikolai Kondratiev, Schumpeter developed a model of business cycles, or "long waves" in which the process of innovation played a central role (cf. Freeman and Louçá 2003).

Schumpeter's ideas have been formative for the ways in which economists and economic historians and, not least, those in the sub-fields of technology management and innovation studies, have since come to characterize and analyze the process of technological change (Jamison 1989). At the beginning of each wave, a cluster of "radical" innovations are seen to propel a new upswing in industrial expansion as they are spread, or diffused in the economy. New companies and branches of industry that are based on the innovations emerge to replace, or "creatively destroy" the companies that had grown up in the previous waves, much as Microsoft, Apple, Nokia, and Google have done in recent decades.

With each of these waves, there has been a change in the kinds of engineering skills and knowledge that have been central. In the nineteenth century, civil and mechanical engineering were important in the "first wave" while chemical engineering and electrical engineering grew in importance in the second. At the turn of the century, the new fields of aeronautic and automotive engineering emerged out of the radical innovations of the airplane and the automobile, while in the fourth wave, after the Second World War, atomic energy and electronics became important fields of engineering. In recent decades, some innovation economists have identified a fifth wave, based on information technology and biotechnology. In the words of Christopher Freeman and Francisco Louçã:

> Even those who have disputed the revolutionary character of earlier waves of technical change often have little difficulty in accepting that a vast technological revolution is taking place, based on the electronic computer, software, microelectronics, the Internet and mobile telephones….From a very much smaller base and on a much smaller scale, bio-technology was also growing very rapidly in the closing decades of the twentieth century. In one sense, it too is a very special form of information technology and it is interacting increasingly with computer technology (Freeman and Louçá 2003, p. 301)

While each wave has played a significant role in refashioning engineering, there has also been a more general shift from a kind of engineering in which science and technology were institutionally distinct to the technosciences of today in which the borders between science and technology have been more or less broken down. In the first two waves of industrialization science and engineering were "mirror-image twins" to borrow a term from Edwin Layton (1971). There was interaction to be sure, but in the United States, as Layton has described, as well as in most of the other industrializing countries, engineers and scientists had different identities, with different forms of education and training.

This started to change in the mid-nineteenth century, especially in Germany, where higher education in engineering was institutionalized primarily in the form of *technische Hochschulen* (renamed "technical universities" in the twentieth century). These institutions combined education in mathematics and natural science with laboratory-based instruction in what started to be called the technical or engineering sciences (cf. Gispen 2002). Some of these theoretically trained engineers contributed to the rise of new industries following the radical innovations in organic chemistry (fermentation, pasteurization, etc.), communications (telephony and wireless radio) and electricity However, for most of the nineteenth century, the practical engineers still dominated industrial development. Even in Germany where the laboratory training at universities was initiated and supported by the creation of research and development facilities in larger corporations, the contribution of engineers in industrial innovation came mostly from their practical experiences and systematic experiments, and only in small part from theoretical, science based knowledge.

Until well into the nineteenth century technological development mostly relied on practical approaches in the crafts and the proficiency and experience of skilled technicians (cf. Lilley 1973). Development and production was characterized by trial and error and a great deal of experience, intuition and working knowledge about materials and mechanisms. Since the late nineteenth century, scientific approaches added a new element to technological development and economic production. As such, in the course of the twentieth century a perception of engineering as technological innovation has grown in importance in many fields, particularly in the newer, more "high-tech" branches of industry. Rather than being either primarily theoretical or primarily practical, the engineer, according to this perception, is an innovator, developing new technology by combining theoretical and practical knowledge both at universities and in private companies, as well as at the interfaces between them.

The Story-Line of Social Construction

While economists and economic historians, and the stories of innovation that they like to tell, tend to dominate both the policy discourses, as well as the academic study of engineering contexts, a second significant story-line or narrative approach has emerged within the field of science and technology studies, or STS.

STS emerged in the 1970s as an acronym for "science, technology and society" and, at many universities around the world, STS courses were developed to offer science and engineering students instruction in the history, philosophy and sociology of science and technology. The idea was to offer instruction about the social and cultural contexts of science and technology, as well as to provide meeting places for natural scientists, engineers, social scientists and humanists for discussion seminars and workshops and eventually for carrying out research projects together. The field of STS, at least at the beginning, was part of a more general interest within universities to foster interdisciplinary studies.

Inspired by the extremely influential book by Thomas Kuhn, *The Structure of Scientific Revolutions*, first published in 1962 and then later revised in 1970, STS courses were meant to provide science and engineering students – and, for that matter, social science and humanities students, as well – insights into how science and technology were actually carried out. Kuhn, and the sociologists, historians and philosophers who established the first STS programs, tried to show that science and engineering were not free from social influence, and that in all fields, the governing paradigm – or what Kuhn later termed the "disciplinary matrix" of methods and theories and concepts – was at least in part, shaped by interests in the broader society. In the course of the 1970s, a number of different approaches, or schools emerged to explore these "social relations" of science and technology, as STS, at least at some universities, started to take on the character of a discipline or academic field of its own and call itself "science and technology studies".

Since the mid-1980s, the French philosopher and anthropologist Bruno Latour and the Dutch engineer turned sociologist Wiebe Bijker have been among the most active in transforming STS into "science and technology studies". In what Latour has called the "proliferation of hybrids" scientists and engineers are seen to bring together, or hybridize human, or social elements with "non-human", or natural, elements in constructing scientific facts and technological artifacts in so-called "actor-networks" (Latour 1993, 2005).

Over the past 25 years, STSers have sought to uncover the ways in which scientists and engineers through their professional activities actually go about carrying out their work (Latour 1986; Bijker et al. 1987; Jasanoff et al. 1995; Hackett et al. 2008). In relation to engineering, Thomas Hughes has contrasted the "networks of power" that were involved in the development of electricity systems in Europe and the United States (Hughes 1983), and Wiebe Bijker has elucidated the social interests and technological frames that were at work in a number of different fields of engineering (Bijker 1995).

Science and technology studies have thus offered a new kind of understanding of how scientific knowledge and engineering projects are actually produced in the contemporary world, which has been both influential and controversial. The engineer according to this story-line is seen as a professional "actor" involved in the construction of a technologically mediated reality. The expertise or professional competence of engineers is thus not seen as purely technical or scientific; there is also a kind of social competence, or social capital that is necessary to cultivate, and an education in this kind of contextual knowledge has thus been seen, as in Aalborg, as an important part of the professional expertise of an engineer.

During the 1990s, however, a number of scientists and engineers launched attacks on science and technology studies for presenting an overly critical and "irrational" view of science and engineering, leading to what came to be called the science wars. Particularly important was the "hoax" perpetrated by the physicist, Alan Sokal, by which he published an article in a special issue on science and technology studies in a cultural journal that purported to show how social interests affected quantum physics that he had simply made up.[2] One of the effects of the science wars is that STS is rarely included any longer in engineering education, and, when it is, it is usually presented in a non-political and academic way.

Indeed, most of the field's practitioners are located at a good distance from science and engineering faculties, for the most studying rather than affecting science and technology. In any case, it seems fair to say that far too many engineering students complete their educations with little or no awareness of the scholarship that has been accumulating during the past 40 years in the relations of science, technology, engineering and society in STS, and, more recently, engineering studies.[3]

The Story-Line of Cultural Appropriation

While the economic meaning of engineering is by far the most dominant, the social, or professional meanings have become ever more influential in recent years, especially in the arenas of policy-making and government. Both focus on the production of science and technology, and have tended to disregard all of the other forms of engineering that involve the "cultural appropriation" or multifarious uses of technology and science in the broader society (Hård and Jamison 2005).

A main source of inspiration for this story-line was the American writer Lewis Mumford, especially his classic work, *Technics and Civilization*, from 1934. Mumford was one of the first to discuss the cultural preconditions for modern science and technology, and to explore the long process of cultural preparation prior to the scientific and industrial revolutions. He was also one of the first to discuss the cultural consequences, and, not least, the forms of cultural resistance and opposition to science and technology. Later in his life, he became one of the main critics of the so-called military-industrial complex in the United States which he saw as a new kind of authoritarian engineering, what he termed the "megamachine" (Mumford 1970).

More recently, the British cultural historian Raymond Williams has written about the relations between technology and broader processes of cultural transformation in a number of books that have contributed to the creation of the academic field of cultural studies. Williams emphasized how the idea of culture, at least in the British context had emerged in the nineteenth century as a "record of our reactions, in thought and feeling, to the changed conditions of our common life… Its basic element is its effort at total qualitative assessment." (Williams 1958, p. 285)

---

[2] See Sokal's book, *Beyond the Hoax* (Sokal 2008) for the full text of his original article and related essays on science and philosophy and critiques of science and technology studies.

[3] Science and technology studies have become popular at many business schools, often in relation to programs in entrepreneurship and innovation. As a possible sign of the times, the 2012 meeting of the European and American societies for science, technology and society (The European Association for the Study of Science in Society, EASST, and the Society for the Social Study of Science, 4S) was held at the Copenhagen Business School.

Another influential writer was the literary historian, Leo Marx, who was a pioneer in investigating the artistic and literary representations of science, technology and engineering in his important study, *The Machine in the Garden* from 1964. Marx's student, David Nye, has been one of the most prolific contributors to the story-line of appropriation, in a series of books on the ways in which electricity and other forms of power have been used in different ways by different people. His recent book, *Technology Matters*, provides a highly readable introduction to this way of discussing engineering contexts (Nye 2006).

This third kind of engineering takes place in very different contexts or social locations than the other two, often in what are characterized as social and cultural movements rather than in established or formalized institutions and organizations. Understanding these contexts of engineering brings out the ambivalence, or mixed meanings of science and technology in human history, and the ways in which engineering has often had to be carried out at the "grass-roots" in informal and temporary public spaces, in order to provide alternatives to the dominant approaches.

Historically, these forms of engineering have been a part of broader political struggles, from the religious struggles of the sixteenth century through the social movements of the nineteenth and twentieth centuries and into the present. One of the founders of interior design, William Morris, was, for example, an active member of socialist organizations, as well as a professional artist and designer. In the anticolonial movements of the early twentieth century, especially in India, Western-trained scientists also joined forces with political activists to resurrect traditional forms of engineering, or what are sometimes now called indigenous technology, that became important parts of the liberation struggle. Similarly, in the environmental movements of the 1970s grass-roots forms of engineering provided "utopian" or radical examples of appropriate technology that have since developed into significant branches of industry (Dickson 1974; Jamison et al. 2011).

Particularly influential was how, within the context of the opposition to nuclear power, many professional scientists and engineers joined forces with environmental activists to experiment with alternative forms of energy. In countries like Denmark, as a part of the movement against nuclear energy, an organization for renewable energy was created that provided a space, or cultural context in which people could learn how to build wind energy power plants and solar panels. Like similar activities in other countries, these forms of grass-roots engineering were an important part of a social movement, and like other movements today, in organic agriculture, alternative health care, sustainable design and architecture, they open engineering to popular, or public participation.

Turning Nanotechnology Green

One of the project groups that we supervised in the academic year, 2006–2007, in the nanotechnology educational program provides a particularly good example of this cultural approach in action. The course on "(nano)technology, humanity and society" for the nanotechnology students took place around the time when Al Gore was conducting his world tour on the "inconvenient truth" of climate change; indeed, he came to Aalborg in January, 2006, and spoke to great fanfare – and student interest – about the importance of taking global warming and the challenges associated with climate change seriously. And so it was perhaps not surprising that one of the nanotechnology groups came up with the idea of relating their technical project work to the climate change debate.

Nanotechnology and climate change were both discovered at about the same time, in the late 1980s. The one emerged from the meeting of Richard Feynman's speculative remarks about there being life at the nanoscale and the development of new scientific equipment, especially electronic microscopes using laser sensors that made it possible to "see" reality at the nanoscale. The other emerged from findings by atmospheric scientists and climatologists, using advanced scientific equipment of their own, that seemed to provide evidence for speculations that scientists had been making for many years that the earth's atmosphere could be affected, and its climatic conditions altered by excessive emissions of carbon dioxide.

The two fields of technoscience, however, have rarely met in the years since, as they have been subjected to very different forms of "cultural appropriation," both institutionally and intellectually The one – nanotechnology – has become a well-funded and rapidly developing field of technoscience, and a much discussed topic for the next big thing in science-based economic development, while the other – climate change – has become a much discussed topic for science-based policy deliberation, with the Intergovernmental Panel for Climate Change (IPCC) providing the subject matter for public debate and controversy, as well as the basis for international agreements, most famously the Kyoto Protocol of 1997.

Perhaps the mention of the missing connection between climate change and nanotechnology, and, more generally, of the relative lack of interest on the part of environmentalists in nanotechnology in one of the course lectures helped set the stage for a student group to attempt to link the two fields in their project work. The list of technical project proposals provided by the main advisers included the "raspberry solar cell", which had been developed, primarily for educational purposes, as a way to teach some of the basic principles of nanotechnology, and not least, nanofabrication to chemistry and chemical engineering students in the US, and learn something interesting and useful, as well.

Under our encouragement, the student group was able to produce a report in which contextual knowledge was successfully integrated into the project work. The group combined an ambitious, and highly enterprising technical research activity with a serious effort to explore the climate change debate and the role that their solar cell, and other kinds of renewable energy, could play in dealing with climate change in a meaningful way. As the students put it in the synopsis of their report:

> This report concerns the problems with global warming and investigates how dye sensitized solar cells (DSSC) might solve some of these. The report starts from IPCC's Fourth Assessment Report and analyzes the current global warming discussion. Next the possible technological solutions to the global warming problem is briefly described, and the DSSC is described in detail.

The students tested other substances than raspberries in order to improve the efficiency of the solar cell (spinach seemed to be more effective in some experiments than raspberries, and at the oral examination at the end of the semester, they discussed the possibilities and technicalities of combining various substances in the dye for the solar cell, coming up with the idea of a "spinberry solar cell," using both spinach and raspberry molecules). The students corresponded with scientists and engineers at companies and research institutes in Denmark and Germany to obtain better materials than were available in the laboratories that are used for the first year students in Aalborg, and produced several alternative products that they displayed and tested at the examination.

They also developed a basic understanding of the quantum mechanical theories that are relevant for nanotechnological fabrication, as well as a familiarity with the basic principles of solar energy and electrical transmission. The project also involved a detailed assessment of how dye sensitized solar cells could be used in society, and, in particular, in energy neutral houses. After reading and discussing a number of different reports, in particular, the so-called Stern report from the United Kingdom on the economic implications of climate change, and the Energy Plan 2030, produced by the Danish Society of Engineers, the students carried out a SWOT analysis of their raspberry solar cells, using established methodologies of technology assessment.

The project on raspberry solar cells is a good example of fostering a hybrid imagination. The students connected two different fields, or problem areas, of science and engineering, and, at the same time, learned different skills and forms of knowledge. They acquired an understanding of the natural scientific theories that are embedded in solar cells, and gained experience in conducting experiments and working with laboratory equipment, as well as obtaining a basic orientation in the climate change debate and some of the key policy documents. Most importantly, the students, in their report, combined the technical-scientific sections with the contextual sections in an integrated fashion. The report began with a discussion of the contextual issues rather than, as is normally the case, the other way around, that is, starting with the scientific theories and technical solutions. The problem, on which the learning process was based, was a contextual, rather than a scientific/technical one.

Engineering Citizenship

Two other groups in the program decided to take a serious interest in public concerns about nanotechnology. Methodologically inspired by studies within the field of "public understanding of science," both groups developed questionnaires to gauge public understanding – in the broadest sense of the word – of nanotechnology. Based on insights from these previous studies on public attitudes to science and technology, the students' surveys concentrated around issues such as public trust in scientists and engineers, citizen knowledge about nanotechnology applications, attitudes and expectations regarding societal implications of nanotechnology, and public engagement in various activities for acquiring information about nanotechnology and for influencing public policies.[4]

In terms of research focus and objectives, the groups worked along similar lines. In relation to data collection, however, they chose rather different designs. One group posted their survey on a number of websites and passed it electronically through their personal networks and to staff and students of several Danish universities. The group was very aware of the implications of web-based data collection for the quality and validity of the data and discussed issues of randomization and representativity at some length in the report. Literally cost-free, this data collection procedure lasted 3 days and resulted in a total of more than a thousand respondents, out of which nearly one in ten had also – encouraged by the group – commented on the survey design and the issues at stake in an open-ended question at the end of the questionnaire.

The second group conducted close to 200 face-to-face interviews with people in a nearby shopping area, randomly selected and screened to meet stratification criteria in terms of basic socio-demographic background variables, age and gender. The sample obviously did not represent the Danish population, as all interviews were done in the city of Aalborg; however, the entire exercise was more about understanding the methodological and practical aspects of different strategies for data collection than about getting a perfect sample.

The efforts of the groups are interesting for a number of reasons. Normally, survey based research on the public understanding of science and technologies is performed by social scientists, who are sometimes – and sometimes not – well-informed about the field of science and technical activities in question, but who are not themselves knowledge producers within the field. The social scientists involved in studying public understanding of science are mediators between science and society, which is obviously a useful task, but none the less, direct interaction between scientists, engineers and citizens is an important part of keeping science and engineering accountable and scientists and engineers informed about society. Or in the words of the groups:

> It should be continuously specified how moral and ethical boundaries are understood in a technological context… dialogues between scientists and citizens can serve to define these boundaries (group 1).

> The more that ordinary citizens are involved, the less fearful they will probably be towards the changes that new technologies bring about. The aim of the dialogue is not that citizens should take a supportive position, but that they will contribute, critically, towards solving problems (group 2).

The nanotechnology students' surveys represent a direct link between producers and (potential) users of nanotechnology. They are not added on to the scientific and technical parts of the engineering education, but were an integrated part of the edu-

[4] Niels Mejlgaard, who was at the time completing his Ph.D. on scientific citizenship, had been working in the field of public understanding of science carrying out public opinion surveys as part of the so-called Eurobarometer program (Mejlgaard 2007, 2009). He served as assistant supervisors for these groups and could thus offer them first-hand knowledge of the details of survey methodology and data collection.

cational process, which contextualizes and situates science and engineering, while it is being produced. They created dialogue and (in the case of one of the groups) direct, physical interaction between engineers (or students, as it were) and citizens. One thing that empirical studies of PUS have persistently shown is that citizen trust in scientists and engineers – "interpersonal identification" – is a main driver for public acceptance of controversial technologies. Basically, most people need to feel able to rely on the engineers behind the technical artefacts and processes they are confronted with in their daily lives. In fact, intersubjective trust is much more important than objective knowledge of the factual, technical aspects of technologies when people make up their minds about how to assess new technologies. Interestingly, both groups came to this conclusion in the course of their surveys, but no less important, they practically experienced the advantages of direct communication with lay persons when collecting the data. In a strange way, their activities and search for public concerns about nanotechnology may in fact have brought about much more public appreciation than ever so glamorous PR activities would have done.

The two projects are also interesting because they manage to combine a technical problem and laboratory work on the one hand with a contextual issue and social science research methods on the other hand, and feed it into a coherent structure and project report, even if it was done in different ways. One group worked, in the technical parts of the project, with carbon nanotubes and used the survey to investigate public perceptions of risk and toxicology, knowledge of nanoparticles, and opinions related to health issues. The other group compared, in a laboratory setting, the strengths and weaknesses of different basic methods for measuring at the nano scale, using scanning probe microscopy and optical spectroscopy, and used their survey experience to discuss parallel methodological questions of how to operationalize and measure in social science research.

In terms of analyses and results from the survey work, both groups came to conclusions which appear entirely plausible, based on what we know from similar research: that public knowledge of nanotechnology is still limited, that attitudes are strongly associated with personal efficacy and trust in scientists and public authorities, and that women and people with less education are more sceptical than other segments of the citizenry towards the societal implications of nanotechnology. Another main result of their learning process, though, was the development of a strong commitment to engaging in dialogue with lay citizens. By fostering their hybrid imagination, the students have themselves developed a hybrid identity, an engineering citizenship which simultaneously embraces scientific-technical competence and social responsibility.

These examples of student project work in the nanotechnology program indicate that there is an enormous potential for bringing contextual knowledge directly and explicitly into the education of scientists and engineers. These examples also show that it is possible to combine contextual knowledge with scientific/technical knowledge in a meaningful and integrated fashion. Obviously, these three projects are exceptional cases. In the other nanotechnology groups, the contextual knowledge, while important and interesting to the students, was quite separate from the scientific/technical part of the project work. One group that investigated the military implications of nanotechnology, and wrote about some research projects financed by the U.S. Department of Defence, made no attempt to bring this contextual knowledge into their report in more than a marginal way. And, as is the case in all of our first-year educational programs in Aalborg, the contextual knowledge was not given equal "weight" in the evaluation of the reports.

Medialogy for Sustainability

Another example comes from the study program in Medialogy, which in itself is a hybrid field combining communication science and computer and media engineering. The program was initiated in 2006 at Aalborg University and has since become a popular subject. Technically, they work with animation, computer games, sound and special effects, graphics and interactive environments. The core of medialogy lies in the design process and creativity is a key skill to master the profession. In the first few years of the program, the course on medialogy, technology and society, introduced students to the story-line of cultural appropriation and the notion of the hybrid imagination. As many of the medialogy students already felt like hybrids in a field that mixed a wide range of skills and competencies, they responded very well to this course.

One group by its own initiative took up the sustainability challenge as early as 2007 by making a computer game to educate children in how to reduce energy consumption in private households. The game was related to an educational initiative in elementary schools that was to be adopted in the obligatory subject in Danish schools called "Nature and technology". In their project, the group took their point of departure in the history of electricity production, while discussing the environmental impacts and possible risks of the related technologies developed. After that they gathered statistical data on electricity use in households in general and the use of electricity by children and teenagers in particular. Based on this statistical data, they estimated the environmental consequences of everyday household practices, taking into consideration energy consumption from electrical equipment in stand-by mode and (what they could relate to in everyday life of their own using their television etc.) electrical equipment turned on without being of any actual use.

By relating to the school context, the group also took an educational point of departure, by specifically addressing computer games as a means to learn, and in more general terms by referring to learning theory e.g. the notion of a "zone of proximal development" discussed in Vygotsky's learning philosophy. However, they also pointed to the weakness of theory by concluding that direct interaction with the target-group was needed. As one of the students stated, they could not themselves remember how it was to be in elementary school – and thereby they had no sense of what children knew at that stage and thereby no sense of their proximal zone of development. This motivated them to make an interview with an elementary school teacher, and observe sessions with first grade pupils.

From the interview with the teacher, students got an impression of the skill level for the pupils, and they found (comparing with their own childhood) that the pupils had more advanced computer skills than expected. Furthermore, the teacher stressed that the game had to have some kind of special figure in order to get the pupils interested; and he gave as example adding a "wise" teddy polar bear to a game about climate change. This example was in fact taken up by the group as they ended up making a game with a polar bear as the cool mascot guiding the members of a family to consume less energy.

From the observations students not only got a chance to recall what they had been like at that age, they in fact realized that they got "a good insight into how children are educated at this level today compared to 10 years ago". So they got a sense of the dynamic nature of contexts as well. Another conclusion from the observations was the realization that the pupils reacted very positively to praise, and that the teacher had a way of making the rather easy tasks (in the view of the group) seem a lot more difficult, so pupils would feel they had made a huge accomplishment when they solved a problem. After these sessions the students decided to make follow-up questioning of a couple of pupils to know more about what they preferred in terms of computer games and formal educational exercises.

The final in-game experience had a simple linear form where at each level a girl or boy had to make the right choices (for example, turning off the lights when moving out of a room) or persuading others to do so by making the right arguments by the help of the teddy polar bear "El-Bear" (in Danish: *Elbjørn*). The pupils could monitor their improvements in the game proportional to the amount of smoke coming from the chimneys outside their room. If the pupils succeed, the home of El-Bear, a flow of ice, will be prevented from melting – if not, the flow of ice will gradually melt – and El-Bear of course is ready to tell them why.

Unfortunately, in both the medialogy and nanotechnology programs the contextual knowledge advising has been significantly reduced in recent years, and has since been removed from the first semester project work in all of the science and engineering programs. The opportunities for fostering a hybrid imagination among science and engineering students at Aalborg University have thus been diminished in recent years, but these examples do indicate that there is a great deal of potential for integrating cultural and social understanding in a meaningful way into student projects in different engineering programs.[5]

[5] There has been a major reconstruction of the entire Aalborg model during the past 5 years that has meant that there has been a shift in the relative share of time and credits devoted to project work. At the same time, the previous "project-oriented" courses such as the one we gave for the nanotechnology students in 2006–2007 have been eliminated. For the first year students a general course in science, technology and society (STS) is now given in the first semester and contextual knowledge is meant to be included in project work in only one semester, with the help of assistant supervisors. All courses, including the STS course, are now subjected to separate examinations, rather than being evaluated as part of the project examinations. The changes have been motivated by the continuing decline in funding from the government, which has meant that the resources devoted to "peripheral" teaching such as ours has tended to be reduced at the expense of the "core" curriculum.

Conclusions

Hybridity, or a hybrid imagination, is something that has to come from within; it requires a student who is interested in obtaining what might be best characterized as a dual competence. But it requires something else, as well: a motivation, a commitment, a sense of engagement in broader processes of social and cultural change. Not all students, by any means, are willing to put in the extra effort involved, but would rather stick to the more traditional specialized competence of an academic discipline, but we have been fortunate through the years to meet enough of them who are willing to put in that effort. The problem has been the tendency to hubris that has crept into our educational programs in Aalborg; as at so many places around the world, where engineering education has largely been transformed into business training.

In the introduction to *Hubris and Hybrids: A Cultural History of Technology and Science,* the tendency to hubris is characterized as the "if only" syndrome:

> the eternal technical fixation that is deeply embedded in our underlying conceptions of reality. If only we could develop an even better instrument of production and destruction, if only we could tame another force of nature to provide us with unlimited energy, then our wealth and our capacities – the values by which we measure progress – would be so much greater. More than two millennia after the sun melted the wings of Icarus for coming too close, we are still under the spell of hubris, trying to fly higher and higher (Hård and Jamison 2005, p. 5).

The Finnish philosopher Georg Henrik von Wright associated the "particular hubris of the modern technological way of life" with an "unreasonable redirection of nature's causality for human purposes" (von Wright 1978, p. 90). In referring to classic tales of hubris such as Prometheus stealing the fires of the gods for human benefit, Francis Bacon's vision of a "New Atlantis" that would be based on his philosophy of useful knowledge, and Mary Shelley's story of Doctor Frankenstein creating human life in his laboratory, von Wright attempted to mobilize cultural history and what he called the humanist attitude to life in order to help reorient the ways in which science and technology are used in society. Von Wright's argument was that by aiming to transcend the limits of reality, science and technology had escaped the bounds of human reason and become threats to the survival of the planet. As such, it had become crucial to find ways to tame the hubris of our technological civilization. The tendency to hubris is not merely a matter of the hype that is so much taken for granted in our commercialized world as a regrettable, but necessary fact of life. The problem is not commercialization as such, but the general lack of awareness and interest in any other possible meanings that science and technology might have.

Even more seriously, in most fields of science and engineering in the contemporary world, there is an absence of appropriate procedures and institutions by which decision-makers, and, for that matter, scientists and engineers can account for their decisions. In most countries, it seems fair to say that there is no public space available any longer for serious discussion and debate of science and technology, no meaningful effort in the media, the schools, the universities or anywhere else in the public sphere to provide opportunities for qualified reflection or cultural assessment of what we, as a species, might actually want to do with all the amazing new discoveries and technologies at our disposal. The "story-line" of economic innovation is so dominant, so hegemonic that it tends to overwhelm all the other possible ways to talk about science and engineering in society (cf. Jamison and Hård 2003).

In Denmark, for example, the active involvement of the government in various activities of "technology assessment" – from research projects at the universities to so-called consensus conferences, bringing lay people and experts together to discuss technological issues of societal importance – have more or less disappeared during the past two decades as the government has instead sought to encourage technological innovations that can help the country's companies compete more successfully on the global marketplace. At the same time, universities have been encouraged to devote ever more effort to training in "entrepreneurship" and the transfer of research results to the private sector. They have also been subjected to ever more explicit quantitative and monetary forms of management and administration with the allocation of financial resources increasingly coupled to "output". The techniques of new public management that were first introduced in Great Britain in the 1980s by Margaret Thatcher have been an active part of Danish higher education policy, especially since 2001 when a conservative-liberal coalition came to power.

This shift – from assessment to promotion – has occurred throughout the world, both at the government policy level, as well as within universities and research institutes. The starting signal was the closing of the US Office of Technology Assessment in 1994, which had been established in the 1970s as an advisory body to the US Congress, and which, for twenty years, served as a model for similar institutions around the world. In many European countries, as well as at the European Commission, governmental offices and agencies for technology assessment were created in the 1980s, such as the Danish Board of Technology, where consensus conferences were organized throughout the 1980s and 1990s, and which had its budget cut significantly when the conservative-liberal government came to power in 2001.

In order to meet the challenges facing engineering and engineering education in the world today, there is a crucial need within all educational programs to foster hybridity or a "hybrid imagination", combining scientific knowledge and engineering skills with cultural understanding. Specifically, this means that there is a need to integrate, much more ambitiously than is currently the case, non-technical or "contextual knowledge" into science and engineering education, as we have exemplified in this article with our experiences in Aalborg.

Acknowledgements This chapter draws on material that has been published in substantially different form in Andrew Jamison and Jette Egelund Holgaard, "The Cultural Appropriation of Contextual Knowledge," in the proceedings of the Engineering Education in Sustainable Development 2008 Conference in Graz, Austria; Andrew Jamison, "The Historiography of Engineering Contexts," in Steen Hyldgaard Christensen, et al. (eds.), *Engineering in Context*, Academica 2009; Andrew Jamison and Niels Mejlgaard, "Contextualizing Nanotechnology Education: Fostering a Hybrid Imagination in Aalborg, Denmark," in *Science as Culture*, 19, 3, 2010; Andrew Jamison, Steen Hyldgaard Christensen and Lars Botin, *A Hybrid Imagination:*

Science and Technology in Cultural Perspective (Morgan & Claypool 2011); and Andrew Jamison, *The Making of Green Engineers: Sustainable Development and the Hybrid Imagination* (Morgan & Claypool 2013). We thank the publishers and editors of these earlier publications for their comments and support. The chapter has been written as part of the Program of Research on Opportunities and Challenges in Engineering Education in Denmark (PROCEED), funded by the Danish Strategic Research Council.

References

Bijker, W. (1995). *Of bicycles, bakelites and bulbs: Toward a theory of socio-technical change*. Cambridge: The MIT Press.

Bijker, W., Hughes, T., & Pinch, T. (Eds.). (1987). *The social construction of technological systems. New directions in the sociology and history of technology*. Cambridge: The MIT Press.

Buch, A. (2012). Governing engineering. In S. H. Christensen, C. Mitcham, Li Bocong, & Yanming An (Eds.), *Engineering, development and philosophy. American, Chinese, and European perspectives*. Dordrecht/New York: Springer.

de Graaf, E., & Kolmos, A. (2003). Characteristics of problem-based learning. *International Journal of Engineering Education, 19*(5), 657–662.

Dickson, D. (1974). *Alternative technology and the politics of technical change*. London: Fontana.

Flyvbjerg, B. (2001). *Making social science matter. Why social inquiry fails and how it can succeed again*. Oxford/New York: Cambridge University Press.

Freeman, C., & Louçá, F. (2003). *As time goes by. From the industrial revolutions to the information revolutions*. Oxford/New York: Oxford University Press.

Gibbons, M., Limoges, C., Nowotny, H., Schwartzman, S., Scott, P., & Trow, M. (1994). *The new production of knowledge. The dynamics of science and research in contemporary societies*. London/Thousand Oaks: Sage.

Gispen, K. (2002). *New profession, old order: Engineers and German society, 1815–1914*. Cambridge: Cambridge University Press.

Hackett, E., Amsterdamska, O., Lynch, M., & Wacjman, J. (Eds.). (2008). *The handbook of science and technology studies* (3rd ed.). Cambridge: The MIT Press.

Haraway, D. (1991). A cyborg manifesto: Science, technology and socialist-feminism in the late twentieth century. In D. Haraway (Ed.), *Simians, cyborgs and women: The reinvention of nature*. New York: Routledge.

Hård, M., & Jamison, A. (2005). *Hubris and hybrids: A cultural history of technology and science*. New York: Routledge.

Hughes, T. (1983). *Networks of power: Electrification in western society, 1880–1930*. Baltimore: Johns Hopkins University Press.

Jamison, A. (1989). Technology's theorists: Conceptions of innovation in relation to science and technology policy. *Technology and Culture, 30*(3), 505–533.

Jamison, A. (2012). Turning engineering green: Sustainable development and engineering education. In S. H. Christensen, C. Mitcham, B. Li, & A. Yanming (Eds.), *Engineering, development and philosophy. American, Chinese, and European perspectives*. Dordrecht/New York: Springer.

Jamison, A., & Hård, M. (2003). The story-lines of technological change: Innovation, construction, innovation. *Technology Analysis and Strategic Management, 15*(1), 81–91.

Jamison, A., Christensen, S. H., & Botin, L. (2011). *A hybrid imagination. Science and technology in cultural perspective*. San Rafael: Morgan & Claypool.

Jasanoff, S., Markle, G., Petersen, J., & Pinch, T. (Eds.). (1995). *Handbook of science and technology studies*. Thousand Oaks: Sage.

Kjærsdam, F., & Enemark, S. (1994). *The Aalborg experiment*. Aalborg: Aalborg University Press.

Kolmos, A. (1996). Reflections on project work and problem-based learning. *European Journal of Engineering Education, 21*(2), 141–148.

Kolmos, A., Flemming, F., & Krogh, L. (Eds.). (2004). *The Aalborg PBL model*. Aalborg: Aalborg University Press.
Krishnan, A. (2009). *What are academic disciplines? Some observations on the disciplinarity vs. interdisciplinarity debate* (ESRC National Centre for Research Methods). Southhampton: University of Southhampton.
Kuhn, T. (1962). *The structure of scientific revolutions*. Chicago: The University of Chicago Press.
Latour, B. (1986). *Science in action. How to follow scientists and engineers through society*. Cambridge: Harvard University Press.
Latour, B. (1993). *We have never been modern*. Cambridge: Harvard University Press.
Latour, B. (2005). *Reassembling the social: An introduction to actor-network theory*. Oxford/New York: Oxford University Press.
Layton, E. (1971). Mirror-image twins: The communities of science and technology in 19th-century America. *Technology and Culture, 12*(4), 562–580.
Lilley, S. (1973). Technological progress and the industrial revolution 1700–1914. In C. Cipolla (Ed.), *The Fontana economic history of Europe. The industrial revolution*. London: Fontana.
Marx, K. (1976/1867). *Capital. A critique of political economy* (Vol. 1, Trans: Fowkes, B.). Harmondsworth: Penguin.
Mejlgaard, N. (2007). *Scientific citizenship: Conceptualisation, contextualisation, and measurement*. PhD Thesis, Aalborg University.
Mejlgaard, N. (2009). The trajectory of scientific citizenship in Denmark: Changing balances between competence and participation. *Science and Public Policy, 36*(6), 483–496.
Mumford, L. (1934). *Technics and civilization*. Jovanovich: Harcourt Brace.
Mumford, L. (1970). *The pentagon of power*. Jovanovich: Harcourt Brace.
Negt, O., & Kluge, A. (1993). *Public sphere and experience. Toward an analysis of the bourgeois and proletarian public sphere*. Minneapolis: University of Minnesota Press.
Nye, D. (2006). *Technology matters*. Cambridge: The MIT Press.
Schumpeter, J. (1928). The instability of capitalism. *Economic Journal, 38*(151), 361–386.
Sokal, A. (2008). *Beyond the hoax: Science, philosophy and culture*. Oxford/New York: Oxford University Press.
von Wright, G. H. (1978). *Humanismen som livshållning*. Stockholm: Bonniers.
Williams, R. (1958). *Culture and society 1780–1950*. London: Chatto & Windus.

Andrew Jamison B.A. in History and Science from Harvard University, Ph.D. in Theory of Science from University of Gothenburg (Göteborg). Docent in Theory of Science, University of Gothenburg. Professor of Technology, Environment and Society at Aalborg University. Coordinator of Program of Research on Opportunities and Challenges in Engineering Education in Denmark (PROCEED), 2010–2013, and author, most recently, of *The Making of Green Knowledge: Environmental Politics and Cultural Transformation* (Cambridge 2001), *Hubris and Hybrids: A Cultural History of Technology and Science*, with Mikael Hård (Routledge 2005), *A Hybrid Imagination – Science and Technology in Cultural Perspective* (Morgan & Claypool Publishers 2011) together with Steen Hyldgaard Christensen and Lars Botin, and *The Making of Green Engineers: Sustainable Development and the Hybrid Imagination* (Morgan & Claypool 2013).

Niels Mejlgaard M.Sc. in Political Science, Aarhus University, and Ph.D. in Development and Planning, Aalborg University. Senior researcher and director, the Danish Centre for Studies in Research and Research Policy, Department of Political Science and Government, School of Business and Social Sciences, Aarhus University. His research interests include science and technology policy, science governance, and public engagement in science and controversial technologies. Presently involved in the strategic research alliance PROCEED, a program of research on opportunities and challenges in engineering education in Denmark.

Jette Egelund Holgaard M.Sc. in Environmental Planning and Ph.D. in Environmental Communication both from Aalborg University. She is Associate Professor within the field of Sustainability, Technology and Organizational learning at the Department of Development and Planning; Aalborg University. Her research is connected to the research group for Problem Based Learning at the department of Development and Planning, section for Technology, Environment and Society, Aalborg University. She is attached to the UNESCO chair in Problem Based Learning and Engineering Education, where she have the special responsibility to make use of a problem based learning framework to enhance engineering education and innovation for sustainability in the center for PBL and Sustainability.

Chapter 15
Constructions of the Core of Engineering: Technology and Design as Modes of Social Intervention

Ulrik Jørgensen

Abstract For a long period of time math and science subjects have undisputedly been seen as the core of engineering education that unifies the field despite the still growing diversity of engineering domains. These disciplines are assigned the role of providing an instrumental, common basis for the development and operation of technologies serving society and human needs. Though the relative part that these disciplines cover has been reduced in the wake of new technical disciplines and the resulting curricula congestion they are still serving as an ideological backbone in discussions of engineering and have made the introduction of other perspectives very difficult as demonstrated in the history of engineering education. The question raised in this chapter is whether new areas of teaching and new disciplines should be considered as alternative candidates to the core curriculum or whether the mere idea of a core should be revised and given up as part of the 'expansive disintegration' observed within the field of engineering. Socio-material design of not only products and services, but also of technological systems takes seriously the important role that technology has in defining social ordering mechanisms in society. This makes socio-material design a potential candidate to become the new core of engineering, coming together with other approaches that emphasize the social part of technology. If accepted on equal footing with the use of models and science, design could serve to moderate the technocratic and instrumental focus that prevails in engineering education due to the dominance of math and science in the core curriculum of engineering education from the very first lectures.

Keywords Core curriculum • Engineering education • Response strategies • Translation of challenges • Technology • Social order • Entrepreneurship • Design as intervention

U. Jørgensen (✉)
Department of Development and Planning, Aalborg University, A.C. Meyers Vænge 15, Copenhagen, SV 2450, Denmark
e-mail: uljo@plan.aau.dk

S.H. Christensen et al. (eds.), *International Perspectives on Engineering Education*, Philosophy of Engineering and Technology 20, DOI 10.1007/978-3-319-16169-3_15

Introduction

The starting point for this exploration is the observation that the core curriculum of engineering education has for a long time been, in somewhat hegemonic fashion, focused on math and science methods, as well as on the instrumental view of engineering as a profession serving human needs through technical means. Attempts to reform and deviate from this established 'classic' view of engineering training, including decades of initiatives to expand or change the core engineering curriculum, have experienced only limited impacts and have been met with resistance.

Several new forms of multidisciplinary educational programs have surfaced combining technical knowledge from engineering disciplines and domains with business economics and organization, entrepreneurial models for innovation, planning of infrastructures, etc. But in general these have not changed in any radical way the nature of what is considered core engineering, as they mostly have operated with an add-on model where the teaching of new disciplines at best have been integrated in some few students projects. The recent strong campaign for adding an entrepreneurial perspective to engineering educational training does not differ from this add-on approach, though it (again) raises questions as to the core values implicitly taught in many engineering disciplines.

The last three decades have seen an explosion in the number of specialty domains represented in educational programs at engineering schools. Together with the large number of educational programs having technology as an important part, but taught outside engineering schools – e.g., in science departments or even from humanities and social sciences – this has been characterized as an 'expansive disintegration' in relation to the 'classic' vision of the core topics of engineering. Rather than changing the core of engineering to reflect new and cross-disciplinary approaches to technology, overly rigid engineering disciplines and domains have requested more and more space within limited curricula, often at the end resulting in handling the expansion by adding new programs. At the same time, this development is paralleled a loss of hegemony over technology by engineering professionals due to the pervasive nature of technology within a large number of societal activities and functions.

Engineering is indeed challenged by several fundamental new problems and ongoing changes, but the strategies have been rather different and lead to quite different developments when it comes to the institutional and educational responses. At the institutional level this is reflected in either a conservationist focus on the 'classic' core of math and science as the basis for engineering approaches and values, an add-on strategy responding to a demand for change in engineering competencies, or a more radical rethinking of technological knowledge and practice. The latter is most often found, e.g., in relation to educational programs focusing on design and introducing new ways of working with technology – sometimes outside engineering schools.

Still other factors influence how teaching and learning can be organized within engineering schools. Societal visions of technology and progress prevail and support the codes of knowledge dominating in engineering, as does the building of student identities already from their time in primary and high school. What is considered core in engineering does not only set the stage for disciplines and teaching, but is interwoven with the recruitment of students and their views of knowledge and use of methods. While most technical and mathematical disciplines are taken for granted as operational, instrumental, and objective, other fields of knowledge challenge these views and become objects of controversy. This has shown in the difficulties met when introducing add-on disciplines into engineering education, often rendered 'soft' not only by the engineering faculty, but also by students preferring the closed world of methods of problem solving within given technical designs.

This chapter will illustrate the challenges, the institutional response strategies, the identity formation process and the different impacts this may have for educational programs and disciplinary approaches to learning about science, technology, and professional practices within engineering. A comparison will be made between the different response strategies in relation to building professional practice, and how different disciplinary approaches and methods influence the problem identification and problem solving heuristics of professionals, giving room for rather different social ordering expectations and implications. Following this, questions will be asked about the potential role of 'socio-material design' in engineering in combination with actor based, ethnographic approaches to problem identification and as a new core of engineering education. The term 'socio-material design' refers to the integrated material and social impact of technology as the condition for engineering designs and problem solving heuristics. Consequently knowledge and experience of both should be a part of the design process.

The empirical material backing this chapter's examples and analysis combines experiences from teaching several disciplines of philosophy, technology studies, professional practice, and design with a number of field studies of engineering identity formation, engineering professional practice, and the construction of disciplinary approaches, models, and methods.

Finally, it may be appropriate to clarify that this chapter is about the role of engineering education and training, and how it affects engineering practices. The criticism presented does not target engineers or engineering for generally being narrow-minded or lacking vision. Though some engineering institutions could fall prone to such criticism, it is not the aim of this chapter. Also, while many engineers may in their practices reflect narrowly what they have been taught in educational programs, others have taken on other visions and perspectives going far beyond the instrumental and often also technocratic views implicit in the majority of engineering educational programs. Without lessening the need for reforming engineering teaching, we must also realize that educational programs are only one part among a multitude of societal influences that shape engineers and the engineering profession.

Math and Natural Science: The Common Core of Engineering Education

Even though the idea of a common core of engineering education has been challenged several times throughout the history of engineering due to the growth in specialties and new technical disciplines, the idea has remained strong that math and science form the backbone not only in the introduction to engineering, but also in the formation of the engineering profession as a science-based endeavor driven by objectivity and rationality. This core has thus not only provided engineering students with a common neutral and rational set of methods to be used to calculate and optimize technical constructions and machines, it has also maintained the idea of a profession basing its work on expertise that tries to maintain independence from human interest and politics.

Following this, one of the intriguing aspects of engineering is the gap between engineering curricula and the categories of employment of engineers very visible in accounts of the labor market for engineers and in the advertisements of new positions. For example, while roles such as sales engineers, technical application specialists, or technical consultants very often are found in conjunction with specifications of the desired technical domains of engineering education, experience, and training, these functions are seen mainly as just as experiences to be learned in practice. They are typically not viewed as an integral component of engineering education on the same level as the technical domain as defined by the big four, civil, mechanical, electrical, and chemical engineering, or later diversifications and new areas of technical expertise such as environmental, computational or biotechnical engineering (Auyang 2004).

Historically, academic engineering institutions were seen as producing the builders of society's technological infrastructure, often in direct relation to nation building activities. More recently, academic engineering institutions have shifted to become more and more entrenched with, and involved in, industrial and business activities that apply technologies to the production of diverse products and services. The self-images of engineers have undergone corresponding changes, though the foundational role of engineering as a profession still has strong roots in the period where engineers were building the backbone of modern society's machinery. In contrast to the rather complex and multi-faceted picture of the drivers of change provided by the history of technology, the idea of math and science as main contributors to modern technology was developed in close connection with the development of the idea of 'polytechnique' – a basic knowledge able to support any technical specialization within the technical universities and engineering schools. The multiple origins of new technologies was kept alive through the two tier system of theoretical- and practice-based engineering educations until the late 1970s, but even the education for vocational practice has been taking over this idea of a common core (Lutz and Kammerer 1975; Reynolds and Seely 1993; Jørgensen 2007).

Controversies within Engineering about the Core Curriculum

In more practical terms, the growing pressure on engineering curricula from new topics and new domains of technology has not left the proportion of math and basic science teaching untouched. These topics have lost terrain in many engineering programs and the common parts to be covered by all programs across technology domains have over time been reduced similarly. The interesting question is then what consequences this has had for the idea of a core curriculum defining the common basis for engineering education?

There has been controversy over the role of math teaching in engineering, with pedagogical questions being asked about the quality of learning abstract math detached from fields of application. Similarly, the role of natural science teaching as abstract and generalized physics and chemistry has spurred controversies within and between different parts of the faculty at the engineering schools and institutions. Typically the experience of teachers of advanced technical subjects was (are) that the students are not well equipped to apply and use math and basic science as ready-at-hand models when drawing upon this knowledge in a later stage of their education, resulting in a de facto repetition of topics. Some of this may directly relate to a misunderstanding of learning, where abstraction does not automatically also lead to a production of student abilities to use this abstract knowledge in specific new settings (Patel et al. 1991; Jakobsen 1994).

Seen from within, the conception of engineering as being the application of natural laws and mathematical principles has not necessarily harmonized well with the different approaches to teaching math and technical sciences. For instance, the reduction and compression of math and physics into courses restricted by less teaching time has led to a compression of these topics into a less open and questioning type of teaching, and a more factual and instrumental presentation of the remaining, reduced curriculum.

Achievements within the field of logistics and control systems in WWII, along with developments in the post-war period, led to a new belief among engineering scholars that an even more science driven development within engineering and the technical sciences would finally bring engineering out of the shadows of the natural sciences and put its new science-based disciplines on par. This thought is intriguing in light of the debts that the natural sciences owe to the technological revolution, and of the progress within engineering of gaining independence from the more speculative fields such as philosophy.

These developments gave systems theory a boost, and for a period in the 1960s and early 1970s the complexity of technologies and their social application resulted in quite challenging problems to engineering. Systems theory provided a new blend of methods that both could be used in analyzing and structuring problems, as well as could provide tools to identify relations even at the most advanced levels of contradiction between different representations and models (Hughes and Hughes 2000; Mindell 2002). But what the new theoretical language did not provide was a set of tools and methods that took into account the actor-based diversity in the ways properties and relations were understood and acted upon. This pushed systems theory back into being another – maybe more complex – tool used by engineers to machinate and order social processes from a technocratic position. Systems theory, despite its open-ended language and pervasive entrance into other disciplines, even within organizational theory and biology, did not fundamentally change the idea of math and science being the common core of engineering.

In the last decade the divide and boundaries between the natural and technical sciences has increasingly been challenged. This has resulted in the coining of the new notion 'techno-science' to cater for the interrelations and blurring boundaries (Latour 1987; Gibbons et al. 1994). This not only demonstrates the transformation and growth of the technical sciences, but also the change in aims and content of the natural sciences, with the latter increasingly being involved in the development of technologies based on constructive interventions in what hitherto might have been seen as the autonomous sphere of 'nature' – a repository of interactions and processes independent of human intervention.

Questions from Practice to the Idea of a Unifying Core

In contrast to the idea of a common math and science core providing engineers with a 'lingua polytechnique', engineers trained in different technological domains often have very different perspectives on what constitutes a problem. They may differ in their repertoire of methods and solutions, and even assign different properties to the objects they work with. This problem has been caught and described in the studies by Louis Bucciarelli (1996) pointing to the existence of rather different 'object worlds', each of which belong to the different specialized branches of engineering, and are reproduced in the educational specializations. This is not just a question that relates to the specific views and objectifications that belong to different engineering professional groups, each looking at different aspects of a technology with their problem solving and optimization strategies resulting from the practical experiences of professionals (Schön 1983). It goes deeply into the ways basic disciplines are taught. For instance, thermodynamics is seen as a theory to optimize the working of energy machines in the mechanics version of the course, while its focus is on chemical processes in the chemistry version.

The consequences of these different and often divided object worlds goes far beyond the problem of communication and differences in the use of notions as it defines both the visible and the black boxed parts of engineering practices. As shown by Louis Bucciarelli (1996), Eugene Ferguson (1992), and Kathryn Henderson (1999), engineering communication extends beyond the formal communication that uses math as the common 'lingua' of technology, and also extends beyond the laws and known principles of physics and chemistry when it comes to the creation of new technological objects and new designs. As their properties are not given but result from the experiments, discussions, and tests that is part of the design process, new notions and ways of describing the new features and objects is in the making as well. The standard view may be that engineers know what properties are relevant and therefore rationally can work with design specification and a 'catalogue' of properties in their design process. But even when it comes to testing already developed prototypes there are open-ended problems that include processes of verification and testing of hypothesis. Design communication builds on a broader 'lingua' that includes analogies, drawings, sketches, and models.

The problem that design and the solving of wicked problems poses to engineering is very clearly demonstrated, but also reproduced, in the foundational book on understanding engineering work and problem solving strategies written by Walter Vincenti (1990). He almost completely black-boxes the generation of engineering design concepts and describes most engineering work to be about the optimization and testing of already-established technological constructions and machines.

The problem with the focus on manipulating already-established technologies and methods was nicely summarized by Gary Downey (2005) in his article on engineering problem solving. He notes the dominant roles played by problem solving based on existing design concepts, and by the application of methods developed and refined within different engineering disciplines. But he also highlights that the process of problem identification and reduction more or less has been left out as an explicit part of the curriculum – maybe with the final thesis project as an exception, at least at some engineering institutions (Downey 2005; Downey and Lucena 2007).

Technology as the Material Means of Social Order

The dominant conception within engineering of technology as the result of an application of natural laws and mathematical principles for societal purposes correlates well with the philosophical idea of technology being a rather autonomous driver of social change. Historically this idea has been the main inspiration for a variety of technocratic movements emphasizing that technology should not be politicized but instead should guide politics.

Though questions could be raised concerning the relevance of historic cases, the birth of the polytechnic ideal in France, and its application in Denmark, for example, was closely related to the idea of an objective and technology driven development devised by government through the utility of an engineering and bureaucratic elite corps of civil servants and managers. In the vision of the Danish first rector of the Polytechnic Learned Institution (established 1826), engineering education should soon be complemented by an education focusing on civil servants with a basis in political science and law – a combination made with reference to the German notion of government chambers called 'Kammeralwissenschaften'.

Such technocratic neutrality and objectivity may not be gained without the construction of an operational base of action that creates the ground for a whole profession, and the conflicts over the role of science and math in engineering is therefore also related to the historic project of constructing engineering as a profession that can present itself as objective in its statements and interventions, and as a servant of society (Williams 2002). Several scholars have demonstrated the role of technology in ordering social practices from forces of production (Noble 1977), military organization (Roe-Smith 1989), professional engineering cultures (Hård 1994, 1999), gendered identities (Faulkner 2007), large technical systems (Hughes 1987), socio-technical ensembles (Bijker and Law 1992; Jørgensen and Karnøe 1995), and socio-technical regimes (Rip and Schot 2002; Geels 2004).

Technologies comprise of a rather varied set of socio-material practices that are made operational in society, ranging from the building and operation of machines to the construction and use of methods and processes governing infrastructure, communication network, security systems, energy provision, etc. Most technologies today are not simply single machines or devices but are operated as parts of larger technological systems that combine, regulate, and govern the individual technical devices within a larger systemic framework that include aspects of control, state shifts, operators, etc. Maintenance and operation as well as continued adjustments and repair works are needed to account for unforeseen problems in the running of these machines and infrastructures.

Theories of technology have evolved from a state where social scripts were seen as properties closely linked to the specific technology, into a more open and interpretative state where domestication (Lie and Sørensen 1996) and interpretative flexibility (Bijker 1995) opens for actors influencing and using technologies in different ways. Yet the institutional settings and governance structures associated with technologies implies that the social order perspective is still relevant for understanding technology in society.

Engineering Objectivity in Constructing Social Order

Having demonstrated the role of technology in delivering the material structures and objects that are crucial for the socio-material ordering of societal activities, addressing the role of engineering in this construction process becomes important. It seem obvious that simplification and black boxing, as demonstrated in the above examples, is a necessity to make technology work, as a certain level of standardization and ordering is needed for the machination process to provide the anticipated outcomes of technological interventions.

The problem is not whether black boxing and standardization have to be avoided, as these processes are an intrinsic part of the design of socio-material constructions, and are necessary to make them become operational. This is also a core finding from the studies of technology. The problem lies with what parts and actor interests are black-boxed, and therefore blinded and left out in the standardization process.

Following the conception of engineering problem solving practices as grounded in specific object worlds stemming from the existing repertoires of technological concepts and solutions, the reproduction of implied social orders become visible, though also at the same time it becomes blurred and black-boxed and therefore requires a thorough analysis to be identified. The ability to overcome the limitations of the object world is crucially related to the ability within engineering to reflect upon both problem definitions (plural) and to transgress the boundaries created by the object worlds. As once stated by research manager of the Danish Learning Lab, the fundamental problem within the field of engineering is the lack of reflection and understanding of the limits to and boundaries of the knowledge within the different disciplines and educational domains.

Experiences with Bringing Social Perspectives into Engineering Education

Attempts to add and/or integrate social perspectives into engineering have been many and have followed different pathways. In recent Danish research in the Project on Opportunities and Challenges in Engineering Education in Denmark (PROCEED), studies have been conducted of how social challenges have been taken up and integrated into engineering education by interpreting and translating these efforts based on what we have identified as institutional response strategies.

Over a rather long period of time engineering education has been reacting to criticism that those responsible for technology and engineering have not taken seriously the critical role that technology plays in society, and consequently have not taken seriously the social responsibility of engineering. In relation to the idea of engineering as a profession serving societal and human needs, most engineering institutions have felt obliged to respond to such criticism. This has been done by including courses that are intended to provide engineers with social and ethical skills, ranging from the idea in the U.S. of having liberal arts requirements to provide engineers with the broad knowledge to make them good citizens, to more specific course requirements that teach about the role of technology in society and how engineers may handle eventual conflicting goals. The latter has included courses in the history of technology, engineering ethics, and, in Denmark, a special mandatory course about the philosophy and practices of engineering.

Along the same line of arguments that have made math and science into common courses, these social science-based courses on the role of engineers and technology have in most cases been provided as courses given to a large number of engineering students across different programs, and very often with only little connection with the engineering 'hard' topics that students take. The very few examples of integration that can be found have demonstrated that while the model with separate courses may in theory provide better teaching from a disciplinary point of view, it also gives rise to many of the problems of disconnectedness that these add-on topics and courses have experienced. In this respect the objectification of problems and solutions, following the core math and science courses, contrasts the discursive and actor-based teaching in the social science courses, and thereby digs even deeper the divide by enhancing the gap in superficial ideologies that does not fit well with how technologies operate in practice.

Besides the attempt to identify why and how it has been so difficult to insert social science-based teaching into engineering curricula, the studies in PROCEED have been carried out in relation to different contemporary challenges to engineering education that can be identified across Europe and the U.S. We have named these the environmental/climate challenge, the entrepreneurial challenge, the globalization challenge, the design challenge and the high-tech challenge. Without claiming that these cover all relevant aspects of what might be challenging the fields of engineering, nor that they are the key drivers of change within the different specializations and disciplines, we have found that these challenges and the identified response strategies have provided us with a quite broad and relevant set of archetypes for ways that engineering institutions choose to tackle the challenges. Four such archetypical strategies can be identified across the mentioned social challenges (Jørgensen and Valderrama 2012; Jørgensen et al. 2013).

1. In the first type of response strategy, an institution may identify a challenge as important for engineering practice, while at the same time denying it any influence on engineering curricula. This may entail highlighting the challenge as one among many important fields of engagement for engineers as responsible citizens and professionals. The challenge may be seen as something that should affect the attitude and orientation of engineers when solving problems, and with respect to this it might even be taken up in advertisements for engineering education and in competitions where students can demonstrate their creativity in problem solving. Still, the divide between engineering as rational problem solving and the politics of, for example, humanitarian design, sustainable solutions, and innovative ideas is maintained. The latter are not made objects of study within engineering, only objects of application.
2. In the second type of response strategy, an institution takes up the challenge by identifying new topics and disciplines that might help students in getting supplementary knowledge and competence as an add-on to their engineering training. These new topics and disciplines may come from fields of natural sciences, as in the case of biology, physiology, and medicine, or they may come from the humanities and social sciences in the form of ethics, organization, economy, business models, or psychology. These add-on contributions generally have their origins in other educational and institutional settings outside of engineering. While the idea of add-on topics has been very dominant in many response strategies, as it provides the flexibility to expand the competences of engineers, it has also resulted in two directly related problems: i) the integration between the engineering problem solving methods and the new approaches has been left to the students, and ii) the teachers of the add-on topics often have been placed in a conflict between adapting their teaching to be a part of engineering curricula and their own background and research options. This response strategy is very often seen when engineering programs include a course in design, a course in entrepreneurship, a course in communication, and/or a course in humanitarian engineering, which in several instances even may be followed by an optional

project assignment where the students can experience the challenge and get some basic, practical experiences.

3. A third type of response strategy operates with a more subtle and disciplinary change process in which the challenge results in the development and assimilation of new problem solving strategies that incorporate and translate aspects of the challenge into the instrumental perspective of engineering. This adds to the repertoire of analytical tools and defined problem solving methods and solutions that are presented to the students. It also contributes to the continued evolution of the technical disciplines in engineering, keeping them up to date with novel methods and cases. The critical aspect of this strategy is that the challenge may appear as a new set of tools and methods, after having been filtered and translated to meet the disciplinary structure and approach in the specific domain of teaching. In response to the environmental concerns in the public in the 1970s, many engineering programs, hitherto focusing on sanitation and water, responded to the challenge by expanding their teaching to include how to deal with pollution through handling the emission to air, soil, and water. They translated pollution threats into handling waste streams. In the field of wastewater treatment, this entailed new advanced processes for mechanical and biological treatment. The origins of pollution – often, extant technologies – were eventually addressed a decade later when cleaner technology strategies were developed that included addressing production processes and their use and handling of materials and energy. These perspectives entered engineering education as, for example, a new course in life cycle assessment methods, and new, less polluting processes added to the repertoire of existing methods of production. But overall, the challenge was effectively reduced to some new parameters in the optimization and choice of technologies, along with a few new courses.
4. The fourth and last type of response strategy combines and goes beyond the two former strategies by seeking new ways of integrating society and nature as an intrinsic part of technology, grounded in the view that technology is much more than just the application of math and technical sciences to subdue nature in service to human needs. Technology, in this perspective, is deeply entrenched in social development and must be understood as a product that integrates the social and the material. As a consequence, new disciplines like 'technology studies' and other interdisciplinary contributions play an important part in providing the new perspectives. As sociology, technical sciences, economy, and other classic disciplines tend to be bound by their framing of the difference between what is considered social and technical, they also have little to tell about technology in practical operation due to their partial view of the workings and impact of a technology. While it is easier to operate the distinction for existing technologies, where an established difference between function and impacts seem obvious, the need for an interdisciplinary approach is promising when design of new products and systems are at the center of engineering work. Here, the actors to be involved, the use qualities of an outcome, the properties and functions in question, as well as the problems to be solved, are less fixed, which suggests a more open-ended process, both creative and analytical.

This general presentation of the four strategies does not render the specific application of these less relevant, as a lot of the impacts and outcomes need to be identified in relation to the detailed transformations that follow from the individual cases. A complete picture is not possible in this context, but hopefully the few examples presented for illustration purposes may help underpin the general lessons learned from the study of response strategies.

The Challenge of 'Expansive Disintegration'

In her book 'Retooling', Rosalind Williams concludes that the field of engineering has gone through a process of 'expansive disintegration' in the recent decades (Williams 2002). This process has challenged engineering schools and educational structures by taking away their dominance as the providers of professionals innovating and working with technology. At the same time it raises questions about the idea of a uniform entity known as engineering education, and built on, for example, the idea of a core curriculum and a specific science-based way of understanding technology in society. The field of technology has been expanding into all aspects of human life and society; at the same time, many new educational programs not belonging to engineering schools provide knowledge about technical topics that seem to be crucial to development in these areas of knowledge. There is a tendency within engineering to maintain a certain resistance to accepting social science and technology studies entering into engineering education in more prominent roles, rather than staying as add-on topics complementing, but not fundamentally changing the approaches in engineering. These factors led Williams to suggest that engineering schools risked losing their status as the institutional and ideological framework that governs engineering.

This perspective is tempting when seen in light of decades of problems with reforming engineering education, though it does pose other problems that often have been back-grounded in discussions of the role of universities in modern society. Engineering has had a continued discussion about the gap between the technical sciences and their relevance to the professional practice of engineers, which raises critical questions about the instrumental focus and narrow framing of engineering science and object worlds, as well as methods that maintain a technological hegemony (Bucciarelli and Kuhn 1997; Sheppard et al. 2009). But the same is true, if only in more limited ways, for a number of university educational programs. Even though academically trained economists, managers, administrators, lawyers, doctors, etc., increasingly dominate societal institutions, their roles in modern everyday life, their professional training, and the power exerted through their disciplinary knowledge still needs to be taken up more critically.

Another problem relates to the practical and material skills that – though increasingly lost to computer-based virtual problem solving – still are part of the training and professional approach in engineering. These critical problems related to the university educational system at large demonstrate that the job is not done by dismantling engineering schools without bringing some of the reform controversies in engineering to the fore.

Engineering schools and institutions have built a formidable institutional network, which leaves the idea of dismantling these institutions as a provocative, but also somewhat idealistic approach. As the engineering hegemony over technology nevertheless has slipped, the challenge and question remains about the direction of future developments of engineering and other technology-focused educations. Several institutional strategies can be observed to point in very different directions. Some tend to follow the idea of techno-science and invest heavily in the new high-tech areas, arguing for these to have huge innovation potentials and to point to futures technology. Others take seriously sustainability challenges, and focus on energy, green technologies, and solving environment and climate problems. Some take up new dimensions of entrepreneurship and/or design as part of reforming their engineering curricula. In this respect, the disintegration is showing in the form of diversity of institutional strategies.

Socio-material Design – A New Core Element of Engineering?

The main argument in this chapter is centered round the new role that socio-material design approaches, which build on lessons from technology studies, can play in a reform of engineering education. This is not just a nice idea, but has been substantiated through a number of recent developments in engineering programs in the U.S. and Europe, for instance the Product Design and Innovation program at Rensselaer Polytechnics, Troy (USA), the Design and Innovation program at the Technical University of Denmark, Lyngby (Denmark), the Engineering Design program at Delft University of Technology, Delft (the Netherlands), and the new Sustainable Design program at Aalborg University in Copenhagen (Denmark).

Taking a design approach may entail rather different pathways for change, as the notion is very open for interpretation and has been taken up in very many different ways in public and professional discussions. A clarification of what is referred to as socio-material design is therefore needed.

First, socio-material design defines the role of the engineering designer as a professional able to stage, and navigate among and with, a number of different actors who have stakes in the processes of designing, producing, implementing, using, and eventually disposing of a technology.

Second, a design is in this perspective not limited to the materialized result, but to the process of involvement and the process of enrollment. A design is not just a product, a service, or a system (of products, operations, maintenance, and services) – it is the outcome of a networking process that brings the design artifact or result into being. In this perspective the design result is clearly not only the material thing, but its socio-material existence and application. It resembles the broad and economic definition of an innovation, but with much more emphasis on the design process as a professional process that involves a set of relevant and necessary actors.

Third, socio-material design builds on a thorough problem identification and analysis. Problems in design do not – even not in the case of an already existing design specification – just operate with problem solving based on already established concepts and methods. A fundamental aspect of a design process is to question specification, design briefs, and pre-selected concepts, as these may correctly state the design script from a single actor perspective but may overlook important problems and challenges with respect to other actors involved. In this perspective, any design process starts with the ability to ask questions and map the sphere of problem statements found among the different actors.

Fourth, design professionalism combines the ability to be creative with the competence of visualization, the ability to analyze a field of use, the employment of a repertoire of models and technical knowledge and known concepts, and the ability to analyze and synthesize the variety of problem-solution relations that define the space of socio-material outcomes.

Many of these dimensions are relevant to quite a large part of engineering education, but they also expand the needed knowledge and experience base that constitute a 'good' and well trained engineer by adding some of the dimensions that, while often defined as crucial to being an engineer, are also often seen as an implicit – almost magic – outcome of education without being addressed in the curriculum. These dimensions are not core to the ways students are taught to analyze and solve problems. Rather, understanding of the social aspects of making designs operational, as opposed to the technical means and methods needed, is assumed to result from a few project assignments and some rather general teaching in social and design topics.

There is a dilemma in re-focusing the core of engineering, as many engineering curricula are crowded with coursework. Consequently, any new topic or project assignment at first glance seems to reduce the math and technical part of the curriculum. This has resulted in a basically hopeless controversy over the loss of quality in engineering education, as measured by the number of topics and by the number of pages the students have to read. In most other professional settings, engineers would argue the need for analyzing and measuring the resulting outputs and competencies that different engineering educational styles produce. But when it comes to engineering education itself, the measure is based on input, not output.

Of course there are reasons for this situation. Such measures of practice are quite difficult to make, and it is even more difficult to relate the measured competencies to the individual, as engineers very often work in teams. Even worse, the coupling of the composition of educational programs with these results makes the measured relationship very complex. Also, engineering educational programs and institutions have a very meager tradition for discussing the relationship between professional practice and educational practice. Many teachers of engineering subjects may never have worked as practicing engineers, but have instead been recruited based on their research work.

What still makes socio-material design a potentially good alternative core of engineering education is not that this perspective is seen as a substitute to math and science in any banal way. Rather it is because it emphasizes competencies that better reflect those needed by engineers in professional practice. The math and science topics are as important as knowledge of the frames and boundaries within which engineering is operating – e.g., organizations, staging processes, ethnographic approaches to field studies, economic valuation, etc.

In all parts of engineering in all its variety, classic functions such as verification of solutions, building trust through references and documentation, and testing, testing, testing of new products, services, and systems are continuously crucial parts of engineering work. These do not lose their importance because more emphasis is oriented towards problem analysis and design processes. On the contrary, more focus on the variety of possible solutions and the open-ended character of design processes will also result in engineers becoming more aware of risks and vulnerabilities, as this will do away with the illusions of the one, objective, best way to solve a problem.

The 'End' of Engineering – Or a Plea for Heterogeneity

There have been several critical contributions, in addition to that of Rosalind Williams, describing a change in engineering and indicating a fall from being the profession ruling technology and providing progress to society. Despite the criticism of such technocratic ideas of technology as an autonomous force guiding societal development, the dominant image among many politicians and in the public still may include some basic assignment of core contributions to be provided by technology. The same holds for engineering institutions when they try to portray the future role of engineering for society and sustainability (Millennium Project 2008; National Academy of Engineering 2004; National Academies 2009). The important role of technological visioning does render the idea of engineering's obsolescence rather problematic. The popular image may have pushed engineers away from the top of the most attractive trades and professions, but engineers are still assigned a number of specific roles nicely captured by the phrase, 'to solve this problem we may need an engineer', though it might remain unclear if this refers to a skilled technician or an engineer trained at a university.

While engineers may have lost their supreme role and influence, and other professions and educational programs, from science to humanities, have taken up technical subjects and produce professionals that both can innovate and operate specific areas of technology, just arguing for the 'end' of engineering would at the same time miss the importance of knowing and handling material objects and integrating the social and the material. Instead of 'ending' engineering, some of these skills, from being able to analyze material objects to knowing about the limits of one's professional models and concepts, are becoming more and more relevant to other fields of education, like economy, management, anthropology, etc. So perhaps engineering's need to embrace non-engineering ideas is complemented by the need for non-engineering fields to embrace engineering ideas.

Alongside the socio-material design approach other ideas have surfaced that try to produce a generally new focus for engineering. At the cognitive and conceptual level, Andrew Jamison has proposed the concept of hybrid imagination as a way to combine rational, analytical thinking with a critical and reflexive perspective. This approach attempts to support a new way of knowing and working for engineering students:

> A hybrid imagination can be defined as the combination of a scientific-technical problem solving competence with an understanding of the problems that needs to be solved. It is a mixing of scientific knowledge and technical skills with what might be termed cultural empathy, that is, an interest in reflecting on the cultural implications of science and technology in general and one's own contribution as a scientist or engineer, in particular (Jamison et al. 2011, p. 4).

This approach takes as a starting point a cultural critique of engineering practice, along with the monolithic reasoning that follows from the math and science based core of engineering. It provides – at a rather abstract level – a program that can be applied in engineering education as a way of thinking and a way to understand the need for combining very different modes of thinking and acting.

In a another proposal that involves more specific considerations of how to organize a new form of engineering programs, Louis Bucciarelli has proposed an engineering program that is grounded in the liberal arts, placing these disciplines in a much more important position in the curriculum and making them stronger and equal to the science topics (Bucciarelli 2011). Also, this vision presents new ways of opening up engineering education to become part of an exchange of knowledge with disciplines outside the field of technical sciences.

There might also exist other way to redefine the core of engineering than the proposed focus on socio-material design. This is still a topic to be explored through discussions and studies that take the gap between engineering practices and the specific and productive role of engineering teaching more seriously. Besides focusing on design as the candidate core of engineering work practices, another large field of engineering is related to technological consultancy, to planning of large technical systems, and to the construction of standardized procedures and measures, which all are fields in which complex social aspects, and their crucial role for engineering problem analysis and problem solving, tend to have been neglected. But to lay the foundation for these new ways of making engineering education a more heterogeneous trade, the approach taken with socio-material design at least provides an exemplary pathway for change.

Acknowledgements This chapter is based on a life-long engagement with engineering education and contemporary empirical studies funded by a grant from the Danish Strategic Research Council (DSF) during 2010–13 to the Program of Research on Opportunities and Challenges in Engineering Education in Denmark (PROCEED). I would like to thank Byron Newberry who has been very supportive in improving language and arguments presented in this chapter.

References

Auyang, S. Y. (2004). *Engineering: An endless frontier*. Cambridge, MA: Harvard University Press.

Bijker, W. E. (1995). *Of bicycles bakelites and bulbs – Toward a theory of sociotechnical change*. Cambridge, MA: MIT Press.

Bijker, W. E., & Law, J. (1992). *Shaping technology/building society: Studies in sociotechnical change*. Cambridge, MA: MIT Press.

Bucciarelli, L. L. (1996). *Designing engineers*. Cambridge, MA: MIT Press.

Bucciarelli, L. L. (2011). *Bachelor of Arts in Engineering*. http://dspace.mit.edu/handle/1721.1/71008

Bucciarelli, L. L., & Kuhn, S. (1997). Engineering education and engineering practice: Improving the fit. In S. R. Barley & J. E. Orr (Eds.), *Between craft and science: Technical work in U.S. settings* (pp. 210–229). Ithaca: ILR Press.

Downey, G. (2005). Are engineers losing control of technology? From 'problem solving' to 'problem definition and solution' in engineering education. *Chemical Engineering Research and Design, 83*(A6), 583–595.

Downey, G., & Lucena, J. C. (2007, June 22–24). Globalization, diversity, leadership, and problem definition in engineering education. *1st International Conference on Engineering Education Research*, Oahu.

Faulkner, W. (2007). Nuts and bolts and people: Gender-troubled engineering identities. *Social Studies of Science, 37*(3), 331–356.

Ferguson, E. S. (1992). *Engineering and the mind's eye*. Cambridge, MA: MIT Press.

Geels, F. W. (2004). From sectoral systems of innovation to socio-technical systems. Insights about dynamics and change from sociology and institutional theory. *Research Policy, 33*(6/7), 897–920.

Gibbons, M., Limoges, C., Nowotny, H., Schwartzman, S., Scott, P., & Trow, M. (1994). *The new production of knowledge – The dynamics of science and research in contemporary societies*. London: Sage.

Hård, M. (1994). *Machines are frozen spirit: The scientification of refrigeration and brewing in the 19th century – A Weberian interpretation*. Frankfurt: Campus Verlag.

Hård, M. (1999). The grammar of technology: German and French diesel engineering, 1920–1940. *Technology and Culture, 40*(1), 26–46.

Henderson, K. (1999). *On line and on paper: Visual representations, visual culture, and computer graphics in design engineering*. Cambridge, MA: MIT Press.

Hughes, T. P. (1987). Evolution of large technological systems. In W. E. Bijker, T. P. Hughes, & T. Pinch (Eds.), *The social construction of technological systems*. Cambridge, MA: MIT Press.

Hughes, A. C., & Hughes, T. P. (2000). *Systems, experts, and computers: The systems approach in management and engineering, world war II and after*. Cambridge, MA: MIT Press.

Jakobsen, A. (1994). *What is known and what ought to be known about engineering work*. Delhi: Studies in Technology and Engineering. Lyngby: Learning Lab DTU.

Jamison, A., Christensen, S. H., & Botin, L. (2011). *A hybrid imagination. Science and technology in cultural perspective*. San Rafael: Morgan & Claypool Publishers.

Jørgensen, U. (2007). Historical accounts of engineering education. In E. Crawley, & J. Malmqvist (Eds.), *Rethinking engineering education: The CDIO approach*. Springer: Springer.

Jørgensen, U., & Karnøe, P. (1995). The Danish wind-turbine story: Technical solutions to political visions? In A. Rip, T. J. Misa, & J. Schot (Eds.), *Managing technology in society – The approach of constructive technology management*. London: Pinter Publishers.

Jørgensen, U., & Valderrama, A. (2012). Entrepreneurship and response strategies to challenges. Engineering and design education. *International Journal of Engineering Education, 28*(2), 407–415.

Jørgensen, U., Valderrama, A., Mathiesen, B. V., & Remmen, A. (2013). How is sustainability incorporated into the engineering curriculum? The case of DTU and AAU. Conference paper for the 8th SDEWES conference in Dubrovnik.

Latour, B. (1987). *Science in action – How to follow scientists and engineers through society*. Cambridge, MA: Harvard University Press.

Lie, M., & Sørensen, K. H. (1996). *Making technology our own? domesticating technology into everyday life*. Oslo: Scandinavian University Press.

Lutz, B., & Kammerer, G. (1975). *Das ende des graduierten ingenieurs? (The end of the 'craft-based' engineer?)*. Frankfurt: Europäische Verlagsanstalt.

Millennium Project. (2008). *Engineering in a changing world – A roadmap to the future of engineering practice, research, and education*. Ann Arbor: The University of Michigan.

Mindell, D. (2002). *Between human and machine – Feedback, control, and computing before cybernetics*. Baltimore: John Hopkins University Press.

National Academies. (2009). 21 century's grand engineering challenges unveiled. Available at: http://www8.nationalacademies.org/onpinews/newsitem.aspx?RecordID=02152008

National Academy of Engineering. (2004). *The engineer of 2020: Visions of engineering in the new century*. Washington, DC: National Academy Press.

Noble, D. F. (1977). *America by design – Science, technology and the rise of corporate capitalism*. Oxford: Oxford University Press.

Patel, V. L., Evans, D. A., & Groen, G. J. (1991). Developmental accounts of the transition from medical student to doctor: Some problems and suggestions. *Medical Education, 25*(6), 527–535.

Reynolds, T. S., & Seely, B. E. (1993). Striving for balance: A hundred years of the American society for engineering education. *Journal of Engineering Education, 82*(3), 136–151.

Rip, A., & Schot, J. (2002). Identifying loci for influencing the dynamics of technological development. In K. Sørensen & R. Williams (Eds.), *Shaping technology, guiding policy: Concepts, spaces and tools* (pp. 156–176). Cheltenham: Edward Elgar.

Roe-Smith, M. (1989). *Military enterprise and technological change: Perspectives on the American experience*. Cambridge, MA: MIT Press.

Schön, D. A. (1983). *The reflexive practitioner: How professionals think in action*. New York: Basic Books.

Sheppard, S. D., Macatanguay, K., Colby, A., & Sullivan, W. M. (2009). *Educating engineers. Designing for the future of the field. A report of the Carnegie foundation for the advancement of teaching*. San Francisco: Jossey Bass.

Vincenti, W. G. (1990). *What engineers know and how they know it, analytical studies from aeronautical history*. Baltimore: John Hopkins University Press.

Williams, R. (2002). *Retooling: A historian confronts technological change*. Cambridge, MA: MIT Press.

Ulrik Jørgensen M.Sc. in Engineering, Ph.D. in Innovation Economics, Technical University of Denmark. Professor at the Department of Development and Planning, Aalborg University Copenhagen, where he is heading a Center for Design, Innovation and Sustainable Transitions. The center is involved in building the curriculum for a new design engineering education focusing on sustainable design of products and systems. In his former job at the Technical University of Denmark, he was responsible for the introduction of all students to the engineering profession and was heading the development of a new interdisciplinary education in design engineering. He has been involved in the management of several Danish and EU research projects. His research is within engineering and design competences, user driven innovation, environmental and innovation policy, the role of experts in public advice, as well as waste and energy systems transition. He is presently involved in the strategic research alliance PROCEED on 'Program of Research on Opportunities and challenges in engineering education in Denmark'.

Chapter 16
Transforming Engineering Education: For Technological Innovation and Social Development

Tony Marjoram

Abstract Engineering, and engineering education, drive innovation, social, cultural and economic development, and are vital in addressing global challenges such as sustainability, climate change and poverty and the other UN Millennium Development Goals.

This chapter examines the urgent need for innovation and transformation amid changing modes of knowledge production, dissemination and application, and to counter declining interest, enrolment and retention in engineering education, the shortage of engineers reported in many countries, brain drain of engineers from developing countries and consequent impact on development. Student-centred, project- and problem-based learning (PBL) plays an important role in this process, together with an emphasis on humanitarian engineering and technology – combining fun and fundamentals, and the need for engineering to be seen as a major factor in development and addressing global issues and challenges. The chapter emphasises the need to develop engineering studies, policy and planning to support and facilitate this process.

Keywords Engineering • Education • Innovation • Transformation • Development • Problem-based learning • Poverty reduction • Sustainable development • Climate change

Introduction

Engineering knowledge and application drives innovation, social and economic development around the world. Our physical infrastructure is designed, built and maintained by engineers, and most innovations derive from engineering (Metcalfe

T. Marjoram (✉)
UNESCO Centre for Problem Based Learning in Engineering Science and Sustainability, Department of Development and Planning, Faculty of Engineering and Science, Aalborg University, Aalborg, Denmark
e-mail: t_marjoram@yahoo.com

S.H. Christensen et al. (eds.), *International Perspectives on Engineering Education*, Philosophy of Engineering and Technology 20, DOI 10.1007/978-3-319-16169-3_16

2009, also 1995). Significant change in knowledge production, dissemination and application has taken place over the last 50 years, driven by and generating associated needs for engineering. Engineering is the most radical profession, in terms of technological, social, economic and cultural change. On the other hand, engineering is conservative, and engineering education has changed very little over this period. This is a factor in the decline of interest, enrolment and retention of young people in engineering, reported shortages of engineers in many countries and brain drain from developing countries.

This is a major challenge for engineering, and occurs at time of two other major global challenges – the need to mitigate and adapt to the effects of climate change and facilitate the sustainable use of resources, and the need to reduce poverty and enhancing the quality of life for the 20 % of humanity who live in poverty, on less than 1$ per day (1.3 billion people, 70 % of them women). Engineering and engineering education are of vital importance in addressing climate change and in improving housing, water supply, sanitation, food, nutrition, transport, communications and employment creation, through the development and application of humanitarian engineering.

Historical Background of Engineering and Engineering Education

Enlightenment thinking was instrumental to and continued into the Industrial Revolution – powered by engineering knowledge, application and education, which developed rapidly in eighteenth century England, transferring to Europe, North America and world. Machines replaced muscle in manufacturing, in a synergistic combination of knowledge and capital. The first Industrial Revolution took place from 1750 to 1850, focused on the textile and related industry. This was the first of the 'Kondratiev waves' of innovation, industrial development and surges in the world economy – periods of alternating sectoral growth, initially of around 50 years duration but decreasing with increasingly rapid knowledge change. Five major waves of innovation have been identified as part of the 'Schumpeter-Freeman-Perez' model. The second wave or revolution focused on steam and the railways from 1850 to 1900. The third wave was based on steel, electricity and heavy engineering from 1875 to 1925. The fourth wave was based on oil, the automobile and mass production, from 1900 to 1950 and onward. The fifth wave was based on information and telecommunications and the post-war boom from 1950. The sixth wave, based on new knowledge production and application in such fields as IT, biotechnology and materials, began around 1980.

This model is generally accepted, although the precise dates, phases, causes and effects of these major changes are subject to debate. A seventh wave would appear to focus on sustainable 'green' engineering and technology, and began, at least conceptually, at the time of the United Nations Conference on Environment and Development held in Rio de Janeiro in 1992, with interest increasing from 2005. Green technology was emphasised at the Rio+20 conference in 2012, although engineering has been overlooked, undervalued and marginalised. It is important to note the six major waves of technological innovation have all been reflected in subsequent innovations and transformations in engineering education (Beanland and Hadgraft 2013). It is therefore timely to be considering transformations based on new knowledge production, application and sustainability.

Development of Engineering Education

The eighteenth and nineteenth centuries were a crucial period in the development of engineering – particularly the Iron and Steam Ages, the second wave of innovation and industrial revolution. Early interest in the development of engineering education began in 1702 in the German mining industry, with the creation of a school of mining and metallurgy in Freiberg. The Czech Technical University in Prague was one of the oldest technical universities, founded in 1707. Engineering education in France developed with the creation of the École Nationale des Ponts et Chaussées (established in 1747) and École des Mines (1783). The École Polytechnique was established in 1794, during the French Revolution, to teach mathematics and science – the revolution in engineering education beginning during a revolution. France developed a formal system of engineering education after the Revolution, under Napoleon's influence, and engineering education has retained a strong theoretical and military character in France. At the beginning of the nineteenth century, the French model influenced the development of polytechnic engineering education institutions around the world, especially in Germany. Between 1799 and 1831 polytechnics were established in Berlin (1810, by Wilhelm von Humboldt, establishing the university model we see today – see below), Karlsruhe, Munich, Dresden, Stuttgart, Hanover and Darmstadt. In Germany, polytechnic schools were accorded the same legal foundations as universities. In Russia, schools of technology were opened in Moscow (1825) and St. Petersburg (1831), based on the model of military engineering education. The first technical institutes appeared at the same time in the USA – including West Point in 1819 (modelled on the École Polytechnique), the Rensselaer School in 1823 and Ohio Mechanics Institute in 1828.

In England, on the other hand, after the early years of the Industrial Revolution, engineering education continued to be based on the model of apprenticeship with a working engineer, and many engineers had little formal or theoretical training. Famous men such as Arkwright, Hargreaves, Crompton and Newcomen, followed by Telford, Maudslay, George and Robert Stephenson, all had little formal engineering education, yet developed technologies that powered the Industrial Revolution and changed the world. Practical activity preceded scientific understanding in many fields, – steam engines preceded thermodynamics, for example, and 'rocket science' is more about engineering than science. England tried to maintain technological lead by prohibiting the export of engineering goods and services in the early 1800s. This was one reason why countries in continental Europe developed their own engineering education systems based on French and German models, with a foundation in science and mathematics, rather than the British model, based on artisanal empiricism and laissez-faire professional development. Through the nineteenth and into the twentieth centuries, however, engineering changed and with it engineering education. England was also obliged to change toward a science and university-based system, partly due to fears of lagging behind the European model in terms of international competitiveness. This reflected the rise of the 'engineering sciences' and increasingly close connection between engineering, science and mathematics.

Toward the end of the nineteenth century, most of the industrialising countries had established their engineering education systems. These were based on the liberal, student-centred model introduced by Wilhelm von Humboldt at the University of Berlin, combining theory and practice, with a focus on scientific research. The German "Humboldt model" went on to influence the development of universities in France and elsewhere, although the emphasis on practice as well as theory was often later overlooked. Over time, emphasis on the practical gave way, largely on professional grounds and the desire to emulate science, with an increasing focus on theory. This has now had a negative impact on the interest and enrolment of young people in engineering, and consequent need for educational approaches for the next generation of engineers based on problem-based learning, projects and real-world needs.

In the twentieth century, the professionalization of engineering continued with the development of learned societies and the accreditation of engineers through qualification and continued professional development. Universities and professional societies facilitated education, research and the flow of information through journals, technical meetings and conferences. This activity continues with the development of international accords, standards and accreditation for engineering education, and the mutual recognition of engineering qualifications and professional competencies. These include the Washington Accord (established in 1989), Sydney Accord (2001), Dublin Accord (2002), APEC Engineer (1999), Engineers Mobility Forum (2001) and the Engineering Technologist Mobility Forum (2003), and, in Europe, the Bologna Declaration in 1999 relating to quality assurance and accreditation of bachelor and master programmes.

Changes in Knowledge Production, Dissemination, Application and Innovation

It is important to note that the waves of technological innovation and industrial revolution reflected in transformation in engineering education, were also reflected epistemologically in changing knowledge production, dissemination and application. The first major wave of technological change in the early 1800s, based on iron and water power, was reflected in a craft mentor-based hands-on approach. The second wave in the later 1800s, based on steam, was reflected in apprenticeships and trades. The third wave in the later 1800s and early 1900s, based on steel, was reflected in the development of technical schools, colleges and universities. The fourth wave in the early/mid 1900s, based on oil, reflected an increasing science, theory and hands-off approach. The fifth wave, post 1950s, based on ICTs, reflects the significant changes in knowledge and technology over the previous 50 years, as does the sixth wave, based on changes in knowledge production, dissemination and application from the 1980s – with "post-industrial science" and the convergence of science and engineering based on interdisciplinarity, networking and a problem-solving, systems approach, with an increasing focus on applications. These changes relate particularly to what has been typified as the change in knowledge production from "Mode 1" (academic/disciplinary) toward "Mode 2" (problem-based/interdisciplinary) (Gibbons et al. 1994 and Nowotny et al. 2001; Etzkowitz and Leydesdorff 2000). Changes from theory toward practice, individual to teamwork, with converging bodies of knowledge, professional practice and employment, need to be reflected by change in science and engineering education toward project- and problem-based, student-centred learning. These changes will be needed in the move into a seventh wave, based on cleaner/greener technology for climate change mitigation and adaptation, where a focus on practice, teamwork, converging knowledge, applications and innovation will be of paramount importance.

Innovation

Innovation and waves of innovation are the history of the world. The Stone Age did not give way to the Bronze Age because they ran out of stones (Yamani 1973). Innovation relates to the introduction, dissemination and use of an idea, method, product or process that is new to the user or user group, but may not be absolutely new. Innovation initially related particularly to technological innovation, although the meaning has now expanded to include broader subjects – such as innovation in education. Epistemologically, innovation was first portrayed as a linear model, where basic science was imagined to lead to applied science, engineering, technology, innovation and dissemination of ideas, products and processes. The linear model of innovation, initially promoted by Vannevar Bush, proved to be deceptively simple and endearing, especially for the science community and policy makers in the post-war period. Although the model has later been shown to be inaccurate and misleading, its simplicity and usefulness for science in the argument for funding in the public policy debate has proved enduring. Academic policy analysis has moved forward, toward systems thinking, "national systems of innovation" and related regional and global models (Freeman 1995; Lundvall 1992). Most recently, the "ecosystem" model of science, engineering, technology and innovation has appeared, as a metaphorical didactic finesse, but offers little in terms of epistemological insight, and may confuse the casual observer. While the systems models may be more accurate for economically developed, OECD member countries (where they were developed, with particular reference to Japan), they can also be misleading in the developing, non-OECD country context, where elements of the innovation system (industry, research, government) are less developed.

In the context of innovation in education, it has been noted that educational practices change slowly and evolve to match cognitive and professional paradigms, requirements and expectations. In engineering education, "engineering science", following the Humboldt model, is the dominant paradigm. Changes in knowledge and technology production, dissemination and application have occasioned the need for change in associated learning approaches – toward cognitive, knowledge- and problem-based learning, and the need for innovation and development in engineering education. Engineering is a problem-based profession, and needs a problem-based, just-in-time approach to learning and continued professional development (UNESCO 2010). It is not easy to identify emerging needs in terms of changing knowledge production, dissemination and application, cognitive and professional paradigms. In the case of engineering, we are fortunate that the needs for the next generation of engineers are reflected in the twelve graduate attributes and professional competencies as identified in the Washington Accord (Washington Accord):

1. Engineering knowledge
2. Problem analysis
3. Design/development of solutions
4. Investigation
5. Modern tool usage
6. The engineer and society
7. Environment & society
8. Ethics
9. Individual & team member
10. Communications
11. Project management & finance
12. Life-long learning

As can be seen, only five or six of these criteria relate to the "core" or "old" engineering curricula, with the other half relating to more modern needs in terms of professional practice (interestingly, along the lines originally advocated by von Humboldt, before practice gave way to theory).

Issues and Challenges Facing the World, and Engineering

The main overall challenges facing the world relate to the Goals of the Millennium Summit (the Millennium Development Goals – MDGs, 2000–2015), and the post-2015 Development Agenda, presently under discussion in the UN (with a spotlight on the role of science, technology and innovation, without overview mention of engineering). The MDGs include, particularly, poverty reduction – enhancing the quality of life for people living in poverty, and sustainability – promoting environmental sustainable development, climate change mitigation and adaptation. These are also the main global issues facing engineering, and engineering is vital in addressing these goals. Engineering also has its own internal issues and challenges, especially the shortages of engineers reported in many countries, and associated decline of interest and enrolment of young people in engineering – which is a major concern for future capacity and addressing international development goals.

The 2000–2015 UN Millennium Development Goals, consist of 8 overall goals, with 18 quantifiable targets measured by 48 indicators. The overall goals are:

1. Eradication of extreme poverty and hunger
2. Achievement of universal primary education
3. Promotion of gender equality and empower women
4. Reduction of child mortality
5. Improvement of maternal health
6. Combating HIV/AIDS, malaria and other diseases
7. Ensuring environmental sustainability
8. Development of global partnership for development

These goals are aspirational and qualitative rather than actual, although the targets and indicators are more quantitative. As may be the case with such goals, success has been limited by the difference between aspiration and reality, and only three of the eighteen quantifiable targets have so far been achieved, boosted by economic development in China and India, but constrained by the Global Financial Crisis. As is also the case with such visionary and aspirational goals, there was little mention of how they might be achieved, or what areas of knowledge might be important or instrumental in achieving them. The role of science and technology was only mentioned in relation to MDG8, target 18 relating to ICTs (the very last target), for example, and there was no mention of engineering. Limited success in achieving aspirational MDGs in a time of economic crisis may be of little surprise, given the scope of the challenge, and emphasizes the need for more realistic, measurable goals and appropriate indicators. It also emphasizes the need for better delineation of how such goals may be achieved, the vital role of engineering and technology in the process, and the generally outdated understanding of the role of engineering and technology in development by the "aid" community and associated policy makers and decision takers.

Issues and challenges facing the world are listed below, in terms of the percentage and numbers of the world population that do not have access to the areas of basic need noted above.

39 %	2.6 billion people	Do not have safe water
35 %	2.3 billion people	Do not have improved sanitation
24 %	1.6 billion people	Do not have electricity
20 %	1.3 billion people	Live in poverty (<1$/day, 70 % women)
15 %	Over 1.0 billion people	Lack adequate housing/live in slums
15 %	Over 1.0 billion people	Lack any ICT connection
13 %	852 million people	Go hungry every day

Life expectancy – poor countries: 52 years; rich countries: 78 years

Addressing basic needs in these areas is an engineering issue, with engineering solutions. Engineering is essential in this process, and engineering education, in developing and developed countries, needs to focus on the development of curricula and learning approaches to graduate engineers with the attributes and competencies to address these challenges. Student-centred, project- and problem-based learning will be vital to address such real and relevant world issues and challenges.

Poverty Reduction and Engineering

Poverty is a major issue and challenge facing the world. Poverty is defined conventionally as living below US$2 per day, and extreme poverty as living below $1.25 per day. Poverty therefore relates particularly to the developing and least developed countries, although not exclusively so – there are examples of relative poverty in most cities and countries around the world. In 2012 the World Bank released data from a study over the period 2005–2008 indicating that, while absolute numbers had increased, the percentage of people living in poverty had declined for the first time since 1981, estimating in 2008 that 2.49 billion people lived on less than $US2 a day and 1.29 billion below US$1.25, down from 2.59 and 1.94 billions in 1981, respectively. The eradication of poverty, especially extreme poverty, is the first of the UN Millennium Development Goals. Poverty depends on social and economic context and such issues as access to land and resources, and is a measure of income and resource distribution and inequality. Poverty is also gendered – 60 % of the world's poor are women, who are also in many countries mainly responsible for family care and services.

Although conventionally considered, measured and indicated financially, poverty relates essentially to the access of people to the resources with which to address basic human needs. This depends on resource availability and population pressure – people living in poverty spend more of their income on basic needs such as food, and are especially vulnerable to increases in the cost of living. Poverty depends on natural factors such as drought and famine, and also on government policies regarding income and resource distribution. In the 1980s, for example, free market policies of economic liberalization and structural adjustment cut government support of social programs, subsidies and public financing in developing countries and lead to an increase in poverty and a substantial increase in inequality within and between countries. In the context of access to resources, poverty is also defined as a denial of basic human rights to food, housing, clothing, a safe environment, health and social services, education and training, decent work and the benefits of science and technology.

While poverty is often thought of economically, it relates primarily to the limited access of people living in poverty to the knowledge and resources with which to address their basic human needs and promote sustainable economic, social and human development. Areas of basic human need include water supply, sanitation, food production and processing, housing, energy, transportation, communication, income generation and employment creation. Addressing basic needs in these areas consists essentially of the transfer, innovation and application of engineering and technology appropriate to the social, economic, educational and knowledge situations in which poor people live. Such engineering and technology has to be appropriate to context – to be affordable, understandable and build upon local knowledge, skills and materials. This requires an understanding of the needs and knowledge systems of people living in poverty and their participation in the identification, development, adaptation, transfer and application of appropriate engineering knowledge and skills and technology. The development of agricultural technologies in the Industrial Revolution revolutionized rural and urban productivity in line with increasing populations, and dramatically reduced poverty. This helped to break the perception that food shortages and poverty were an inevitable fact of life.

Engineering and technology consists of 'hardware' tools, equipment and infrastructure, and 'software' knowledge that develops the technology that surrounds and supports people around the world. The application of engineering and technology helps address poverty at all levels. At the macro level, neo-classical and later economic growth theories paid increasing reference to technology and innovation as the main drivers of economic development and growth, and emphasise economic growth as the main factor in the reduction of poverty, despite criticism of the 'trickle down' effect. Recent research also indicates that growth does not necessarily reduce poverty, but also requires government policies that reduce inequality, with infrastructure playing a key role. At the middle level, many businesses in developed and developing countries are medium and small-scale enterprises, employing less than 250 or 50 employees, and many more businesses are at the micro level with less than 10 employees. Around the world, especially in developing and least developed countries, micro, small and medium scale enterprises (MSMEs) represent the vast majority of companies and jobs, up to 50 % of GDP, and higher growth compared to larger industries. Many MSMEs are focused on particular technologies and innovations.

Engineering and technology is most vital and visible in addressing basic human needs and improving the quality of life of ordinary people in direct applications at the community and family level – in both villages in rural areas and in urban communities. Engineering and technology is vital for the provision and development of water and food supply and other areas of basic need. Examples include domestic food processing tools, the construction of wells, water tanks and improved toilets, equipment and techniques, animal- and engine-powered farm machines, better housing and cooking stoves, low-cost roads and mobile phones. Enterprise and technology helps create income and jobs, but technology for the poor does not have to be poor technology, or low technology. One of the greatest challenges for the next generation of engineers will be to continue to address poverty. Engineering and technology needs to be appropriate to the social, economic, educational and knowledge situations of people living in poverty in order to facilitate and enable them to address their own basic needs, alleviate poverty and promote sustainable livelihoods and development. This requires effective policy formulation, implementation, and the integration of engineering and technology into such debt relief and aid qualifiers as Poverty Reduction Strategy Papers (PRSPs). It also requires effective capacity and capacity building, and the education and training of young engineers, particularly those in developing countries, to be aware and sensitive to the role of engineering and technology in poverty reduction. Government ministries and departments, donor agencies, universities, NGOs and other relevant organizations need to be encouraged and supported in this process with the transfer of information and experience.

The identification, development, adaptation, transfer and application of engineering and technology also requires the provision of information, learning and teaching material using multimedia approaches and ICTs for human and institutional capacity building, and associated support services, particularly micro-finance, to promote technological innovation and application. Technology can then empower the poor by helping them to address their basic needs to reduce poverty – this is a human right and this approach should therefore be central to a rights-based approach to poverty eradication. Specific regional and social dimensions of poverty and poverty eradication require reference to particular areas and issues – including urban and rural poverty, the problems of young people, the elderly, women and gender issues and the 'feminisation' of poverty. The poverty divide is therefore closely connected to the knowledge and technology divide, and the world can be seen to be divided into technology innovators, technology adopters, and the technologically excluded (Sachs 2000). The number of scientific research papers and patents per capita population, for example, is in absolute reverse correlation to measures of poverty. It is the responsibility of engineering and engineering educators to address and reduce the knowledge and technology divides.

Sustainable Development, Climate Change, Engineering Capacity and Education

There is an increasing global challenge regarding the need for development to be environmentally sustainable and for mitigation and adaptation to climate change, especially sea level rise. The use of resources needs to be sustainable for future generations, and better protection from pollution and degradation will be needed. The use of natural resources has approached and in some cases exceeded critical limits, natural and man-made disasters are more frequent, and the economic gap and "knowledge divide" between the rich and many poor countries continues to widen. These issues are a major threat to global prosperity, security, stability and sustainable development.

Engineering is at the heart of sustainability, and sustainability is a major challenge for engineering. Most countries now recognize the need for sustainability and agree that there is an urgent need to reduce emissions and use resources more efficiently, if we are to mitigate and minimize the catastrophic effects of climate change. The question is how to achieve this, amid increasing population pressure and consumption? This question was first raised in 1972, with the publication of "Limits to Growth" by the Club of Rome, and created major interest, concern and debate, which has unfortunately declined since that time. Many countries recognized the need for policy instruments and initiatives to mitigate climate change prior to the 2009 United Nations Climate Change Conference in Copenhagen, and similarly for sustainability prior to the UN Conference on Sustainable Development in 2012. Unfortunately, both COP15 and Rio + 20 failed to deliver any binding agreements and were broadly disappointing, especially for the science and engineering communities – with engineering hardly mentioned at Rio + 20 and in associated documents. Addressing these issues, and the specific outcomes and follow-up to COP15 and Rio + 20, will be a challenge for engineering. This includes the development of environmental engineering, the greening of engineering, and the need for the engineering community to ensure that engineering and technology are on the agenda for sustainable development and climate change mitigation and adaptation.

The Intergovernmental Panel on Climate Change (IPCC) has emphasised the importance of technology and finance in climate change mitigation and adaptation. Despite this, the role of engineering in sustainable development is often overlooked. At the same time, there is a declining interest and enrolment of young people, especially young women, in engineering. This will have a serious impact on capacity in engineering, and our ability to address the challenges of sustainable development, poverty reduction and the other MDGs. The most pressing challenge for engineering is to ensure that there are enough appropriately qualified and experienced engineers to meet this demand. This will require the development of new, more interesting and hands-on courses, training materials and systems of accreditation featuring sustainability. Young people will hopefully be attracted to such courses, which will raise overall awareness of the role and importance of engineering in sustainable development, at the centre of building a carbon-freer future.

Significant investment in technology and infrastructure will be required to enhance sustainable development and climate change mitigation and adaptation. The use of coal may double by 2030, and so will the need to develop carbon capture and sequestration and related technologies – this will be a challenge on a technological scale similar to that of the petrochemical industry. Many countries were looking to develop nuclear power generation, which abated in the shadow of Fukushima, although seems to be returning, even though the nuclear industry has declined over the last decades. Renewable energy has developed over the last decade, and will need further development, marketing and incentives. Similar comments apply to other sectors, such as housing and transportation, and many new engineers will be required as the demand for knowledge and technology increases. While market demand attracts young people into engineering, it takes over 5 years to develop courses and graduates, and over 10 years to train and accredit an engineer – so urgent government support will be required for course development and associated R&D and innovation. Although investment in current technology is a pressing issue, R&D for new technology is also urgent, and governments need to invest now to stimulate R&D and industry in the direction of the coming wave of sustainable technological development, which will need to be at the centre of the engineering agenda.

How can the public understanding of engineering and application of engineering in sustainability be promoted? Public understanding and interest in engineering is facilitated by an understanding of engineering as a vital part of the solution to sustainable development, climate change mitigation and adaptation. University courses can be made more interesting with the transformation of curricula and pedagogy and use of less formulaic approaches that turn students off – with more activity, project and problem-based learning, just-in-time approaches and hands-on applications relating to sustainable development. These approaches promote the relevance of engineering, address contemporary concerns and help link engineering with society in the context of sustainability, and need to build upon rather than displace local and indigenous knowledge. These challenges are linked in a possible solution to promote sustainability and enrolment – many young people are concerned about sustainable development, climate change and other international issues such as poverty, and are attracted to engineering when they see engineering as part of the solution, rather part of the problem. Engineering has changed the world, but is professionally conservative and slow to change – there is a need for innovative examples of schools, colleges and universities around the world that have pioneered activity in such areas as problem-based learning. Engineers introduced just-in-time techniques in industry, and need to do the same in education.

Engineering education needs to be transformed to respond to rapid change in knowledge production and application, with the emphasis on a cognitive, problem-solving approach, synthesis, awareness, ethics, social responsibility, experience and practice in national and global contexts. There is a need to learn how to learn, to emphasize the importance of lifelong and distance learning, continuous professional development, adaptability, flexibility, interdisciplinarity and multiple career paths, with particular reference to engineering and sustainability. While the need for holistic and integrated systems approaches in engineering has been recognised, there is still a need to share information on what this means in practice, and to share pedagogical approaches and curricula developed in this context. This is particularly important for universities and colleges in developing countries, with serious constraints of human, financial and institutional resources to develop such curricula and learning/teaching methods. The transformation of engineering and engineering education will be essential if engineering is to catch the seventh wave of technological revolution in innovation for sustainability.

Knowledge development and application in engineering is vital for sustainable social and economic development, climate change mitigation and adaptation. To promote international cooperation and bridging the "knowledge divide", however, engineering needs to be more closely positioned at the centre of the sustainable development and climate change debate and policy agenda. Sustainable development and climate change also need to be positioned at the centre of the engineering agenda. An important contribution in this context and the "Limits to Growth" debate was the publication of Ernst von Weizsäcker's "Factor Four: Doubling Wealth, Halving Resource Use" in 1997. More recently, Von Weizsäcker and the Natural Edge Project have shown that engineering and innovation can improve resource use and wealth creation by a factor of 5 – an 80 % improvement in resource productivity (von Weizsäcker et al. 2009). At a time of increasing concern over climate change and ongoing economic focus on growth, such contributions help focus attention on engineering and the wider "Limits to Growth" debate, "a green new deal" wave of sustainable engineering and technology. It is apt to recall the native American saying attributed to Alanis Obomsawin, "Your people are driven by a terrible sense of deficiency. When the last tree is cut, the last fish is caught, and the last river is polluted; when to breathe the air is sickening, you will realize, too late, that wealth is not in bank accounts and that you can't eat money."

Humanitarian Engineering, Technology and Development

Engineering applications and innovation for humanitarian development include all levels of technology, from low to high. Technology should reflect social need, affordability, operability, maintainability, sustainability – for example from higher tec solar PV systems to medium and lower tec, such as foot-operated water pumps for African farmers (an innovation is a technology that is new to the user or user-group). The crucial consideration is that technology should be appropriate to social, economic, environmental, engineering and technological context. For a background to appropriate technology see "Small is Beautiful" (Schumacher 1973). Interest in engineering and technology for development has waxed and waned over the last 50 years, with increasing interest in appropriate technology in the 1960/70s. Interest declined in the 1980s/1990s with changing politics, cuts in aid in many Western countries and linkage to policies of structural adjustment. There was a reemergence of interest in appropriate technology in the 2000s, reflected, for example with the publication of "Small is Working" (UNESCO, ITDG, TVE 2004), establishment of Appropedia (2006) and the development of Engineers Without Borders groups around the world. Appropriate technology is not therefore dead (Paul Polak 2010), but resting. Appropriateness is also a feature of new modes of knowledge generation and dissemination, networking (sixth wave of innovation), sustainability, greener engineering and cleaner technology for climate change mitigation and adaptation (seventh wave).

Engineering and engineering education is vital to the sixth and seventh waves of innovation, in developed and developing countries – where much of the technological, economic and social change will take place. This relates to a "political economy of engineering and development" – in developed and developing countries engineering applications and technology depend on knowledge, resources and funding, and in less developed countries also includes development assistance. The contribution of engineering and technology to development depends on various considerations, internal and external to engineering. Considerations external to engineering include awareness of the role of engineering/technology in development and the need for appropriate policy and implementation by policy-makers and decision-takers. Considerations internal to engineering include the need for information and advocacy regarding the role of engineering in development, and the inclusion of development issues in engineering education.

Various factors for success relate to the application of engineering and technology for humanitarian development, these include the need for:

- Information, advocacy, resources, leadership
- Appropriate policy, need for commitment, implementation of policy
- Technologies to be appropriate to local social and economic needs conditions affordable, operable, maintainable, sustainable
- Engagement and involvement of local community and engineers
- Drive by the engineering and technology community, popular champions
- Focus on various communities: engineering organisations and education institutions policy, planning, development in government and private sectors
- NGOs, international and intergovernmental organisations

Issues, Challenges and Opportunities – Fun and Fundamentals

As noted above and in the UNESCO Engineering Report (Marjoram 2010, in UNESCO Report 2010), particular issues and challenges for engineering include:

Decline of interest and enrolment of young people in engineering
Shortages of engineers, technologists and technicians
Brain-drain of engineers from developing countries
Need for investment in infrastructure, capacity and R&D
Climate change, mitigation and adaptation, move to lower-carbon future

The decline of interest and enrolment of young people (especially women) in engineering appears mainly due to negative perceptions that engineering is uninteresting, unappealing, uncool and boring, that university courses in engineering education are difficult and hard work, that engineering jobs are not well paid, and that engineering has negative environmental impact. There is evidence that young people turn away from science at age 10–12, that good science education at primary/secondary level is vital, and that teachers can turn young people on and off science and engineering (National Science and Technology Centre Australia 2007). The image of the nerdy engineer is epitomised in the "Dilbert" newspaper cartoon, and by Mr Bean (although Rowan Atkinson has a degree in engineering). The overall message is that engineering is uncool. Many countries report shortages of engineers, and many Western countries solve this problem through immigration from developing countries, although from the developing country perspective this is brain-drain, and has a serious impact on capacity and development in those countries. Such brain-gain may therefore be considered unethical, where a better 'engineering' solution is to enhance enrolment in developed countries. The need for investment in infrastructure, capacity and R&D following the Global Financial Crisis in 2007–8 was emphasized by Barack Obama in the run-up to the 2008 and 2013 elections (Obama 2008, 2013).

The importance of engineering in climate change, mitigation and adaptation, and the move to lower-carbon future, elsewhere in this chapter.

To address this situation there is a need to counter specific negative perceptions of engineering as unappealing, boring and uninteresting, and a need to promote public awareness and understanding of the important role of engineering in development. To counter the perception that engineering education and university courses are hard work there is a need to make education and university courses more interesting and relevant for problem-solving, that emphasise a problem-based learning (PBL) approach. To counter the perception that engineering jobs are not well paid there is a need to promote the perception of pay parity with similar professions and levels of qualification (although, following supply and demand, salaries are increasing in areas of shortage). Finally, to counter the perception that engineering has a negative environmental impact, there is a need to promote engineering as a part of the solution to sustainable development, climate change reduction and mitigation, rather than part of the problem. To sum up, there is an ongoing need to address and present an overall picture of engineering to:

- Emphasize engineering as the driver of social/economic development to get engineering on the development agenda
- Develop public and policy awareness of engineering
- Develop information on engineering highlighting the need for better statistics and indicators on engineering
- Promote change in engineering education, curricula and teaching to emphasize relevance and problem-solving
- More effectively apply engineering to global issues such as poverty reduction, sustainability and climate change
- Develop greener/sustainable engineering and technology – the next wave of innovation

The promotion of relevance and engineering problem-solving to address global issues such as poverty, sustainability and climate change is exemplified in such initiatives as the Daimler-UNESCO Mondialogo Engineering Award that ran from 2003 to 2010, attracting 10,000 student participants from 100 countries (Mondialogo 2010). The Mondialogo Engineering Award was a problem-based, project-design exercise involving international student cooperation focused on global issues. The interest in such issues is also reflected in the rapid growth of Engineers Without Borders (EWB) groups at universities around the world over the last 20 years. EWB groups have been shown to attract students, and several universities have supported EWB groups in the enrolment and retention of students. Such activities promote engineering and appropriate technology as highly relevant in addressing global issues, ensuring positive feedback, promoting public interest and understanding and conveying the important overall message that engineering is cool.

It is also useful to note that these remarks regarding engineering are part of the wider picture regarding perceptions of recent trends in academia relating to declining standards and funding, and the increasing overloading of academics. These trends have been linked to increasing bureaucracy, corporatisation, and focus on public relations, revenue, efficiency, profile and position – based on indices of academic ranking (Hill 2012).

Many of the above issues, challenges and opportunities facing engineering are linked in the provision of positive solutions. When young people, the public and policy-makers see that engineering is a major part of the solution to global issues, their attention and interest is raised and young people are attracted to engineering. They are also attracted by innovative pedagogical approaches, such as problem-based learning, and to relevance in relation to addressing global issues, such as the use of appropriate technologies to enhance sustainability and reduce poverty. There is therefore a need to promote transformation and innovation in engineering education, that includes theory and practice that was a core of the original Humboldt model – to promote fun and fundamentals.

Innovation and Transformation of Engineering Education

The main goals of innovation and transformation in engineering education to address the issues and challenges noted above are to respond to rapid change in knowledge production, dissemination and application, and the need to move from the traditional, formulaic, engineering curricula and pedagogy toward a cognitive, knowledge-based approach. This approach emphasizes experience, problem-solving and insight, with a more just-in-time, hands-on approach, and is exemplified by project and problem-based learning. This also responds to the changing need for engineers to be better attuned to knowledge change in terms of synthesis, awareness, ethics, social responsibility, experience, practice, applications and intercultural sensitivity. Due to rapid change in knowledge production and application, there is an increasing need for engineers to learn how to learn, in terms of lifelong and distance learning, continued professional development, adaptability, flexibility, interdisciplinarity and multiple career paths. There is also the need for relevance regarding pressing global issues and challenges – including poverty reduction, sustainability (environmental, social, economic and cultural), climate change mitigation and adaptation. As noted above, these needs are reflected in the graduate attributes of the Washington Accord.

Engineers are problem-solvers and innovators, and need to innovate in engineering education toward a curricula focused on project and problem-based learning, with particular reference to real world, relevant issues and problems, cleaner and greener engineering and technology appropriate to social, economic, environmental and cultural context. Curricula need to reflect formal and informal learning trends, especially the use of ICT resources for student-centred learning, with limited lectures and staff acting more in a role of learning facilitators. There should be a focus on development and the assessment of graduate attributes, and the provision of suitable learning and work space to facilitate student interaction. The focus on real world, relevant issues and problems also serves to promote engineering as essential, exciting and a rewarding career (Beanland 2012).

Innovation and transformation is a socio-political as well as a technical process, and as such may encounter barriers and resistance to change. In general, universities and academics have a focus on research, rather than education, are conservative and resist change, and have a culture and space for lecturing, rather than learning. Universities focus on staff performance in terms of papers published and grants gained, and have higher rewards for researchers than effective educators, and university leaders rarely see the need for transformation. Other barriers and resistance relate to accreditation authorities, who also tend to be conservative, slow to change, often averse to an output-oriented, graduate attribute approach, and may not effectively enforce attribute achievement at the individual student level. This is not always the case, however, and accreditation authorities may lead and be instrumental to and noteworthy champions of change, as is the case with many members of the Washington Accord. Despite the rhetoric of excellence, quality, innovation and creativity noted above, however, there are also real concerns regarding declining standards in these areas.

Transforming Engineering Education

There is an increasingly urgent need to transform engineering education to address points raised above – to address shortages of engineers reported in many countries, to move with changing modes of knowledge production, dissemination and application, and in recognition of changing needs for engineers, in terms of knowledge, learning, graduate attributes and professional competencies. These include a problem-solving, problem-based learning approach and link to global issues – poverty, sustainability, climate change. There is an associated need to promote information, evidence, examples of good practice and advocacy on the need to transform engineering education, targeted at engineers, engineering organisations, accreditation bodies, universities, decision takers, policy makers and governments, to emphasise the need for change, facilitate support and enlist champions for change and transformation.

Various 'transformative actions' are of vital importance for change, and it is possible to identify areas of transformative action that are crucial for change in the transformation of engineering education (UNESCO Report 2010; Beanland and Hadgraft 2013). These relate particularly to:

- Knowledge systems – in engineering, science, technology
- Data and information – in/on engineering, science, technology
- Ethical issues – in engineering, science, technology
- Engineering and science education and educators
- Engineering profession and associated institutions
- Engineering industry, employers and associations
- Engineering and government policy and policy makers
- Society and social context for engineering, science, technology

Transformative actions require guidelines, and in the above areas this includes the need to develop and disseminate a better understanding of the knowledge system of engineering – how knowledge in engineering is produced, disseminated and applied in academic, industrial and consultancy settings, and associated social, economic and ethical contexts. This relates particularly to and underlines the need to develop engineering studies to better understand engineering, as a partner to science studies and input to policy. This requires data and information on engineering, in this case to support evidence-based advocacy for change. This needs to be directed toward engineering and science educators (at tertiary and secondary level), the engineering profession, institutions and industry, policy makers and politicians. Particular guidelines for transformative actions include the following:

- Use of the Washington Accord graduate attributes as overall objectives for engineering education, with assessment based on these attributes.
- Design and use of curricula based on Washington Accord graduate attributes to establish student goals and develop professional competencies
- Emphasis on student-centred, problem-based learning and ICT resources, as an alternative to lectures, to encourage motivation and engagement, especially in the first year
- Use of student learning rooms, personal learning environments and e-portfolios, staff operating more as learning facilitators than lecturers
- Use of projects focused on real-world needs to develop design skills, teamwork and communication (such as the Mondialogo Engineering Award, EWB Challenge in Australia).
- Development of university-industry cooperation to facilitate project activity, work and professional experience, staff exchange and promotion of engineering as a career.

Barriers and resistance to change may be overcome with various strategies. The university and academic focus on research needs to be addressed with more emphasis on and reward for educational activity. The conservative nature of universities in relation to pedagogical change can be addressed with information and advocacy for change. One of the main concerns here relates to the belief of many academics that problem-based learning takes more time and effort than conventional lecturing – which is not necessarily the case. The university focus on lecturing persists, to the extent that some academics regard lecturing as synonymous with learning. The validity of such a perception can be reviewed by research and information on learning. PBL emphasises learning, and Aalborg PBL graduates, for example, are sought after by industry for their initiative and innovation. University space has generally been designed for lecturing, although many universities are realising and addressing the need for student learning areas. While some accreditation authorities may still be conservative and reluctant to change, many recognise the importance of the graduate attributes of the Washington Accord and are leading champions of change.

Engineers and educators can help facilitate change by recognising, supporting and promoting the transformation of engineering education to universities and government through example, research, information and advocacy. They can work with accreditation authorities and universities to implement Washington Accord graduate attributes, professional competencies and development. They can also work with industry on projects, professional experience and staff exchange to facilitate transformation.

Concluding Remarks

Transformation and change in engineering education is required to attract and retain young people to engineering, to address reported shortages of engineers around the world, and associated brain drain from developing countries, and to keep up with changing needs for engineers, changing modes of knowledge production and application and changing global needs. These include the increasing need for sustainability, climate change mitigation and adaptation, and humanitarian engineering to reduce poverty and promote social and economic development – challenges that concern and appeal to many young people, and attract them to engineering. The transformation of engineering education needs to be student-centred, with a focus on graduate attributes, professional competencies and relevance. This transformation will not only benefit students and engineering, but also universities, industry and the wider public. Other professions, such as medicine, have transformed toward 'patient-based' learning, when there was no enrolment need to do so, whereas engineering has enrolment and retention issues that transformation will address. These issues are internal and external to engineering, and require internal and external incentives to change, including a move from teaching to learning, and a better balance of reward between learning and research at universities. Student-centred, problem- and project-based learning has been shown to facilitate such transformation at universities around the world (including Aalborg, Olin College and Singapore University of Technology), with many other universities taking increasing interest. Accreditation authorities and governments need to recognise, support and help facilitate the output-oriented, graduate-attributes approach and transformation of engineering education.

There is a particular need to recognise the changing context of knowledge production and application, and changing needs for engineers in terms of learning, graduate attributes and professional competencies, as indicated in the Washington Accord. These include a problem-solving, problem-based learning approach and link to global issues – especially poverty, sustainability and climate change. There is also a need to develop and promote information, evidence, examples of good practice, and to enlist champions for advocacy regarding the transformation of engineering education, focusing on engineering organisations, accreditation bodies and universities, with the goal of facilitating government and private sector support for transformation. To conclude, it is useful to consider the consequences of failure in addressing the need to transform engineering education – continued and increasing shortages of engineers around the world, continued brain drain and impact on social and economic development, especially in developing countries, a world of increasing borders without engineers. This is the backdrop to the need for engineering education to transform itself to interest, promote enrolment and retention of young people, reflecting changing knowledge, production, dissemination and application, changing societal and economic conditions and needs.

References

Appropedia – the first internet portal on appropriate technology for development, established in 2006.

Beanland, D. (2012). *Engineering education: The need for transformation.* Presentation to Engineers Australia, Melbourne, 19 July 2012.

Beanland, D., & Hadgraft, R. (2013). *Engineering education: Innovation and transformation.* Melbourne: RMIT Publications.

Etzkowitz, H., & Leydesdorff, L. (2000). The dynamics of innovation: From national systems and "mode 2" to a triple helix of university-industry-government relations. *Research Policy, 29*(2), 109–123.

EWB challenge – annual project of engineers without borders, Australia, established in 2007, to provide students with an opportunity to learn about design, teamwork communication through real, inspiring, cross-cultural development projects. http://www.ewb.org.au/whatwedo/institute/ewb-challenge#sthash.VwPCjSKi.dpuf

Freeman, C. (1995). The national system of innovation in historical perspective. *Cambridge Journal of Economics, 19*, 5–24.

Gibbons, M., Limoges, C., Nowotny, H., Schwartzman, S., Scott, P., & Trow, M. (1994). *The new production of knowledge. The dynamics of science and research in contemporary societies.* London, Sage: Sage.

Hill, R. (2012). *Whackademia: An insider's account of the troubled university*. Sydney: University of New South Wales Press.

Lundvall, B. A. (Ed.). (1992). *National innovation systems: Towards a theory of innovation and interactive learning*. London: Pinter.

Marjoram, T. (2010). UNESCO report, 2010.

Metcalfe, S. (1995). The economic foundations of technology policy: Equilibrium and evolutionary perspectives. In P. Stoneman (Ed.), *Handbook of the economics of innovation and technological change*. Oxford: Blackwell Publishers.

Metcalfe, S. (2009). Keynote presentation at the OECD-UNESCO international workshop, Innovation for development: Converting knowledge to value, OECD, Paris, 28–30 Jan 2009, co-hosted by the OECD and UNESCO.

Mondialogo. (2010). *Daimler-UNESCO Mondialogo Engineering Award*. UNESCO report.

National Science and Technology Centre, Australia. (2007). Personal communication, Brenton, H., World conference on science and technology education, Perth.

Nowotny, H., Scott, P., & Gibbons, M. (2001). *Re-thinking science: Knowledge in an age of uncertainty*. London: Polity.

Obama, B. (2008, 2013). Presidential inaugural addresses. http://www.nytimes.com/2009/01/20/us/politics/20text-obama.html?pagewanted=all&_r=0; http://www.nytimes.com/interactive/2013/01/22/us/politics/22obama-inaugural-speech-annotated.html#/?annotation=490a0fc13

Polak, P. (2010). The death of appropriate technology: If you can't sell it, don't do it. Blog: "*Out of poverty*". Available at: http://www.paulpolak.com/the-death-of-appropriate-technology-2/

Sachs, J. (2000, June 24). Globalisation: A new map of the world. *The Economist*, pp. 99–101.

Schumacher, E. F. (1973). *Small is beautiful: Economics as if people mattered.* London: Blond and Briggs.

UNESCO, ITDG, TVE. (2004). *Small is working: Technology for poverty reduction*, video and booklet available at: http://upo.unesco.org/details.aspx?Code_Livre=4133

UNESCO Report. (2010). *Engineering: Issues, challenges and opportunities for development.*

Von Weizsäcker, E., Hargroves, K., Smith, M., Desha, C., & Stasinopoulos, P. (2009). *Factor five: Transforming the global economy through 80 % improvements in resource productivity.* London: Earthscan.

Yamani, A. Z. (1973). Quotation attributed to Sheikh Yamani, Saudi Minister for Oil, at OPEC.

Tony Marjoram Guest Professor at Aalborg University, Honorary Fellow at Melbourne University and *Manchester Institute of Innovation Research.* Responsible for the Engineering Program in the Division of Basic and Engineering Sciences at UNESCO from 2001 to 2011. Ph.D. from the University of Melbourne on Technology and Development in the Pacific Islands, M.Sc. on Science and Technology Policy from Manchester University, B.Sc. Hons in Mechanical Engineering from UMIST. Conceived, edited and produced the UNESCO Report *Engineering: Issues, Challenges and Opportunities for Development* in 2010 – the first UNESCO Report on engineering. Worked for UNESCO for 18 years, first at the Regional Office for Southeast Asia and the Pacific in Jakarta. Prior to UNESCO was a Senior Research Fellow at the International Development Technologies Centre of the University of Melbourne, Senior Development Fellow at the University of the South Pacific, and Research Fellow at Manchester University. Published over 50 papers, books, articles and reports.

Chapter 17
Appropriate Curricula for Engineering Management Programmes: A South African Approach

Alan Colin Brent

Abstract Education in the field of engineering management is rapidly increasing worldwide, and particularly in developing countries that are industrialised. In South Africa, Stellenbosch University and the University of Pretoria respond to the need for these skills through dedicated postgraduate programmes. The programmes link the concept of adult learning with a web-based environment. Furthermore, project-based learning activities are typically used in the modules of the programmes. The research summarised in this chapter set out to determine whether the web-based platform is a constraint for project-based learning on the programmes, and to ascertain whether the learning styles on the programmes are conducive for these learning activities; to then understand better how learners develop appropriate knowledge and competencies in engineering management. An action research approach and a questionnaire research methodology were combined, focusing on a specific module that is offered on the programmes. It is concluded that the web-based platform is not a constraint to the programmes. Furthermore, the learning styles of the typical educators and learners are conducive to project-based learning, although the principles of flexible learning, through reflective practice, must be incorporated for students of different characteristics. Finally, the chapter introduces a conceptual model for curricula design to facilitate effective and flexible learning on the programmes. Further research is required to determine the practicality of the conceptual model for the postgraduate engineering management programmes.

Keywords Engineering management • Project management • Technology management • E-learning • Situational learning • Project-based learning • Problem-based learning • Learning styles • Learning model

---

A.C. Brent (✉)
Department of Industrial Engineering, Stellenbosch University,
Stellenbosch 7600, South Africa
e-mail: acb@sun.ac.za

S.H. Christensen et al. (eds.), *International Perspectives on Engineering Education*,
Philosophy of Engineering and Technology 20, DOI 10.1007/978-3-319-16169-3_17

Introduction

Education in Engineering Management (EM), Technology Management (TM), and Project Management (PM) has grown substantially at a global level (Brent and du Toit 2006). In South Africa, Stellenbosch University and the University of Pretoria address the increasing need for postgraduate education in the EM, TM and PM fields. The aim of Education, Training and Development (ETD) in these fields is to give intellectual development and systematic instruction, growth, and advancement to professional engineers with management responsibilities in a technical and business environment (Duhaney 2005). The challenge of ETD in these fields is then to incorporate the separate, dominant epistemologies of the engineering and management sciences in an appropriate manner for the engineering management oriented discipline.

Linking the concepts of 'adult learning' with the web-based ETD environment of the Masters programmes is paramount, and a number of associated challenges have been highlighted (Brent and du Toit 2006). Most notably, all of the modules of the different programmes are offered via an online platform, such as WebCT (2013) – now Blackboard, or Moodle, with limited contact periods; and academic staff have access to training on the platform, but little attention is currently given to the curricula development of the modules from a web-based learning perspective. Also, the students that participate in the modules are of extremely diverse backgrounds with different professional career paths, and they either represent different parts of South African society, or they are from other countries in Africa or other continents; for example Europe and China. A situation analysis relating to the typical learners and learning processes of the postgraduate programmes is summarised in Table 17.1.

Seven learning perspectives or principles have been analysed in the context of the postgraduate programmes (Brent and du Toit 2006): action learning, experiential learning, reflective learning, constructivism, transformational learning, cooperative learning, and culture and learning, which specifically address indigenous aspects and the concept of situated learning. It was concluded that all of the learning principles are applicable to the programmes and modules in one way or another, although some of the learning approaches might be more suitable for the specific learning environment and context. An investigation was subsequently undertaken on the engineering management programmes to prioritise the learning principles from the perspectives of academic staff that are responsible for the programme modules and the students that participate in the programmes. The observations indicated that a shift is required in how the curricula are designed for the programme modules (Brent and du Toit 2006). It was concluded that the action, experiential, and reflective learning perspectives should receive the highest priority for the curricula design of the modules, with some regard of constructivism, and transformational and cooperative learning. Aspects of culture and situation learning are deemed of lesser importance in the context of these engineering management programmes.

Table 17.1 Assumptions of the postgraduate engineering management learner and learning process with consequences

Assumption	Consequence
The web-based ETD technology WebCT is the most appropriate learning platform for the programme modules	The educators must be knowledgeable of and comfortable with the WebCT platform, as the astute and informed use of these types of technologies have been shown to be key success factors associated with learning experiences with such technologies (Oliver 1998)
All learners have affordable access to the web-based technology	Access and equity has been noted as a problem when technology leads the learning process (Davison 1996), and access efficiencies will influence the learning rates or progresses of different learners
All learners are comfortable with the web-based technology	It has been noted that e-learning experiences amongst students range from inspiration to frustration, with the latter too often being the case. The educator therefore has to do much to engage all learners actively in the learning process and limit frustrations (Sharpe and Benfield 2005)
The web-based technology provides a platform for faster turn-around of information	The learning content of an effective educational process should always be timely, which is especially true in the fast-growing field of engineering management, and places additional pressure on the educator; it has been illustrated that speed is the essence of the digital society and time can either be a competitive asset or liability (Rosenberg 2001)
The postgraduate programmes specifically require group work and associated team skills, which are increasingly desired by employers and graduates	In an online learning environment there is enormous potential for cooperative or collaborative work within groups, particularly among geographically diverse student populations, but there is as yet no adequate mechanism for identifying and formally assessing group skills in an online environment (Underhill 2006); this should be considered by the educators of the postgraduate programmes during curricula design

Objectives of the Chapter

Many of the learning activities of the engineering management Masters programmes are based on the known reiterative problem-based learning (PBL) approach that was designed to develop the skills of effective reasoning and participative learning (Ryan 1999). The approach has also been used for the learning and teaching of sustainability concepts to engineers and scientists (Mitchell et al. 2004), and is then appropriate for one of the modules of the programmes, which this chapter focuses on in terms of understanding how the engineering learners develop knowledge and competencies. Accordingly, and given the challenges that are faced in the programmes (Brent and du Toit 2006), the South African Universities aim to establish a generalised model whereby effective and flexible learning curricula can be developed for all the engineering management modules. A collaborative project-based learning approach forms the core of the generalised model, namely a constructivism-learning environment (Carr et al. 1998), which addresses all levels of learning as described through Blooms taxonomy (Atherton 2005), namely: knowledge, comprehension, application, analysis, synthesis, and evaluation.

To reach the goal of the Universities, the primary objective of this chapter is to determine if the web-based (WebCT-Blackboard or Moodle) platform – hereafter referred to as WebCT, as part of a 'blended learning' environment (Konrad 2003), is a constraint for project-based learning in the postgraduate Masters programmes. As a secondary objective the chapter also ascertains whether the learning styles of the typical educators and students on the engineering management programmes are conducive to project-based learning in a 'blended learning' environment. Finally, the chapter introduces a conceptual model for curricula design to facilitate effective and flexible learning on the engineering management programmes.

Research Methodology

Research based in the social sciences is often 'quantitative', which is typically based on statistical analyses. Such research often follows a quasi-scientific model with a hypothesis tested through the gathering of data from many individuals (Monaghan 2003). Other research is described as 'qualitative'. While generalisations from qualitative research are difficult to make, such research typically examines a small number of cases in depth rather than attempting to summarise numerical information (Monaghan 2003). A combination of research methods has been suggested, referred to as triangulation (Myers 1997). To address the primary objective, the research subsequently combined the action research (qualitative) and questionnaire (quantitative) methods. The research methods were applied on a specific module of the engineering management programmes; as a main stream instrumental case.

The Sustainable Life Cycle Management (SLCM) Module as a Case Study

Management structures for projects and the business are essential for engineering practices. Apart from the general issues of management, namely planning, organizing, leading and control, the aspects of safety, health, and the environment (SHE), and sustainability in general, form an integral part of decision support (Brent 2012). A sound basic knowledge of SHE related issues are therefore required at all levels. It is increasingly important that an engineering manager can apply the various principles of SHE management to ensure the future sustainability of the business (Labuschagne et al. 2005). The tremendous change in the importance of sustainability issues strongly influences Life Cycle Management (LCM), which is of increasing importance as a management concept (Labuschagne and Brent 2005). LCM is a multi-disciplinary study field and integrates, for example, risk

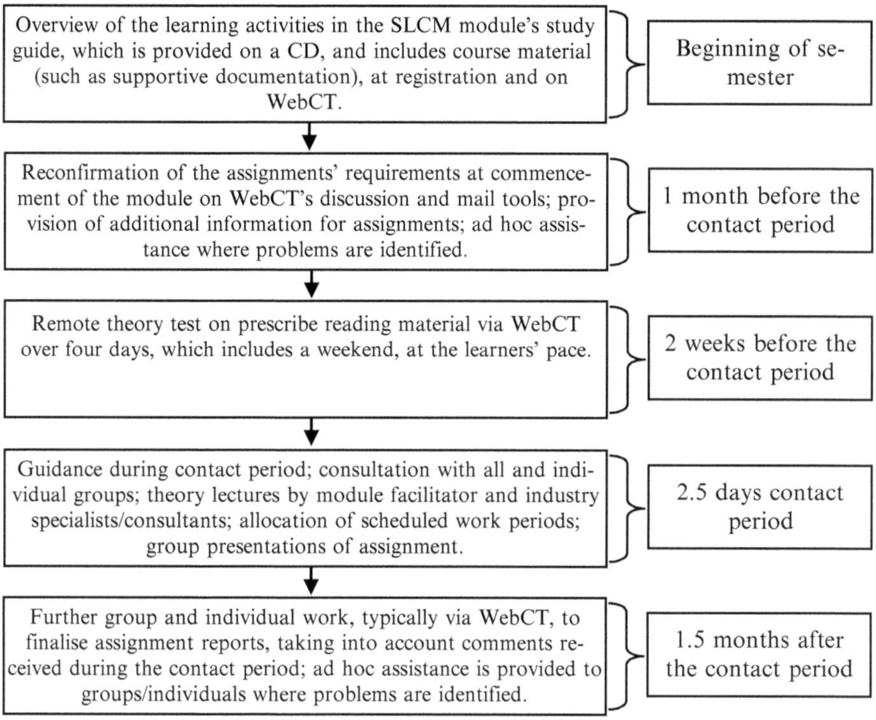

Fig. 17.1 Typical schedule of the SLCM module for engineering management programmes

management, supply chain management, maintenance management, and logistics management, among other disciplines (Brent 2012).

The SLCM module, which forms part of the engineering management programmes, is thus structured as one unit to deal with all the issues that are essential to ensure the sustainability of a project or technology and a business overall.

The learning activities of the module occur over one semester through the web-based platform technology WebCT and a limited contact period of 2½days. The typical schedule of the module on the engineering management programmes is summarised in Fig. 17.1. Over 30 students participate in the module every year from different engineering and sciences backgrounds, and with various employment experiences.

Action Research: Problem Identification and Corrective Action on the SLCM Module

In previous years very little engagement between the educator and the students occurred through WebCT. The study guide was posted, ad hoc communications occurred via the discussion and mail tools, and the assignment tool was used for the

Table 17.2 Study guide information on the individual assignment

Introduction	Sustainable development is a vast concept from an engineering perspective, which has been demonstrated through a number of cases in industry (see Part II of the prescribed textbook). Students must be able to critically analyse such cases and highlight how the concepts are applicable to engineering management and could be applied to **another** industry sector or product in South Africa (as described in the textbook cases). For this module one of the case study chapters of the prescribed textbook will be allocated to each student
Description	Provide an overview of the described case and identify the key concepts that are put forward. Consider the questions that the authors pose in each chapter to guide you through the process. Then reflect on the importance of the key concepts from an engineering management perspective. Finally, synthesise the key concepts into a proposed framework that can be used by decision-makers for another industry sector or product (as appropriate) in the South African context
General	The assignments must be submitted in MS Powerpoint format, suitable for presentation to decision-makers as applicable, and not exceed 20 slides. Do not overburden the slides with unnecessary text and graphically illustrate the concepts, elements and conclusions. This presentation must be prepared for presentation during the contact period
	The assignment must be submitted by using the WebCT Assignment tool prior to the contact period

submission of individual and group assignments. This follows the normal practice of many postgraduate modules, which has been noted (van Ryneveld 2004). Subsequently, no information is available in terms of the student profiles of the programmes (and modules), and the related constraints that the WebCT environment may pose to learning, and specifically to collaborative group learning activities. Specific WebCT related interventions were taken from the 2006 academic year onwards in the module. Apart from the assignment information provided in the module's study guide (see Tables 17.2 and 17.3), an emphasis was placed on the participation of the discussion tool of WebCT, which has been noted as a valuable way to contribute to learning and also to continually evaluate a module (Keat and Watts 2003), through concise, but intentionally directing, postings at the beginning of the semester.

The lecturer visited the discussion and mail tools of WebCT on a daily basis for the 5 weeks from the initial posting and leading up to the contact period. The announcement tool of the WebCT platform was also used to direct students to the discussion tool and to motivate them on a continual basis to participate in the online discussions.

Participation Index and Questionnaire: Hypotheses of the Learner Profiles and Related WebCT Constraints

The participations of 136 students were tracked over the typical 5-week period of the SLCM module, over 4 years, through the WebCT platform in terms of:

Table 17.3 Study guide information on the group assignment

Introduction	Techniques are best learned by implementing. In practice, management problems are normally solved by groups rather than by individuals. Furthermore, small groups afford the opportunity to learn from one another's experience. In order to learn from the experience and insight of other students, discussions in small groups form an important part of the course. Although each student should interact with the other members of his/her group right through the semester, time for group discussions will also be made available during the contact period. Reaching conclusions on *all* of the study themes constitutes an important part of your preparation for the examination. Students should study the material before discussions. Since there will not be sufficient time available in class to discuss all the aspects of SLCM listed in the Course Notes, only unresolved issues should be discussed in class. For this module students will be allocated to different groups that will deal with only **one** of two types of assignments
Description 1	Compile from the relevant literature as well as any other relevant source, a list of environmental or environmentally related elements (evaluation framework) that can be used during the execution of an Environmental Impact Assessment (EIA). Select a development project where such an EIA must be performed; for example, a new or expanded manufacturing facility, the construction of new infrastructure, etc. Apply the evaluation framework and analyse all the relevant aspects, as EIA specialists, according to the possible impacts the development project may have on these elements. Rate the possible impact on a defined scale. Perform this analysis for at least the physical and the socio-cultural environments, i.e. positive and negative impacts on the natural environment and society. The results should be presented in a table
	Discuss these findings briefly and clearly indicate how possible negative impacts may be eliminated or at least be minimised together with cost implications. Also, clearly describe (to the decision-making board of the company considering to undertake the project) the contribution the EIA has made or could make to the execution of the project or the functioning of the organisation
Description 2	The group will need to investigate issues relating to a specific product manufactured in industry, which can be a material, service, value-added consumer-based product, etc. Sustainability problem areas must be highlighted in the (cradle-to-grave) life cycle of the product, namely in the supply chain, during normal manufacturing operations, and during usage and final recycling, reuse or disposal, and remedial proposals should be made. Real life situations should be considered so that the solved problems that are highlighted will not only benefit the student, but also benefit a company that supplies the product with respect to minimising liabilities and improving the company image
	The current situation must be investigated and compared with life cycle systems and strategies used in other product life cycle systems, as well as the efficiency and cost effectiveness thereof. In compiling this presentation, it is expected of you to highlight shortcomings in the current arrangements. To make this evaluation meaningful, it will be expected of you to make proposals of rectification with regards to the shortcomings that you have highlighted, which will benefit the company that manufactures the product. The report should specifically address the four phases of a Life Cycle Assessment as specified by the international standard (ISO 14040); with associated sub-headings

(continued)

Table 17.3 (continued)

General	The assignments must be submitted in MS Powerpoint format, suitable for presentation to decision-makers as applicable, and not exceed 20 slides. Do not overburden the slides with unnecessary text and graphically illustrate the EIA or LCA concepts, elements and results. This presentation must be prepared for presentation during the contact period
	The assignment must be submitted by using the WebCT Assignment tool directly after the contact period

- WebCT logged-in sessions;
- Read and sent mail messages; and
- Read and posted discussion messages.

A simple Participation Index was derived as follows:

$$PI = \frac{D_P}{S} \tag{17.1}$$

Where:

PI = Participation Index.
D_P = Discussion messages posted by learner participants.
S = Sessions logged in by learner participants.

A questionnaire was then distributed amongst the 136 students during the contact periods. The questionnaire aimed to address the following hypotheses:

- The Participation Index and background cultures of the participants are independent;
- The Participation Index and work experiences of the participants are independent;
- The Participation Index and web-access type of the participants are independent;
- The Participation Index and regularity of the web-access of the participants are independent; and
- The Participation Index and the number of visits to the discussion tool of WebCT by the participants are independent.

Results

Student Participant Profiles

A total of 92 students completed the questionnaire in total, representing just over two thirds of the total 136 students that participated in the module over the sampling period. The student profiles and Participation Indexes are summarised in Table 17.4. The majority of participants were from African or European-African decent. This is

Table 17.4 Profiles and Participating Indexes of the participating students

Learners' profiles		Number of participants	Percentage of sample	Participating Index[a]	
				<1	>1
Culture	**African**	45	48.9	0.4	2.0
	European-African	29	31.5	0.2	1.5
	Indian-African	14	15.2	0.4	1.9
	Coloured-African	4	4.4	0.4	4.0
Work experience	**<2 years**	19	20.7	0.2	1.7
	2–5 years	35	38.0	0.4	1.7
	5–10 years	23	25.0	0.3	2.0
	>10 years	15	16.3	0.4	2.7
WebCT-access location	**Work LAN**	77	83.7	0.3	1.8
	Home high-speed	7	7.6	0.5	2.1
	Home slow-speed	6	6.5	0.4	2.0
	Roaming	2	2.2	0.9	3.7
Regularity of WebCT-access	**Daily**	70	76.1	0.4	1.8
	2–4 days	13	14.1	0.4	1.7
	5–7 days	3	3.3	0.1	2.4
	>Weekly	6	6.5	0.1	3.3
Visits to the discussion tool	**Daily**	34	37.0	0.5	1.9
	2–4 days	41	44.5	0.3	1.8
	5–7 days	10	10.9	0.3	2.7
	>Weekly	7	7.6	0.1	1.3

[a]Arithmetic mean for the participating learning in a specific profile

not a reflection of the South African demographics, but is reasonably representative of the typical student-profiles on the specific postgraduate programmes. The work experiences of the students were widespread, but at least one third had between 2 and 5 years of working experience in the private or public sectors.

The overwhelming majority of the participants had high speed LAN internet access at their respective places of work, and subsequently accessed the WebCT site daily. However, only one third visited the discussion tool of the specific module on a daily basis.

Testing of the Research Hypotheses

The data was arranged in contingency tables and the guidelines for analysis of r-by-c tables (Johnson et al. 1994) were used to perform hypothesis, Chi-Square testing on the data. Five tests were performed, as set out in the Research Methodology section of the chapter. Table 17.5 summarises the results.

The hypotheses tests show that at a level of significance of 0.01 the obtained Participation Indexes do not reflect the culture, work experiences, web-access type, regularity of web-access, and the number of visits to the discussion tool, of the participants.

Table 17.5 Summary of the hypotheses testing

	Null hypothesis (H_0)	Alternative hypothesis (H_1)	Results for χ^2	Conclusions[a]
Hypothesis 1	The Participation Index and background cultures of the participants are independent	The Participation Index and background cultures of the participants are dependent	**0.130**	The null hypotheses can likely not be rejected for all hypotheses
Hypothesis 2	The Participation Index and work experiences of the participants are independent	The Participation Index and work experiences of the participants are dependent	**0.069**	
Hypothesis 3	The Participation Index and web-access type of the participants are independent	The Participation Index and web-access type of the participants are dependent	**0.033**	
Hypothesis 4	The Participation Index and regularity of the web-access of the participants are independent	The Participation Index and regularity of the web-access of the participants are dependent	**0.654**	
Hypothesis 5	The Participation Index and the number of visits to the discussion tool of WebCT by the participants are independent	The Participation Index and the number of visits to the discussion tool of WebCT by the participants are dependent	**0.192**	

[a]Level of significance $\alpha = 0.01$; reject the null hypothesis if $\chi^2 > 11.34$ for 3 degrees of freedom

Discussion

Analyses and Reflection on the Learning Opportunities

For the group-learning activities that have been defined the educator, as part of an education, training and development (ETD) activity, acts as an 'Honest Broker' (see Table 17.6) (Mitchell et al. 2004). The group-learning activities were analysed according to the requirements of the model as specified in Table 17.6. The analyses are summarised in Table 17.7. A number of shortcomings are highlighted that should be addressed to improve the learning activities and therefore the curriculum of the SLCM module:

A. Meeting the problem situation. The students are not familiarised sufficiently with expected processes in terms of PBL, team management, and reflective practice. It is proposed that this should occur at programme level for all modules. The groups do not consider the impact of cumulative knowledge, values and reasoning abilities on their engagement of the problem. This aspect should be raised in the study guide and/or through the WebCT discussions.

Table 17.6 Stages in problem-based learning (PBL) for the honest broker

Educator's role	Student or project group/team's role
A. Meeting the problem situation	
Educator introduces the problem "as defined by the client" (or learners are given guidance on selecting an appropriate problem)	Students evaluate their own knowledge, values and reasoning abilities associated with the problem
Details of problem are presented, as they would be in professional practice	Project team "takes stock" of cumulative knowledge, values and reasoning abilities
Familiarises learners with expected processes, e.g. reiterative PBL, team management, reflective practice	Team judges likely impact of above on their engagement with the problem (reflective practice)
B. Characterising the problem situation	
Educator leads discussion on assessing levels of uncertainty, and identifying stakeholders or "extended peer community" (EPC)	Project team evaluates the level of uncertainty associated with the problem situation, and identifies an EPC
Educator provides a critique and feedback on team's judgements of uncertainty and identification of EPC	Team judges likely impact of individual and team values, assumptions, skills, reasoning abilities and knowledge of the problem on judgement of uncertainty and identification of EPC (reflective practice)
C. Positioning the extended peer community	
Educator provides guidance on investigating the values perspectives and priorities of the EPC (mechanisms might include interviewing, role-play, literature review, internet enquiry)	Students synthesise, justify and represent various EPC positions on the problem (mechanisms might include positioning papers, oral presentations)
Critiques position papers, etc	Team uses reflective practice to judge the quality of their EPC positioning
D. Framing the problem	
Educator provides guidance on defining objectives of the problem solution and criteria by which various problem solutions might be judged	Project teams describe the objectives of the problem solution as defined by the various EPC positions (learners may be asked to reconcile or trade-off competing objectives)
Facilitates, guides or informs processes for reconciling or trading-off competing objectives	Teams describe the criteria by which various problem solutions will be judged based on EPC positions
E. Identifying and investigating potential solutions	
Educator provides guidance on the means of investigating and evaluating a range of technical and non-technical solutions (might include providing information resources, acting as technical consultant, critiquing design or analysis calculations)	Project team identifies and investigates a preliminary range of technical and non-technical solutions appropriate to satisfy the EPC's objectives
	Preliminary evaluation of the costs and benefits of each according to the EPC's criteria
	Team reflects on identification, investigation and evaluation processes

(continued)

Table 17.6 (continued)

Educator's role	Student or project group/team's role
F. Communicating potential solutions	
Educator creates realistic setting for feedback and discussion on proposed solution, e.g. oral presentation to other learners	Project team communicates the preliminary costs and benefits of each solution in medium and appropriate language to EPC; for example, flow diagrams, written and oral reporting, sketches
Facilitates diverse feedback on proposed solutions (could be by technical critique, consultant, critiquing design or analysis calculations, strategic questioning, role-play)	Project team gathers comments and questions or proposed solutions
	Identify a smaller field of "preferred options" (these may include options from outside the range of potential solutions offered by the project team)
	Team reflects on communication and feedback processes
G. Reiterating the investigation into preferred options and communicating potential solutions	
Educator repeats the previous two steps providing closer critique	Project team repeats (E) at a greater level of detail and accuracy
	Team once again communicates costs and benefits of preferred options and seeks feedback from EPC (F)
H. Reflective summary	
	Individual learners review personal and group reflection to generate appreciation of how and what they have learned, and a sense of what they still need to learn

B. Characterising the problem situation. The groups do not engage adequately in reflective practice, which the educator should emphasise.
C. Positioning the extended peer community. Again, reflective practice is lacking.
D. Framing the problem. This stage of the learning task is adequate.
E. Identifying and investigating potential solutions. The groups do not reflect on identification, investigation and evaluation processes.
F. Communicating potential solutions. This stage of the learning task is adequate.
G. Reiterating the investigation into preferred options and communicating potential solutions. No provision is made for reiteration. This would require additional class interaction through WebCT, which is considered altogether impractical for this type of student.
H. Reflective summary. This is currently not encouraged in the learning task. The educator should prompt responses from students through WebCT.

From the above it is clear that reflective practice, of both the groups and the individual students, is ill represented in the defined learning activities. At the very least this should be promoted in the module's material and through the interactions between the educator and students, but the postgraduate programmes as a whole should also address reflective practices.

Table 17.7 Analyses of the learning activities of the SLCM module

Educator's role	Student or project group/team's role
A. Meeting the problem situation	
√×	√×
B. Characterising the problem situation	
√	√×
C. Positioning the extended peer community	
√	√×
D. Framing the problem	
√	√
E. Identifying and investigating potential solutions	
√	√×
F. Communicating potential solutions	
√	√
G. Reiterating the investigation into preferred options and communicating potential solutions	
×	×
H. Reflective summary	
	×

Influence of the Learning Styles of Typical Educators and Students on the Programmes

The educator of the SLCM module evaluated his learning style to establish whether characteristics could be highlighted to improve the learning opportunities. The evaluation is shown in Table 17.8 and Fig. 17.2, according to the Kolb Learning Style Inventory, which has been used before for the evaluation of learning styles in online learning environments (Terrell and Dringus 2000). The educator prefers a combination of the Abstract Conceptualisation, Active Experimentation, and Concrete Experimentation of the Kolb (1984) learning cycle. However, the Reflective Observation step is not a preferred way of learning, which emphasises the findings of the analyses of the learning opportunity. The educator therefore needs to adapt the learning opportunity, through a process of flexible learning, to accommodate the needs of students with Diverger and Assimilator type characteristics, although only one third of the students strongly display such characteristics.

These outcomes where further highlighted in the feedback from students on the module as a whole. Although a questionnaire was not provided to assess the learning opportunities specifically, the students did indicate in the general assessment of the module that more time should be spent to discuss the assignments' presentations and associated learning thereof. This is addressed in the conceptual model for the curriculum design of the module.

Table 17.8 Learning style questionnaire as executed by the educator

Mark a cross in the appropriate block between the two descriptions on both sides of the following tables on the position, which describes you best.

Scale:

A or 1 = Identify strongly with the word (s) on the left

B or 2 = Identify not so strongly with the word (s) on the left

C or 3 = Identify more with the word (s) on the right but not very strongly

D or 4 = Identify strongly with the word (s) on the right

	A	B	C	D	
Talking		×			**Listening**
Acting		×			**Reacting**
Taking small steps			×		**Observing overall picture**
Being quick		×			**Being deliberate**
Experimenting			×		**Digesting**
Carrying out ideas			×		**Thinking up ideas**
Changing	×				**Remaining constant**
Being animated		×			**Being reserved**
Doing		×			**Watching**
Being goal-orientated		×			**Being process-orientated**
Being practical		×			**Seeing ideals**
Changing as I go		×			**Mapping out in advance**
Finding solutions			×		**Identifying problems**
Formulating answers			×		**Formulating questions**
Total	1	8	5	0	

	1	2	3	4	
Intuitive			×		**Logical**
Personally involved		×			**Impersonally objective**
Emotional			×		**Intellectual**
Supportive		×			**Critical**
Eager to discuss with others		×			**Prone to analyse myself**
Interested in new experiences			×		**Interested in new ideas, models**
A believer in opinion		×			**A believer in theory**
Accepting		×			**Questioning**
Feeling			×		**Thinking**
A quick risk taker			×		**A slow risk taker**
Prone to trial and error		×			**Prone to planning and organising**
People-orientated			×		**Task-orientated**
Ready to jump in		×			**Wanting facts first**
Dependent			×		**Independent**
Total	0	7	7	0	

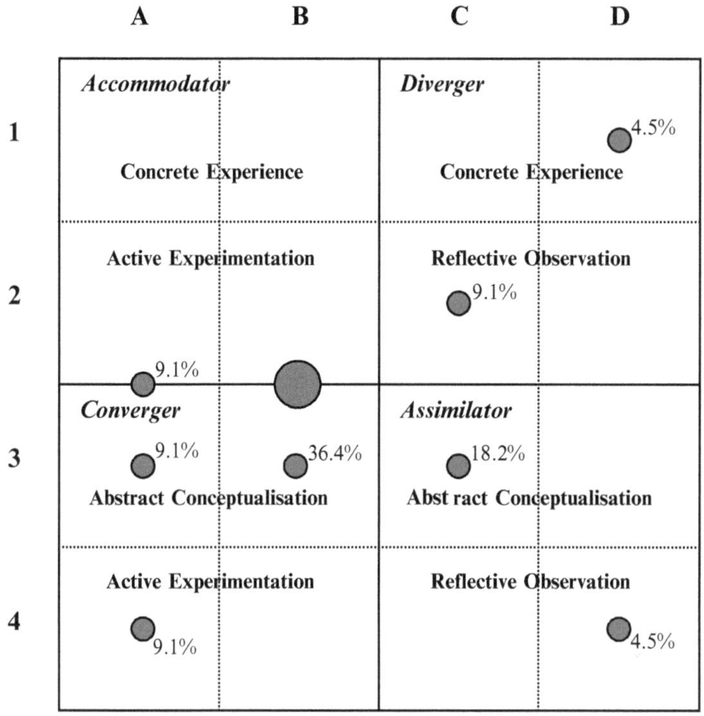

Fig. 17.2 Learning style grid associated with the educator/students

Indicates the intersect, which identifies the preferred way of learning of the educator

Indicates the intersect, which identifies the preferred way of learning of the students

Conceptual Model for the Curriculum Design of the Module

A variety of models for designing modules in higher education have been developed (Toohey 1999; Biggs 1999). However, many of the same issues are relevant in the context of designing modules and a framework that integrates these issues has been proposed (Donnelly and Fitzmaurice 2005a). The framework (see Fig. 17.3) provides an overview of the process, highlighting the important variables in module design and illustrating the relationships between them. However, it is not considered to be linear process (Donnelly and Fitzmaurice 2005a). It is proposed to apply the model as basis for curricula designs on the engineering management programmes.

Figure 17.3 emphasises three main areas that are key to design an effective module:

- Situation analysis or constructive alignment of the module;
- Prioritising learning perspectives or theories for the module; and
- Defining an assessment strategy for the learning outcomes of the module.

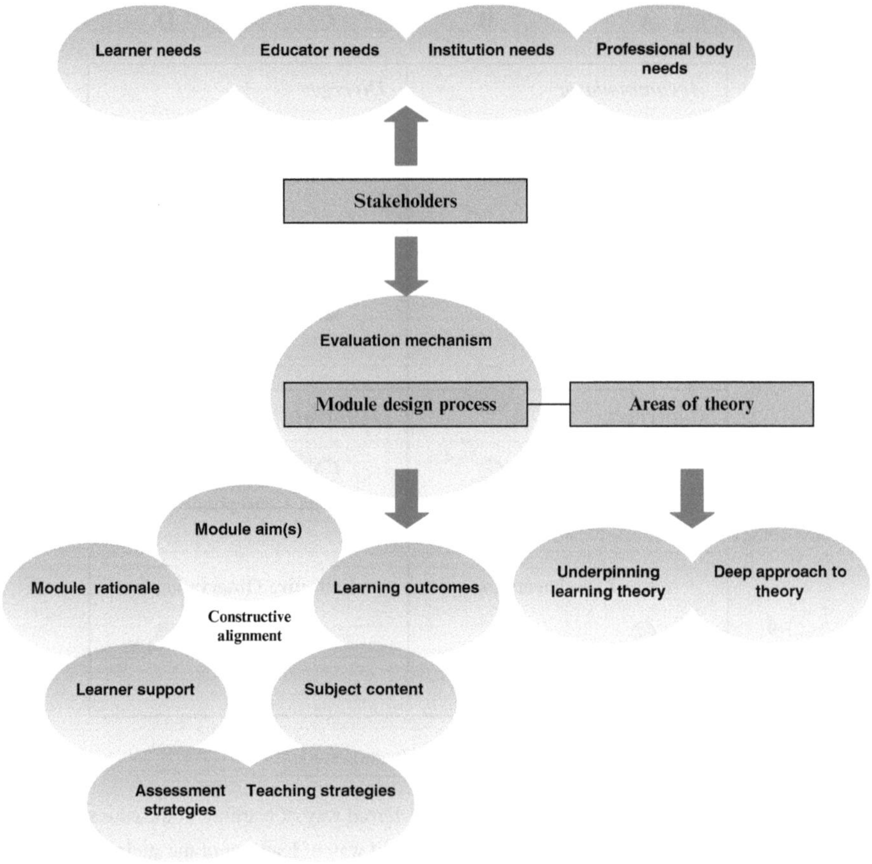

Fig. 17.3 Conceptual model for module design and development in postgraduate engineering management programmes

The first two areas have been addressed in this chapter and elsewhere (Brent and du Toit 2006). The later calls for further discussion in the context of the engineering management programmes.

Defining Assessment Procedures in the Context of the Engineering Management Programmes

An assessment strategy should be based on key learner attributes. It is proposed that the engineering management programmes should focus on 12 learner attributes that has been emphasised in literature (Holzl 2000):

- Communication. The ability to convey ideas and information clearly and fluently, in both written and spoken form (could also include reading, listening, using electronic media, quantified data etc.).

- Computer literacy. The ability to use computers for information retrieval, processing and presentation, to a level comparable with workplace expectations.
- Critical thinking. The ability to identify issues, think independently, apply critical reasoning and make informed judgements.
- Cultural and historical appreciation. The knowledge of other cultures and times which fosters intercultural communication and an appreciation of cultural diversity, historical consciousness and a global perspective.
- Ethics. Knowledge of ethics and ethical standards in relation to their major discipline area/s.
- In-depth knowledge. Deep understanding of at least one field of knowledge, including its methodology.
- Information management. The ability to collect, analyse and organise information.
- Interdisciplinary perspective. Possession of a wide general knowledge, including an appreciation of the philosophical and social context of their major discipline/s.
- Lifelong learning. A propensity to continue learning, including self-management, an ability to adapt to a changing environment and learn new skills.
- Problem solving. The ability to identify, define and analyse problems, create solutions, evaluate opinions, innovate and improve current practices.
- Scholarship. Experience in the scholarship process through which knowledge is gained and disseminated.
- Team work. The ability to interact effectively with others in order to contribute to a common outcome, and to take a leadership role when necessary.

The overall assessment policies and strategies for the engineering management programmes can subsequently be defined as follows:

- The primary focus of assessment is to encourage, direct and reinforce learning; assessment should be designed to assist students in their learning.
- Assessment should be capable of indicating achievement, maintaining standards and providing certification.
- The assessment system should be as transparent as possible.
- Assessment requirements should be communicated clearly, accurately, early and in some detail to all students at the beginning of each module.
- The assessment methods employed should reflect the variety of module and programme goals.
- Well-constructed self-assessment and peer assessment exercises have the potential to provide valuable learning experiences and encourage lifelong learning.
- Where possible, assessment should be based upon a programme of study rather than individual modules.

All of these aspects should be incorporated in the assessment strategy and approaches of the SLCM (and other) modules. Considering the characteristics of the engineering management programmes and the typical students that participate in the programmes, a collaborative project-based learning approach (Donnelly and Fitzmaurice 2005b) is followed in most modules, which can be defined as: an

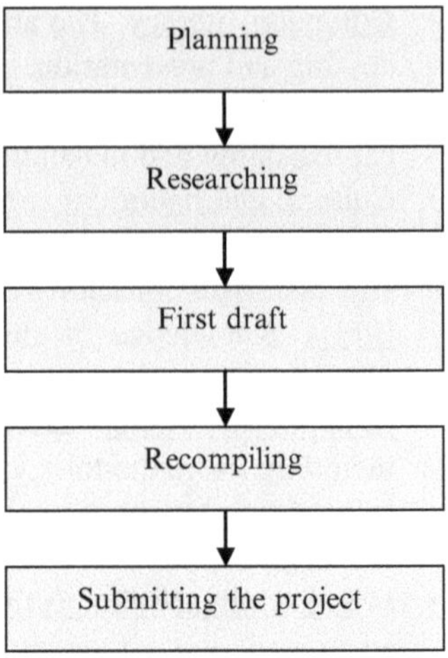

Fig. 17.4 Stages in doing a project-based task

individual or group activity that goes on over a period of time, resulting in a product, presentation or performance. Project-based learning typically commences with the end product in mind, the production of which requires specific content knowledge or skills and typically raises one or more problems that students must solve together. A good project brief of the different project stages is then essential, which is illustrated in Fig. 17.4 and described elsewhere (Donnelly and Fitzmaurice 2005b).

The roles of the educators and students on the engineering management programmes in general have been addressed (see Table 17.6). However, the importance of the educator's role in the design stage of a project-based learning and assessment approach must be emphasised:

- Strong guidance is needed on how to tackle project work at the outset in order to reduce the likelihood of students attempting to undertake overly ambitious projects.
- Project specifications should be more detailed that they would be in 'face-to-face' teaching.
- Careful piloting and testing of proposed projects should be undertaken in advance of the first delivery of a module in order to establish reasonable estimates of the time required for successful student completion.
- Sample projects should be provided to indicate to students the scope of the project that is expected, in order to assist students to form a realistic picture of what they are expected to achieve.
- Module teams or groups should be aware of the importance of Project Guides (documents containing guidelines for undertaking the relevant project) and strive to make it as clear and as helpful as possible.
- It should be recognised that extra demands are made upon educators, both in terms of personal involvement and of time commitment in evaluating and assessing projects.

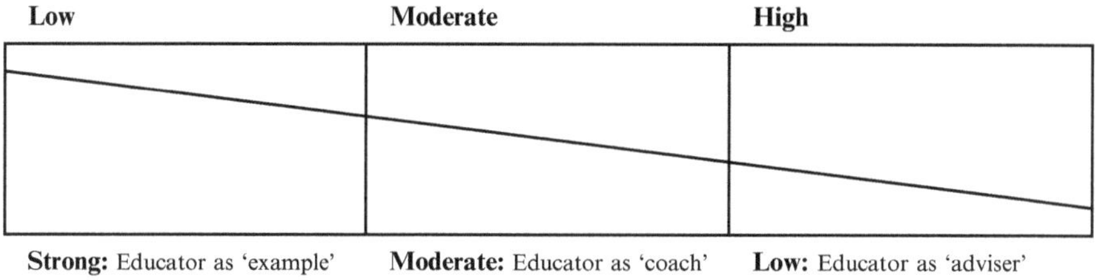

Fig. 17.5 The independence of the learner and support of the educators

In summary, and considering the context of the engineering management programmes, it is expected that the level of independence of the students participating in the respective modules would be moderate to high, whilst the support of the educators be moderate to low (see Fig. 17.5).

Conclusion

The postgraduate Masters programmes of engineering management of Stellenbosch University and the University of Pretoria typically rely on project-based learning activities through an e-learning platform such as WebCT. The research summarised in this paper set out to determine if the WebCT platform, as part of a 'blended learning' environment, is a constraint for project-based learning in the postgraduate programmes; in the South African context. A qualitative action research methodology and a quantitative questionnaire methodology were combined for one module on the programmes, namely Sustainable Life Cycle Management (SLCM). It is shown that the culture, work experiences, web-access type, regularity of web-access, and the number of visits to the discussion tool, are independent of the students' participation on the module. Therefore it is concluded that the WebCT platform is not a constraint to the learning activities of the module (and the programmes). It is further concluded that the learning styles of the typical educators and students on the engineering management programmes are conducive to project-based learning, although the principles of flexible learning must be incorporated for students of different characteristics. Specifically, more opportunities should be provided, as part of learning activities, for reflective practices. Finally, the chapter introduces a conceptual model for curricula design to facilitate effective and flexible learning on the programmes. Certain aspects of the model have been addressed in the chapter, although in the context of the one module only; other aspects have been discussed in general for the programmes. However, further action research needs to be undertaken to determine the practicality of the conceptual model for the postgraduate engineering management programmes.

References

Atherton, J. S. (2005). Learning and teaching: Bloom's taxonomy. Available at: http://www.learningandteaching.info/learning/bloomtax.htm. Accessed 27 Nov 2013.

Biggs, J. (1999). *Teaching for quality learning at university*. Buckingham: SRHE/OU Press.

Brent, A. C. (2012). An overview of environmental management and sustainable development concepts for management practices. In W. Nel (Ed.), *Management for engineering and technology scientists*. Cape Town: Juta.

Brent, A. C., & du Toit, P. (2006). Evaluation and prioritisation of learning principles for modules of postgraduate engineering and technology management programmes. *Engineering Education for Sustainable Development, Proceedings of the 3rd African Regional Conference on Engineering Education & the 4th South African Conference on Engineering Education*, pp. 235–246.

Carr, A. A., Jonassen, D. H., Litzinger, M. E., & Marra, R. M. (1998). Good ideas to foment educational revolution: The role of systemic change in advancing situated learning, constructivism, and feminist pedagogy. *Educational Technology, 38*(1/2), 8–9.

Davison, T. (1996). Distance learning and information technology: Problems and solutions in balancing caring, access and success for student. *Distance Education, 17*(1), 145–158.

Donnelly, R., & Fitzmaurice, M. (2005a). Designing modules for learning. In: G. O'Neill, S. Moore, & B. McMullin (Eds), *Emerging issues in the practice of university learning and teaching* (pp. 99–110). Dublin: AISHE Readings.

Donnelly, R., & Fitzmaurice, M. (2005b). Collaborative project-based learning and problem-based learning in higher education: A consideration of tutor and student roles in learner-focused strategies. In: G. O'Neill, S. Moore, & B. McMullin (Eds.), *Emerging issues in the practice of university learning and teaching* (pp. 87–98). Dublin: AISHE Readings.

Duhaney, D. C. (2005). Technology and higher education: Challenges in the halls of academe. *International Journal of Instructional Media, 32*(1), 7.

Holzl, A. (2000). How do we assess graduate attributes. *TEDI Teaching and Learning Conference*, Brisbane.

Johnson, R. A., Miller, I., & Freund, J. (1994). *Miller and Freund's probability and statistics for engineers* (5th ed.). Englewood Cliffs: Prentice-Hall International, Inc.

Keat, R., & Watts, P. (2003). *The development and use of WebCT in interdisciplinary social science courses*. School of Social and Political Studies, C-SAP Project 2001–2, University of Edinburgh. Available at: http://www.c-sap.bham.ac.uk/resources/project_reports/admin/extras/21-SAP_02.pdf. Accessed 27 Nov 2013.

Kolb, D. A. (1984). *Experiential learning: Experience as the source of learning and development*. Eaglewood Cliffs: Prentice Hall.

Konrad, J. (2003). Review of educational research on virtual learning environments: Implications for the improvement of teaching and learning and access to formal learning in Europe. *European Conference on Educational Research*, University of Hamburg.

Labuschagne, C., & Brent, A. C. (2005). Sustainable project life cycle management: The need to integrate life cycles in the manufacturing sector. *International Journal of Project Management, 23*(2), 159–168.

Labuschagne, C., Brent, A. C., & van Erck, R. P. G. (2005). Assessing the sustainability performances of industries. *Journal of Cleaner Production, 13*(4), 373–385.

Mitchell, C. A., Carew, A. L., & Clift, R. (2004). The role of the professional engineer and scientist in sustainable development. In A. Azapagic, S. Perdan, & R. Clift (Eds.), *Sustainable development in practice: Case studies for engineers and scientists* (pp. 29–55). London: Wiley.

Monaghan, P. (2003). *Interdisciplinary research design*. Committee on personnel and professional development, School for New Learning. Available at: http://www.learningace.com/doc/2799880/f650b0197f7bf5ca6c364276ea450bd0/ird. Accessed 27 Nov 2013.

Myers, M. D. (1997). Qualitative research in information systems. *MIS Quarterly, 21*(2), 241–242.

Oliver, R. (1998). Maintaining accessible and equitable education programs using technology: A Western Australian perspective. In J. M. Barker (Ed.), *Learning together: Collaboration in open learning forum* (pp. 99–102). Springwood: Conference Publications.

Rosenberg, M. J. (2001). *E-learning: Strategies for delivering knowledge in the digital age*. New York: McGraw-Hill.

Ryan, G. (1999). Ensuring that students develop an adequate, and well-structured, knowledge base. In D. Boud & G. Feletti (Eds.), *The challenge of problem-based learning* (2nd ed., pp. 125–136). London: Kogan Page.

Sharpe, R., & Benfield, G. (2005). The student experience of e-learning in higher education: A review of the literature. *Brookes eJournal of Learning and Teaching, 1*(3).

Terrell, S. R., & Dringus, L. (2000). An investigation of the effect of learning style on student success in an online learning environment. *Journal of Educational Technology Systems, 28*(3), 231–238.

Toohey, S. (1999). *Designing courses for higher education*. London: Routledge.

Underhill, A. F. (2006). Theories of learning and their implications for on-line assessment. *Turkish Online Journal of Distance Education, 7*(1), 165–174.

Van Ryneveld, L. (2004). *Surviving the game: Interaction in an adult online learning community*. PhD thesis, Department of Teaching and Training Studies, University of Pretoria.

WebCT. (2013). *Blackboard*. Available at: http://www.webct.com. Accessed 27 Nov 2013.

Alan Colin Brent B.Sc. in Chemical Engineering, B.Phil. in Sustainable Development, and M.Phil. in Sustainable Development. M.Sc. in Technology Management, and a Ph.D. in Engineering Management. He is a professor in the engineering management program of the Faculty of Engineering at Stellenbosch University, and the Associate Director of the Centre for Renewable and Sustainable Energy Studies. He is also a part-time professor in the Graduate School of Technology Management at the University of Pretoria. He has been active in the development of engineering management programs since 2002. His research revolves around sustainable technology management with over 50 papers in international scientific journals, books chapters, and conference proceedings.

Part IV
Innovative Approaches and New Pathways

Introduction

Dean Nieusma and Louis L. Bucciarelli

The chapters in this part share unease over the traditional science-based, technical-problem-centered core of undergraduate engineering education and sketch out a range of pathways to reform. While different in structure and approach, the new pathways proposed have key characteristics in common. First, they largely deny that education can be reduced to the transmission of knowledge nuggets, discrete bits of material-like substance that can be passed from instructor to student in a way that is divorced from students' experiences or their active participation in knowledge generation. Second, they dismiss simplistic proposals for "broadening engineering education" by, say, adding in occasional supplements from the arts, humanities, or social sciences, since such approaches rest on the presupposition that engineering and culture/society are somehow separable, distinct phenomena. Third, they reject reform strategies that neatly demarcate between teaching and research, showing how teaching and research not only inform one on the other but also how they are co-constituted in robust reform initiatives.

As such, these chapters describe deficiencies in engineering education that go beyond the popular concerns about lack of graduates' ability to communicate, to work more effectively in teams, to better understand the "impacts" of technology on society (ugh!). Rather, they ask: Are we preparing our students to live a full life, as citizens, as professionals? Or are we preparing them to be merely "technical functionaries in support positions" – to be "guns for hire." How are our graduates going

D. Nieusma (✉)
Department of Science and Technology Studies, Rensselaer Polytechnic Institute,
110 8th Street, Troy, NY 12180, USA
e-mail: nieusma@rpi.edu

L.L. Bucciarelli
School of Engineering & School of Humanities, Art & Social Science,
Massachusetts Institute of Technology, 77 Massachusetts Av., Cambridge,
MA 02139-4307, USA
e-mail: llbjr@mit.edu

to meet the challenges of engineering practice today, dealing with the uncertainties and ambiguities of complex tasks that require the collaboration of individuals with different expertise and distinct ways of seeing the world, if their main educational occupation is finding the single right answer, always quantitative, to a well posed problem wrenched free of any real-world context, and all this while working alone in competition with their peers?

In response to such concerns, these chapters also share a common sensibility regarding what ought to be considered in the quest to reform engineering education. The answers are as much about values and attitudes, about norms and beliefs – of students and instructors and administrators – as they are about the educational structures, assumptions, and methods that we ought reconsider or the knowledge, concepts, and principles we seek to "transmit" to our students. In fact, the reader is advised to leave behind prevailing suppositions surrounding engineering education as transmission of knowledge bits, misdirecting attention as it does to always-inadequate refinements in sequencing or delivery of existing components.

One new pathway leads through problem- and project-based learning (PBL) to a strategy of engineering education change that is enabled by research-cum-practice. In Chap. 18, Anette Kolmos reports on a research project carried out by the UNESCO Chair in Problem Based Learning in Engineering Education at Aalborg University in Denmark. Drawing on Dewey's insistence of the centrality of experience and experimentation in educational reform, this project has Aalborg Ph.D. students, who come originally from different Asian universities, carrying out educational research on PBL initiatives within their home universities. Part of the challenge is transforming traditionally hierarchal educational environments to ones that are student-centered. Kolmos asks, "how can a western developed pedagogy like PBL be implemented outside Europe where the educational system practices other types of cultural values?" Her answer is to create Ph.D.-student change agents employing design-based research – an educational research strategy that mixes research with reform-oriented practice in a way similar to action research. Kolmos characterizes these change agents as boundary workers, who must simultaneously cross cultural, disciplinary, and theory-practice boundaries. While she reports early success with the project, she also notes that the pressures and contingencies of practice in each context tend to push aside theoretical reflection.

Chapter 19 follows on with the theme of education research, this time looking at the contribution engineers and engineering can make. Jonte Bernhard challenges the ordinary split of research from (undergraduate) teaching and of engineering technical research from engineering education research. Following a design-based research approach similar to Kolmos, Bernhard suggests three ways engineers can contribute to education research: (1) Through their ability to handle both general (i.e., analytic tools) and particular (i.e., local contextual) aspects of their profession; (2) through their facility moving between theory and practice via design; and (3) through their proficiency in the use of material objects. For Bernhard, the promise of this last item remains under recognized. He sees the role of technology, instruments, and artifacts in the development of engineering thinking and practice, including education, as a fertile yet neglected domain of research. In this respect, Bernhard argues, engineering knowledge is not wholly contained in texts, or even the cognitive domain, but is also embodied in the instruments and artifacts they develop and deploy.

Several chapters turn more fully toward various educational initiatives and other efforts to elevate contextual awareness among engineers. In Chap. 20, Dean Nieusma reports on Rensselaer's Programs in Design and Innovation, a dual-major undergraduate initiative, which he directs. This set of programs entails two constituent majors: one in Design, Innovation, and Society (a liberal-arts take on design inspired by STS) and one in a technical area, usually mechanical engineering. The program challenges the pervasive compartmentalization of knowledge into the social and technical so common in engineering thinking and, as importantly, curricular structure. It does this in several ways, including most notably through a sequence of design studios with varying foci and interdisciplinary instruction (usually engineering and STS), where students are challenged to integrate technical, social, and formal analyses with creative synthesis in response to specific problems in the world. Through a judicious outlay of possible projects as well as the program's supplementary STS course requirements, the programs aim "at enhancing students' ability to engage – explicitly and productively – the many contexts of engineering design work." Notably in this approach, context is treated as a set of variables at play alongside the technical, not merely as static constraints. Nieusma too can be seen to be doing a type of design-based research, seeking not only to produce scholarship that informs engineering educational reform but also to apply those insights in the classroom and curriculum.

Gary Downey addresses the dominance of mathematical problem solving in engineering sciences explicitly, challenging those who advocate retaining such a position by asserting "that visible leadership for engineers will likely not come through claims of technical genius and technological heroism when engineers do not have jurisdiction over technology in the first place." In Chap. 21, Downey offers an alternative vision of engineering as problem definition and solution, or PDS. PDS entails stepping back from problem solution to problem formulation and problem definition in a way that is authentic to engineering practice by accommodating, even mediating, participation by diverse stakeholders who bring divergent perspectives to the problem formulation process. Downey provides three strategies for integrating "practices of collaborative problem definition" into engineering education, each working with a different element of the curriculum. Notably, Downey rejects the "breadth" approach to reform because it legitimates the core/periphery distinction, which Downey's strategies aim to break down. This seemingly simple movement from problem solution to problem definition, if authentic, makes a world of difference, enabling students to confront and deal with the interests and power of alternative stakeholders, contend with alternative definitions of the problem, and alternative views about what is taking place in the problem definition and solution process. Through this approach the true power and limitations of the engineer's contribution to problem solving might be overcome, not through claims of "technical genius, but through "the hard work of including collaborative problem definition in engineering work as a core competence, responsibility, and set of practices [that] offer a more realistic pathway [than] hanging onto a declining image."

Chapter 22 takes a similar a big-picture view of both the opportunities for expanding the social commitment of engineering graduates and the barriers to such reforms. Javier Cañavate, Manuel José Lis Arias, and Josef M. Casasús follow Downey closely in their understanding of the problem and their proposed solutions, but explore those insights with very different empirical material. Cañavate et al. start with a brief survey of engineering education accreditation requirements of various countries, identifying a set of "competencies that are outside what traditionally was considered the core of engineering knowledge." This set of competencies is broken down into five categories: environmental risk prevention, soft skills (e.g., communication), ethics, sustainability (which includes but transcends environmental risks), and social integration. Cañavate et al. then identify three major strategies for providing these competences: adding new courses (e.g., in STS), modifying existing (technical) courses, and changing teaching methods. While they recognize potential in each of these approaches, Cañavate et al. explore in some detail a major dilemma: whether the traditional focus on problem solving might ever move faculty to take seriously the calls for "broadening" their programs in accord with the social-commitment criteria of accrediting agencies. They see a "need to change engineering studies at a deeper level" and refer to the notion of *habitus*, which they define, quoting Pierre Bourdieu, as "'a system of durable and transposable dispositions,' [that] 'functions as a matrix of perceptions, appreciations and actions.'" They conclude by reviewing how these issues are responded to in the Spanish context, highlighting the need to modify, not merely courses and curricula, but "the *habitus* of the profession."

In Chap. 23, Matthias Heymann reviews the historical development of wind turbine technology in Denmark and shows how this particular case can be used to highlight more general characteristics of engineering practice that tend to be missed in educational approaches. Such characteristics include especially the complex interlinking of various facets of engineering processes and technology-society relations. Heymann presents a close analysis of how the Danes came to dominate in wind technology. He directs attention to the importance of practical experience, the tacit and personal, in the generation of knowledge and in technical design, but that's just one factor that explains. Just as important is the emergence of a social network of individuals and groups sharing a common interest in furthering the development of the technology. Power relations among corporate interests, the public, and government agencies were just as important, as were visions and ideologies about the proper form of Denmark's entire energy system. Contra historians of technology, Heymann takes issue with a way of speaking about technology as having a context, or cultural ambience, which suggests that engineering is an activity that is surrounded by a set of given and unchanging conditions external to it. He sees this boundary between engineering and culture as artificial, and implicitly supports Downey's emphasis on collaborative problem formulation as an opportune place to reassess engineers' actual and ideal contributions.

Anders Buch and Louis Bucciarelli fear that engineering education reforms centered on matters of curriculum and didactics rarely address "fundamental questions about the nature and character of knowledge and learning." In Chap. 24, they interrogate the instrumental focus of traditional engineering curricula and the reified conception of "engineering knowledge." Specifically, Buch and Bucciarelli see the teaching of engineering as an act of decontextualizing through and through. This approach engenders and sustains a vision of engineering knowledge as well defined, compartmentalized, wholly cognitive, and value free – in other words "stuff" that individuals can either possess or lack. Such a vision fails to recognize the essential role of context not just in problem formulation but also in the very process of learning. Whereas Downey rejects "an idealized mathematical space free of human difference and conflict" as an insufficient basis for effective problem formulation, Buch and Bucciarelli turn the table by understanding the decontextualization process as a critical learning opportunity in itself. They show how narrowness of the domain, its instrumentalist approach, "frees the engineer from social concerns." "[U]ncontaminated by human foibles…, one can dream of reinventing and saving the world…[while] oblivious to what goes on in the world." Arguing that the harm of decontextualization will not be mitigated by the addition of a few courses in the humanities or social sciences, always positioned as they are as peripheral degree requirements, Buch and Bucciarelli suggest that technical knowledge be "intrinsically interwoven with other types of knowledges in our stream of experiences." Instead, educational experiences should start from "authentic engineering situations where problems emerge, …comprised [as they are] by technical, organizational, social, political, etc. elements."

Finally, in Chap. 25, Lars Botin and Tom Børsen describe the Techno-Anthropology programs at Aalborg University Copenhagen. Techno-Anthropology is a recent undertaking but with some of the same goals as Rensselaer's Programs in Design and Innovation. Techno-Anthropology programs go beyond conventional science and engineering programs in that they integrate different disciplinary approaches – anthropology and social studies, philosophy and ethics, natural and technical sciences of instrumental character – in the programs' modules. The focus is on technology as it relates to people's social and cultural needs from an anthropological point of view. The chapter refers to the programs' core competencies – namely hybrid imagination, meta-reflection on technological practices, social responsibility, and interactional expertise – and describes how these connect to and can extend conventional engineering skills. Ultimately, they anticipate that techno-anthropologists will work alongside traditionally trained engineers and anthropologists, as well as philosophers and scientists, on collaborative teams to address contemporary technological problems. In this context, the techno-anthropologists – by virtue of their hybrid imagination – will play a special role in mediating tensions between techno-science/engineering and culture/humanities.

If these various programs and proposals are to gain traction among faculty of engineering and their allies in the humanities and social sciences, it is clear that attitudes toward the social and political realities of engineering practice – and the engineering classroom – must change. These features need to be given the respect they are due, no longer ignored or discounted but seen as important ingredients of engineering research, of problem definition, and of collaborative thought and action in fixing upon a "solution" – as important as the technical skills we teach our students. This is no light matter: Our engineering mindset, assumptions, and dispositions are all conditioned to see the world as a world of things and events that can be measured and quantified, that interact in trustworthy ways once modeled correctly, a world that is devoid of ambiguity, uncertainty, and social exchange. The chapters that follow provide a range of new pathways to understanding, teaching, and practicing engineering in context.

Dean Nieusma B.S. Mechanical Engineering and Bachelor of General Studies, University of Michigan. M.S. and Ph.D. in Science and Technology Studies, Rensselaer Polytechnic Institute. Director of Programs in Design and Innovation and Associate Professor of Science and Technology Studies, Rensselaer. Editor of *International Journal of Engineering, Social Justice, and Peace*. Research interests: engineering professional and educational reform; interdisciplinary collaboration in design; appropriate technology design; design pedagogy; expertise and democratic theory.

Louis L. Bucciarelli B.S. in Mechanical Engineering. M.S. in Aeronautical Engineering from Cornell University, and Ph.D. in Aeronautics and Astronautics from MIT. Professor Emeritus in Engineering & Technology Studies. Current research concerns innovation in engineering education.

Chapter 18
Design-Based Research: A Strategy for Change in Engineering Education

Anette Kolmos

Abstract There is a need for more complex collaborative and innovative competences in engineering education all over the world and more student-centred curricula are means to meet this. However, research has shown that it is challenging to implement student-centred learning, and there is a need for change agents that are familiar with the context, culture, the subject area and new teaching and learning methods, and who have the ability to facilitate the transformation of practice in collaboration with local academic staff. In this chapter, I will present design-based research (DBR) as a research methodology that is suitable in the change process in engineering education. Three Ph.D. students at the UNESCO Chair in Problem Based Learning in Engineering Education at Aalborg University in Denmark have tried out this methodology as part of their Ph.D. study. A framework for DBR as a combined research and change strategy will be presented in this chapter together with experiences from the three Ph.D. projects in which DBR methods have been utilized, adjusted and experienced in an Asian context when implementing student-centred problem- and project-based learning (PBL) at their home institutions. The conclusion of the chapter is that the DBR method can definitely be used as part of the process of curriculum change; however, there are a lot of issues concerning academic standards, educational change and individual courage to work on cultural boundaries.

Keywords Educational change • Engineering education • Design-based research • Problem-based learning • Project-based learning • PBL

Introduction

> In short, at present, both students and teachers of education are excessively concerned with trying to evolve a body of definite, usable, educational directions out of the new body of science. The attempt is only too natural. But it is pathetic. The endeavor to forestall experiment

A. Kolmos (✉)
Department of Development and Planning, Aalborg University,
Vestre Havnepromenade 5, Aalborg 9000, Denmark
e-mail: ak@plan.aau.dk

S.H. Christensen et al. (eds.), *International Perspectives on Engineering Education*,
Philosophy of Engineering and Technology 20, DOI 10.1007/978-3-319-16169-3_18

> and its failures and achievements, the attempt to tell in advance how successfully to do a new kind of things ends, at most, in rectification of old ways and results, plus a complacent assurance that the best methods of modern science are employed and sanction what is done. This sense of being scientifically up-to-date does endless harm. It retards the creation of a new type of education, because it obscures the one thing deeply needful: a new personal attitude in which a teacher shall be an inventive pioneer in use of what is known, and shall learn in the process of experience to formulate and deal with those problems which a premature "science" of education now tries to state and solve in advance of experience (Dewey 1922, p. 273).

During the last decades in the Western world, there has been an explicit formulated requirement for new types of engineering competences such as project management, entrepreneurship, innovation, and global collaboration. These competences are formulated by a wide range of actors from the politicians (e.g., Bologna process and Outcome Based Education), governments (such as the accreditation bodies), engineering societies (National Academies), managers and academic staff (such as conferences and workshop activities on new ways of teaching and learning). In Europe, several ministerial communiqués have emphasized new innovative and creative competences like in the Leuven Communique (Europa 2009; Rocard et al. 2007). Several reports on new engineering competences state that the universities have to pay much greater attention to real-life problems and to societal needs in order to address the employability agenda including collaboration with companies (National Academy of Engineering 2004; Royal Academy of Engineering 2007). Industry organizations point out the need for graduates who are able to participate in engineering project organizations and to collaborate (Chinowsky 2011; Kolmos and Holgaard 2010). Accreditation bodies have defined transferable or professional skills as an important part of the curriculum (ABET, EURACE).

So there is no doubt that there is a need for a change in engineering education – and the growing number of research publications on new types of teaching and learning methods and management of change witness both the research society's concern in this matter and the dissemination of new innovative educational practices.

Research on change in engineering education indicates that more active student centred learning methodologies increase students' motivation for learning, deep learning, and achievement of professional and process competences (Prince and Felder 2006; Graham 2012). Especially problem- and project-based learning (PBL) has been applied as one of the core institutional response strategies to the change agenda, and over the last 20 years there is an increasing number of places that utilize these type of learning principles (De Graaff and Kolmos 2007; Graham 2009; Beddoes et al. 2010; Kolmos and De Graaff 2013; Schmidt and Moust 2000).

During the last five years, the same demand for change and new types of innovative learning methods have emerged in Asian countries. Higher education and in particular engineering education has started to experiment with new educational approaches to achieve new types of competences. In India, there are several studies indicating that engineer graduates are not employable (Blom and Hiroshi 2010; Shinde 2011). In Thailand, there is a concern for regional development and in

Malaysia there is a process of implementing outcome-based education at all levels in the educational system to emphasize the development of competences (Borhan et al. 2012). This is fully in line with what is happening in the rest of the world (e.g. South America) where there also is a growing awareness of the need to educate new types of engineers and academics who are able to participate in global, collaborative and sustainable innovation processes.

The question is how can a western developed pedagogy like PBL be implemented outside Europe where the educational system practices other types of cultural values? And how can this development become a research-based development? Most of the research reported is based on retrospective empirical research reporting on data collected either by qualitative or quantitative methods. How can research and curriculum development be linked with a research paradigm in order to also plan ahead? Graham (2012) argues that research results are rarely the cause of, or the justification for, educational changes, however, I would argue that it is an implicit driver for the managers and staff to believe in new educational practices and to make decisions for establishing new practices. But there is a missing link between the existing research and theories on the one side and the design and implementation of new practice on the other side. I.e. knowing how an innovative educational system could be and how to get this in place from where you are now – is not an easy or a trivial problem.

In this chapter I will report on why and how we at the UNESCO Chair in Problem Based Learning in Engineering Education (UCPBL) at Aalborg University have tried to overcome the obstacles of the missing link between research and development of new innovative practices in Asian institutions by applying design-based research. I will start by addressing the challenges in educating change agents that are able to work across cultural boundaries followed by a short introduction to design-based research and experiences by utilizing this methodology in three Ph.D. studies where PBL is implemented at Asian institutions.

Educating the Integrative Change Agent

The concept of a change agent originally comes from the organizational learning theories and is defined as an organization or a person that acts as a catalyst for conceptual and organizational change. This function involves knowledge about the organizational conditions and a framework for how to establish change in organisations together with a vision on the directions of change. A change agent is a person that manages to carry out cross-boundary work – that can be cultural boundaries, discipline boundaries or theory-practice boundaries (Balogun et al. 2005). In this sense the change agent is comprehensively knowledgeable and a competent person that manages to understand and link diverse perspectives in order to change practice. This involves an integrative personal commitment that might go beyond the values in the academic university. Dewey emphasized this point more than 100 years ago in an article on the engineering of education.

> I do not underestimate the value of the guidance which some time in the future individuals may derive from the results of prior collective experience. I only say that the benefit of such an art cannot be had until a sufficient number of individuals have experimented without its beneficial aid in order to provide its materials. And what they need above all else is the creatively courageous disposition. Fear, routine, sloth, identification of success with ease, and approbation of others are the enemies that now stand in the way of educational advance. Too much of what is called educational science and art only perpetuate a regime of wont and use by pretending to give scientific guidance and guarantees in advance (Dewey 1922, p. 273).

Dewey argues that scientific knowledge is not enough – educational change also has to derive from practitioners' practice and is a result of experience with change of practice. So the experience is important along with the research and the courage to establish new practices. That is quite a radical standpoint as he indicates that change might happen even without any relation to research (Biesta 2009).

Barnett and Coate (2004) emphasize the growth of the individual person as an aim for any curriculum and state that a curriculum is not a firm construction, but it should be a place for learning of three aspects: knowing, acting and being. Knowing can be compared to learning of scientific knowledge, acting with the learning of competences, and being to do the boundary work between disciplines and in relation to practice. Similar to this understanding of curriculum as a much more dynamic learning process, the training of a researcher should address scientific knowledge, competences and not least the transformative element related to the development of professional and personal identity. The being has to be interpreted as an integrative individuation process that enables the individual to work on boundaries and encourage the establishment of alternative practices that involves not only scientific knowledge and competences, but goes far beyond these notions and points at identity and the social formation of the individual's courage to try out new alternatives. Letting Ph.D. students gain experience from designing and experiencing practice can be one way of achieving this goal.

This understanding of curriculum has to be seen in relation to the understanding of universities. In a recent article (Jamison et al. 2014; also Jamison 2013), different types of universities are linked to the concepts of modes 1, 2 and 3. The mode 1 university is seen as the traditional academic and disciplinary university and the mode 2 university as the innovative and collaborative university in terms of company collaboration (Gibbons et al. 1994). Barnett (2011) indicates that it is depressing that the mode 2 concept has been announced as something new and radical. The fact that only two types of knowledge modes are defined limits science:

> Why should it be thought that just two modes of knowledge offer a complete understanding of the world? In a globalized world, in a world in which cultures and traditions are colliding with each other, any number of knowledges will arise. The question is: what is to be the stance of the knowledge towards multiple … can the knowledge university become epistemologically generous, such that no mode of knowledge is especially favoured? (p. 25)

Both Dewey and Barnett represent a critical view on the definition of science and knowledge in general and argue for a more open approach and at the same time an integrative approach where one of the aims in education is that there is intrinsic continuity (Dewey 1916). This is in line with the mode 3 concept that is based on both academic knowledge (mode 1), a collaborative approach (mode 2), and a community orientation together with identity development (Jamison et al. 2014).

To educate people that can work on boundaries is to educate people to learn to live with risks. Boundary shakers are individuals that have the courage to work across existing organisational frames that involves an organisational refiguring (Balogun et al. 2005). To educate Ph.D. candidates that academically reach a high level and relate their research to practice is an ambitious goal in itself. So adding an element of educating change agents is an even higher level and combining scientific research training with competences of designing and facilitating change is very ambitious.

Change Agents Across Cultural Boundaries

Since its establishment in 2007, the UCPBL at Aalborg University has been an organizational agent for change and facilitating change in engineering education towards more innovative, team-based and student-centred learning (UNESCO 2013). UCPBL builds on previous organizations on problem based and project based learning (PBL) and right from the beginning we were involved in faculty development within engineering education by running a Master's program for academic staff and running training workshops. More than 50 workshops have been given all over the world, however, the impact on actual educational change was doubtful.

In 2008, the strategy as an organisational change agent was re-considered as we concluded that if we wanted to have an impact on educational change, we needed to educate change agents who can lead changes in local contexts. Therefore, we decided to educate the change agent that can establish and facilitate change and we decided to build in the change aspect in our Ph.D. program. This should not only establish new local practice but the research training should also go beyond cultural boundaries. Most of our Ph.D. students are employees at their home institutions and they have been asked to learn about PBL and become able to implement PBL in the curriculum at their home institutions. Therefore, the starting point and motivation derives from practice.

In the process of developing the research scope, we have dealt with at least three challenges: (1) meeting a high academic research standard across the discipline boundaries, (2) combining theory, research and change of curriculum practice, and (3) researching and practicing across cultural boundaries. To address all three aims is quite ambitious and loaded with constraints, but also possibilities.

The first challenge concerning the academic level is challenged by cross-disciplinary qualifications as many of the Ph.D. students do not have an educational background in education but they do have a science or an engineering background. The double qualification of knowing both the subjects and educational research is an important aspect that has to be addressed in one way or another within subject-specific educational research. As the research is on engineering education, we associate with the engineering education research that is a growing international field

(Jesiek et al. 2009). In many of the US engineering education research programmes (at Purdue and Virginia Tech), there is a requirement of having formal credits from engineering education together with formal credits from educational research. In the present Ph.D. programme at the UNESCO Chair, there are no requirements of formal credits from either engineering or education; however, this is addressed in the specific study plan for the Ph.D. fellow.

The second challenge is the relation between research and practice, which is an on-going issue in educational research. There is a dilemma that on the one side we would like to base educational development on research evidence and certainly avoid random beliefs and experiments. On the other side, educational research – even the subject-specific educational research – moves away from practice by both phrasing the practice problems differently from practitioners' understanding of the problems, and communicating in an academic language that might be hard to understand for academic staff from other academic fields. Furthermore, knowledge is published in journals that the teachers do not read. So even if researchers do have intentions of researching relevant problems for practitioners, the academic research communities most often act on internal criteria. This is exactly what Dewey criticizes in the introduction quotation and maybe even indicating that having research knowledge is not even a necessary condition for development (Biesta 2009). However, Dewey regards the educational development process as an engineering process and as a creative way bringing elements together from theories, practices and ideas.

The third challenge concerns research and practice across cultures. Any PBL practice is a social construction and what might work in one cultural or institutional setting, might not work in another. Before implementation, the Ph.D. students need to have an idea of alternative practices and maybe not only an idea, but certainly experiences. The practice experiences have to cross culture and national educational policies – and what might be possible in Denmark, might not stand a chance in Malaysia. For development of practice, the contextual conditions (culture, national, institutional) create even more dilemmas, as the research results on efficiency of educational practice cannot be transferred directly. The element of cultural practice might be more dominant in a student-centred learning model compared to a more instructional model based on textbook approaches, as the learners are more actively involved.

Figure 18.1 intends to illustrate the balancing act of the boundaries in the middle layers and for the UNESCO Chair's Ph.D. training, the goal is to enter the hidden squares at the rear in the figure. Student-centred learning will not work in a hierarchical learning environment where the academic staffs decide on all aspects. Student-centred learning is about empowering the students to be able to make their own choices, to analyse contexts and carry out transparent judgements.

However, Fig. 18.1 is a far too simplistic picture of the wealth of boundaries that change agents will have to deal with. For example, culture is not only a question of

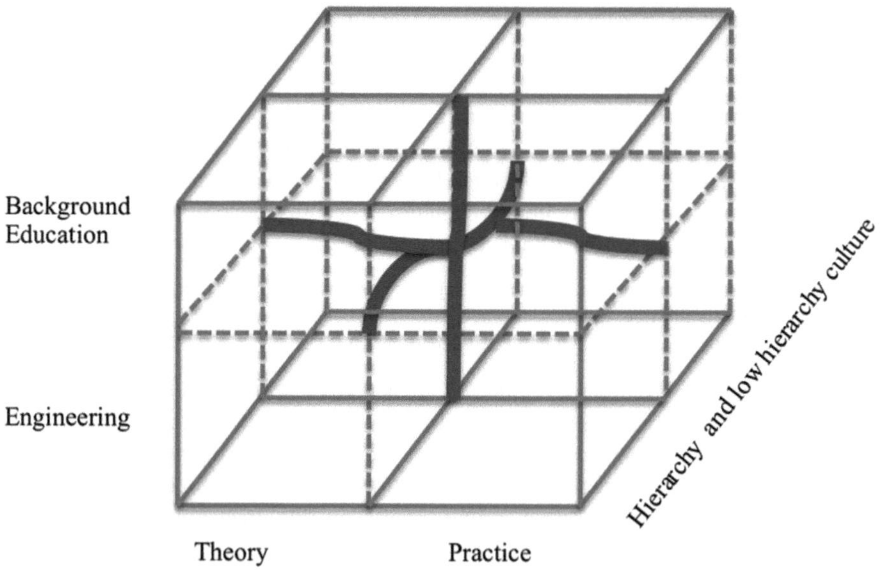

Fig. 18.1 Three integrated global educational innovation challenges

national culture and language – but indeed also a question about institutional and academic culture. Student-centred learning pedagogy belongs to a new mode of universities that either addresses the need for academic competence development or the need for community oriented learning.

Openness is needed when dealing not only with crossing the disciplines, theories and practice boundaries, but indeed also the cultural boundaries where the practices that are to be implemented are far from the tradition in a given educational culture. Therefore, we deal with challenges and dilemmas that can be very complex and where the reaction from the academic community can be harsh in defending existing boundaries.

Design-Based Research

The UNESCO Chair has applied a new methodology by the following reasoning: if we want to establish new educational practice in another culture and at the same time use this for data collection and a core part of a Ph.D. project, we need to be very conscious of the criteria for designing a new practice and for how to research the effect of the new practice.

We choose design-based research (DBR), which is a relatively new research methodology that has emerged as a response to the mentioned challenges. The community that has discussed and developed this methodology is primarily educational researchers of which many are working on ICT and learning. The methodology has been developed in Western countries and has not been used for cross-cultural projects.

During the last 15 years, a growing number of researchers report on DBR both in terms of methodological considerations and in particular on results (Anderson and Shattuck 2012). In the literature DBR is defined as both a systematic and flexible research framework aiming to understand the messiness of educational practice and trying to enhance learning by iterative analyses, designs, developments and implementation. Barab and Squire (2004) point out that:

> Design-based research is not so much an approach as it is a series of approaches, with the intent of producing new theories, artifacts, and practices that account for and potentially impact learning and teaching in naturalistic settings. (p. 2)

Intentionally, DBR was developed to (1) address learning theories, (2) to study learning in context, (3) to develop measures of learning, and (4) to contribute to new designs and learning theories (Reimann 2011). DBR tries to reach out to both development of theories and practices at the same time and this research and development process is based on collaboration among researchers and practitioners in real world settings where they interact as partners (Wang and Hannafin 2005; Barab and Squire 2004). Summing up the key features of DBR would result in the following list of characteristics:

- Deriving from and developing learning theories simultaneous with the development of innovative practices.
- Situated in real world context and problems as the starting point for the research process.
- Collaborative partnership between researcher and practitioner.
- Focus on the design involving comprehensive considerations for the construction of the design both in terms of previous experiences and theories. This involves a starting point in literature review, theoretical and methodological considerations as a basis for the design of experiments that can be implemented in a new educational and cultural context.
- Multiple iterations in the implementation phase and further development of the educational practice to overcome the boundaries that might exist between theory and practice. These iterations are based on reflection and immediate adjustments as well as results from analyses of data collections.
- Utilizing formative evaluation.
- Focus on the effect of designs by collection of data during the process by using mixed methods approach.
- Multiple iterations in the development of the research design.
- Contribution to the development of instruments for data collection and models for changes in practice.
- Development of new design principles.
- Contribution to theoretical development.

DBR shares a lot of similarities with action research and especially the pragmatic research paradigm. A comparison between DBR and action research on a more specific level is complicated as there are many different "schools" within the heading action research ranging from action research in general to participatory action research, critical action research and action learning. One could claim that in higher education practitioners have "hijacked" action research in some communities and especially action research has been used as a strategy for academic staff to experiment and improve their practice. Anderson and Shattuck (2012) point at two important differences, partly the theoretical contribution and partly the partnership between researchers and teachers. The fact that the theoretical contribution is highlighted in DBR may be one of the really core differences. However, in reality action research and DBR will be and should be combined in many variations depending on the contextual factors. Andriessen (2007) combines action research and DBR by using action research methods for the implementation phase containing the iterations and reflections and the data collection and findings, whereas DBR adds the theoretical component and the design development.

In the following, I will explain how we have used DBR at the UNESCO Chair and the experiences from applying the methodology in three Asian institutions in India, Malaysia and Thailand.

Important Phases in DBR

In the DBR literature, there are numerous number of phase definitions going from three to six phases. Reimann (2011) describes three phases for DBR: (1) preparation of the experiment, (2) the experiment phase and (3) finally the phase of retrospective analyses. Collins et al. (2004) define six phases: (1) implementing a design, (2) modifying a design, (3) multiple ways of analyzing the design, (4) measuring dependent variables, (5) measuring independent variables, and (6) reporting on design research. We found that it was beneficial to work with four phases and an extra component that plays a central role during all the phases: supervision and peer discussions. In our definition of the phases, we found that the design phase with preparation of various elements was extremely important to create awareness of what and how we change. Furthermore, we have worked with the implementation and the data collection phase simultaneously, but again to create awareness of practice on the one side and research on the other side, as it has been important to separate the phases.

Table 18.1 illustrates these four phases: design, implementation, data collection, and findings. For each of the phases, several important elements have been identified and in the following I will present core issues that we have had to deal with.

In each phase there are issues that are difficult to handle and these issues are contextual and depend on local conditions. So the list of issues in Table 18.1 can look very different in other contexts.

Table 18.1 Issues in DBR phases

Phases	Elements
1. Design	Research questions: Initial problem analysis and educational background
	Theory: Literature review and theories of learning and curriculum
	Practice: Experiences with new practices and empirical studies
	Local context: Collaboration with partner and identify constraints and needs of the local context
2. Implementation	Re-design phase
	Collaboration with partners and home institution
	Daily iterations and adjustments
3. Data collection and analysis	Research design with mixed methods
	Amount of data
	Analysis
4. Findings and conclusions	Empirical findings
	Theoretical findings
	New designs
	Organizational development

Design

The development of design experiments and design studies is well rooted in early learning theories and at certain times influenced by clinical studies. DBR is very different from clinical experiments and designs as here we address design of practice (Confrey 2006). The design phase encompasses several sub-phases: research questions, theories, experiences from innovative practices and local context (see Fig. 18.2). We have chosen to cluster these elements in the design phase as they all interact and influence each other in the first period of the study. In the process the elements are intertwined and mixed and influence each other and even in the design phase there are several iterative processes and plenty of reflection loops going on.

To design a new practice seems to be an easy act, but educational change is a process loaded with constrains and drawbacks as educational change really has to be understood as a cultural change involving new approaches and perceptions to learning, education, students and practice. Even more challenging is the cultural element.

Research Questions: Initial Problem Analysis and Educational Background

The first element is when the Ph.D. students arrive from a different culture to a Danish culture and particularly to Aalborg University. They have heard about the Aalborg PBL model and they arrive with a clear expectation from their institution that they are going to implement PBL at their home institution in one way or another. This brings

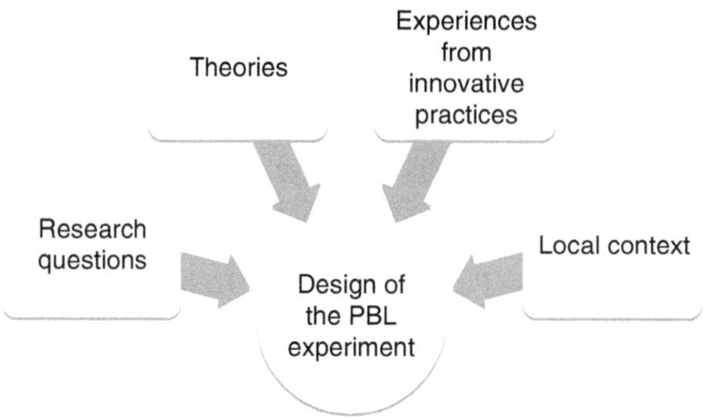

Fig. 18.2 Elements in the design phase of DBR

us to the initial problem and research question: How can PBL be implemented in a Thai/Indian/Malaysian context? This "how question" is very much directed towards a toolbox of what kind of measures should be used in the implementation of PBL.

Educational background is also a variable in this first phase. Ph.D. students with an educational background in education or humanities are more into learning theories than Ph.D. students with an engineering degree. Educational background gives either a preference to the educational literature or a contextual understanding and the specific course plans will have to be adapted accordingly.

In the UCPBL program, a special course plan is set up for each Ph.D. student as a result of negotiation between the supervisor and Ph.D. student. The literature review will shed light on the relevance and the scope of the initial research questions – and along the entire project the research questions will develop as in all Ph.D. studies. But the DBR approach requires focus and overview from the very beginning, as the research question will guide all the phases. Thus, special training on this is provided in the interaction between state-of-art knowledge and the formulation of research questions. As a result of this type of training, the character of the questions shift from the "how questions" to "what kind of impact questions" in a given context.

In general, the experiences from these three DBR cases can be used as general considerations. There is a constant interaction between practice and theory or between the "how to do questions" and the "what kind of impact questions" and there is a need for focus on the research questions from the very beginning. Joseph (2004) emphasizes that the formulation of research questions is a continuous activity during the entire research process.

Theory: Literature Review and Theories of Learning and Curriculum

The second element is to carry out a literature review on PBL and especially on why practice PBL and the impact of PBL. PBL might be a very efficient learning methodology in some cases – but not in all cases. Another difficulty is that the PBL concept has become very broad and covers a wide range of student-centred practices from real participatory and collaborative knowledge processes to more task-oriented group work. Using DBR in non-PBL contexts would mean that the students would need to focus their literature review on other student centred learning methodologies. The research results will have to be interpreted in relation to the practice as well as the theoretical scientific framework for the study. Thus it is important to develop a critical stance to what is PBL and why, what and how does various forms of PBL work. Furthermore, the PBL literature has increased enormously during the last 10 years, and therefore selection criteria are important.

The theories that these DBR projects contribute to, will be the learning theories and the curriculum theories combined with a cultural dimension. What kind of implications will emerge when a more participatory learning approach is practiced in a traditional and very hierarchical culture? Basically, the DBR experiments will prepare for a lack of alignment between the students' pre-conceptual understanding and the new PBL curriculum – and it is the implications of the emergence of new cultural practices that is the new curriculum contribution.

The challenge in the DBR process is to balance the amount of theoretical reading. This is a challenge in all empirical studies, but for DBR this is an even more complex issue as the empirical part also contains a design of a new practice. In general, the linking between theories, research questions and practice is a huge challenge in the DBR research approach. One way of handling this is to create parallel thinking by phrasing the research questions from a theoretical angle and a practice angle at the same time and by moving backwards and forwards between the two perspectives to include theories into the design of practice. Awareness on how to deal with theory and practice is needed to strengthen the theoretical component in the design phase and this might be one of the areas that needs much more development in the DBR community. The theories can be used as input for research questions, design, and a lens into the analyses of data and a reference in the summing up of results (Joseph 2004), but the data should also contribute to a further theoretical understanding of the learning taking place in practice.

Practice: Experiences with New Practices and Empirical Studies

In order to develop a new unknown practice, the Ph.D. student has different possibilities of imagining curriculum practices from theories or getting ideas from practice. We have done both – reflecting on implications of different theories, but indeed what has been most beneficial are the experiences with and studies on the PBL curriculum at Aalborg University. For this purpose, we have used case-study methodology. These smaller case studies partly provides initial research training, partly a very close interaction with practice by observing what students are doing. At the UNESCO Chair we have found these very close interactions with new types of practices very valuable as it creates a concrete mind-set of how a new curriculum practice can be effectuated.

Local Context: Collaboration with Partners and Identification of Constraints and Needs of the Local Context

The local context is the last major element in the design phase. The national and institutional policy level is the first step – how much freedom is there in the curriculum, e.g., in India the project was carried out at an institution that had licensed their curriculum from a nearby university and all exams have to follow a certain pattern. From the beginning of the Ph.D. study, there has been contact with the home institution. Contact with the home institution and identification and collaboration with the local partner (practitioner) has been challenging in all cases. The challenging part is that organizations are not stable. When a rector/dean has decided to send one of their staff members to Aalborg University for a Ph.D. on PBL, it is actually a desire to educate a change agent. But in most cases, it gets difficult, e.g., change of managers, explaining to the partners that are going to run the classes what PBL is, managing PBL training, and maintaining motivation, etc.

In the pre-design phase collaboration with local partners was sporadic as it was difficult to maintain contact and therefore, the pre-design would undergo a phase of re-design of the entire process when the Ph.D. student returned to the home institution for implementation. In some cases, the Ph.D. student even had to take over teaching the course and therefore played double roles as a practicing teacher and a reflective researcher. We have had to compromise along the way and have done that by reflecting on the consequences. But an important learning is that the pre-design needs to be loosely structured, as a lot of re-design will take place.

Implementation

The implementation phase consists of a long series of adjustments and negotiations with both the students and the colleagues.

Re-design Phase

When the Ph.D. students return home, new issues emerge. The local partner often has other ideas, the top management changed since the start of the project, requirements for teaching new courses emerged, and a lack of resources or other types of issues materialize. In the three cases that we have been involved in, nothing has turned out exactly as it was planned in the pre-design phase and a re-design has been necessary. In the re-design DBR can include the contextual elements although there is a tendency that practice and pragmatic choices will determine. The Ph.D. students are under pressure to come up with a design that can work and the practicalities lead the work. This is unavoidable and it is important that learning from this phase might be very hard to highlight the theoretical considerations and therefore, the pre-design should clearly contain more theoretical learning, otherwise this will disappear.

Collaboration with Partners and Home Institution

The re-design is done in collaboration with local partners. In two out of the three cases, there have been established faculty development courses on PBL in order to train the faculty. As advisor to the Ph.D. students, I visited all three institutions after the Ph.D. students returned for implementation and in some cases contributed to the faculty training.

Visiting the institutions was necessary in order to understand the local conditions and especially to give advice in the re-design phase. Furthermore, it was possible to have meetings with vice-chancellors, deans and other responsible academic staff members. These meetings were beneficial for all involved partners – both the local managers, the Ph.D. students and the advisor. Furthermore, it was decided to look for a local advisor that could join the team.

Daily Iterations of Adjustment

In the implementation phase there will be daily iterations of adjustment. Perhaps the instruction to each PBL case or the PBL project was not explicit enough – maybe the students needed a lecture on project management and writing reports. These adjustments are part of a reflective practice and in this case in the DBR process and the negotiation with the partners. The learning from this element is that it is hard to write down all these iterations as there are so many, but on the other side it is quite important to have created awareness of how the practice differed from the pre-design and re-design phases.

Data Collection and Analysis

This phase addresses the research design and research methods. Along with the design of the experiment, we have discussed research design for collection of data and strategies for data analysis. A research design can be considered as part of the preparation and indeed of the entire design, however, we have worked with this phase as an independent element in the DBR approach.

Research Design and Mixed Methods

Common practice within the DBR community is to use mixed methods design, which is in line with a more pragmatic world view (Cresswell 2009). The scientific discourse in the literature reveals that DBR argues with quantitative and positivistic experimentation (Barab and Squire 2004; Dede 2005) more than social constructivism. But both worldviews are needed. The social constructivism for conceptualization and understanding of new practices and positivism for large-scale studies and identification and evaluating the effect of diverse variables (Dede 2005). But it should be clearly indicated that to evaluate the effect of a design would be complicated as it would not be possible to isolate variables. Therefore, mixed methods are needed in the research design, which is in line with the pragmatic worldview based on Dewey and pragmatism (Barab 2006; Biesta and Burbules 2003). DBR has a pluralistic scientific approach both in terms of theories and methodologies and emphasizes a more coherent methodology linking theory, methods and practice (Bell 2004; Hoadley 2004).

Table 18.2 presents the data collection methods for the three Ph.D. studies using mix methods approach. In each Ph.D. study, the amount of data collection methods increased during the implementation phase as possibilities emerged. From the beginning none of the projects have thought about the project reports or videotaping project presentations. So in most cases, new methods were added during this phase.

Methodological triangulation has been a core criterion for choosing methods to capture uniqueness in learning. The collection of data can be experienced overwhelming when it is happening simultaneously with all the daily iterations of implementation loops and we learned that more focused data collection could be needed to avoid the amount of data.

Table 18.2 Methods for data collection in the three Ph.D. studies

	India	Malaysia	Thailand
Surveys on effect of faculty development activities	X		X
Student pre-questionnaires	X	X	X
Student questionnaires	X	X	X
Student group interviews	X	X	
Student individual interviews		X	
Student observations	X	X	X
Student log-books		X	
Project reports	X	X	X
Grades	X	X	X
Presentations	X	X	

Data Analysis

In the literature, the high amount of data is pointed out as a weakness in DBR. The real challenge that emerged in the process has been the amount of data using multiple methods and trying to examine it within a methodological frame such as grounded theory, discourse analysis, etc. This is a challenge in the analysis of the effect of the design and will involve research questions that are formulated for a feasible study (Dede 2004). This has also been the case in the three Ph.D. studies – the wealth of data has been enormous and it has been difficult not to collect the data when opportunities were there.

Findings

The three Ph.D. projects are still in the data analysis phase and the findings are about to be published and the results of these processes are developing into four different dimensions.

New Curriculum Designs

In all three cases, totally new designs for PBL in an Asian context are developed and implemented. The fact that it is possible to practice this design and to get positive results will be a source of inspiration for other Asian institutions.

Empirical Findings

The empirical findings in all three cases indicate that the students' learning outcomes as well as the satisfaction with the learning process are improved. In India two designs have been developed and the initial results indicate that students improved their learning outcomes and were positive with their learning experience. In the Malaysian and Thai cases, there are similar results showing significant improvement on a series of variables: motivation, collaboration and communication skills, problem-solving skills and critical thinking.

Organizational Development

What is not an obvious result is the impact on the institutional level and if the institutions will use the results to proceed with change. There is no doubt that for all three institutions, there are positive attitudes towards change to more student-centred and innovative-learning methods; however, as management in some cases change, the landscape and directions might change.

For the Indian institution, there will be changes at a more comprehensive level (across disciplines) at a Master programme level. At the Malaysian institution, the DBR project has boosted the documentation for new PBL practices and might feed into long- term planning. At the Thai institution, there is a plan for institutional change. However, this only represents the period of study and may change in 3 months' time. At least what can be concluded is that the three cases contribute to institutional change.

Theoretical Findings

The last finding to mention here is the contribution to theory which is also the most complex finding as learning and curriculum theories in general lack contextual and cultural sensitivity. However, this is really where the three DBR projects add to the theoretical landscape as they document it is possible to practice a new curriculum design in more hierarchical-oriented cultures and even get positive results concerning learning outcomes such as new competences and comprehensive learning.

These findings also raise issues if the global media culture reaching out to young people almost all over the world will be the link in the missing cultural alignment between hierarchical school systems on the one side and student-centred learning on the other side. In other words, the misalignment is between Asian university tradition and the student-centred learning approach and not between Asian students and student-centred learning. However, the conclusion concerning this point in the three Ph.D. projects is still pending.

Perspectives

I started out by describing some of the challenges in educating change agents across cultures and how DBR has been chosen as a research strategy to overcome these challenges. It is challenging as the DBR process requires rigorous research, understanding of the relation between research and practice, courage to design and implement new learning practices and courage to go through with the implementation despite resistance from students, occasionally from colleagues and sometimes lack of support from the organization. The Ph.D. students have been trained in boundary work – both cultural boundaries, and academic boundaries between disciplines and theory/practice. For the UNESCO Chair the academic training of change agents that can do this boundary work is essential.

In the three cases discussed in this chapter, PBL has determined the curriculum change. It is crucial to mention that PBL should never be an end goal in itself. PBL is a means to change engineering education in a direction of more participatory learning processes, which in combination with social and sustainable contexts can educate engineers with a higher degree of complex understanding and competences.

In the same way, DBR is not an end goal in itself. It is a means for establishing change processes in engineering education based on research and documentation.

The three cases indicate overall success in all parts, although not without pitfalls in the process. One of the main pitfalls derives from the theory-practice relation with the risk of becoming more practice oriented and less theoretical focused. The educational background adds to this issue, as the Ph.D. students with engineering background often will be more oriented towards practical solutions. Furthermore, the expectations from the home institutions might move in the direction of practice as it is expected to try out PBL implementations. So the theoretical contribution is in risk unless this is explicitly addressed. In the DBR community this is not an explicit concern, instead it is discussed as defining the relevance of the design to the practitioners (Dede 2004). Maybe the potential dominance of practice in the referred cases are due to cultural boundaries, as Western inspired designs cannot be implemented directly in an Asian practice. Iterations and adjustments are constant variables in the process.

But change takes place – maybe at a minor level but as seeds for further inspiration – and is well documented. This change of practice has been directed by a researcher that has been to the "engine room" or curriculum practice and at a personal level has experienced both the scientific approach and the often-muddy practice. As an added value, this personal growth should not be underestimated, as confidence is very important in leading any process at universities or for enhancement of character. And it provides valuable research-driven development.

As advisor/supervisor or facilitator of this process, it requires more than intellectual support and dialogue. It also involves visits to the engine room, personal support, practical response as well as scientific guidance. Visiting the local environment is a must in this process for support and organizational matters, but also for understanding the conditions under which university educations are run in different cultures. And it is an eye-opener of the different cultural perspectives that have to permeate PBL models in different cultures. Even if there might be a trend among young people towards more openness and democratic participation influenced by the Internet, learning models have to be adjusted to context and culture. For all these purposes, DBR is a methodology that needs development but also has the potential to link theory and practice or research and change.

Acknowledgement Thanks to Andrew Jamison for encouraging me to write this contribution and to Mohamad Termizi Borhan, Prarthana Coffin and Vikas Shinde for their contributions and collaboration during the DBR process.

References

Anderson, T., & Shattuck, J. (2012). Design-based research: A decade of progress in education research. *Educational Researcher, 41*(1), 16–25.

Andriessen, D. (2007, August 22–24). *Combining design-based research and action research to test management solutions*, the 7th World Congress Action Learning and Process Management, Groningen.

Balogun, J., Gleadle, P., Hailey, V. H., & Willmott, H. (2005). Managing change across boundaries: Boundary-shaking practices. *British Journal of Management, 16*, 261–278.

Barab, S. (2006). Design-based research – a methodological toolkit for the learning scientist. In R. K. Sawyer (Ed.), *The Cambridge handbook of the learning sciences*. Cambridge: Cambridge University Press.

Barab, S., & Squire, K. (2004). Design-based research: Putting a stake in the ground. *The Journal of the Learning Sciences, 13*(1), 1–14.

Barnett, R. (2011). *Being a university*. London: Routledge.

Barnett, R., & Coate, K. (2004). *Engaging the curriculum in higher education* (1st ed.). Maidenhead/New York: Open University Press.

Beddoes, K. D., Jesiek, B. K., & Borrego, M. (2010). Identifying opportunities for collaborations in international engineering education research on problem- and project-based learning. *Interdisciplinary Journal of Problem-Based Learning, 4*(2), 7–34.

Bell, P. (2004). On the theoretical breadth of design-based research in education. *Educational Psychologist, 39*(4), 243–253.

Biesta, G. (2009). Building bridges or building people? One the role of engineering in education. *Journal of Curriculum Studies, 41*(1), 13–16.

Biesta, G. J. J., & Burbules, N. C. (2003). *Pragmatism and educational research*. Oxford: Rowman and Littlefield Publishers.

Blom, A., & Hiroshi, S. (2010). *Employability and skill set of newly graduated engineers in India*. Policy research working paper 5640, World Bank. Available at: http://www-wds.worldbank.org/servlet

Borhan, M. T, Shinde, V., & Coffin, P. (2012). *Addressing the contextual elements in designing PBL curriculum: Lessons learned from three Asian universities*, presented at the ICED Conference in Bangkok.

Chinowsky, P. (2011). Engineering project organization: Defining a line of inquiry and a path forward. *The Engineering Project Organization Journal, 27*(3), 170–178.

Collins, A., Joseph, D., & Bielaczyc, K. (2004). Design research: Theoretical and methodological issues. *The Journal of the Learning Sciences, 13*(1), 15–42.

Confrey, J. (2006). The evolution of design studies as methodology. In K. R. Sawyer (Ed.), *The Cambridge handbook of the learning sciences*. Cambridge: Cambridge University Press.

Cresswell, J. (2009). *Research design*. Thousand Oaks: Sage.

De Graaff, E., & Kolmos, A. (2007). *Management of change implementation of problem-based and project-based learning in engineering*. Rotterdam: Sense Publishers.

Dede, C. (2004). If design-based research is the answer, what is the question? *The Journal of the Learning Sciences, 13*(1), 105–114.

Dede, C. (2005). Why design-based research is both important and difficult. *Educational Technology, 45*(1), 5–8.

Dewey, J. (1916/1998). Aims in education. In L. A. Hickman & T. M. Alexander (Eds.), *The essential Dewey, Volume 1. Pragmatism, education, democracy*. Bloomington: Indiana University Press.

Dewey, J. (1922/1998). Education as engineering. In L. A. Hickman & T. M. Alexander (Eds.), *The essential Dewey, Volume 1. Pragmatism, education, democracy*. Bloomington: Indiana University Press.

EUROPA. (2009). Communiqué of the conference of European ministers responsible for higher education, Leuven and Louvain-la-Neuve, 28–29 April 2009. Retrieved from http://europa.eu/rapid/pressReleasesAction.do?reference=IP/09/675&format=HTML&aged=0&language=EN

European Commission. (2013). *Europe 2020*. Retrieved from http://ec.europa.eu/europe2020/index_en.htm

Gibbons, M., Limoges, C., Nowotny, H., Schwartzman, S., Scott, P., & Trow, M. (1994). *The new production of knowledge – The dynamics of science and research in contemporary societies*. London: Sage.

Graham, R. (2009). *U.K. approaches to engineering project-based learning*, white paper. Retrieved March 29, 2012 from http://web.mit.edu/gordonelp/ukpjblwhitepaper.pdf

Graham, R. (2012). *Achieving excellence in engineering education: The ingredients of successful change*. London: The Royal Academy of Engineering.
Hoadley, C. M. (2004). Methodological alignment in design-based research. *Educational Psychologist, 39*(4), 203–212.
Jamison, A. (2013). *The making of green engineers. Sustainable development and the hybrid imagination*. San Rafael: Morgan & Claypool.
Jamison, A., Kolmos, A., & Holgaard, J. (2014). Hybrid learning: An integrative approach to engineering education. *Journal of Engineering Education, 103*, 2.
Jesiek, B. K., Newswander, L. K., & Borrego, M. (2009). Engineering education research: Field, community, or discipline? *Journal of Engineering Education, 98*(1), 39–52.
Joseph, D. (2004). The practice of design-based research: Uncovering the interplay between design, research, and the real-world context. *Educational Psychologist, 39*(4), 235–242.
Kolmos, A., & De Graaff, E. (2013). *Problem and project based learning – Merging models in handbook for engineering education*. Cambridge: Cambridge Publishers.
Kolmos, A., & Holgaard, J. (2010). Responses to problem based and project organised learning from industry. *International Journal of Engineering Education, 26*(3), 573–583.
National Academy of Engineering. (2004). *The engineer of 2020 – visions of engineering in the new century*. Washington, DC: National Academies Press.
Prince, M. J., & Felder, R. M. (2006). Inductive teaching and learning methods: Definitions, comparisons, and research bases. *Journal of Engineering Education, 95*(2), 123–138.
Reimann, P. (2011). Design-based research. In L. Markauskaite, P. Freebody, & J. Irwin (Eds.), *Methodological choice and design: Scholarship, policy and practice in social and educational research* (pp. 37–50). New York: Springer.
Rocard, M., Csermely, P., Jorde, D., Lenzen, D., Walberg-Henriksson, H., & Hemmo Valerie. (2007). *Science education now: A renewed pedagogy for the future of Europe*. European Commission, Directorate-General for Research, Directorate L: Science, Economy and Society, Brussels.
Royal Academy of Engineering. (2007). *Educating engineers for 21st century*. Retrieved March 29, 2012 from http://www.raeng.org.uk/news/release/pdf/Educating_Engineers.pdf
Schmidt, H. G., & Moust, J. H. C. (2000). Factors affecting small-group tutorial learning: A review of research. In D. H. Evensen & C. E. Hmelo (Eds.), *Problem-based learning: A research perspective on learning interactions*. Mahwah: Lawrence Erlbaum Associates.
Shinde, V. (2011). Relevance of the problem and project based learning (PBL) to the Indian engineering education. In J. Davies, E. de Graaff, & A. Kolmos (Eds.), *PBL across the disciplines: Research into best practice* (pp. 489–502). Aalborg: Aalborg Universitetsforlag.
UNESCO Chair in Problem Based Learning in Engineering Education. (2013). Retrieved from http://www.ucpbl.net
Wang, F., & Hannafin, M. J. (2005). Design-based research and technology-enhanced learning environments. *Educational Technology Research and Development, 53*(4), 5–23.

Anette Kolmos Professor of Engineering Education and PBL and Chairholder for UNESCO in Problem Based Learning, Aalborg University, Denmark. She was also president of SEFI 2009–2011 (European Society for Engineering Education). Founding Chair of the SEFI-working group on Engineering Education Research. During the last 20 years, Dr. Kolmos has researched the following areas, primarily within Engineering Education: development and evaluation of project based and problem based curriculum, change from traditional to project organized and problem based curriculum, development of transferable skills in PBL and project work, and methods for staff development. She is Associate Editor for the *European Journal of Engineering Education* and was Associated Editor for *Journal of Engineering Education* (ASEE). A member of several organizations and committees within EER, national government bodies, and committees in the EU.

Chapter 19
Engineering Education Research as Engineering Research

Jonte Bernhard

Abstract Engineering Education Research (EER) has recently emerged as a field of research worldwide. In this context one could focus on the conceptual difficulties experienced by engineers learning about educational research. However, in this chapter I explore the *contributions* that engineering and engineers can make to education research, based on the view, drawn from John Dewey's essay "Education as engineering", that EER could be regarded as engineering research. My first point is that engineers have learned to handle both general aspects (in the case of bridge building: engineering mathematics, solid mechanics, materials science, geology etc.) and particular aspects (the local situation of particular bridges) of their profession. Hence, it is not possible in engineering to simply apply knowledge from science to practice and Dewey points out that this also applies to education. My second point is that engineers are trained to acquire proficiency in design and both understanding and improving complex systems. Similarly, in "design-based research" or "design experiments" in education, insights from design and engineering are employed to address the complexity of educational activities and the need, as known from engineering, for theory as well as tinkering. My third point is related to the role of technologies in promoting engineering students' learning in, for example, laboratories. Diverse technologies (artifacts) are crucial in engineering for collecting and processing data from experiments and/or real environments for numerous applications, for example controlling and monitoring production processes and monitoring the environment. Thus, engineers have high proficiency in the use of technologies and materiality, strong awareness of their impact on human perception, and hence can make valuable contributions to their application in educational contexts.

Keywords Engineering education research (EER) • Engineering research • Dewey • Bildung • Didaktik

---

J. Bernhard (✉)
Engineering Education Research Group, Department of Science and Technology (ITN), Linköping University, Campus Norrköping, Norrköping 601 74, Sweden
e-mail: jonte.bernhard@liu.se

S.H. Christensen et al. (eds.), *International Perspectives on Engineering Education*, Philosophy of Engineering and Technology 20, DOI 10.1007/978-3-319-16169-3_19

Introduction

> When I first came across Dewey's wonderful piece "Education as engineering" … it proved to be very helpful in my day-by-day work at the Norwegian University of Science and Technology (NTNU) at Trondheim. Most of my partners in projects and university affairs came from engineering; indeed, one of my closest colleagues was a bridge-builder in the original sense of the word, having designed and built bridges across Norway and elsewhere.
>
> He often complained about the misconceptions of engineering that many of my social science colleagues seemed to have. "Of course you've got to do the math right", he said: "but when it finally comes to building a bridge, you've got to understand the uniqueness of the site you are approaching. This requires a deep understanding of what this bridge will be, how it will fit into the landscape and to the needs of its customers, an understanding, which we can't teach at universities, which only can be acquired by doing bridges".
>
> For him the argument in Dewey's short essay about the nature of knowledge in engineering and about the short-comings of an educational theory not firmly rooted in preceding practical improvement seemed to be a perfect fit – and to confirm his everyday experience of my education colleagues at NTNU ("all theory, no practice", he would say) (Hopmann 2009, p. 7).

Engineering education research (EER) has recently emerged as an important field of research within schools of engineering and at technical universities worldwide (Baillie and Bernhard 2009; Borrego and Bernhard 2011). There are three fundamental components of the term *Engineering Education Research*, namely *Engineering*, *Education* and *Research*. This chapter discusses some aspects of the relationships between these *three* terms and combinations of them, such as *Engineering Research*, *Education Research*, *Engineering Education* and finally *Engineering Education Research* (EER).

Some researchers in EER have (like myself) a background in engineering and have conducted "pure" engineering research. Conducting research in engineering education is, of course, not *exactly* the same as pure engineering research. Maura Borrego (2007, p. 99, my italics) even claims that "Engineering education [research] is just beginning to emerge as a discipline", and makes the strong claim that "its research is *fundamentally different* from engineering research". By contrast, in the wonderful essay, "Education as engineering" John Dewey (1983) called for the development of an "art of educational engineering" and in France the term "didaktik engineering" [ingénierie didactique] is used to denote a certain approach to educational research (e.g., Artigue 1988). Furthermore, in a discussion of "design research in education" Anthony Kelly (2004) claimed that "design studies [in education research] usually involve *engineering* a broader 'learning environment'". Thus, despite the claim that EER is fundamentally different from engineering made by Borrego (2007) several scholars in education research use engineering metaphors to describe their approaches and/or thinking (Artigue 1988; Dewey 1983, 1984; Hopmann 2009; Kelly 2004). I propose that these apparently conflicting views can be traced back to different views on the nature of engineering. However, these issues are not deeply addressed here. Instead, I explore the *contributions* that engineering and engineers can make to education research, focusing on similarities between engineering and EER, taking Dewey's essays "Education as engineering" and "The sources of a science of education", together with my experience as an engineer, as points of departure. Thus, I focus on similarities rather than differences between engineering and engineering education research.

The field of EER is defined in various ways around the world, as discussed by Borrego and Bernhard (2011). In accordance with the central and northern European *Didaktik*[1]- and *Bildung*-traditions (e.g. Comenius 1657; Hopmann and Riquarts 1995b; Westbury et al. 2000; Christensen et al. 2006) the field as a subject in doctoral training at Linköping University, Sweden, is defined as follows:

> The subject of engineering education [Ingenjörsvetenskapens didaktik] deals with learning, teaching and the formation of knowledge in the art and science of engineering[2] in a broad sense. In focus stand fields of knowledge of relevance for the practice of, and education for, the engineering profession and its relation to the advance of knowledge in techniques and technology.[3] This leads to a special interest in students' and practicing engineers' acquisition of and further growth of knowledge of techniques, technology and about different fields of technology, selection of content in and the arrangement of education, knowledge about and insights in the relation between techniques, technology, the evolution of technology and changes in society and the development of the ability of solving problems with the help of technology (Linköping University 2004).

According to this definition, EER should address not only *how* a given topic is best taught or a learning environment best designed, but also *what* should be taught and *why* certain topics should be included or excluded. These *what* and *why* questions are core elements of the didaktik-tradition (Hopmann and Riquarts 1995a, p. 26), indeed Künzli (2000, p. 46, italics in original) holds that the "fundamental question of Didaktik is *Why is the student to learn the material in the first place?*" Borrego and Bernhard (2011, p. 32) claim that "professional appropriateness (e.g., complex, authentic problems similar to real engineering problems) is emphasized [in the didaktik tradition]". Furthermore, they note that "*Bildung* is a Northern and Central European concept that extends beyond knowledge and skills – it is a forming of the individual as a person and as a professional

---

[1] The spelling, "didaktik" is deliberately used to distinguish the European "didaktik-tradition" from the English term "didactics," which has a different meaning.

[2] The original Swedish word "ingenjörsvetenskap" (cf. German "Ingenieurwissenschaften") has been translated as "the art and science of engineering". The Swedish word "vetenskap" (and, for example, the German word "Wissenschaft") for "science" does not have the same restricted connotation as "science" in modern English usage, i.e. (natural) science using *the* (positivist) scientific method. In Swedish (and in German) the original, broader, usage of "science" is retained, meaning any body of systematic knowledge such as, for example, engineering, geography or history (cf. Layton 1976, pp. 689–690). This double meaning can lead to confusion when reading texts since the term "engineering science" is used in its broader meaning in parallel with its usage in, for example, German and Swedish as well as to denote, as **it** sometimes is in the United States, a special approach to engineering as an applied (positivist) science using *the* scientific method.

[3] The original Swedish terminology used here reflects Alfred Espinas' distinction between "*techniques* (skills in some particular activity), *technologie* (systematic organization of some technique), and *Technologie* (generalized principles of action that apply in many cases)." (See, for example, Mitcham 1994, p. 33).

[and] no objectives are seen as worth teaching or learning that are not also seen as directly, or indirectly, contributing to Bildung" (Borrego and Bernhard 2011, p. 32). Rudolph Künzli (2000, p. 45) states: "The concept of Bildung has proved a stable source of orientation for this approach. Bildung is not determined by separate academic disciplines, but by life as a whole and the individual's share in this whole." Lars Henriksen (2006, p. 55) relates the concepts of Bildung to professional appropriateness by arguing "Bildung is an essential part of the engineering profession, and engineering is much more than the application of science, scientific facts, and scientific methods".

Didaktik Analysis and Knowledge of Engineering

> This form of opening up, of rendering the learners open to contents and values, can be achieved only by what we call contents of education because they have a particular characteristic: *They are always specific contents, are examples that represent a larger set of cultural contents*. A content of education must always make fundamental problems, fundamental relations, fundamental opportunities, general principles, laws, values, and methods understandable. Such elements that effect understanding of the general in or through the medium of the specific are conveyed in the term *educational substance* (*Bildungsgehalt*). Any specific content thus contains general substance (Klafki 2000, p. 150).

Wolfgang Klafki (2000) sees a *didaktik analysis* of content, which should "bring out the *substance* of the *objects of learning*" (Willman 1957, p. 460, cited by Klafki 2000, p. 150, my italics), as the first and most important step in preparation for teaching. Accordingly, the definition of EER at Linköping University (see above) calls for research into "formation of knowledge in the art and science of engineering" and studies of what constitutes "knowledge of relevance for the practice of, and education for, the engineering profession and its relation to the advance of knowledge in techniques and technology". As can be easily seen, the *what*- and *why*-questions are regarded as highly important. Klafki (2000, pp. 150–157) describes the task of a *didaktik analysis* as addressing five general questions:

1. What wider or general sense or reality does this content exemplify and open up to the learner? What basic phenomenon or fundamental principle, what law, criterion, problem, method, technique, or attitude can be grasped by dealing with this content as an "example"?
2. What significance does the content in question, or the experience, knowledge, ability, or skill to be acquired through this topic already possess in the minds of the [students] in my class? What significance should it have from a pedagogical point of view?
3. What constitutes the topic's significance for the [student's] future?
4. How is the content structured (which has been placed in a specifically pedagogical perspective by Questions 1, 2 and 3)?
5. What are the special cases, phenomena, situations, experiments, persons, elements of aesthetic experience, and so forth, in terms of which structure of the content in

question can become interesting, stimulating, approachable, conceivable, or vivid for [students] of the stage of development of this class?

Clearly, profound understanding of engineering practice and engineering research is required to answer these questions in the context of engineering education and EER. Indeed, any researcher should be able to answer these questions or include attempts to answer them in research aims. Hence, in my mind, *didaktik analysis* is at the very core of both engineering practices and engineering research since such questions address the very fundamentals of (rather than something fundamentally different from) the "objects of learning" in engineering education, i.e. what engineering is about. It is important to understand that content is the central unit of analysis in *didaktik analysis* and not, for example, frame factors, social functions or dynamics of classroom settings. However, teaching is not just "conveying content, but it is… also education *by* content, the *Bildungsgehalt*" (Hopmann 2000, p. 198, italics in original), i.e. skills, norms and values are also fostered by education as, for example, argued by Henriksen (2006) in the above quotation.

Furthermore, it is important to understand that it is *not a one-way process*: the subject-matter-specialist (in the case of EER the experienced engineer or engineering researcher) contributes his or her domain-specific knowledge, but a *didaktik analysis* and didaktik research *can also contribute to the development of knowledge in a domain*. An important aspect of the model of *educational reconstruction* [Didaktische Rekonstruktion] (e.g. Kattmann et al. 1997) is the "clarification of science content matter" resulting from analysis of content structure starting with a *didaktik analysis*. This is important because specialists' understanding is often tacit and not clearly expressed. For example, in Physics Education Research David Hestenes (1992, p. 733) found that that the zeroth law "which specifies the primitive kinematic properties of position and motion, thus defining the Newtonian concepts of space and time" is necessary to understand Newtonian mechanics. However, Hestenes (1992, p. 733) notes that "the Zeroth law is tacitly taken for granted in conventional physics textbooks [and] there is no justification for such an omission … since Einstein has shown that the most profound deficiency of Newtonian theory resides in the Zeroth law". In a similar vein, in the context of Chemistry Education Research, Helge Strömdahl (1996) showed that chemistry experts held slightly different views on the concept *amount of substance* and the definition of the *mole*. The scientific community later clarified the definitions of these concepts. Similarly, as an example from EER, González Sampayo (2006) found that teachers at technical universities did not have a unified view on subject-matter such as the role of the Laplace transform in relation to, for example, electric circuit theory. Furthermore, James Trevelyan (2007, 2009, 2013), who has published several empirical studies of engineering practice, reported that the "predominant aspect of engineering practice is informal technical coordination"[4] (Trevelyan 2013) rather than design and technical problem solving, as emphasized in most engineering education

[4] It is important to understand that "technical coordination" is *not* the same as "management"; even newly graduated engineers are heavily involved in informal "technical coordination".

textbooks. Although many practicing engineers express a belief that they do not do any "real engineering", Trevelyan (2013, my italics) maintains that this is "contradicted by observations from the workplace that reveal that *the content* of engineers' social interactions, especially in the workplace, is *dominated by technical issues*". Trevelyan (2009, 2013) and Bill Williams et al. (2013) argue that their findings regarding engineering practice have implications for the theoretical framework of engineering and should prompt changes in engineers' education. According to the doctoral thesis presented by Roger Malmberg (2007), a practicing engineer, engineers who have newly graduated from universities do not apply the knowledge of electric circuit theory they have learned at technical university when they design circuits, because of the gap between what is taught at universities and engineering practice. Furthermore, he argues, this neglect of theory in practice results in inefficient design processes. He concludes "that both the industry and universities need to adapt their work and teaching methods, respectively, for the engineer to get a more unified and effective learning situation, enabling continuous learning throughout the career" (Malmberg 2007, p. i).

As stated above, my intention in this chapter is to explore the *contributions* engineering and engineers can make to education research. The most obvious potential contribution of engineers to EER is knowledge of engineering, as in *didaktik analysis* and *educational reconstruction.* However, although this is very important, by virtue of their experience and training engineers have (or at least should have), specific knowledge and skills that could contribute specifically to education research and development.

As noted in a previous publication, ideas learned during my engineering education include the following, which I suggest are also important to understand in the context of educational research (cf. Dewey 1983; Artigue 1988):

- Humans are part of any technical system.
- Most systems are complex.
- Scientific theories do not provide sufficient knowledge for successful design, but they can be used as a starting point.
- Designing is always a contextualized practice and must address the possibilities and restrictions in the actual context.
- Designing should take account of diverse and sometimes conflicting aims.
- Designing is not neutral, but is a value-laden practice.
- There is no "best" design and many different solutions are possible.
- Adherence to details is critical for successful design.

(Bernhard 2009)

The following three sections of this chapter explain my reasons for believing that these understandings based in engineering are important for educational research, focusing on: The nature of knowledge in engineering in relation to what is needed in (engineering) education research; design and design-based-research in education; and the role of physical artifacts[5] (i.e. technologies) in human perception.

[5] Also commonly spelled artefact.

The Nature of Knowledge in Engineering and Education Research

> The notion of didaktik engineering [ingénierie didactique] emerged in mathematics didactics in the early 1980s. The term was used to denote a form of work in didaktiks, *comparable to that of an engineer*, who *bases efforts to realize* a specific project *on the scientific knowledge* in his field and submits it to a form of scientific testing, but *at the same time* must address *much more complex objects than the ideal objects considered in science* and to *address problems in a practical way*, with the *means at his disposal*, although science does not or cannot yet handle them (Artigue 1988, p. 283, my translation and italics).

Michèle Artigue (1988) and Dewey (1983) use engineering metaphors to describe aspects of education research because knowledge in the subject area is messy, complex and situated. In the words of Artigue (1988), those engaged in education research are "forced to work on much more complex objects than the pure ones in science" and Dewey (1984, p. 9) states that "no conclusion of scientific research can be converted into an immediate rule of educational art". Rather ironically, according to views represented by Borrego (2007), "scientists and engineers are trained to expect that once a fact is proven or discovered, it is universally true" and because knowledge in education research is messy, complex and situated it is "fundamentally different from engineering research".

At play here are, I suggest, different understandings of the nature of engineering knowledge and engineering. Borrego (2007, pp. 91–93) argues that the level of consensus is an important characterization of a disciplinary field. She describes the view in engineering research as paradigmatic: "the theoretical framework ... is often the traditional scientific paradigm based on the scientific method. ... Scientific and engineering theories are so universal, they need never be mentioned among adherents" (Borrego 2007, p. 92). Furthermore, she holds that Anthony Biglan (1973) "found the disciplines with the highest levels of consensus to be the physical sciences and engineering". However, critical reading of the cited work shows that Biglan did not explicitly study the level of consensus, instead he made an assumption. He argues that "Kuhn specifically designates physical and biological sciences as paradigmatic. He *does not* discuss agricultural and engineering areas, but they may also be *considered to be paradigmatic*, since they are *grounded in their related pure fields*" (Biglan 1973, p. 202, my italics).[6]

Clearly, Biglan (1973) sees engineering as an applied science. By contrast, in his essay "The Structure of Thinking in Technology" Henryk Skolimowski (1966, p. 372) maintains that "it is erroneous to consider technology as being an applied science". Sylvain Lavelle (2009, p. 88) has compared the nature of knowledge in science and engineering, summarizing the findings as shown in Table 19.1.

Clearly, as illustrated in Table 19.1, Lavelle (2009) holds different views, regarding engineering as a poly-paradigmatic field in which theory and pragmatic values are used eclectically (in accordance with Dewey's perception of pragmatism). Louis

[6] In contrast, Kuhn (1970, p. 161) sees "profound differences between science and technology".

Table 19.1 Nature of knowledge in science and engineering according to Lavelle (2009)

	Science	Engineering
Delimitation of objects	Idealized, isolated objects Causal mechanisms	Real entities and artifacts
Epistemic and ontological	Essential	Adopted from pure science
Theory structure	Hierarchical structure of nomological systems Mainly mono-paradigmatic	Theory adapted to problems. Poly-paradigmatic. Eclectic use of theory
Methods	Derived from theory	Methods more fundamental than theory
Values	Explicit justification Truth is important	Implicit justification *Efficiency and practical usefulness.* Pragmatic

Bucciarelli (1994, p. 113, my italics) argues against "reductionist, mythical, object-world representation [that] misses the *uncertainty* and *ambiguity* of what really goes on in designing. Unlike the kinematics of particles, designing is not *lawlike* or *deterministic*". Thus, the rigor in engineering research described by Borrego (2007) is absent in the accounts of engineering practice represented by Bucciarelli (1994), Lavelle (2009) and Trevelyan (2009, 2013). In a similar vein, Donald Schön (1987, p. 42, my italics) argues that "although some design products may be superior to others, there are *no unique right answers*". Schön (1983, p. 76) describes "design as a reflective conversation with the situation" and as expressed by the engineering professor in the quote in the introduction, "Of course you've got to do the math right, but when it finally comes to building a bridge, you've got to understand the *uniqueness of the site*" (Hopmann 2009, p. 7, my italics). Artigue (1988, p. 283, my italics) uses the label *didaktik engineering* because, like an engineer, education researchers are "…forced to work on *much more complex objects than the pure ones in science* and to *address problems in a practical way*".

Looking back on my Swedish education as an engineering student this describes very well what I learned; in retrospect the practical epistemology of the training was not positivist, but pragmatic as Lavelle explains. Re-reading my master's thesis (Bernhard 1978) it does not have the characteristics of post-factum reconstruction. It was based on an engineering fluid mechanics project, involving the design of a new system to control the flow (and re-use) of water in a pulp and paper mill. The report also included a critical evaluation of modeling and the limits of models. I remember feeling it was necessary to include this section as the older engineers at the mill admired the work too uncritically, because I had used a computer to simulate the system (this was in 1978 and at that time I had to write all the code myself) and therefore believed that the results were true more or less *a priori* because they had been generated by a computer.

Layton (1976, p. 695, my italics) explains the differences between engineering and (basic) science as follows:

> Engineering science often differs from basic science in important particulars. Engineering sciences often drop the fundamental ontology of natural philosophy, though on *practical* rather than metaphysical grounds. Thus, in solid mechanics, engineers deal with stresses in

continuous media rather than a microcosm of atoms and forces. Engineering theory and experiment came to differ from those of physics because it was concerned with *man-made devices* rather than directly with nature. Thus, engineering theory often deals with idealizations of machines, beams, heat engines, or similar devices. And the results of engineering science are often statements about such devices rather than statements about nature. The experimental study of engineering involves the use of models, testing machines, towing tanks, wind tunnels, and the like. But such experimental studies involve *scale effects*. From Smeaton onward we find a constant concern with comparing the results gained with models with the *performance of full-scale apparatus*. By its very nature, therefore, engineering science is less abstracted and idealized; it is much *closer to the "real" world of engineering*. Thus, engineering science often differs from basic science in both style and substance. Generalizations about "science" based on one will not necessarily apply to the other.

The call for rigorous application of scientific theory in engineering (and EER) is rooted in a restricted view of knowledge as only consisting of *episteme* (see below) and the positivist tradition of the philosophy of science. The idea of technology as applied science is a common *received* view among many scientists and engineers. According to this view "technics and engineering practice are kinds of activity, not kinds of knowledge" (Mitcham 1994, p. 197). However, I propose that is important to distinguish between the *received* view (sometimes learned from theoretical or philosophy of science courses) and the *practical* epistemology engineers use in their praxis.[7] By contrast, as illustrated by the quotation above, Layton (1974) and others definitely regard "technology as knowledge".[8] I maintain that *how* we understand technology is not only important for a *didaktik analysis*, but also for how we understand the relationship between EER and engineering research.

Malmberg (2007, p. 64, italics in original), an engineer with extensive industrial practical experience before starting on his Ph.D. studies on circuit design, points out that we should not neglect human aspects:

Due to the human learning aspect, no approach should be applied consider *only* the technical side of knowledge. It takes us to the area of human understanding philosophy, instead of the area of detailed technical knowledge.

In a similar vein, my former colleague Rune Hedberg (1995) strongly emphasized that "humans are always part of any measurement system, and this must be taken into account" (personal communication).

---

[7] Recent studies in the field of "science studies" has revealed that in their praxis scientists, even "pure scientists", do not encompass a positivist epistemology or follow a linear sequential structure in their experiments. See for example Bruno Latour and Steve Woolgar (1986), Latour (1987) and David Gooding (1990). Gooding (1990, pp. 4–8) notes that "philosophers and AI researchers tend to consult only published papers or text-books" and "take scientists' own narratives as realistic accounts … ignoring the extent to which scientists' accounts are reconstructions rather than records" hence they miss what scientist *do* in reality.

[8] It could also be noted that, for example, Heidegger (1954) argued that modern science could be seen as "applied technology": "It is said that modern technology is something incomparably different from earlier technologies because it is based on modern physics as an exact science. Meanwhile, we have come to understand more clearly that the reverse holds true as well: modern physics, as experimental, is dependent upon technical apparatus and upon progress in the building of apparatus" (pp. 17–18).

Malmberg (2007, p. 64, italics in original) continues by suggesting that reasoning about knowledge in (circuit) engineering should be based on "Aristotle's five concepts of knowledge: *episteme*, *techne*, *phronesis*, *sophia* and *nous*." He presents the following definitions for use in technical engineering:

- **Epistéme:** Knowledge of science, based on *general rules and structure*. It is often theory formulas, verified by experiments once and for all.
- **Téchne:** Knowledge of practice, tied to a *specific production situation*. ...
- **Phrónesis:** Wisdom of practice from own experience, how to handle (make decisions in) *new situations* similar to previously experienced situations. ... Each decision in each new design situation is preceded by a weighing process in our minds, where we weigh relevant fact knowledge such as our episteme and techne into each new decision situation [and thus] phronesis [is] *situation dependent* (i.e., not *generally* valid, as is episteme).
- **Sóphia:** Wisdom on *combining science* and *intelligence*. We need sophia, for example, when comparing design areas to each other, which possess properties so different that they are hard to compare directly. We also need sophia when developing new episteme as during research work. ... In general sophia is needed (and learnt) when we make intelligent judgements ... on design and research information.
- **Nóus:** *Intelligence*, *holistic comprehension*, and *intuition* that borders on divine reason, all combined to make *holistic judgements* on design and research situations. ...

(Malmberg 2007, pp. 66–68, bold and italics in the original)

Although fascinating topics in their own right, the subtleties of Aristotle´s concepts of knowledge and Malmberg's interpretation are beyond the scope of this chapter.[9] However, it is worth noting Malmberg's emphasis on *phronesis* for making "good judgment", especially in new and unknown situations to judge "what knowledge (techne and episteme)...is applicable in this new situation" (p. 66). Malmberg (2007, p. 83) criticizes Bloom's taxonomy (Bloom et al. 1956; Anderson and Krathwohl 2001), which is commonly used (even in technical universities), for assuming that knowledge consists solely of "theory-based knowledge (episteme), maybe with some added application experience on this theory (episteme-related phronesis)" not recognizing techne or similar knowledge. A critique on similar grounds has been voiced by González Sampayo (2006, chapter 8). I share the critical views expressed by Malmberg and González Sampayo. It is also apparent that Malmberg does not agree with the views about the nature of engineering ascribed to members of the engineering faculty surveyed in Borrego's (2007) study (see above).

The differences and tensions between "science" and "technology", as well as those between "episteme", "techne" and "phronesis", discussed above are also present within educational research. Some researchers investigate education and student learning as a "given reality" and see teaching praxis or the design of teaching and learning environments as only a matter of applying educational theory. However, as any engineer knows and pointed out above by Malmberg (2007), knowledge of Maxwell's equations is not sufficient for constructing an amplifier circuit and knowledge of Newton's laws is not sufficient for building a bridge. Similarly, Dewey (1984) states in "The sources of a science of education" that "no conclusion of scientific

[9] Interesting discussions can be found in, for example, Hickman (1990) and Mitcham (1994).

research can be converted into an immediate rule of educational art" (p. 9) and compares this with the use of scientific theory in engineering:

> When, in education, the [researcher] in any field reduces his findings to a rule which is to be *uniformly adopted* then, only, is there a result which is *objectionable* and *destructive* … But this happens not because of scientific method but because of *departure* from it. It is not the *capable engineer* who treats scientific findings as *imposing* upon him a certain course which is to be *rigidly* adhered to: it is the third- or fourth-rate man who adopts this course (Dewey 1984, p. 6, my italics).

For Dewey (1983), therefore, engineering is not simply about applying knowledge from science in practice. Rather, he maintains that knowledge is gained, in engineering, by *doing* things differently and in this sense engineering knowledge is practical or in his words "there was … no definite art or science of modern bridge-building until *after* bridges of the new sort had been constructed" (p. 324). In a similar vein Trevelyan (2013) states that "the knowledge of practice is created by practice". Dewey continues by pointing out that theory developed as a result of a new achievement cannot precede the achievement. However, Dewey (1983) also pointed out that it is not fruitful to take the approach of "blindly trying one's luck or messing around in the hope that something nice will be the result" (p. 326). Rather, he observes, "pioneers in the educational field need an extensive and severe intellectual equipment" (p. 326), they need "imagination, courage and the desire to experiment and to learn from its results" (p. 325), but nevertheless there is a "certain amount of dependable knowledge" that can be relied upon and used to proceed with any endeavor. The problems lie in identifying how this knowledge should be used in "new social conditions" and avoiding limiting imagination to what is already familiar (p. 325). He stresses that engineering is a human enterprise by stressing that "the essential need was thus human rather than scientific" (p. 325). Education is a human enterprise and is a domain well worth the attention of engineers and appropriate for the application of engineering knowledge.

> The control of conditions demanded by laboratory work leads to a maximum of *isolation of a few factors* from other conditions. The scientific result is *rigidly limited to what is established with these other conditions excluded.* In education individualities, *no such exclusion can be had.* The number of variables that enter in is enormous. … Judgement in such matter is of qualitative situations and must itself be qualitative (Dewey 1984, p. 33, my italics).

I believe that any practicing engineer would recognize the truth in Dewey's statement (as is apparent in Malmberg's thesis, discussed above) and claim that in *real-world* engineering "no such exclusion can be had". In the earlier quote from Layton (1976, p. 695) scale-effects are mentioned and both engineers and engineering researchers know (or should know) the problem of transferring laboratory results into a design that *functions* in the *real*-world. The same holds true for education and there are ample examples from engineering as well from education of findings that worked well in the laboratory but failed in the *real* messy world. As pointed out by Dewey (1984, pp. 10–11, my italics) "no genuine science is formulated by *isolated conclusions*, no matter how scientifically correct the technique by which these isolated results are reached, and no matter how exact they are" and "the physical sciences … deal with subjects that are intrinsically *less complex*, involving *fewer variables*".[10] This is the reason why Layton (1976, p. 695) claims that "engineering science often differs from basic science in both style and substance". This does not mean that these laboratory experiments under controlled conditions and scientific theories are of no value. As previously noted, Dewey stated that you cannot proceed by "blindly trying one's luck or messing around in the hope that something nice will be the result", you need theory as a point of departure.

Engineers have learned (through necessity) to handle both general aspects (in the case of bridge building: engineering mathematics, solid mechanics, materials science, geology etc.) and particular aspects (the local situation of particular bridges) of their profession. This is reflected in Dewey's understanding of engineering as an "art" and a "science". Similarly, in his investigation of engineering practice Trevelyan (2013) concluded:

> The results of engineering practice strongly depend on localized social, economic factors, even though the underlying principles of the natural science are universal … The influence of local factors requires us to understand how social, cultural, philosophical, even religious beliefs affect practice, the sharing and distribution of human knowledge, and the financial capital on which it depends.

I maintain that such an understanding of the relationship between the general and particular, as advocated in works cited above by Artigue (1988), Dewey (1983, 1984), Hopmann (2009), Malmberg (2007) and Trevelyan (2013), is also needed in education to make theoretical as well as practical progress. In this respect, as well understood by Dewey, education has something to learn from engineering.

Design has already been mentioned several times in this section and it is often claimed that "artifact design is what constitutes the essence of engineering" (Mitcham 1994, p. 147). Thus, I maintain, engineers could also make valuable contributions to design-based educational research, as outlined below.

Design and Design-Based Research

> Usually, in their professional careers, engineers are often involved in design projects. Their training prepares them to approach design problems in a systematic way. In most cases they learn to apply a systems approach … analysing the objectives and planning alternative solutions to reach the desired goal … Surprisingly, engineers seldom put their design skills into practice when they are faced with the task [of] curriculum design in engineering education (Rompelman and De Graaff 2006, p. 215).

As already mentioned, Skolimowski (1966) maintains "it is erroneous to consider technology as being an applied science" (p. 372) because "in science we *investigate*

---

[10] Colleagues and I have proposed that investigations of students' understanding and learning of common *single* concepts in science education research are not sufficient for engineering education research. We propose that learning in engineering should be regarded and examined as the learning of *complex* concepts (Bernhard et al. 2010).

the reality that is given; in technology we *create* a reality according to our designs" (p. 374, italics in the original). He summarizes this as "science concerns itself with what *is*, technology with what *is to be*" (p. 375, italics in the original). A "technological object" is, according to Skolimowski, any "artifact produced by man to serve a *function*" (p. 375, my italics). Therefore, I maintain that the design of a learning environment could, indeed, be seen as the design of a "technological object", i.e. an artifact. Similar reasoning can be found in Kelly's (2004) discussion of *design research in education*: "A design is not [a] design without some form of designated artifact" (p. 116). He continues, "in my opinion, design studies should produce an artifact that outlasts the study and can be adopted, adapted and used by others … The design of such artifacts usually involves *engineering* a broader 'learning environment'…" (pp. 116–117, my italics). Hence, as previously proposed by Dewey, education can be regarded as "engineering", indeed, he called for the development of an "art of educational engineering" (Dewey 1983). The complexity of technological systems and learning environments has already been mentioned and Schön (1987) notes that designing involves handling this complexity, which involves many variables, constraints and also conflicting values:

> Designing, in its broader sense involves *complexity* and *synthesis*. In contrast to analysts or critics, designers put things together and bring new things into being, dealing in the process with *many variables* and *constraints*, some initially known and some discovered through designing. Almost always, designers' moves have consequences other than those intended for them. Designers juggle variables, reconcile *conflicting values*, and manoeuver around constraints – a process in which, although some design products may be superior to others, there are *no unique right answers* (Schön 1987, pp. 41–42).

An important point made by Schön is that there are "no unique right answers", i.e. numerous designs for almost any object may meet specified functional and aesthetic criteria. What is regarded as "the best" design also strongly depends on how different aspects of the outcomes are valued (cf. phronesis discussed earlier). Dewey's stance in "Education as engineering" and "The sources of a science of education" (Dewey 1983, 1984), as well as the French notion of *didaktik engineering* (Artigue 1988) are highly consistent with an emergent approach called *design-based research* or *design experiments*. According to the Design-Based Research Collective (2003), design-based research "must account for how designs function in *authentic settings*. It must not only document success or failure but also *focus on interactions* that refine our understanding of the learning issues involved." Paul Cobb et al. (2003) described this shift as follows: "Prototypically, design experiments entail both 'engineering' particular forms of learning and systematically studying those forms of learning within the context defined by the means of supporting them. This designed context is subject to test and revision, and the successive iterations that result play a role similar to that of systematic variation in experiment".

A statement by Mun Ling Lo et al. (2004), that *inter alia*, the main "benefits of design experiments [in education] are that [they] will … contribute to theory development, and improve practice at the same time" seems very similar to the position taken by Dewey that the development of practice and theory is closely and synergistically related. The same holds true for engineering. In the "design-based research" or "design experiments" approach, insights from design and engineering are employed to address the complexity of educational activities and the need, as known from engineering, for theory as well as tinkering.

My personal design-based research (e.g. Bernhard 1999, 2000, 2003, 2005, 2010, 2011a, b; Bernhard and Carstensen 2002; Bernhard et al. 2007; Carstensen and Bernhard 2004, 2007, 2009) has, for example, contributed to the design of environments that foster learning, the development of theories regarding the role of technologies in labs (e.g. Bernhard 2007, 2008, 2012, 2013), critical factors for learning in labs (e.g. Bernhard 2003, 2010, 2011a, b; Bernhard et al. 2005; Carstensen 2013; Carstensen and Bernhard 2009), threshold concepts theory (e.g. Bernhard et al. 2011; Carstensen and Bernhard 2008, 2013; Carstensen 2013) and a model of the learning of complex concepts (e.g. Bernhard et al. 2010, 2011; Carstensen 2013; Carstensen and Bernhard 2004, 2013; Carstensen et al. 2005).

I do not further consider the subtleties of design-based educational research in this chapter, because the preceding chapter by Anette Kolmos in this volume is devoted to "Changing the learning paradigm by design-based research". My point is that engineers should receive training in designing that could be utilized in EER for designing engineering education learning environments and I would claim that my engineering training has been personally beneficial.

As already mentioned, Mitcham (1994, p. 147) proposes that "artifact design is what constitutes the essence of engineering" and Kelly (2004) maintains that design studies in education involve "some form of designated artifact". Tools play an important role in Dewey's philosophies of education and technology. This leads to the third contribution that I maintain engineers could make to education, namely improving understanding of the role of physical artifacts (i.e. mediating technologies) in human perception and the application of appropriate artifacts in the design of learning environments.

Learning Through and with Artifacts (Technologies)

The production of knowledge in science and engineering in modern society is technologically embodied. This means that science not only uses instruments (technologies), but also uses them in innovative, informative ways. According to Alfred Whitehead (1963, p. 107, my italics):

> The reason we are on a higher imaginative level [in modern science] is not because we have a finer imagination, *but because we have better instruments.* In science, the most important thing that has happened in the last forty years is the advance in instrumental design…a fresh instrument serves the same purpose as foreign travel; *it shows things in unusual combinations. The gain is more than a mere addition; it is a transformation.*

Learning is described by Ference Marton and Amy Tsui (2004) as developing a vision: "Arranging for learning implies arranging for developing learners' ways of seeing or experiencing, *i.e., developing the eyes through which the world is perceived*". If so, an important issue for educational research is how students and professionals in a specific discipline acquire a "professional vision" (Goodwin 1994). As mentioned above, a central characteristic of learners' and professionals' experience of our world in engineering and most sciences is that experience should not be seen as a direct *human-world* experience, but as a mediated experience shaped by the use of physical and symbolic tools, i.e. artifacts. The concept of mediation and mediating tools could be represented schematically as:

Human – Mediating Tools (Artifacts) – World

Questions about the role of technology (artifacts) in everyday human experience include:

- How do technological artifacts affect the existence of humans and their relationship with the world?
- How do artifacts produce and transform human knowledge?
- How is human knowledge incorporated into artifacts?
- What are the actions of artifacts?

Both the structure of an artifact and learning to use an artifact changes the structure of human interaction with the world. This has profound implications for learning, thus the role of technologies in learning warrants extensive investigation. However, the role of instrumental technologies in student learning in laboratories has been rarely studied and is generally either neglected or taken for granted. Consequently, research generally focuses mainly on instructions, concepts, and ideas or the organization of labs. This is in line with the "[traditional belief] that … instruments and experimental devices … *per se* … has no cognitive value" (Lelas 1993, pp. 423–424, italics in original), i.e. in traditional beliefs about science the *technological means* by which nature is perceived leave no trace in our conceptions of nature (e.g., Kroes 2003). Karl Popper (1972, p. 118), for example, restricted his epistemology to the "world of language, of conjectures, theories, and arguments". Hence, the role of instruments is often neglected or taken for granted and the emphasis is placed only on concepts and ideas. However, neglecting the role of instruments (i.e. technological artifacts) in science leads to naïve realism or naïve idealism (Ihde 1991; Ihde and Selinger 2003). Two common views regarding technology in education are summarized in Fig. 19.1 below.

The theoretical framework in education research is commonly based on cognitivist and mentalist ideas that could be described as based on "the presumption that all psychological explanation must be framed in terms of internal mental representation" (Still and Costall 1987). Hence, in cognitivist theories "*technology is nearly invisible*", however in "postcognitivist" theories such as activity theory, distributed cognition, actor-network theory, and phenomenology "a major point of agreement … is the *vital role of technology* in human life [and a criticism] of

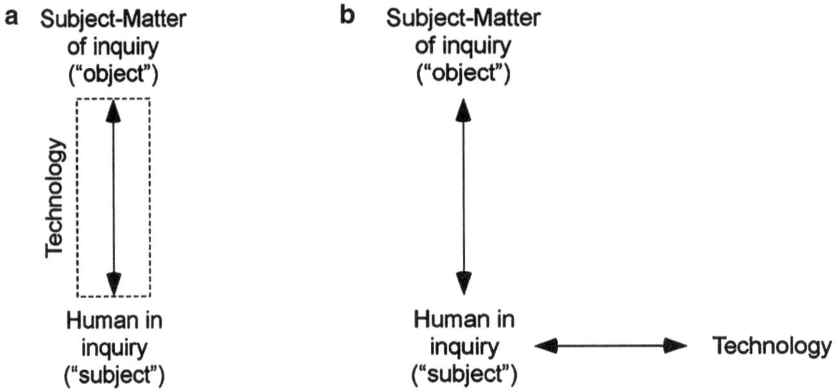

Fig. 19.1 (**a**) A "transmissive" view of technology as merely a neutral vehicle for transportation of information. (**b**) An "auxiliary" or "supportive" view of technology as merely a provider or source of information or support (Adapted from Bernhard (2008)

mind-body dualism" (Nardi and Kaptelinin 2006, pp. 195–197, my italics; cf. Cole and Derry 2005; Bernhard 2008).

Criticism of the neglect of experimental technologies in analysis can also be found in the emergent field of philosophy of scientific experimentation. An early critic was Ian Hacking (1983), who argued that "philosophers of science constantly discuss theories and representation of reality, but say almost nothing about experiment, technology, or the use of knowledge to alter the world". Twenty years later Hans Radder (2003, pp. 1–8) advocated "a more developed philosophy of scientific experimentation" and claimed (cf. Harré 2003; Kroes 2003; Baird 2004; Gooding 1990):

> The fact that many scientists … spend most of their time doing experiments of various kinds is not reflected in the basic literature in the philosophy of science. … Thus, the philosophy of experimentation is still underdeveloped …There has been a strong tendency to take the production of empirical knowledge for granted. … In sum, if philosophers keep neglecting the technological dimension of science, experimentation will continue to be seen as a mere data provider for the evaluation of theories.

However, by neglecting the active role of technology, humans are also quite ironically excluded from consideration and turned into passive recipients of information. Peter Kroes (2003) summarizes this view as follows:

> In [the traditional] view, the physicist is essentially a passive observer in experiments: once the stage is set he just observes (discovers) what is going to happen.

As noted by Ihde (1991) and Kroes (2003), for example, observation is not generally regarded as problematic in positivist approaches and from the anti-positivist perspective the praxis-ladenness of observations tends to be overlooked. The importance Hedberg (1995) perceived of humans as part of any measurement has already been mentioned. The important role of technology in learning has also been pointed out by Nora Sabelli (1995, my italics), who claims:

> What and how we learn have always depended on the *tools available* to students and teachers and should change with significant changes in the tools available. … Educators [are] responsible for exploring the profound *pedagogical implications of the changes brought about by technology on the practice of science*.

Therefore, I consider that the role of technology in education requires careful analysis (cf. Waltz 2004), and has not been sufficiently addressed. A theoretical foundation for this can, for example, be found in the philosophy of technology presented by Ihde, which integrates non-foundational phenomenology and pragmatism in an approach dubbed postphenomenology (Selinger 2006). According to Ihde, all science in its production of knowledge is *technologically embodied* and perception is co-determined by technology. In science, instruments do not merely "mirror reality", but mutually constitute the reality investigated. The *technology used* places some aspects of reality in the *foreground*, others in the *background*, and makes *certain aspects visible* that would otherwise be *invisible* (e.g. Ihde 1979, 1991, 1998, 2009).

My own research in the field of engineering education (Bernhard 2010, 2012) corroborates the assertion that technologies (artifacts) play important roles in student learning by placing some aspects of reality in the foreground, others in the background, and making certain aspects visible that would otherwise be invisible. Different technologies have different affordances for discernment and hence the possibilities for learning different objects of learning are dependent upon the technologies available or made available to students.

Karen Barad (2007) maintains that discourse cannot be seen solely as language and that materiality cannot be neglected, coining the term *material-discursive practice* (cf. *material hermeneutics* [Ihde 2009]). An engineer knows that materiality cannot be neglected and is used to applying instrumentations and technologies to investigate and model the world and is therefore well-equipped for *material-discursive* analysis of both the world and learning.

Conclusions

In this chapter I have explored some aspects of the relationships between: (i) technology, engineering and engineering competencies, and (ii) education and the design of education. I have presented examples of knowledge and skills founded in the art and science of engineering that could contribute to the development of both engineering education and education in general. According to Cobb et al. (2003, p. 13) an aim of design experiments should be to develop theories that do "real work in practical educational contexts". Engineering research and design have similar aims – theories should be useful and do "real work in practical contexts".

Engineers have developed theories for design including awareness of the tensions involved and for dealing with complex systems. In addition, the contingencies and particularities of practical situations have been proficiently reconciled in the art and science of engineering. Such proficiencies could be of great value in the development of educational theories and education practices to engineer modes of thought that enhance both education and education research.

Furthermore, I conclude that to exploit the full potential of technologies as learning tools in education we must understand their cognitive role(s). As previously mentioned, the design of artifacts is an engineering specialty, thus both the philosophy of technology and engineers can make essential contributions to our understanding by encouraging critical reflections about *what* kind of skills and awareness are important for sound engineering practice, and improving understanding of *how* technologies can be used in education and modulating human perception.

Furthermore, educating engineers is a social discipline with an engineering content. Indeed, Trevelyan (2009) describes engineering as a "technical and a social discipline at the same time: the social and technical are inextricably intertwined".

Engineering and knowledge of technologies could contribute to the development of the "art of educational engineering" by deepening insights regarding design, the role of humans in socio-technical systems and awareness of technology. Research in engineering education has a great potential to contribute to the "art of educational engineering" and the knowledge and skills founded in the art and science of engineering has a great potential, as argued in this chapter, to contribute to the development of engineering education in particular as well as the development of education in general. Hence, I maintain, engineering education research should be seen as engineering and as engineering research.

Acknowledgements This chapter draws in part on research funded by the Swedish Research Council and the Council for Renewal of Higher Education.

References

Anderson, L. W., & Krathwohl, D. R. (2001). *A taxonomy for learning, teaching and assessing*. New York: Pearson Education.

Artigue, M. (1988). Ingénierie Didactique. *Recherches en Didactique des Mathématiques, 9*(3), 281–308.

Baillie, C., & Bernhard, J. (2009). Educational research impacting engineering education. *European Journal of Engineering Education, 34*(4), 291–294.

Baird, D. (2004). *Thing knowledge: A philosophy of scientific instruments*. Berkeley: University of California Press.

Barad, K. (2007). *Meeting the universe halfway: Quantum physics and the entanglement of matter and meaning*. Durham: Duke University Press.

Bernhard, J. (1978). *Simulering av planerat bakvatten och avloppsvattensystem i Grycksbo pappersbruk*. Uppsala: Teknikum/Uppsala University.

Bernhard, J. (1999). Hands-on experiments in advanced mechanics courses. In G. Born, H. Harreis, H. Litschke, & N. Treitz (Eds.), *Hands on-experiments in physics education* (pp. 175–177). Duisburg: Didaktik der Physik, University of Duisburg.

Bernhard, J. (2000). Teaching engineering mechanics courses using active engagement methods. Paper presented at *Physics Teaching in Engineering Education (PTEE) 2000*. Budapest.

Bernhard, J. (2003). Physics learning and microcomputer based laboratory (MBL): Learning effects of using MBL as a technological and as a cognitive tool. In D. Psillos, P. Kariotoglou, V. Tselfes, E. Hatzikraniotis, G. Fassoulopoulos, & M. Kallery (Eds.), *Science education research in the knowledge based society* (pp. 313–321). Dordrecht: Kluwer.

Bernhard, J. (2005). *Experientially based physics instruction – using hands on experiments and computers: Final report of project 167/96*. Stockholm: Council for Renewal of Higher Education.

Bernhard, J. (2007). Thinking and learning through technology – mediating tools and insights from philosophy of technology applied to science and engineering education. *The Pantaneto Forum, 27*. http://www.pantaneto.co.uk/issue27/Bernhard.htm

Bernhard, J. (2008). Humans, intentionality, experience and tools for learning: Some contributions from post-cognitive theories to the use of technology in physics education. *AIP Conference Proceedings, 951*, 45–48.

Bernhard, J. (2009). Learning through artifacts in engineering education: Some perspectives from the philosophy of technology and engineering science. Paper presented at *SEFI 2009*. Rotterdam.

Bernhard, J. (2010). Insightful learning in the laboratory: Some experiences from Ten years of designing and using conceptual labs. *European Journal of Engineering Education, 35*(3), 271–287.

Bernhard, J. (2011a). Investigating student learning in two active learning labs: Not all "active" learning laboratories result in conceptual understanding. Paper presented at *ASEE Annual Conference*. Vancouver.

Bernhard, J. (2011b). Learning in the laboratory through technology and variation: A microanalysis of instructions and engineering students' practical achievement. Paper presented at *SEFI/WEE 2011*. Lisbon.

Bernhard, J. (2012). Learning through artifacts in engineering education. In N. M. Seel (Ed.), *Encyclopedia of the sciences of learning* (pp. 1983–1986). New York: Springer.

Bernhard, J. (2013). What matters? Learning in the laboratory as a material-discursive practice. Paper presented at *European Association for Learning and Instruction (EARLI), 15th biennial Conference*. Munich.

Bernhard, J., & Carstensen, A.-K. (2002). Learning and teaching electrical circuit theory. Paper presented at *PTEE 2002: Physics Teaching in Engineering Education*. Leuven.

Bernhard, J., Carstensen, A.-K., & Lindwall, O. (2005). Highly structured open inquiry labs. Paper presented at *EARLI 2005*. Nicosia.

Bernhard, J., Carstensen, A.-K., & Holmberg, M. (2007). Design-based educational research and development of engineering education – examples from courses in mechanics and electrical engineering. Paper presented at *ASEE Global Colloquium on Engineering Education*. Istanbul.

Bernhard, J., Carstensen, A.-K., & Holmberg, M. (2010). Investigating engineering students' learning – 'learning as the learning of a complex concept'. Paper presented at *IGIP-SEFI 2010*. Trnava.

Bernhard, J., Carstensen, A.-K., & Holmberg, M. (2011). Analytical tools in engineering education research: The "learning a complex concept" model, threshold concepts and key concepts in understanding and designing for student learning. In W. Hernandez (Ed.), *Proceedings of research in engineering education symposium 2011* (pp. 51–60). Madrid: Universidad Politécnica de Madrid (UPM).

Biglan, A. (1973). The characteristics of subject matter in different academic areas. *Journal of Applied Psychology, 57*(3), 195–203.

Bloom, B. S., Engelhart, M. D., Furst, E. J., Hill, W. H., & Krathwohl, D. R. (1956). *Taxonomy of educational objectives*. New York: David McKay.

Borrego, M. (2007). Conceptual difficulties experienced by engineering faculty becoming engineering education researchers. *Journal of Engineering Education, 96*(2), 91–102.

Borrego, M., & Bernhard, J. (2011). The emergence of engineering education research as a globally connected field of inquiry. *Journal of Engineering Education, 100*(1), 14–47.

Bucciarelli, L. L. (1994). *Designing engineers*. Cambridge, MA: MIT Press.
Carstensen, A.-K. (2013). *Connect: Modelling learning to facilitate linking models and the real world through lab-work in electric circuit courses for engineering students*. Linköping studies in science and technology dissertation no. 1529, Linköping.
Carstensen, A.-K., & Bernhard, J. (2004). Laplace transforms – too difficult to teach, learn and apply, or just matter of how to do it. Paper presented at *EARLI sig#9 Conference*. Gothenburg.
Carstensen, A.-K., & Bernhard, J. (2007). Critical aspects for learning in an electric circuit theory course – an example of applying learning theory and design-based educational research in developing engineering education. Paper presented at the *First International Conference on Research in Engineering Education*. Honolulu.
Carstensen, A.-K., & Bernhard, J. (2008). Threshold concepts and keys to the portal of understanding: Some examples from electrical engineering. In R. Land, E. Meyer, & J. Smith (Eds.), *Threshold concepts within the disciplines* (pp. 143–154). Rotterdam: Sense Publishers.
Carstensen, A.-K., & Bernhard, J. (2009). Student learning in an electric circuit theory course: Critical aspects and task design. *European Journal of Engineering Education, 34*(4), 389–404.
Carstensen, A.-K., & Bernhard, J. (2013). Make links – learning complex concepts in engineering education. Paper presented at the *Research in Engineering Education Symposium (REES)*. Kuala Lumpur.
Carstensen, A.-K., Degerman, M., González Sampayo, M., & Bernhard, J. (2005). Interaction in labwork – linking the object/event world to the theory/model world. Paper presented at *ESERA2005*. Barcelona.
Christensen, J., Henriksen, L. B., & Kolmos, A. (2006). *Engineering science, skills, and bildung*. Aalborg: Aalborg University Press.
Cobb, P., Confrey, J., diSessa, A., Lehrer, R., & Schauble, L. (2003). Design experiments in educational research. *Educational Researcher, 32*(1), 9–13.
Cole, M., & Derry, J. (2005). We have met technology and it is us. In R. J. Sternberg & D. D. Preiss (Eds.), *Intelligence and technology: The impact of tools on the nature And development of human abilities* (pp. 209–227). Mahwah: Lawrence Erlbaum.
Comenius, J. A. (1657). Didactica magna. In J. A. Comenius (Ed.), *Opera didactica omnia* (pp. 5–190). Amsterdam.
Design-Based Research Collective. (2003). Design-based research: An emerging paradigm for educational inquiry. *Educational Researcher, 32*(1), 5–8.
Dewey, J. (1983). Education as engineering. In J. A. Boydston (Ed.), *The middle works of John Dewey, 1899-1924* (pp. 323–328). Carbondale: Southern Illinois University Press.
Dewey, J. (1984). The sources of a science of education. In *John Dewey, the later works, 1925-1953* (pp. 1–40). Carbondale: Southern Illinois University Press.
González Sampayo, M. (2006). *Engineering problem solving: The case of the laplace transform as a difficulty in learning electric circuits and as a tool to solve real world problems*. Linköping studies in science and technology dissertation no. 1038, Linköping.
Gooding, D. (1990). *Experiment and the making of meaning: Human agency in scientific observation and experiment*. Dordrecht: Kluwer.
Goodwin, C. (1994). Professional vision. *American Anthropologist, 96*(3), 606–633.
Hacking, I. (1983). *Representing and intervening: Introductory topics in the philosophy of natural science*. Cambridge: Cambridge University Press.
Harré, R. (2003). The materiality of instruments in a metaphysics for experiments. In H. Radder (Ed.), *The philosophy of scientific experimentation* (pp. 19–38). Pittsburgh: University of Pittsburgh Press.
Hedberg, R. (1995). *Mätteknisk system*. Lund: Studentlitteratur.
Heidegger, M. (1954). Die Frage nach der Technik. In M. Heidegger (Ed.), *Vorträge und Aufsätze*. Stuttgar: Klett-Cotta.
Henriksen, L. B. (2006). Engineers and Bildung. In J. Christensen, L. B. Henriksen, & A. Kolmos (Eds.), *Engineering science, skills and Bildung* (pp. 43–60). Aalborg: Aalborg University Press.
Hestenes, D. (1992). Modeling games in the newtonian world. *American Journal of Physics, 60*(8), 732–748.
Hickman, L. A. (1990). *John Dewey's pragmatic technology*. Bloomington: Indiana University Press.

Hopmann, S. (2000). Klafki's model of Didaktik analysis and lesson planning in teacher education. In I. Westbury, S. Hopmann, & K. Riquarts (Eds.), *Teaching as a reflective practice: The German Didaktik tradition* (pp. 197–206). Mahwah: Lawrence Erlbaum.

Hopmann, S. T. (2009). Mind the gap: Dewey on educational bridge-building. *Journal of Curriculum Studies, 41*(1), 7–11.

Hopmann, S. T., & Riquarts, K. (1995a). Didaktik and/or curriculum: Basic problems of comparative Didaktik. In: S. T. Hopmann, & K. Riquarts (Eds.), *Didaktik and/or curriculum* (pp. 9–40). Kiel: Institut für die Pädagogik der Naturwissenschaften an der Universität Kiel (IPN).

Hopmann, S. T., & Riquarts, K. (1995b). *Didaktik and/or curriculum*. Kiel: Institut für die Pädagogik der Naturwissenschaften an der Universität Kiel (IPN).

Ihde, D. (1979). *Technics and praxis*. Dordrecht: Reidel.

Ihde, D. (1991). *Instrumental realism: The interface between philosophy of science and philosophy of technology*. Bloomington: Indiana University Press.

Ihde, D. (1998). *Expanding hermeneutics: Visualism in science*. Evanston: Northwestern University Press.

Ihde, D. (2009). *Postphenomenology and technoscience: The Peking university lectures*. Albany: State University of New York Press.

Ihde, D., & Selinger, E. (2003). Chasing technoscience: Matrix for materiality. In D. Ihde (Ed.), *Indiana series in the philosophy of technology*. Bloomington: Indiana University Press.

Kattmann, U., Duit, R., Gropengießer, H., & Komorek, M. (1997). Das Modell der Didaktischen Rekonstruktion: Ein Rahmen für naturwissenschaftsdidaktische Forschung und Entwicklung. *Zeitschrift für Didaktik der Naturwissenschaften, 3*(3), 3–18.

Kelly, A. (2004). Design research in education: Yes, but is It methodological? *Journal of the Learning Sciences, 13*(1), 115–128.

Klafki, W. (2000). Didaktik analysis as the core of preparation of instruction. In I. Westbury, S. Hopmann, & K. Riquarts (Eds.), *Teaching as a reflective practice: The German Didaktik tradition* (pp. 139–159). Mahwah: Lawrence Erlbaum.

Kroes, P. (2003). Physics, experiments, and the concept of nature. In H. Radder (Ed.), *The philosophy of scientific experimentation* (pp. 68–86). Pittsburgh: University of Pittsburgh Press.

Kuhn, T. (1970). *The structure of scientific revolutions* (2nd ed.). Chicago: University of Chicago Press.

Künzli, R. (2000). German Didaktik: Models of re-presentation, of intercourse, and of experience. In I. Westbury, S. Hopmann, & K. Riquarts (Eds.), *Teaching as a reflective practice: The German Didaktik tradition* (pp. 41–54). Mahwah: Lawrence Erlbaum.

Latour, B. (1987). *Science in action*. Cambridge, MA: Harvard University Press.

Latour, B., & Woolgar, S. (1986). *Laboratory life: The construction of scientific facts*. Princeton: Princeton University Press.

Lavelle, S. (2009). Technology and engineering in context: Analytical, phenomenological and pragmatic perspectives. In S. H. Christensen, B. Delahousse, & M. Meganck (Eds.), *Engineering in context* (pp. 75–95). Aarhus: Academica.

Layton, E. T. (1974). Technology as knowledge. *Technology and Culture, 15*(1), 31–41.

Layton, E. T. (1976). American ideologies of science and engineering. *Technology and Culture, 17*(4), 688–701.

Lelas, S. (1993). Science as technology. *The British Journal for the Philosophy of Science, 44*(3), 423–442.

Linköping University. (2004). Allmän studieplan för forskarutbildningsämnet Ingenjörsvetenskapens didaktik [Studyplan for the PhD education in engineering education]: Approved (FoFudel 04–21) by The Institute of Technology, Linköping University, on 29 Sept 2004.

Lo, M. L., Marton, F., Ming Fai Pang, & Wing Yan Pong. (2004). Toward a pedagogy of learning. In F. Marton & A. B. M. Tsui (Eds.), *Classroom discourse and the space of learning* (pp. 189–225). Mahwah: Lawrence Erlbaum.

Malmberg, R. (2007). *Analog circuit topology development: Practice methods for technology and teaching based on comprehensible transistor models*. Gothenburg: Chalmers University of Technology.

Marton, F., & Tsui, A. B. M. (2004). *Classroom discourse and the space of learning*. Mahwah: Lawrence Erlbaum.

Mitcham, C. (1994). *Thinking through technology: The path between engineering and philosophy*. Chicago: The University of Chicago Press.
Nardi, B. A., & Kaptelinin, V. (2006). *Acting with technology: Activity theory and interaction design*. Cambridge, MA: MIT Press.
Popper, K. R. (1972). *The logic of scientific discovery*. London: Hutchinson.
Radder, H. (2003). Toward a more developed philosophy of scientific experimentation. In H. Radder (Ed.), *The philosophy of scientific experimentation* (pp. 1–18). Pittsburgh: University of Pittsburgh Press.
Rompelman, O., & de Graaff, E. (2006). The engineering of engineering education: Curriculum development from a Designer's point of view. *European Journal of Engineering Education, 31*(2), 215–226.
Sabelli, N. (1995). For our Children's sake, take full advantage of technology. *Computers in Physics, 9*(1), 7.
Schön, D. A. (1983). *The reflective practitioner: How professionals think in action*. New York: Basic Books.
Schön, D. A. (1987). *Educating the reflective practitioner: Toward a New design for teaching and learning in the professions*. San Francisco: Jossey-Bass.
Selinger, E. (2006). *Postphenomenology: A critical companion to Ihde*. Albany: State University of New York Press.
Skolimowski, H. (1966). The structure of thinking in technology. *Technology and Culture, 7*(3), 371–383.
Still, A., & Costall, A. (1987). Introduction: In place of cognitivism. In A. Still, A. Costall, & A. Costall (Eds.), *Cognitive psychology in question* (pp. 1–16). Brighton: Harvester Press.
Strömdahl, H. (1996). *On mole and amount of substance*. Göteborg: Acta Universitatis Gothoburgensis.
Trevelyan, J. (2007). Technical coordination in engineering practice. *Journal of Engineering Education, 96*(3), 191–204.
Trevelyan, J. (2009). Engineering education requires a better model of engineering practice. Paper presented at the *Research in Engineering Education Symposium*, Palm Cove.
Trevelyan, J. (2013). Towards a theoretical framework for engineering practice. In B. Williams, J. Figuereido, & J. Trevelyan (Eds.), *Engineering practice in a global context: Understanding the technical and the social* (pp. 33–60). CRC Press.
Waltz, S. B. (2004). Giving artifacts a voice? bringing into account technology in educational analysis. *Educational Theory, 54*(2), 157–172.
Westbury, I., Hopmann, S., & Riquarts, K. (2000). *Teaching as a reflective practice: The German Didaktik tradition*. Mahwah: Lawrence Erlbaum.
Whitehead, A. N. (1963). *Science and the modern world*. New York: New American Library.
Williams, B., Figueiredo, J., & Trevelyan, J. (2013). Engineering practice as an emergent field of study – implications for engineering educators. Paper presented at the *Research in Engineering Education Symposium*. Kuala Lumpur.
Willman, O. (1957). *Didaktik als Bildungslehre nach ihren Beziehungen zur Socialforschung und Zur Geschichte der Bildung*. Freiburg im Bresgau: Herder.

Jonte Bernhard M.Sc. in Engineering and Ph.D. in Solid State Physics both from Uppsala University. Professor in Engineering Education at Linköping University. Before this appointment he was an associate professor in Electronics and docent in Physics at Linköping University and he has taught courses in electrical engineering and in engineering physics for 30 years as well as courses in education for 15 years. Coordinator for the *Nordic Network in Engineering Education Research* (NNEER) funded by NordForsk and associate editor for the *European Journal of Engineering Education*. He has published extensively on topics related to Engineering and Physics Education Research as well in Material Science. He is presently coordinating the research project 'Investigating learning in engineering and in techno-science as a material discursive practice' funded by the Swedish Research Council.

Chapter 20
Analyzing Context by Design: Engineering Education Reform via Social-Technical Integration

Dean Nieusma

Abstract This chapter describes and analyzes one model of engineering education reform aimed at enhancing students' ability to engage wide-ranging contexts of engineering work. Rensselaer's Programs in Design and Innovation (PDI) use a dual-major strategy to engage engineering students in contextual analysis in a way that is continual and dynamic. The chapter summarizes PDI's approach, assesses its significance, and identifies its limitations. The bulk of the chapter is dedicated to characterizing PDI's approach, first by reviewing its curricular structure and then by describing some of the didactical strategies used in its core courses. Next, PDI's achievements are assessed in terms of how they work to transform students' experiences with engineering education in a way that enhances their ability (and willingness) to engage contextual matters. Finally, prominent limitations of PDI are identified.

Keywords Programs in design and innovation • Engineering and liberal education integration • Curriculum reform • Science and technology studies (STS) • Design studios • Project-based learning

Introduction

As this volume attests, there are many ways to interpret and engage the various contexts of engineering work, both in professional practice and in educational environments. This chapter describes and analyzes one model of engineering education reform aimed at enhancing students' ability to engage – explicitly and productively – the many contexts of engineering design work. The initiative is called the Programs in Design and Innovation (PDI) at Rensselaer Polytechnic Institute. While

D. Nieusma (✉)
Department of Science and Technology Studies, Rensselaer Polytechnic Institute, 110 8th Street, Troy, NY 12180, USA
e-mail: nieusma@rpi.edu

S.H. Christensen et al. (eds.), *International Perspectives on Engineering Education*, Philosophy of Engineering and Technology 20, DOI 10.1007/978-3-319-16169-3_20

PDI has an established history and is now institutionally embedded at Rensselaer, its approach to engineering education reform remains innovative and, in certain respects, is effective in responding to a range of challenges faced by traditional engineering curricula. There have been relatively few publications describing PDI, and those that exist (Bronet et al. 2003; Steiner and Winner 2005; Nieusma 2008) have become dated. Hence, this chapter seeks to take a fresh look at an established yet innovative and promising educational reform effort that focuses squarely on the question of context in engineering practice and education.

Because context in engineering can mean so many things (Christensen et al. 2009; Jamison 2009), it will be helpful to summarize at the outset the overarching approach taken by PDI. In its curricula, didactics, and organizational formation, *PDI seeks to integrate technical, social, and formal*[1] *analysis with creative synthesis* – all in response to specific problems in the world. In other words, PDI students and faculty move among the technical, social, and formal dimensions of a given problem and its proposed solution at each stage of development of the solution. A core component is to investigate the various social contexts of a given problem, including user needs and experiences as well as the full range of economic, organizational, political, and cultural conditions existing around the problem. This contextual framing is probably a critical starting point for any systematic attention to context in engineering design work, yet such an approach is surprisingly difficult to do well – both conceptually and in practice (Kroes and van de Poel 2009).

Despite the fundamental importance of engaging the social contexts of engineering problems, direct inclusion of contextual variables within traditional engineering problem solving approaches may not be the most effective way of connecting with engineering students or of conveying the elusive, expansive, evolving nature of social context (Leydens and Lucena 2009). In some sense, insofar as various dimensions of social context are integrated into traditional engineering problem solving – as a "given" opportunity or constraint (i.e., as *fixed*) – the engineering problem solving process need not change much. As an illustration of this point, consider engineering students' facility with responding to *economic* constraints or even to the technological possibilities arising within a given economic system. In integrating economic considerations, the real variable at play in such engineering problem solving remains *technology*; the economic context remains static.

A more promising opportunity for engaging the complex nature of context lies beyond the reading of and responding to existing contextual conditions. It entails *treating context as a set of variables at play alongside technology*. Not only does this offer a truer picture of the changing nature of context, but it highlights the dynamic interplay among technology-society relationships, where neither can be neatly separated as independent or dependent variables (MacKenzie and Wajcman 1999). This is the approach taken in PDI, and this chapter seeks to elaborate how that is achieved.

[1] In this chapter, *formal analysis* refers to examination of the structure or arrangement of material and symbolic elements of a given design, including aesthetic characteristics.

To convey how PDI teaches engineering students to engage social context in a continual and dynamic way, the chapter summarizes the program, assesses its significance as an engineering education reform effort, and addresses limitations to the PDI model of reform. First, I review PDI according to its curricular structure, some of the pedagogical strategies used within its core courses, and the educational culture that results. Second, I analyze salient features of PDI that are potential levers for engineering education reform, both generally and in response to questions of context. Third, I identify several limitations of our approach. Before moving into a summary of the program, however, I will reflect briefly on the context of this analysis – my own role in PDI and the methodology underlying the claims put forward in this chapter.

Methodological Note

I currently serve as director of Rensselaer's Programs in Design and Innovation. Before that appointment, I served as a core member of PDI's faculty steering committee and have taught core courses in the program for 8 years. Earlier still, as a graduate student in Rensselaer's Science and Technology Studies Department, faculty members involved me in creating the program's early foundations with experimental STS "companion courses" that paralleled the required capstone engineering design course. Hence, it is fair to say I am both deeply invested in the program and committed to its success.

This investment risks projecting my own intellectual priorities and goals onto the educational reform initiative described and critically analyzed here. This acknowledgement is in part a qualification of the analysis that follows. But it is also a declaration: Following Downey (2009), I seek not merely to produce scholarship that informs critical participation in engineering educational reform, but to actively participate in such reform efforts and to reflexively analyze and refine that participation.

While I strive to make clear my situatedness in the analysis that follows, it is also important to acknowledge upfront that the voices and perspectives of other key stakeholders are largely absent from my analysis, except as they are filtered through my own understanding and experiences. Absent voices include those of other instructors participating in the program, a subset of the current students (especially those relatively disengaged or otherwise distant from the PDI community), most former students (except for the few who have maintained contact), and Rensselaer's administrators.

PDI Curriculum

PDI seeks to integrate technical, social, and formal analysis with creative synthesis. Two high-level educational dynamics are at play in this effort. First is the *integration of technical, social, and formal analytic approaches*. Most generically, this integration parallels broad disciplinary approaches to problem identification and solution: the types of technical analysis done by engineers; the types of social analysis done by social scientists and some humanists; and the types of formal analysis done by "creatives" – architects, designers, and artists – and associated humanists – historians and critics of architecture, design, and art.

Of course, problems in the world cannot cleanly be divided into technical, social, and formal dimensions. After all, a central contribution of STS and engineering studies scholarship is that the social-technical binary is untenable, notwithstanding continued reliance on the demarcation (Bijker et al. 1987; Downey and Lucena 1995). Nevertheless, academic disciplines and their respective undergraduate programs have maintained a surprisingly tidy demarcation among these areas, with the creatives perhaps most comfortable transgressing traditional disciplinary boundaries. Integrating across technical, social, and formal domains is, obviously, easier said than done. How it is done in PDI will be elaborated below.

The second educational dynamic at play in PDI is the *integration of analysis and creative synthesis*: After breaking a complex problem down and understanding its components, a proposed solution is built up in a way that responds to each major problem component. (In practice, the process is not so linear, but instead oscillates between analysis and synthesis. Nevertheless, there is a progressive shift in emphasis from the former to the latter (Nieusma 2004a).) In making this educational move, again there is precedent in the approach taken by designers of all kinds, starting with architecture, creative design (e.g., industrial design, graphic design, fashion design), and some art but extending to engineering design, policy design, and business planning among other areas that employ different types of design problem solving.

Creative solution posing is nothing new – and it usually entails a degree of systematic analysis at the front end and throughout the process – but integrating solution development with systematic, interdisciplinary analysis across technical, social, and formal domains sets a very high bar. The challenge is mostly due to the fact that integration requires continually transgressing disciplinary boundaries and, hence, moving beyond existing disciplinary educational arrangements (Winner 1995). PDI seeks to reach this high bar for integration without being naïve to the complexity of real-world problems or the impossibility of arriving at a "complete" analysis (Cañavate et al. 2009).

PDI strives to achieve this double integration in part through its curricular design. From the perspective of traditional engineering curricula, at least in the US context, the PDI curriculum has three major innovations, each of which will be elaborated in turn: (1) a dual-major structure, (2) a multi-year curriculum core of interdisciplinary design studios, and (3) a complementary sequence of traditional STS courses.

Dual-Major Strategy

The Programs in Design and Innovation are, somewhat confusingly, constructed in the grammatical plural, because they include a broad set of degree options. All PDI students are required to enroll in the "Design, Innovation, and Society" Bachelors of Science degree program. The Design, Innovation, and Society program is offered by Rensselaer's STS Department, alongside a more traditional Science, Technology, and Society program and a new Sustainability Studies program. In PDI, it is the Design, Innovation, and Society program that carries most of the innovative educational reform elements reviewed in this chapter.

The umbrella term, Programs in Design and Innovation, is retained because the Design, Innovation, and Society program is most effective as an educational reform when combined as a dual major. It is also carefully designed to combine seamlessly with the traditional mechanical engineering curriculum at Rensselaer. The vast majority of PDI students pursue Design, Innovation, and Society as a dual major with an engineering program, usually Mechanical Engineering. Many fewer students pursue Design, Innovation, and Society as a single major or as a dual major with wide-ranging program combinations, from Business Management to Communication/ Graphic Design to Computer Science.

As a degree offering (and hence an optional dual major), Design, Innovation, and Society can stand on its own; it is a New York State-approved Bachelors of Science degree program. This helps ensure the depth and coherence of the program in terms of its STS/social sciences content. But because the Design, Innovation, and Society program has been designed to couple with engineering curricula in particular, it is reasonable to treat it as an engineering education reform effort. Accordingly, this analysis will focus primarily on the coupling of Design, Innovation, and Society (DIS) with Mechanical Engineering, while recognizing that this particular coupling is more powerful because other students, who are pursuing other degree combinations, are also present.

PDI's Spring 2013 enrollment was 112 students, with 25–30 students per cohort. This enrollment profile is shown in Table 20.1.

The dual-major curricula for the three largest enrollment groups – Mechanical Engineering, Business Management, and Communication/Graphic Design – have been harmonized to integrate seamlessly both administratively and, to the degree possible, in terms of course content and sequencing. Each of these three dual-major combinations also imposes curricular constraints on the non-DIS program side as well. For example, a PDI student combining DIS and Mechanical Engineering will have specific PDI-requirements on the Mechanical Engineering side of the curriculum, requirements that other Mechanical Engineering students do not share. Specifically, these students are required to satisfy their Mechanical Engineering technical options with courses in design and manufacturing processes. Hence, while DIS contains most of the educational innovation, the Mechanical Engineering program requirements have been modified slightly as well to maximize the effectiveness of integration with DIS. The same situation applies to Business Management and Communication/Graphic Design students.

Table 20.1 PDI Enrollment breakdown by major, Spring 2013

Degree combination	
DIS + Mechanical engineering	72 %
DIS + Business management	8 %
DIS + Communication/Graphic design	8 %
DIS only (stand-alone major)	5 %
DIS + All other engineering	4 %
DIS + All other non-engineering	3 %

There are many reasons for building our reform efforts upon a dual-major strategy. As suggested above, requiring an entire dual major enables achieving both the breadth and depth of content necessary for responding to wide-ranging contexts of engineering work. Requiring the dual major also ensures a high number of instructor-student contact hours, enabling repeated, in-depth feedback as engineering students practice skills in social and formal analysis and integrative design. Specific strategies for developing the design experience and depth of analytic skills will be reviewed in the following sections of this chapter.

Besides these educational motivations, a host of more pragmatic goals are achieved with the dual-major strategy. One is educational certification. Important for many of our students is formal recognition of both their extra work and their identity as "not just an engineer" (also elaborated below): PDI students' Bachelors of Science diploma lists both their engineering concentration area and Design, Innovation, and Society. This double credential provides many PDI graduates with non-traditional and highly competitive engineering job opportunities.

Another pragmatic motivation for the dual-major strategy is to provide an institutional incentive for Rensselaer's STS Department and some of its faculty members to participate. Because PDI students are "our majors," as well as engineering majors, it is easier to justify investing considerable energy and departmental resources in interdisciplinary didactics and advising. This opportunity is magnified by the particular institutional context at Rensselaer, where the vast majority of all students – undergraduate and graduate – are engineering and science majors. Offering dual majors in humanities or social sciences is an effective recruiting strategy for attracting students who have chosen to attend what is overwhelmingly an engineering and science school.

A third pragmatic if indirect motivation for our approach is that it can operate relatively independently of ABET accreditation requirements and protocols. Such requirements – both as formally stated and as practiced via program reviews – produce reluctance among many engineering faculty members and administrators to experiment in the way and to the extent PDI does (Splitt 2003). The engineering curricula ensure ABET requirements are addressed, leaving the Design, Innovation, and Society curriculum free to focus on interdisciplinary integration and the complex and dynamic nature of context.

Design Studio Core

The main curricular mechanism by which PDI achieves its educational objectives is a sequence of interdisciplinary design studios – PDI's hallmark. The "PDI Studio" sequence provides a stem within each of the various curricula, starting in the first semester and reaching across all 4 years of undergraduate coursework. While each studio has a different emphasis, all PDI Studios offer interdisciplinary instruction and respond to broad, real-world problems, typically with a user, social, or environmental focus. PDI Studios are aligned with already-existing design courses within students' dual-major programs, resulting in a total of seven required studios and one optional studio over eight semesters. PDI Studios provide interdisciplinary instruction, through either multiple instructors representing different disciplines or instructors with interdisciplinary backgrounds. Table 20.2 provides an overview of the PDI Studio sequence for PDI students pursuing a dual-major in DIS and any area of engineering.

PDI Studios, those created specifically for the DIS program to serve PDI students exclusively, are inserted around existing engineering-program design studios. In addition to its senior-year capstone design course (what we label as "Studio 7"), which is a typical requirement of ABET-accredited engineering programs in the US (Dutson et al. 1997), Rensselaer has added a second-year design course for all engineering programs (labeled "Studio 4"). These two courses entail the total extent of design coursework for most engineering students at Rensselaer. PDI adds Studios 1, 2, 3, 5, and 6, which are restricted to PDI students only. An optional Studio 8, offered as a Mechanical Engineering advanced elective, is selected by many PDI students (including non-engineering dual majors) but is open to all Rensselaer students.

With the optional studio, the eight-course sequence entails 32 credit hours of a total 132 credit hours – almost a quarter of our students' total coursework. Each PDI Studio also entails six instructor-student contact-hours per week for 4 credit hours (instead of the typical four contact-hours). This results in PDI students experiencing considerably more instructor contact than with typical courses, and *a full one-third*

Table 20.2 PDI Studio sequence for DIS-engineering dual majors (Studios 4 and 7 meet generic program requirements in engineering)

Studio	Focal area	Primary disciplinary affiliation of instructors
1	Creativity & visualization	Engineering, STS, & graphic design
2	Design process	Engineering & STS
3	User experience design	STS (with engineering background)
4	*Engineering design*	*Interdisciplinary engineering*
5	Participatory design	STS (with cybernetics background)
6	Organizational design	STS (with engineering background) & business
7	*Capstone design*	*Interdisciplinary engineering*
8	Optional: inventors' studio	Engineering (with business background)

of their total instructor contact time is with their studio instructors compared with the approximately 5 % for typical engineering students.[2]

Beyond the amount of student-instructor contact, PDI Studios benefit from interdisciplinary instruction and, in some studios, multiple instructors. PDI is fortunate to have programmatic support from Rensselaer's School of Engineering, even as the Design, Innovation, and Society program is offered by the STS Department in the School of Humanities, Arts, and Social Sciences. Engineering professor Mark Steiner directs Engineering's Multidisciplinary Design Lab, which coordinates both of the engineering design studio courses (Studios 4 and 7). Steiner and other engineering instructors also participate in PDI Studios 1 and 2, and Studio 8 is led by engineering instructor Burt Swersey. As a result of this overlap, PDI benefits considerably from instructional participation by those engineering faculty members with expertise in design.

Several faculty members from Rensselaer's STS Department also participate in PDI Studio instruction, including myself (Studios 3 and 6), ethnomathematics scholar Ron Eglash (Studio 5), political philosopher Langdon Winner and feminist anthropologist Linda Layne (Studio 2), and occasional others. Additional instructors over the past few years have come from backgrounds in communications and graphic design, business management, and industrial design.

Co-teaching by interdisciplinary instructors with widely divergent approaches to problem solving provides a unique educational opportunity, especially for our students who are not accustomed to knowledge claims being negotiated in the classroom. This opportunity was created deliberately. When PDI was first implemented, all PDI Studios included multiple instructors bridging technical, social, and creative design disciplines, with the explicit goal of *forcing the negotiation of knowledge claims across disciplines* – all in front of the students (Bronet et al. 2003).

One of PDI's founders, anarchist philosopher John Schumacher, saw such real-time negotiation in the classroom as a critical method for helping students understand the disciplinary differences that need to be resolved in design. More ambitiously, however, Schumacher saw such negotiation as a way to convey to students the constructed nature of knowledge claims as well as the contingency of the authority structures underlying them. Many new students in the program experience the disagreement among their instructors with considerable consternation: "Whom should I believe? Which articulation of the assignment should I satisfy?" But, over time, the practice teaches them the need to exercise their own judgment in the complex and sometimes contradictory domain of design.

Despite precedent for design-centered curricula in creative design disciplines, design continues to play a minor role in most US engineering curricula (Dym et al. 2005). This is largely due to the existence of many competing program requirements, such that proposing additional content of any sort is met with the inevitable

[2]Contact hours calculations are based on hours in class under the guidance of instructors and teaching assistants, which inevitably will be higher than actual student-instructor interaction times. However, because PDI Studios do not have breakout sections assigned to teaching assistants, it is likely that the actual relative contact time is even higher in PDI.

chorus, "Where will it fit? What will be left out?" But the modest emphasis on design in engineering education is also due to an educational model founded primarily on technical analytical skills: Mathematics and basic sciences first, engineering sciences second, and design as a culminating exercise. In a sense, PDI seeks to invert this model by placing design front and center in students' educational experience, thereby using real-world problems as context of and motivation for learning mathematics, the sciences, and, eventually, the engineering sciences. In this way, PDI is an attempt to renegotiate "the fundamentals" of engineering education, making human problems the starting point for educational activities and placing technical analysis – and its underlying math and science – at the service of that problem solving approach.

Based on similar insight and ongoing concerns over engineering program retention, the past decades have witnessed an explosion of interest in design didactics (Eder and Hubka 2005), problem- and project-based learning (Mills and Treagust 2003), and service learning in engineering programs (Tsang 2000), as well as pre-college design education (Brophy et al. 2008). Amongst the most radical of such experiments in the US is Olin College, an engineering university created from scratch with a curriculum built around interdisciplinary problem-based learning and student-centered project work (Olin College 2012). In terms of disciplinary breadth and social analytic depth achieved through design studios, however, I am aware of no other program that goes as far as PDI.

STS Sequence

In addition to the interdisciplinary studio sequence, the Design, Innovation, and Society program requires a five-course-sequence in traditional STS areas. Here, the requirements include an introduction to STS, a course on design history and culture, two STS advanced topics courses (e.g., politics of design, sustainability problems, Internet culture, etc.), and a senior STS research project. The goal of these courses is to develop social analytic breadth and some depth, specifically around technology-and-society themes, using traditional social sciences and STS frameworks and approaches (i.e., not design per se). Also at 4 credit hours each, these courses taken together constitute 20 credit hours, or about 15 % of the dual-major total course load.

This STS course sequence, in combination with an engineering leadership sequence, satisfies Rensselaer's humanities and social sciences (H&SS) requirements for engineering majors. Even without the PDI Studios, the DIS STS sequence makes two improvements over the generic H&SS requirements for engineers. First, it structures students' H&SS course selection in a logical way that provides good breadth and some depth. Breadth is achieved because STS is constituted by several traditional H&SS disciplines: anthropology, history, political science, sociology, etc. Depth is achieved with one survey course, one topical introduction, two advanced topical courses, and a senior research project.

In terms of cumulating and refining social analytic skills, this approach offers considerable advantages over the *a la carte* H&SS course-selection process followed by most engineering students at Rensselaer and many elsewhere. Under the *a la carte* approach, students tend to take the entirety of their H&SS courses based on general interest, scheduling convenience, or perception of relative ease (Ollis et al. 2004). With notable exceptions – such as when pursuing minors in H&SS disciplines – engineering students do not identify and select H&SS courses in a structured way that facilitates the accretion of a social-analytic skillset.

Second, given that PDI students satisfy their H&SS requirements exclusively with STS courses – and not generic H&SS selections – the course content tends to connect more directly with students' career aspirations around technology design and the social analysis thereof. For example, taking a course on the politics of design, instead of, say, a more generic course on political theory, helps PDI students see clearly the relevance of political theory to their other coursework as well as to their future as practicing designers and engineers.

PDI Didactics and Pedagogy

The curricular structure of PDI is perhaps the program's major innovation and most notable achievement, because it structures and institutionalizes the integration of technical, social, and formal analysis and the integration of systematic analysis and creative synthesis. It does this with an organizationally sustainable model that has persisted apart from any individual's role in shepherding the program. (In fact, all three original faculty founders of the program have long departed; a formal director and two informal directors have cycled through the program; and literally dozens of faculty members have been involved in teaching studios and supporting courses.) Yet, despite the importance of the program's curricular structure, a look at classroom-level activities is needed to provide a clearer picture of how context is engaged by PDI students. Hence, this section of the chapter reviews didactics and pedagogy used in the program, where didactics are understood as the classroom-level activities used for educating students and pedagogy as their underlying theoretical foundations.

Certainly, numerous didactical strategies and pedagogical models operate in PDI classrooms, and they work together to create the distinct educational experience of PDI students. While my understanding of the full diversity of educational strategies at play in the program is admittedly limited, my direct experience collaborating with many PDI instructors and my direct or indirect experience with all of the PDI Studios provides some comprehension of the teaching approaches used across the program. I will identify several overarching approaches used in PDI, focusing particularly on those strategies that contribute most directly to student learning about and engagement with the matter of context in engineering and technology design.

Social-Technical Decompartmentalization

Given PDI's motivating goals around integration, it makes sense to elaborate on how, in general terms, disciplinary integration is achieved in the PDI Studios. While the logic of engineering capstone design courses may be said to be integration through accumulation of skills gained over prior years, PDI's main logic of integration – at least analytically – is decompartmentalization. PDI provides an alternative to disciplinary compartmentalization in skills building, and it does so both structurally and pedagogically.

PDI Studios overcome what is perhaps the most pernicious problem of engineering education as commonly structured: the systematic and pervasive compartmentalization of social (e.g., H&SS) and technical (e.g., "core engineering") educational content and experiences. The segregation is *temporal*, with separate courses addressing the technical and social in turn; it is *spatial*, with the H&SS coursework often taken across campus or in the "humanities building"; and it is *cognitive*, with engineering students and their instructors able to rapidly discriminate between "real engineering" and all the "soft," "fuzzy," "humanities" coursework that provides a "break" from the rigors of engineering courses (Ollis et al. 2004).

Importantly, PDI Studios do not necessarily attempt to *resolve* the social-technical binary. The separation is simply too deeply embedded for most students (and their instructors) to overcome. Plus, it provides a useful if simplistic conceptual tool for identifying and addressing the marginalization of "the social" within engineering. Instead of resolving the binary, PDI Studios decompartmentalize (i.e., integrate) students' work around the technical and social dimensions of the problems being addressed and the solutions being proposed.

Several specific mechanisms are used to treat technical and social dimensions together, but perhaps the single most important one is that both are considered the proper domain of the PDI Studio *as a course*. In other words, PDI Studios do not neatly fall into either side of the social-technical binary, since both dimensions of any given problem are addressed in all studios. This is not to say that PDI Studios are identified as "real engineering courses," even if they do include "hard technical" engineering content. But the status of PDI Studios as "not engineering" has as much to do with the rigid boundaries drawn around engineering courses as it does with students' level of engagement with technical content.[3]

As with the PDI Studio courses, the design projects students work on are similarly integrated across technical (or material) and social dimensions. Most PDI Studio projects seek to respond to "real-world" problems faced by specific populations, communities, or social groups – usually marginalized social groups. In contrast to the contextual simplifications commonly made in engineering problem

---

[3] For accounting purposes, some PDI Studios are labeled as engineering courses and others as H&SS courses, even though all might be more appropriately labeled part engineering and part H&SS.

solving, PDI students are taught from the very first semester that understanding the context of a given problem is the most important initial step of their design process.

Student engagement with a problem's context is open-ended and expected to be far-reaching, investigating wide-ranging dimensions such as: user needs and experiences; current solutions or work-arounds; organizational resources available in the local context that may be drawn upon in implementing a solution; financial constraints and resource pools; legal regulations applying to particular products, services, or industries; opportunities for engaging targeted users directly in conceiving of or implementing a solution; and potential environmental impacts including those created through material specification, energy requirements, and embodied energy.

Students also engage higher-order contexts that serve to exacerbate the problems they attempt to solve or to preclude consideration of certain types of solutions altogether. They consider cultural practices and values surrounding the problem, both dominant and competing alternatives; social power imbalances that create or exacerbate the problem; diverse stakeholder experiences that highlight differential perspectives on the problem and desired solution; and, importantly, limitations of consumerist-oriented innovation strategies themselves for addressing market-based inequities. Ultimately, consideration of social power imbalances – including around gender, race, socio-economic status, disability, age, etc. – provides much of the analytic foundations for addressing these higher-order contexts.

While learning about the context of a given problem is an early step in PDI students' design work, it is not neatly constrained within the first week or two of their design process. PDI Studios encourage the problem definition to iterate and evolve, both with students' deepening understandings of the problem as they engage it over time and with what becomes possible to achieve with a given solution. PDI students do not learn problem definition as engineers do: the translation of a story problem (i.e., a textual narrative of a problem that embeds a range of technical, financial, and sometimes social constraints) into a set of technical specifications that are precisely quantified. Instead, they learn it as the translation of messy and complex non-optimal real-world circumstances with incomplete understanding to a set of broad performance goals and then, ultimately, a type of product, service, or system that is likely to achieve those goals. Both social opportunities and constraints and material opportunities and constraints are included as the problem is redefined and refined throughout the design process.

Another mechanism by which PDI projects integrate across technical and social domains is by treating "the social" itself as a variable to be manipulated in the context of problem solving. It is not just that PDI students must understand the social context in order to better specify the technical solution, although that is certainly true. Also, PDI students are encouraged to explore changes to the social context – for example, regulatory changes – that might better enable a given technical solution – such as product reuse or recycling. This encourages deeper engagement with context, as students explore existing regulations in the given context, alternative regulations in different existing contexts, and then compare the contexts to determine the conditions under which regulatory change might be feasible. Treating context as variable also encourages students to grapple with the dynamic interplay between social forces and material conditions surrounding technological change: If technology and society are mutually shaped, then both can be understood as amenable to deliberate shaping through design.

Integration of STS Frameworks

In addition to the mutual shaping framework underlying PDI's social-technical integration, other conceptual frameworks from STS and Engineering Studies motivate and enable a more integrated educational experience for PDI students. Some of these concepts include: technological determinism and momentum, technological fix, large-scale sociotechnical systems, real-time and anticipatory technology assessment, politics of artifacts, actor-networks and lash-ups, macro-ethics in engineering, appropriate technology and design, technology appropriation, and others (see, e.g., Jasanoff et al. 1995; Hackett et al. 2008).

Students are taught many of these concepts explicitly in their STS concentration courses and sometimes in the PDI Studios, but PDI Studios have students applying the underlying analytic logic of the various concepts irrespective of using the particular label. So, for example, students learn the liabilities and limitations of the technological fix by confronting the shortcomings of employing a "technology only" solution in response to complex social problems (such as clean water access in remote, impoverished regions). Similarly, they learn about technology appropriation by creating educational technology prototypes that are inserted in a local grade school and then closely studying how the grade-schoolers actually engage the prototypes, sometimes aligning with the designers' intentions and sometimes not.

The primary educational dynamic at play is driven less by the STS concepts themselves and more by the presence of STS instructors continually probing students' assumptions and expectations surrounding their understandings of the problem as well as their proposed solution concept. In fact, this is one of the primary educational mechanisms of the studio model: continual feedback on student proposals for responding to open-ended problems. Having STS scholars, or similarly trained instructors, in the design studio encourages the application of STS frameworks in an anticipatory manner, which usually challenges students and instructors alike. Students are challenged to think more deeply, more systematically, and more theoretically about the relationships between people and things, and they are motivated to do so because their own proposed "solution" concept is at stake. STS instructors are challenged to move from the more straightforward domain of after-the-fact analysis and critique to suggesting tangible alternative approaches that can preempt such critique.

Because STS provides powerful conceptual frameworks for interpreting diverse contextual factors surrounding engineering design, STS seems particularly relevant for inclusion in an interdisciplinary design studio. However, the same basic logic should certainly work for a range of social sciences and humanities, not least including psychology (of user experience), social history (of any given technology), sociology (of consumption), feminist theory (on the gendering of technology and design), development studies (on Western economic assumptions undergirding most development logics), and so on.

In the context of PDI Studios, as alluded to above, a major challenge is to get the social-content expert to contribute to solution formation and not allow social criticism to degenerate into an infinite regress of problem identification. This usually requires some level of alignment between the social analysis and the technical content of the course. Where there are STS and engineering co-instructors in the PDI Studio, they need not agree on any given problem formulation – and often times they do not – but their feedback must intersect in a way that allows students to respond simultaneously to both in their design work. The worst outcome is when instructors talk past one another and students respond only one or the other of them in their design work.

Project Identification

PDI Studio projects are as wide ranging as the educational mechanisms for integrating across social and technical domains, but some patterns in project selection emerge as characteristic of the program, including especially projects that respond to enduring social inequities, the specific problems faced by marginalized social groups, and environmental problems. Different studio instructors have different motivations when it comes to project selection, but alignment around various facets of social inequity – e.g., challenges uniquely facing the global poor, the elderly, the disabled, women, children, etc. – offers repeated opportunities for students to confront a range of social forces that provide broad context for technology solutions.

In the most general terms, students confront problems of inequity in both their social and technical dimensions, including the social (economic, political, cultural, historical) structures that lead to or reinforce inequities (Nieusma 2004b). Whereas some systemic inequities are connected with contextual variables that elude students' efforts to productively rein them in – especially some of the projects around global poverty – others are responded to quite deftly by students – particularly when the cultural gap between user and designer is not too wide.

The scope of PDI design projects is equally important as their target users. In most cases, PDI projects are grounded in a particular empirical context, so that students have a specific place to go to research the problem in context. Projects are open-ended in two ways: First, the exact problem definition is determined by students, not instructors; second, the nature of the solution is generally quite open. Such open-endedness has considerable risks in the design studio, especially with some engineering students who are accustomed to being given considerable structure within which to design. The major risk is that students spend too much time imagining "what if…" and avoid focusing on a particular dimension of the problem, a particular target audience, or a particular solution concept. To offset this risk, PDI projects usually have students identify a specific empirical context early and then work from there. Examples of such contexts include: the grade-school classrooms mentioned above; a local nursing home; a local community fitness center; and any of numerous existing entrepreneurial initiatives in the region. When projects are set in non-local contexts, specificity is still prioritized. Rather than addressing foot-borne disease across Africa, say, students focused on Cameroon.

Significance of PDI Achievements

In the previous sections, some of the specific educational innovations employed in PDI were explained. The rest of the chapter assesses these innovations. This section reflects on the significance of what has been achieved, and the next section takes on limitations of our approach.

Perhaps the greatest achievement of PDI is the enthusiasm by which it has been received by students. Admittedly, the program attracts only a small cohort of students, but these students, by and large, are deeply engaged with the program, its courses, and its approach and goals. Program instructors may provide educational context and content, but students energize the program and, in particular, the studio environment. What is significant here, however, is not that students are *pleased* with the program, but that they are *engaged* with it. In fact, students are very vocal about the many shortcomings of the program, which is another dimension of their engagement. Their engagement is even reflected in how they identify themselves. Most PDI students identify as "PDI students" first and "engineering students" second. Some of these even identify as "designers" more than as "engineering designers," even as they identify also as "engineers."

The high level of student engagement with the program certainly makes recruitment easier and program management more rewarding, but it also provides a tangible educational benefit, namely, that the students are willing to grapple with complex, underspecified problems as well as the constant barrage of questions and criticism by instructors and still put forward one concept proposal after another. The high level of engagement with the program creates willingness to engage critical social analysis in particular, especially as compared to modal engineering students, and for some students it even enhances their performance in technical engineering courses (Patterson et al. 2011).

Despite their identification as not-typical engineers and their attraction to social-problem solving, PDI students nevertheless remain engineers and retain a strong affinity for technological solutions to human problems. The social-material interplay present across the program enables PDI students to go beyond the technological fix and grapple with the social systems that contextualize technology innovation, but *without rejecting, dismissing, or ignoring the productive role technology can play* in social problem solving. Balancing attention to both the technological/

material and the social as both problem and solution is perhaps PDI's signature educational achievement.

PDI has also been successful in inverting the "fundamentals first" (Baillie 1998) approach to education dominant in engineering programs, which in its simplest formulation entails math and science first, followed by engineering science, followed by engineering analysis, and finally "application" via design. Starting with design from the very beginning ensures complex problem solving – and in particular the weighing of competing possibilities (i.e., design trade-offs) – is part of how engineering students learn to think *as engineers*. This approach provides an antidote to the tendency in engineering problem solving to emphasize one correct answer to a given problem. The design-first approach is also effective as a retention tool in that early design experiences provide many students an intellectual framework within which to situate the "fundamentals" they are learning in their other courses (Knight et al. 2007). Put more precisely, PDI does not actually invert the fundamentals first approach. Instead, PDI raises the question of what is, in fact, most fundamental to engineering and then answers the question: "solving complex social-technical problems."

In a similar vein, PDI has unsettled STS's educational role as well. It has challenged STS instructors to apply their conceptual tools in new ways, anticipating how the impacts of a given design decision will play out in order to contribute more directly to students' visions of a future improved with technology. As with engineering approaches, PDI demands STS tools respond to and serve students' solution formulation, not artificially drive it. STS instructors must continually make the case for the relevance of their frameworks, rather than assume such relevance and enforce their application.

In this way, STS instructors can teach students how to question "the rules of the game" at the same time as the students are learning how to play the game. For instance, they can teach students how to critique consumer culture even as students are designing what would become another consumer product. Thus, PDI teaches students not only to learn how to read context, but also how to engage it and participate in its creation. Ultimately, this leads to a degree of hybridity among PDI students that instructors, perhaps, cannot fully comprehend, grounded as we are in our own disciplinary, and sometimes interdisciplinary, training (Jamison et al. 2011). One may hope that such hybridity among our students supersedes and supplants our analytic categories, especially those built around the social-technical binary.

Limitations of the PDI Approach to Engineering Education Reform

The thrust of this analysis has been overwhelmingly affirmative, and I am confident that PDI has, in fact, succeeded in important ways in providing a compelling alternative to engineering education as typically formulated. However, I am also clearly aware of limitations and shortcomings evident in the PDI approach.

Most notably, despite our successes around social-technical integration, the program remains, for the most part, *institutionally outside engineering*: outside Rensselaer's School of Engineering, outside "engineering" classrooms, outside ABET's accreditation oversight, and devoid of participation by the vast majority of engineering faculty members (including many of those who otherwise support our efforts). Because PDI reforms engineering education from the outside, the dominant educational experience of our students with the engineering side of their dual major is a point of frustration and, often, contention. While most PDI students openly favor what they do in PDI Studios over what they do in their "real engineering" courses, they also recognize the cultural authority of engineering, however narrowly bound. They are "PDI students" and "engineering students," but they are not "PDI engineers" because they recognize contradictions between the two identities.

Another limitation of the program is that more than a few students retain a degree of discomfort with, and sometimes even hostility toward, the critical social analysis conveyed primarily by STS instructors. This is true sometimes in PDI Studios but more often in traditional-format STS courses. Above, I claimed that a major achievement of the program was getting engineering students to be comfortable with and competent in critical social analysis, which remains valid. Nevertheless, a non-trivial minority comes to the program and stays in it less for the social analytic skills and more for the opportunity to engage in creative design work. Some students even state that they are in the program because it is the closest Rensselaer has to product design, and that they wish the program focused "less on STS and more on industrial design." The extent of this problem is not perfectly clear, but the perceived sexiness of design professions and the allure of consumer product innovation (e.g., the next Apple iGadget) are palpable among most PDI students. While PDI instruction sometimes highlights the productive tension between social/technical criticism and technology innovation – much as the art critic advances art through criticism – at least a handful of students appears to find the tension unsettling.

A third limitation of the PDI model for engineering education reform is scalability. In a variety of ways, Rensselaer's particular institutional context is distinctive: a well regarded engineering and science school seeking to become a "world-class technological university" by diversifying program offerings; a top-ranked STS department administratively housed apart from engineering programs but which teaches courses primarily to engineering students and has several faculty with design and engineering studies expertise; an administrative environment that strongly incentivizes attracting majors regardless of institutional location; and relatively low walls for interdisciplinary collaboration among faculty members across the institution. PDI Studio courses are resource intensive: they require 50 % more time in the classroom for instructors as well as students, and they require instructors who can either provide coverage of both technical and social domains or are willing to co-teach with instructors who have radically different training and perspectives.

Many elements of PDI could be introduced easily in a variety of engineering programs, and the past decade has witnessed plenty of related experimentation: interdisciplinary design studios, project-based service learning programs, H&SS courses directed explicitly to the interests and proclivities of engineering students.

But implementing such initiatives simultaneously, with deep penetration into the curriculum over all 4 years, and for a large number of engineering students is a considerable challenge. This challenge has been addressed uniquely if partially at Rensselaer, and our approach may not translate well to other institutions.

Finally, one very practical limitation of PDI – relevant especially in the context of this chapter – is the lack of systematic, empirical assessment of the program and its students. This analysis has been concerned with relatively high-level program strategy and performance, in part because there is no good data on how our students rate the program (apart from what they communicate directly with me), how they compare with non-PDI engineering students, or how they perform in their careers 5 or 10 years out. While being careful to avoid speculation where possible in my appraisal of PDI's effectiveness, without assessment data, this analysis is necessarily limited.

Conclusion

This chapter has described Rensselaer's Programs in Design and Innovation (PDI) as one model of engineering education reform aimed at tackling the challenges of contextual awareness and analysis. Using a reflexive approach, the chapter reviewed the dual-major structure of the program, its educational strategy, its most significant achievements, and some important limitations of the PDI model. In each of these areas, a social-technical dualism played an important role, both in framing the conceptual challenge surrounding engineering education reform and in providing strategies for responding to those challenges.

Ultimately, PDI might be understood as a far-reaching experiment in boundary work around the social-technical divide in engineering education, where the simple division between technical content and social context fails to capture how engineering design actually works in the world and how engineering students might be educated to prepare for it. Rather, by assuming technology and society are mutually constitutive, PDI takes the technical and the social as simultaneously both content and context, and thereby challenges common understandings of what engineering is and what engineers ought to know.

Acknowledgements I would like to thank participants of the 2012 Engineering in Context workshop at the Massachusetts Institute of Technology for productive discussion that helped direct this chapter's analysis as well as the editors of this volume for helpful feedback on the chapter's presentation. I would also like to thank PDI instructors and PDI students for creating, together, a stimulating environment in which to observe, practice, and theorize social-technical integration.

References

Baillie, C. (1998). Addressing first-year issues in engineering education. *European Journal of Engineering Education, 23*(4), 453–465.

Bijker, W. E., Hughes, T. P., & Pinch, T. (1987). *The social construction of technological systems: New directions in the sociology and history of technology*. Cambridge, MA: The MIT Press.

Brophy, S., Klein, S., Portsmore, M., & Rogers, C. (2008). Advancing engineering education in P-12 classrooms. *Journal of Engineering Education, 97*(3), 369–387.

Bronet, F., Eglash, R., Gabriele, G., Hess, D., & Kagan, L. (2003). Product design and innovation: Evolution of an interdisciplinary design curriculum. *International Journal of Engineering Education, 19*(1), 183–191.

Cañavate, J., Casasus, J. M., & Lis, M. J. (2009). Integrating public context perception in engineering design. In S. H. Christensen, B. Delahousse, & M. Meganck (Eds.), *Engineering in context* (pp. 277–289). Århus: Academica.

Christensen, S. H., Delahousse, B., & Meganck, M. (Eds.). (2009). *Engineering in context*. Århus: Academica.

Downey, G. L. (2009). What is engineering studies for? Dominant practices and scalable scholarship. *Engineering Studies, 1*(1), 55–76.

Downey, G. L., & Lucena, J. (1995). Engineering studies. In S. Jasanoff, G. E. Markle, J. C. Petersen, & T. Pinch (Eds.), *Handbook of science and technology studies* (pp. 176–188). Thousand Oaks: Sage.

Dutson, A. J., Todd, R. H., Magleby, S. P., & Sorensen, C. D. (1997). A review of literature on teaching engineering design through project-oriented capstone courses. *Journal of Engineering Education, 86*(1), 17–28.

Dym, C. L., Agogino, A. M., Eris, O., Frey, D. D., & Leifer, L. J. (2005). Engineering design thinking, teaching, and learning. *Journal of Engineering Education, 94*(1), 103–120.

Eder, W. E., & Hubka, V. (2005). Curriculum, pedagogics and didactics for design education. *Journal of Engineering Design, 16*(1), 45–61.

Hackett, E. J., Amsterdamska, O., Lynch, M., & Wajcman, J. (2008). *Handbook of science and technology studies* (3rd ed.). Cambridge, MA: MIT Press.

Jamison, A. (2009). The historiography of engineering contexts. In S. H. Christensen, B. Delahousse, & M. Meganck (Eds.), *Engineering in context* (pp. 49–60). Århus: Academica.

Jamison, A., Christensen, S. H., & Botin, L. (2011). *A hybrid imagination: Science and technology in cultural perspective*. San Rafael: Morgan & Claypool.

Jasanoff, S., Markle, G. E., Petersen, J. C., & Pinch, T. (1995). *Handbook of science and technology studies*. Thousand Oaks: Sage.

Knight, D. W., Carlson, L. E., & Sullivan, J. F. (2007). Improving engineering student retention through hands-on, team based, first-year design projects. In *Proceedings of the 31st international conference on research in engineering education*, Honolulu.

Kroes, P., & van de Poel, I. (2009). Problematizing the notion of social context of technology. In S. H. Christensen, B. Delahousse, & M. Meganck (Eds.), *Engineering in context* (pp. 61–74). Århus: Academica.

Leydens, J. A., & Lucena, J. C. (2009). Knowledge valuation in humanitarian engineering education. In S. H. Christensen, B. Delahousse, & M. Meganck (Eds.), *Engineering in context* (pp. 147–162). Århus: Academica.

MacKenzie, D., & Wajcman, J. (Eds.). (1999). *The social shaping of technology* (2nd ed.). Philadelphia: Open University Press.

Mills, J. E., & Treagust, D. F. (2003). Engineering education: Is problem-based or project-based learning the answer? *Australian Journal of Engineering Education*, online publication 2003–2004. Available at http://www.aaee.com.au/journal/2003/mills_treagust03.pdf. Accessed 5 Mar 2013.

Nieusma, D. (2004a). *The energy forum of Sri Lanka: Working toward appropriate expertise*. Dissertation, Rensselaer Polytechnic Institute.

Nieusma, D. (2004b). Alternative design scholarship: Working toward appropriate design. *Design Issues, 20*(3), 13–24.
Nieusma, D. (2008). Integrating technical, social, and aesthetic analysis in the product design studio: A case study and model for a new liberal education for engineers. In *Proceedings of the 2008 annual conference and exposition of the American Society for Engineering Education*, Pittsburgh.
Olin College. (2012). *Curriculum. Franklin W. Olin College of Engineering*. Available at http://www.olin.edu/academics/curriculum.aspx. Accessed 5 Mar 2013.
Ollis, D. F., Neeley, K. A., & Luegenbiehl, H. C. (2004). *Liberal education in 21st century engineering*. New York: Peter Lang Publishers.
Patterson, E. A., Campbell, P. B., Busch-Vishniac, I., & Guillaume, D. W. (2011). The effect of context on student engagement in engineering. *European Journal of Engineering Education, 36*(3), 211–224.
Splitt, F. G. (2003). The challenge to change: On realizing the new paradigm for engineering education. *Journal of Engineering Education, 92*(2), 181–187.
Steiner, M., & Winner, L. (2005). Thoughts and reflections on Rensselaer's product design and innovation program. In *Proceedings of the engineering and product design education conference*, Napier University, Edinburgh.
Tsang, E. (Ed.). (2000). *Projects that matter: Concepts and models for service-learning in engineering*. Washington: American Association for Higher Education.
Winner, L. (1995). Political ergonomics. In R. Buchanan & V. Margolin (Eds.), *Discovering design: Explorations in design studies* (pp. 146–170). Chicago: University of Chicago Press.

Dean Nieusma B.S. Mechanical Engineering and Bachelor of General Studies, University of Michigan. M.S. and Ph.D. in Science and Technology Studies, Rensselaer Polytechnic Institute. Director of Programs in Design and Innovation and Associate Professor of Science and Technology Studies, Rensselaer. Editor of *International Journal of Engineering, Social Justice, and Peace*. Research interests: engineering professional and educational reform; interdisciplinary collaboration in design; appropriate technology design; design pedagogy; expertise and democratic theory.

Chapter 21
PDS: Engineering as Problem Definition and Solution

Gary Lee Downey

Abstract All of us who teach engineers share at least one common problem: the continuing dominance of an image of engineering formation that places highest value on mathematical problem solving in the engineering sciences. The image grounds a claim of jurisdiction over technology through design. This essay offers an alternative image of engineering as problem definition and solution (PDS) and takes initial steps toward facilitating its travel. The analysis outlines four contemporary challenges to the engineering claim of jurisdiction: changes in the work of scientists, mass production of engineers for technical support, credentialing by exam alone, and shared jurisdiction through teamwork. It then explains that PDS avoids incorporating the image of "breadth" because it lacks an organized vision. Four sets of PDS practices include early involvement in problem definition, collaboration with those who define problems differently, assessing alternative implications for stakeholders, and leadership through technical mediation. Three sets of strategies for enabling the PDS image to travel include adapting pedagogies in engineering science courses, adapting pedagogies in peripheral courses, and adapting curricula to produce more than one thing. What might engineers be if a PDS image gained acceptance across the terrains of engineering formation? Could integrating PDS practices into your teaching work for you?

Keywords Engineering education • Engineering sciences • Engineering problem solving • Problem definition • Dominant images

---

This chapter revises and updates a keynote address to the 2005 World Congress of Chemical Engineering (Downey 2005). Excerpts from the original appear with permission. Thanks much to the editors for their work organizing this volume. Thanks much to Carl Mitcham for his close reading and editing of two drafts of this manuscript. This work is also based upon work supported by the National Science Foundation under grant #DUE-1022898. Any opinions, findings, and conclusions or recommendations expressed in this material are those of the author(s) and do not necessarily reflect the views of the National Science Foundation.

G.L. Downey (✉)
STS Department 0247, Virginia Tech, Blacksburg, VA 24061, USA
e-mail: downeyg@vt.edu

S.H. Christensen et al. (eds.), *International Perspectives on Engineering Education*, Philosophy of Engineering and Technology 20, DOI 10.1007/978-3-319-16169-3_21

Introduction

Do you teach engineers? Are you perhaps an instructor in the heart of the engineering sciences, taking care to make sure your students know how to recognize and solve the difficult technical problems they will encounter on the job? Or perhaps your thing is engineering design, helping students learn how to draw on the engineering sciences to develop new technologies. Either way, might it trouble or frustrate you that prestigious reports, changing accreditation regulations, outside critics, and, perhaps, globalizing employers seem to advocate cramming all sorts of new content into engineering curricula to address their concerns? Are you feeling crowded and overwhelmed with impossible demands?

Are you perhaps an activist engineering educator, testing out new curricular strategies to help engineering students develop a broader range of skills? Do you find your opportunities limited primarily to the first and last years of degree programs? Are you feeling supplemental?

Are you perhaps an instructor in the humanities or social sciences, like me, dedicated to helping engineers become better critical thinkers and practitioners? Does it trouble or frustrate you that centers of power in the making of engineers seem to keep it focused primarily on technical capabilities? Does the force of resistance to incorporating new questions and commitments into engineering formation (Downey 2014) and engineering work still seem overwhelming despite quality critical analysis? Are you feeling left out?

All of us who teach engineers, along with others who care about or are affected by the outcomes of engineering education, share at least one common problem. It is the continuing dominance across many countries of an image of engineering formation that places highest value on mathematical problem solving in the engineering sciences. That image pictures students acquiring diverse practices of mathematical problem solving in order to apply them in the design and construction of new technologies. It links problem-solving to technological development through design. This image of engineering problem solving not only dominates the making of engineers. It has also long grounded claims by engineers to have jurisdiction over technology. By jurisdiction I mean, following Andrew Abbott, intellectual and social control over an arena of expert practice (Abbott 1988, p. 20).

The problem for those who teach in the engineering sciences or engineering design is that this image is woefully incomplete. Since at least the early 2000s, limitations in the engineering claim to jurisdiction over technology have become too obvious to ignore. It is no longer exclusive. The dominant image of engineering problem solving and technological design is scaling down.[1]

The problem for those seeking to expand engineers' skills is that it asserts a distinction between the technical and nontechnical dimensions of engineering work. That distinction makes it difficult for their work to achieve both force and coherence.

[1] For theoretical elaboration of dominant images and scale, see Downey (2009).

The problem for those who critique and resist the dominant image, claiming it is flawed, is that we have not been able to move our practices from the periphery of engineering curricula. We have not been able to successfully challenge the dominant image by offering an alternative that can scale up among engineers, let alone gain dominance. How does one overcome marginality in the face of what feels like immoveable resistance?

This essay offers the image of engineering as problem definition and solution as an alternative to the dominant image of engineering problem solving and technological design. I call it PDS for short. My purpose is to use it to enable practices of critical self-reflection to travel within and alongside practices of mathematical problem solving. The PDS image seeks to hire in, i.e., to participate critically in and inflect the dominant image rather than attempt to undermine and replace it entirely. It accepts the twin risks of co-optation and social engineering (Downey and Lucena 1997, p. 120).

The argument below elaborates the PDS image and outlines a set of practices for facilitating its travel across arenas of engineering formation. To advocates of engineering sciences and design, it claims, only partly tongue-in-cheek, that current curricula teach but half of what engineers need to know to be effective practitioners and leaders. "Collaborative problem definition" is my label for the other half. To activists in engineering education, PDS offers an alternative to "breadth" as an organizing image for new competencies. To those of us who critique engineering's continuing core emphasis on mathematical problem solving and its extrapolation into design, the argument is that PDS can provide an organizing image for integrating practices of critical self-analysis into the making of engineers. It argues that skills and speaking can function more effectively with questions and listening.

I have no illusion or expectation that integrating practices of collaborative problem definition into engineering education would be sufficient to produce technical practitioners who routinely question and thoughtfully adapt their normative commitments as everyday practices of expertise. I am suggesting, however, that such may be a necessary step, achievable by integrating the questions "What is engineering for?" and "What are engineers for?" into engineering practice at every moment. Getting there would radically reframe the next steps.

I begin by outlining four contemporary challenges to the engineering claim of jurisdiction over technological innovation. Other fields have begun claiming jurisdiction in practices of technological development. To the extent engineers acknowledge such claims, continuing to place primary emphasis on solving technical problems amounts to accepting a significant reduction in the status and value of engineering work. How can engineers claim to be unique when others do technology too?

Seeing through the PDS lens depends upon avoiding or abandoning the desire for breadth in formal engineering education. The next section argues that what that image hides far outweighs what it makes visible. The balance of the essay elaborates the PDS image by identifying four sets of constitutive practices and three sets of strategies for enabling it to travel. As analysis, it invites you to reflect on the question – "Could engineers be for more things if an image of engineering as problem definition and solution successfully gained substantial acceptance across schools of engineering?" As an attempt at critical participation, it asks if integrating PDS practices into your teaching might work for you.

The Lost Claim of Jurisdiction

Let's now examine four threats to the claim that engineers have exclusive jurisdiction over the creation or design of new technologies.

Scientists and Technology

The first and seemingly most threatening set of challenges to the engineering claim of jurisdiction over technology has come from significant changes in the work of scientists. The dominant image of engineering problem solving and technological design has long depended upon an image of science and scientists as upstream and, hence, out of the way.

The U.S. National Academy of Engineering's 2004 report *The Engineer of 2020: Visions of Engineering in the New Century* began, for example, with the simple, definitive jurisdictional claim: "Technology is the outcome of engineering" (2004, p. 7). It went on to explain that science lay upstream in the realm of unrestricted inquiry and discovery. "It is rare," asserted the report, "that science translates directly to technology, just as it is not true that engineering is just applied science. Historically, technological advances, such as the airplane, steam engine, and internal combustion engine, have occurred before the underlying science was developed to explain how they work" (p. 7).

The image of science upstream had some plausibility through the mid-twentieth century. As economic historians David Mowery and Nathan Rosenberg found in analyzing time delays between discovery and application from the late nineteenth and to the mid-twentieth century, "technological exploitation of new scientific understanding often require[d] considerable time because of the need for additional applied research before the economically useful knowledge [could] be extracted from a new but abstract formulation" (1989, p. 25). They further found, however, that by the 1980s, "scientific research was [now] loosely tied to [technological] innovation" (p. 28).

Much evidence exists of a turn toward technology among scientists, especially after the Cold War. Consider the expansion in the numbers and character of patents awarded to universities, the traditional centers for basic, unrestricted research. The U.S. National Science Board reported as early as 2004 that "[p]atenting by academic institutions has markedly increased over the past three decades, rising from about 250–350 patents annually in the 1970s to more than 3,200 patents in 2001" (2004, pp. 5-53–5-54). The number of academic institutions receiving patents nearly tripled and the share of patents granted to them increased from 1.5–4 %. Critically, this growth centered not in engineering but "occurred primarily in the life sciences and biotechnology" (p. 5-55) The disciplines experiencing the fastest growth were chemistry, molecular biology, and microbiology.

Another indicator lay in changes in the scope of funding for scientific research. In the early 1980s the U.S. National Science Foundation both acknowledged and contributed to an increasingly blurred distinction between basic and applied science when it stopped designating applied science as a separate funding category (Lucena 2005). Also, in 1987 NSF introduced funding for multi-institutional, multidisciplinary "Science and Technology Centers" with the aggressive economic goal of "respon[ding] to rising global competition by "mount[ing] an innovative, interdisciplinary attack in important areas of basic research" (Graphics and Visualization Center 2004).

Beginning in the 1990s, NSF dramatically increased the number of programs linked directly to technological outputs, expanded programs that encouraged direct collaborations with industry, and rewrote virtually all science program descriptions to include technological development as a desirable outcome alongside contributions to knowledge, education, and training. It also began requiring all project summaries to demonstrate not only the "intellectual merits" of the project but also its "broader impacts" (National Science Foundation 2012). One clear way to demonstrate broader impacts is to posit links between research and potential new technologies.

The delay Mowery and Rosenberg found lay in a research world in which physics provided the dominant image of scientific knowledge production. Images blurring the claimed boundary between science and technology began to scale up with the shift toward the life sciences and information technology. In the much-celebrated field of tissue engineering, for example, the interdisciplinary collaborations of practitioners from biophysics, developmental biology, materials science, biochemistry, genomics, and several braches of medicine with chemical and mechanical engineers demonstrate the increasing comfort scientists have in associating themselves with fields that might be labeled "engineering" (Hogle 2003; Williams 2002) The same can be said for the more recent emergence of synthetic biology – the engineering of biology (Schyfter et al. 2013). Note also that many cutting-edge nanoscientists judge themselves as having fully established their professional reputations only after founding successful start-up companies (Baird and Shew 2004).

The increased degree of comfort among scientists with technological development can be found in the U.S. National Research Council's 2003 report *Beyond the Molecular Frontier: Challenges for Chemistry and Chemical Engineering.* Strikingly, the report "departs from the earlier practice of treating chemistry and chemical engineering as separate disciplines," instead lumping them together under the more general term "chemical sciences." The stated goal was to present "the entire spectrum of activities in the chemical sciences," a spectrum that now includes not only "research" and "discovery" but also "invention." All this was justified, the report held, by "strong couplings" between chemists and chemical engineers in universities and industries (2003, p. 2). In short, invention and technological development no longer distinguished chemical engineering from chemistry, and it was not the label "engineering" that was being celebrated and extended.

Mass-Produced Engineers for Technical Support

A second challenge to the engineering claim of jurisdiction over technological innovation emerged from the mass production of engineers trained only in the engineering sciences. In a 2004 interview, a senior engineering official and influential government consultant from Cairo University in Egypt complained that while the Faculty of Engineering judged itself to have a capacity of 4,000 students, its enrollments typically exceeded 15,000 students in any given semester. Staff members necessarily structured classwork around large lectures and annual exams, testing students' knowledge of relevant engineering sciences (confidential interview, June 2004). The implications go far beyond Egypt since that country has long been a major producer and exporter of engineering graduates trained almost entirely in the engineering sciences to countries across the Middle East.

At the 2004 annual meeting of the U.S. National Academy of Engineering, President William Wulf claimed that "new U.S. engineers account for only about 7.5 % of the world total" (Wulf 2004). He was drawing data from the *2020* study, which highlighted the "rapidly improving educational capabilities in countries like China and India" and asserted that China along was producing "more than three times the graduates in all fields of engineering than is the United States" (2004, p. 33). As Gereffi et al. (Bracey 2006; Gereffi et al. 2008) have shown, what gets reported as engineers in other countries would often be classified as technicians in the United States, with vocational certificates or associates degrees. Yet whether these are engineers or engineer-technicians, the increasing numbers in China, India, Egypt, Philippines, and other countries suggest they are scaling up an image that may well fit what engineers across the planet are perhaps increasingly becoming – technical functionaries in support positions.

The *2020* report identified two key features of this emergent image. These workers are "highly skilled… with engineering and science backgrounds," and they were "willing and able to work for wages well below those in the developed nations" (2004, p. 33). In 2005, I placed four telephone calls for technical support for a Palm Pilot. Two were answered in India, two in the Philippines. All four technicians claimed to hold bachelor's degrees in computer engineering.

One can argue that producing engineers for technical support is an American export, an industrial system that seeks low-wage workers to fuel low-cost production for mass consumption. At the same time, a key implication just may be a reverse flow of influence in what can be claimed as the jurisdiction of engineering – the scaling up of an image in which engineers are valued more for their work as technical problem solvers and less as technology creators.

Credentialing by Exam Alone

The historian of technology Rosalind Williams points toward a third, related challenge to the identities of engineers in an insightful and engaging account of institutional transformation at MIT during the late 1990s and early 2000s. Williams found that "[a]ll engineering departments are becoming, in some form or other, to a greater or lesser extent, departments of applied-information technology" (2002, p. 46). Increasing reliance on a common digital language, she argued, "lifts engineering, once the most down-to-earth of professions, from its familiar ground of materiality, endowing it with a ghostly lightness of being" (p. 47). This dematerialization of engineering work was pulling at least some engineers into a densely populated world of information technology workers, millions of whom had already gained "engineering" credentials by passing exams rather than completing curricula.

In the 2000 U.S. Department of Education report *A Parallel Postsecondary Universe: The Certification System in Information Technology*, longtime education researcher Clifford Adelman mapped out the contours of a system that between 1997 and 2000 produced over two million information technology certifications worldwide, while operating as an "international guild" almost entirely outside of government-operated systems of data collection and accreditation (Adelman 2000). Armed with such titles as Accredited Systems Engineer (Compaq), Certified Novell Engineer, Microsoft Certified Systems Engineer and Red Hat Certified Engineer, students "assemble valises of special knowledge and skills, apply them in different work-organization contexts, and modify them by (1) personal predilection, (2) personal perception of potential 'work-life' paths, and (3) labor market changes" (p. 30). These new adaptive, flexible workers realized that "work life mobility demands the transparent and portable evidence of a certification" (p. 3). This challenge may not have affected most engineers. But the easy use of the term "engineer" in such contexts illustrates the potential risk of devaluation associated with defining engineering work as technical problem solving for clients.

Shared Jurisdiction Through Teamwork

Finally, a fourth source of challenge emerged from a phenomenon that is to this day frequently characterized as a site of promise and opportunity for engineers (which it could be) – the institutionalization of teamwork in industry. Through a succession of movements including total quality management, business process re-engineering, knowledge management, and a variety of other practices, industrial organizations have worked to restructure themselves into flexible mazes of product and process development teams.

Teamwork puts engineers at the table with business managers, marketing and sales-people, researchers, labor representatives, information technology specialists, etc. Effective teamwork necessarily affords all participants some measure of responsibility over and, hence, identification with technological developments. Placing greater emphasis on teamwork in formal engineering education makes it increasingly difficult for engineers to claim jurisdiction over technology for themselves.

Indeed, to the extent engineers may be the participants most inclined to understand the problem at stake in exclusively technical terms, they might very well be least likely to respond to such shared responsibilities in other than defensive terms.

Might becoming a good team member occur in spite of core engineering training rather than because of it?

Overall, changes in the work of scientists, the mass production of narrowly-trained engineers, the rise in engineering certifications through exams, and an increased emphasis on teamwork combine to make visible a unique vulnerability in engineers' identification with technological development and dominant understanding of themselves as technical problem solvers. By claiming jurisdiction over the solving of technological problems, engineering has positioned itself as society's technological consultant, there to help but only when asked. The claim to creativity in technological development is now contested directly by both research scientists and teammates in industry. The re-visioning of engineering into technical support may be modeled by the mass production of engineers in poorer countries and easy appropriation of the label by those who certify engineers with a single exam.

Is it not now obvious to all what has long been clear to scholars in technology studies – that engineering does not (and likely never did) have jurisdiction over technological development? Many fields of engineering have been attempting to integrate bio-, info-, and nanotechnologies, in particular, into their jurisdictions by redefining the engineering sciences at their core. The replacement of unit operations in chemical engineering with multi-scale analysis is a good example (Gillett 2001). But might such efforts misdiagnose and fail to respond adequately to a more fundamental challenge? Might the main challenge facing the making of engineers in the present be to re-imagine and re-define in its entirety the obligatory core and essential heart of engineering identities?

The Limitations of "Breadth"

A key prerequisite to re-theorizing the core of engineering learning and work is to move beyond a geometry of "narrowness" and "breadth." For one thing, the critique of narrowness in engineering education has a long history without resolution. MIT professor Henry Talbot was writing in 1911 when he offered a thoughtful defense of the engineering curriculum against "the general charge of 'narrowness' and inadequacy which is directed against our courses" (Talbot 1911, p. 118).

But the main problem with the critique of narrowness is that it necessarily posits breadth as the solution. As Williams explained, the 1949 Lewis Report at MIT, authored by Warren K. Lewis, a founder of chemical engineering and her grandfather, labeled its central recommendation "A Broader Educational Mission." It asserted that "we recognize especially a need to develop a broader type of professional training that will fit engineers to assume places of leadership in modern society… (2002, p. 67). Likewise, the *2020* report called for engineers "who are broadly educated, who see themselves as global citizens, who can be leaders in business and public services, and who are ethically grounded" (2004, p. 5). Between these two reports, and since, are hundreds of other examples.

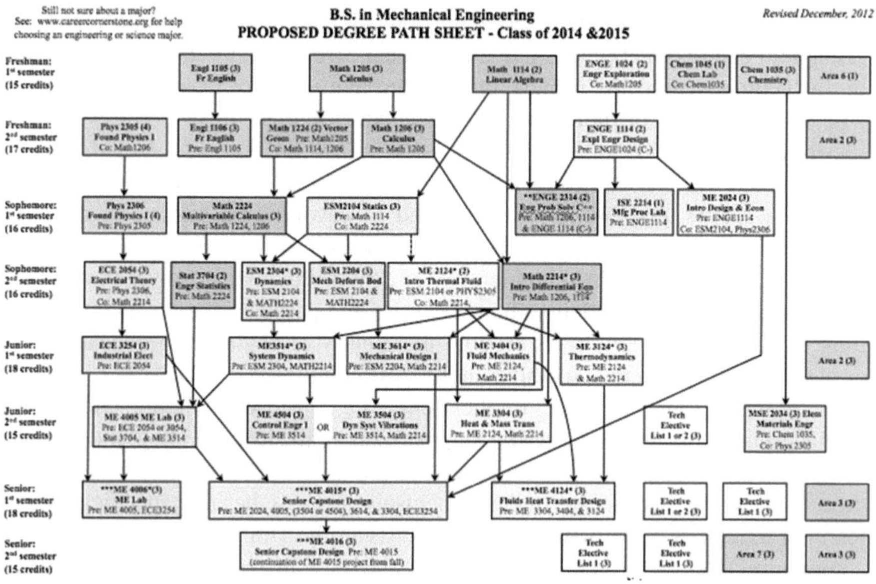

Fig. 21.1 M.E. degree path sheet

The broadly-trained engineer is an attractive image. One can make a plausible case that broadening the training of engineers could help educators address several vexed problems, including ameliorating European difficulties in attracting quality students, U.S. difficulties in recruiting and retaining women and underrepresented minorities, the general invisibility of engineers, lack of public understanding of what engineers do, and, particular to Europe, difficulties in contributing affirmatively and collectively to the Bologna process (designed to make credits and degrees interchangeable).

The image of breadth is problematic, however, because it tends to preserve a distinction between core and periphery, with technical problem solving at the core and everything else at the periphery. Figure 21.1 offers a current example of how this works. The diagram is a flowchart of a U.S. mechanical engineering curriculum distributed to students to guide them in course selection. Similar diagrams could be constructed of other curricula.

Readers need not examine the course titles and numbers inside the boxes. The diagram's key feature is the array of vertical and horizontal lines that constitute the curricular core in an interlocking network of prerequisites and co-requisites. Sitting directly above them are important preparatory experiences in the basic sciences. However, the main broadening experiences, elective courses in the humanities and social sciences ("areas" 2 and 3), sit off to the side on the right, connected neither to one another nor to anything else. They are peripheral.

In the vast majority of engineering curricula, breadth is supplementary. While a given field can reasonably legislate its technical core, it cannot do so with breadth. In this geometry, students achieve breadth through mixes of classes they select at will and integrate, or not, on their own according to their preferences and sensibilities.

The image of breadth thus lacks an organized vision. Discussions about how to overcome narrowness through breadth tend to devolve into arguments over the appropriate distribution of credits between the required core and elected peripheries. For engineering faculty who identify (or contextualize) themselves through the technical core, using it to define their identities and passions, the prospect of whittling away at core credits risks eroding the quality of engineering education and even transforming it into something entirely different.

In a move with dramatic implications, the U.S. Accreditation Board for Engineering and Technology (ABET) in 2000 shifted the locus of integration among the technical and nontechnical dimensions of engineering education from credits on the student's transcript to the students themselves. These became specifications of learning outcomes, greatly energizing activists in the U.S. engineering education community. What started in the 1990s as significantly increased attention to design and information technology now included (at least in principle!) curricular interests in professional ethics, oral and written communication, teamwork, international experiences, continuing education, and more, as well as the legitimation of research on engineering education (see the January 2005 issue of the *Journal of Engineering Education*).

The long-term success of enterprises such as this one will depend upon leaving behind the critique of narrowness and its call for breadth. One reason is that technical education in every engineering field has long been itself both broad and multidisciplinary. A commitment to technical breadth is the reason why each engineering field defines itself not as a discipline but as a collection of disciplines.

A second reason for moving beyond a geometry of narrowness and breadth is that the dominant image of mathematical problem solving limits itself not by being narrow but by being incomplete. It is insufficient as a label or description. Engineering problems do not solve themselves. They are always solved by people. As soon as one introduces people into problem solving, the human dimensions of the process become obvious. When it imagines mathematical problem solving as technical work alone, formal engineering education abstracts out what it counts as human dimensions and defines these as extraneous and irrelevant. It can do so no longer. The long-claimed jurisdictional space for engineering has eroded.

Those of us who teach engineers need a dominant image that both encourages and guides competition over the panoply of potential changes facing engineering curricula and engineering work. The *2020* report pointed in this direction when it observed, "In many ways the roles that engineers take on have always extended beyond the realm of science and technology" (2004, p. 37).

PDS: Adding Problem Definition

One way of formally recognizing the core human dimensions of engineering work is to acknowledge that engineering problem solving has always included activities of collaborative problem definition. In carrying out their work, engineers necessarily negotiate and re-negotiate the definitions of technological problems both among themselves and with non-engineers. They do so in ways that go far beyond laudable, but still limited, efforts to expand the umbrella of technical "specifications" and "needs assessment" in engineering design. One potentially promising way of remapping the jurisdiction of engineering work to adapt effectively to the challenges of the present may be to redefine it in terms of both problem solving and problem definition.

An image of engineering as Problem Definition and Solution, or PDS, would have at least four key sets of practices. To illustrate these, consider an extrapolation from a well-argued proposal by Geoff Moggridge and Ed Cussler (2000) to build chemical product design into chemical engineering curricula. The case involves a hypothetical printing company grappling with a pollution problem from a lithographic ink that contains the carcinogenic solvent methylene chloride (CH_2Cl_2). This solvent is also used in the cleaning process. By entering the air through evaporation, the solvent poses health risks to workers and the company risks censure from environmental regulators.

Focusing on product design, the chemical engineers involved proceed systematically through a procedure that includes (a) identifying needs, (b) generating ideas, (c) rationally selecting among available ideas, and (d) identifying how to put solutions into operation, including building and testing prototypes and estimating costs. The proposed procedure is attractive because it explicitly pushes chemical engineers beyond the purely technical decisions that are typical in conventional models of process design, e.g., batch vs. continuous processes, inputs and outputs, reactors and recycles, and separations and heat integration. Also, even though "obviously a major simplification" (p. 8), the design procedure differs from business management models of product development by insisting that technical knowledge is crucial to sound decision making.

In the hypothetical case, following the procedure yields the short-term solution of substituting the solvent toluene for methylene chloride, for toluene has a similar solubility parameter, is inexpensive, and although "still toxic" has not been banned by environmental authorities. The longer-term solution that appears most desirable is to change the resin chemistry to make the ink solvent-free but water soluble through a chemical trigger.

Early Involvement in Problem Definition

The first set of practices in a PDS image of engineering is that engineers involved in technology development would always expect to participate in activities of problem definition and, equally importantly, would be expected by others to participate. In this design case, the process begins with the pollution problem clearly defined and focuses on translating it into engineering terms in order to provide solutions.

Implementing a PDS image would focus the engineers' attention much earlier, before the problem has been negotiated, described, and, in perhaps a minority of cases, defined clearly. Issues involving emissions and health hazards are notoriously unclear and contested. Who decides initially that methylene chloride poses a danger, through what mechanisms, and at what concentrations? Is this knowledge developed outside the company, appearing through a list of hazardous chemicals published by the environmental authority?

PDS engineers committed to the work of problem definition would possess knowledge about what the environmental authority is, how it makes its decisions, and how methylene chloride showed up on its radar screen. Or perhaps the issue emerges through complaints from workers. PDS engineers would have knowledge about what workers know about the relevant production and cleaning processes, what are their customary work practices, and what has been the history of relationships among workers, between workers and management, etc. Or perhaps someone from management quietly expresses a concern about the future of the cheque-printing business. PDS engineers would have knowledge of various management positions gained by learning about the distinct responsibilities of company managers and the competing visions of the company's past, present, and future that live in management circles.

The key point here is that engineers trained to integrate problem definition into mathematical problem solving would involve themselves early in processes of problem solving, prior to the point at which a clear design problem emerges or can be claimed. These engineers would participate by bringing to bear valuable technical knowledge about chemical process development, product development, and manufacturing, but also substantial knowledge of the nontechnical dimensions of those processes. As PDS engineers, they would include in their work exercises in mapping the positions, interests, and visions of all those groups who have stakes in the industrial processes of the company. Indeed, PDS engineers would be the only participants who expected and were expected by others to explicitly address both the technical and nontechnical dimensions of the processes at the same time.

Collaboration with Those Who Define Problems Differently

A second set of practices in the PDS image involves collaborative work among people who define problems differently than one another. Engineers trained in conventional problem solving know that the first step in solving an engineering problem is to draw a boundary around it so that it can be analyzed in mathematical terms. Equally important is the fact that by successfully defining a problem one also takes possession of it, gaining control over what will count as desirable solutions. Instruction in the quantitative dimensions alone extracts engineers from this real-world condition, enabling them to pursue sound technical solutions to the problem as defined but only by also transporting them into an idealized mathematical space free of human difference and conflict. As such, it provides engineers with no strategies for solving problems when people disagree with one another about how to define the problem in the first place.

In the cheque-printing case, the chemical engineers take an important step by involving other people in the design process. They identify needs by interviewing management, workers, and the company's environmental consultants and health and safety administrators, and they generate ideas by meeting with expert consultants, analyzing the experiences of competitors, and organizing brainstorming meetings. As PDS engineers, their work in interviewing stakeholders would include the additional responsibility to learn and explicitly map how all stakeholders understand the problem, what addressing the issue appears to mean to their future positions and identities, and how they understand their responsibilities. PDS engineers would investigate the history of the relationship between the company and the regulatory authority, knowing if such relations have been positive or not. They would examine the evolution of relationships among managers, engineers, affected workers, and local residents. They would find out if workers were worried about their jobs and trusted engineers and management sufficiently to participate in problem-solving experiences. PDS engineers would learn which managers might fear potential loss of the cheque-printing business and which might see it as a step forward for the company and for themselves.

Creative participation in collaborative problem definition thus includes but extends beyond figuring out how to translate a societal problem into a design problem for the engineering sciences. It can include but also extends beyond the use of systems analyses to link some economic and social dimensions to the technical problem solving process.

The key move in collaborative PDS work involves investigating and assessing other perspectives. Its success depends upon the prior knowledge and conviction that one occupies only one point of view among many in negotiations of technological developments. Also, disagreement is likely, even to the extent that agreement about a single definition of the problem may not be possible. PSD engineers would be important contributors to the collaborative definition of technical problems not only because their technical knowledge would enable them to understand the technical issues at stake. They would also strive to understand these technical issues from different points of view and critically recognize and examine the limitations of their own perspectives.

Assessing Alternative Implications for Stakeholders

The third set of practices in the PDS image involves assessing the implications of alternative solutions for stakeholders. Such work, which has both technical and non-technical dimensions, includes anticipating the possibility that engineers may not possess the knowledge crucial to the most desirable solutions.

In the cheque-printing case, for example, the short-term solution of substituting toluene for methylene chloride works because it has a similar solubility parameter, is inexpensive, and is not banned by environmental authorities. It is still toxic, however. Engineers who defined their work as problem definition and solution would include in their jurisdiction responsibility for analyzing from workers' points of view the implications of substituting a still-toxic solvent for one that has been banned. Would participating workers interpret this option as evidence that engineers are siding with management against them? If so, would they deem this to be an exception or part of a long-standing pattern? Would workers agree that substituting a different solvent is preferable to shutting down the cheque-printing process? What steps might be taken to mitigate these effects? Finally, might attending directly to workers' concerns lead to deliberation over solutions that fall outside of chemical engineering, e.g. introducing breathing apparatus to protect workers from either solvent or even building a room for the presses in which gaseous methylene chloride could be collected, concentrated, and disposed of through other means? PDS engineers would accept responsibility for exploring similar questions with each set of stakeholders.

Solving technological problems typically changes the relationships among participants in one way or another. While one participant may gain additional contacts, status, and/or power, another participant may lose contacts, status, and/or power. Participants tend to weigh alternative solutions in both purely technical terms and in terms of the implications these solutions have for their identities. Indeed, in a given situation, the non-technical dimensions of the process, e.g., the interests of senior managers, may be not only significant but also a key determinant of a desirable outcome. Rather than avoiding such dimensions or rejecting them as politics that falls outside of engineering, PDS-trained engineers would know that technological problem solving always includes such power dimensions and would draw on their training to find ways of dealing with both at the same time.

Leadership Through Technical Mediation

The fourth set of practices in the PDS image involves exercising engineering leadership through a seemingly novel but actually quite common path – technical mediation. In conventional definitions of engineering work, engineers have to make difficult trade-offs among alternative needs or design specifications. In the PDS image, engineers may also have to make difficult trade-offs among alternative stakeholders, alternative definitions of the problem, and alternative perspectives about what is taking place, including their own. Mediating among the positions of stakeholders, whether between employer and regulatory agency, between employer and others affected by the employer's work, between workers and management, among workers, among managers, etc., engineers would continue seeking solutions to meet technical needs but also add the work of reconciling differences in defining them.

Technical mediation by PDS engineers would still be engineering work. Most importantly, it would differ from the business management of people or knowledge management of a firm in that the scope of its vision would continue to extend beyond the identity of the firm. In the cheque-printing case, the new product design engineers discard the idea of changing the presses because "the company does not want to make the enormous capital investment involved" (p. 10). Also, if electronic data processing replaced hand-written cheques, "the company may decide that… printing cheques is like making buggy whips" (p. 10). PDS engineers would certainly have to understand and fulfill their responsibilities as employees. But the jurisdiction of their actual work would, by definition, leave open the boundaries that defined stakeholders, recognizing that these take shape in each case. Engineers would bear a continuing professional responsibility to juxtapose employer considerations with considerations drawn from and attributed to others elsewhere.

Technical mediation is neither absolute subordination nor resistance to management. Nor is it a search for often unattainable consensus judgments. Rather the process takes into account the fact that final decisions affect the next round of decision-making, for technical deliberations necessarily begin with the outcomes of previous deliberations. Reconciling definitional differences as much as possible maximizes the possibility that the process is easier next time around.

Some engineers have told me that labeling engineering work "mediation" would appear to demote it. But the purpose is to avoid the explicit demotion to technical support, as outlined above. Quality engineering work already involves mediation even when it privileges creative technical genius. Engineers already see genius in design as requiring difficult but clever trade-offs among alternative needs or specifications. The PDS image makes visible the fact that creative engineers also make difficult trade-offs among alternative stakeholders, alternative definitions of the problem, and alternative perspectives about what is taking place. Technical mediation can be creative work indeed.

When advocates of engineering position it as waiting for society to ask it for help or give it problems to solve (including via the narrower interests of employers), they fail to fulfill a responsibility to bring its technical knowledge to bear in the definition of problems in the first place. They also deprive others of the opportunity to look to engineers for leadership in problem definition. The *2020* report romantically pictured engineering "strengthen[ing] its leadership role in society" and envisioned engineers working "as leaders who serve in industry, government, education, and nonprofit organizations" (2004, p. 48). Perhaps it is even more romantic to picture engineering identities and responsibilities extending beyond the interests of employers. David Noble (1977) certainly made that case while characterizing engineers as lackeys for capitalism. But I maintain that engineers in fact routinely imagine problem definitions and service outcomes that extend well beyond the boundaries of the firm, even if also commonly through it and not always consciously.

The point here is that visible leadership for engineers will likely not come through claims of technical genius and technological heroism when engineers do not have jurisdiction over technology in the first place. Visible leadership qua engineers may never come. But might the hard work of including collaborative problem definition in engineering work as a core competence, responsibility, and set of practices offer a more realistic pathway than hanging onto a declining image?

Integrating Problem Definition into Engineering Education

A key criterion for identifying and assessing pedagogical strategies to integrate problem definition into engineering education is to ask: How does this learning activity prepare engineering students to work with people who define problems differently than they do?

In any policy-making process, effective travel toward some new desired state of being must always start "here," in the present and at this location. Engineering curricula virtually everywhere tend to include a technical core and non-technical periphery. The most difficult challenge in the work of integrating problem definition into engineering education is to locate and champion both technical and non-technical knowledge practices at the core – education in the engineering sciences. The efforts required include, minimally, three categories of initiatives: (1) adapting pedagogies in engineering science courses to emphasize the limitations of the knowledge they convey along with their strengths; (2) adapting pedagogies in peripheral courses to translate their knowledge practices in ways that engage the practices of mathematical problem solving while also promising to help engineers understand and critically engage diverse technical perspectives on the job; and (3) adapting engineering curricula in ways that legitimize and encourage students to become more than one thing.

Adapting Pedagogies in Engineering Science Courses

How can one teach engineering science courses so that students come to understand what they are not learning? The main challenge to a PDS instructor or PDS textbook author is to teach not only the main mechanisms of analysis but also their boundaries. In his 1994 book *Designing Engineers*, MIT engineer Louis Bucciarelli offered a helpful tool for addressing this issue with the image of "object worlds." Bucciarelli's point was that each engineering science creates and lives in one or more object worlds into which engineers must enter to do their analyses. The mathematical objects in these worlds are both crucial to quality engineering work and a significant source of difference and disagreement among engineers.

"In the simplest terms," Bucciarelli wrote, "design is the intersection of object worlds" (1994, p. 20). Systematically examining three design projects that experienced high levels of uncertainty, Bucciarelli found that "[t]he apparent incoherence and uncertainty of the process[es]... derives in large measure from the differing interests and viewpoints of different parties to the design" (p. 51). He observed how engineers and other professionals working in different object worlds "will construct different stories according to their responsibilities and... technical, professional interests" (p. 71). As a result, because "the authors of these stories display full confidence in their construction" (p. 72), the key issue in defining the engineering problem at stake is not overcoming uncertainty but reconciling different perspectives.

Without overemphasizing the concept of object worlds, which some engineering educators may find too ethereal, engineering science courses could be adapted systematically to present their material as introductions to abstract mathematical arenas that only partly overlap with one another. Engineering sciences, from thermodynamics to heat transfer, build ideal mathematical arenas that are useful and, indeed, beautiful. Each posits a unique configuration of theoretical entities and processes. Engineering science faculty who devote their careers to advancing and improving the abstractions that constitute these arenas often build powerful personal commitments to their promise and value, which includes understanding their boundaries and relations to abstractions in other such arenas. To gain a pedagogical responsibility not only to deliver the mechanisms to students but also to help students learn to articulate the value of those mechanisms and how they are distinct from other mechanisms could very well provide faculty with welcome opportunities to share both their knowledge and their passions.

Given the currently dominant structure of engineering science courses as lectures, problem sets, and exams, the faculty involved in, for example, a chemical engineering thermodynamics class would have to be creative in addressing such questions as: What are the key entities and processes in this thermodynamics course and how do they relate to one other? How are these entities and processes similar to or different from those in the heat transfer course? How do thermodynamics and heat transfer connect to one another, or not? What is different about how thermodynamics and heat transfer are taught in chemical engineering and in mechanical engineering, and why?

The challenge to the faculty trying to help students learn to work with people who define problems differently than they do would bring to classrooms the types of discussions about the relative positioning and value of thermodynamics that often appear in meetings of department faculty, curriculum committees, conferences, and world congresses. But such activities would also carry one key additional dimension, the responsibility to move beyond the defense of strengths to include acknowledging and articulating limitations. Engineering students who are being trained to become leaders who listen will have to learn what they do not know.

One practical strategy for working toward this end is to require students to routinely classify problem sets in addition to solving them. Students would have to examine textbooks in a new way, with the goal of understanding how chapters and sections differ from one another, yet are related. Consider the implications of asking students in a heat transfer course not only to solve conduction and convection problems but to be able to explain what makes these different from one another, what sorts of assumptions each makes, and what sorts of considerations get left out when one uses them in practical applications.

Learning to explain the definition and significance of the mathematical tools they gain in engineering science courses is a crucial step for engineering students to become critical analysts of their own knowledge. Furthermore, rather than diminishing the significance of that knowledge, the acquisition of such critical capabilities is arguably more likely to deepen engineers' commitments to it by enabling them to better articulate and understand what they know in relation to what co-workers know.

A more ambitious strategy would be to develop a separate course experience focused specifically on the issue of problem definition in engineering. Such a course would make visible and analyze examples of disagreement and conflict among the technical perspectives of engineers and non-engineers. Building such a course would require significant effort preparing case studies. Yet students who will later find themselves in senior design courses, which tend to focus on object or product outcomes, could benefit greatly from a second- or third-year "define" course that applied methods of case analysis to instruction in problem definition. Such a course could also better prepare students for the increasingly common inclusion of problem definition activities in senior design.

Adapting Pedagogies in Peripheral Courses

The unique burden on the traditionally peripheral courses would be to mold their critical contributions to advance the knowledge practices of engineers in collaborative problem definition and solution.

It is important to acknowledge that bodies of abstract knowledge originating in the social sciences, humanities, or business management typically do not exist in a form ready for easy and uncontroversial incorporation into the heart of formal engineering education. Faculty from liberal arts disciplines can be inflexible themselves, especially when they seek to reproduce themselves in students rather than to adapt modes of knowledge and practical reasoning to student trajectories.

Substantial communities of scholar/teachers committed to "integrated" liberal arts education for engineers were heartened by Engineering Criteria 2000 in the United States (Ollis et al. 2004) and their analogs in other countries. Once again, a key criterion for facilitating their movement toward the center of engineering curricula is whether or not their contributions help students learn to work with people who define problems differently than they do. In the case of technical communication, for example, an important contribution is to help students recognize, understand, and act on the presence of "audiences" for their work (Winsor 1996). Engineering ethics training calls attention to multiple roles, schemes, or mental models through such concepts as "moral imagination," which involves learning to critically assess one's own point of view and evaluate alternative courses of action (Gorman et al. 2000). Those of us who seek to move our practices from peripheral positions toward the center may have to formulate and focus our critical analyses in ways that maximize the possibility of informed and effective critical participation.[2]

[2] The PDS image evolved from pedagogical strategies in my Engineering Cultures course, an elective that seeks critical participation from the periphery (Downey 2008, 2009, 2011b).

Adapting Curricula to Produce More than One Thing

A third type of adaptation lies at the level of the curriculum. One crucial way to better prepare engineers to work amid differences among co-workers is to acknowledge, accommodate, and even promote differences among themselves. Engineering curricula tend to picture students as acquiring the same core or essence. Although students supplement this core with technical and nontechnical electives, most schools of engineering claim that all graduates from a particular field have a specific configuration of core knowledge and expertise, and, hence, core identity.

Must a degreed engineer be just one thing? After graduation, students set out on pathways that turn them into many different things, yet the focus on a single essence remains. It grounds, for example, the common but highly questionable claim that once engineers become managers they are no longer engineers. Scaling up a PDS image would shift the emphasis away from the minimum requirements to become an engineer and toward the diversity of practices that constitute quality engineering. Working as an engineer would mean that one brings to the field arrays of practices in both mathematical problem-solving and the mapping of perspectives and personnel in relation to one another.

Much research and experimentation would be required to sort out which configurations of knowledge and expertise better prepare students to work with people who define problems differently than they do. Yet it is reasonable to expect that more than one type of knowledge practice and, hence, more than one type of practitioner identity would be essential.

One way to facilitate this shift is to reposition current curricula as tracks inside degree programs that also include other, new tracks.[3] For example, a current curriculum that places highest emphasis on engineering science training could become an engineering science track, structured to prepare students for research positions or graduate school. An engineering design track could include coursework in industrial design, architecture, or other design disciplines, preparing students for careers emphasizing design work. An engineering and management track would specifically help students prepare for the work of problem definition in private industry, especially by training them to analyze the types of knowledge other non-engineering managers possess and use. An engineering and policy track or engineering and society track would prepare students for problem definition work beyond the firm, e.g., in government or non-profit sectors. Extrapolating the idea, a multi-field general engineering track, degree, or possibly advanced degree program could introduce students sufficiently to a range of fields to enable them to function effectively as mediators among different types of engineering specialists.

One benefit from developing alternative pathways to an engineering degree is that faculty would have to compete more for students, thus encouraging them to share both knowledge and passions in the classroom. Also, because every track

[3] A version of this proposal to develop tracks in engineering departments also appeared in Downey (2011a).

would be part of a larger set, each would clearly have both strengths and limitations. What a given track lacked in depth or breadth in a particular area could be supplemented through continuing education depending upon the student's career trajectory. Importantly, the introduction of diversity to curricular structures is made theoretically possible by the shift in accreditation policies from credits to capabilities. If review teams were trained to expect diversity, engineering departments could likely develop and defend alternative ways in which their programs meet outcomes criteria.

In general, strategies at any level to integrate problem definition into engineering education would count as formal moves to claim technical mediation as part of the jurisdiction of engineering work. Such moves could not only help engineers recognize they do not have jurisdiction over technology, but also enable practices of engineering formation to better prepare students for what has always counted as quality work by the best engineers.

References

Abbott, A. (1988). *The system of professions: An essay on the division of expert labor*. Chicago/London: The University of Chicago Press.

Adelman, C. (2000). *A parallel postsecondary universe: The certification system in information technology. office of educational research and improvement*. Jessup: U.S. Department of Education.

Baird, D., & Shew, A. (2004). Probing the history of scanning tunneling microscopy. In D. Baird, A. Nordmann, & J. Schummer (Eds.), *Discovering the nanoscale*. Amsterdam: IOS Press.

Bracey, G. (2006, May 21). Heard the one about the 600,000 Chinese Engineers? *Washington Post*.

Bucciarelli, L. L. (1994). *Designing engineers*. Cambridge, MA: MIT Press.

Downey, G. L. (2005). Keynote address: Are engineers losing control of technology?: From "problem solving" to "problem definition and solution" in engineering education. *Chemical Engineering Research and Design, 83*(A8), 1–12.

Downey, G. L. (2008). The engineering cultures syllabus as formation narrative: Critical participation in engineering education through problem definition. *St Thomas Law Journal (Special Symposium Issue on Professional Identity in Law, Medicine, and Engineering), 5*(2), 101–130.

Downey, G. L. (2009). What is engineering studies for?: Dominant practices and scalable scholarship. *Engineering Studies: Journal of the International Network for Engineering Studies, 1*(1), 55–76.

Downey, G. L. (2011a). Epilogue: Beyond global competence: Implications for engineering pedagogy. In G. L. Downey & K. Beddoes (Eds.), *What is global engineering education for?: The making of international educators* (pp. 415–432). San Rafael: Morgan & Claypool Publishers.

Downey, G. L. (2011b). Location, knowledge, and desire: From my two conservatisms to engineering cultures and countries. In G. L. Downey & K. Beddoes (Eds.), *What is global engineering education for?: The making of international educators* (pp. 385–414). San Rafael: Morgan & Claypool Publishers.

Downey, G. L. (2014). *The (Professional) formation of engineers*, Keynote delivered at NSF EEC Engineering Education Awardees' Meeting, September 29. National Science Foundation, Arlington, VA. Available at www.downey.sts.vt.edu

Downey, G. L., & Lucena, J. C. (1997). Engineering selves: Hiring in to a contested field of education. In G. L. Downey & J. Dumit (Eds.), *Cyborgs and citadels: Anthropological interventions in emerging sciences and technologies* (pp. 117–142). Santa Fe: School of American Research Press.

Gereffi, G., Wadhwa, V., Rissing, B. A., & Ong, R. (2008). Getting the numbers right: International engineering education in the United States, China, and India. *Journal of Engineering Education, 97*(1), 13–25.

Gillett, J. E. (2001). Chemical engineering education in the next century. *Chemical Engineering and Technology, 24*(6), 561–570.

Gorman, M. E., Mehalik, M. M., & Werhane, P. H. (2000). *Ethical and environmental challenges to engineering*. New York: Peter Lang.

Graphics and Visualization Center. (2004). Available at http://cs.brown.edu/stc/allstc.html

Hogle, L. (2003). Life/time warranty: Rechargeable cells and extendable lives. In S. Franklin & M. Locke (Eds.), *Remaking life and death: Toward an anthropology of the biosciences*. Santa Fe: School of American Research.

Lucena, J. C. (2005). *Defending the nation: U.S. Policymaking to create scientists and engineers from Sputnik to the 'War against Terrorism'*. Lanham: University Press of America.

Moggridge, G. D., & Cussler, E. L. (2000). An introduction to chemical product design. *Chemical Engineering Research and Design, 82*(A12), 1525–1532.

Mowery, D. C., & Rosenberg, N. (1989). *Technology and the pursuit of economic growth*. Cambridge/New York: Cambridge University Press.

National Academy of Engineering. (2004). *The engineer of 2020: Visions of engineering in the new century*. Washington, DC: National Academies Press.

National Research Council. (2003). *Beyond the molecular frontier: Challenges for chemistry and chemical engineering*. Washington, DC: National Academies Press.

National Science Board, National Science Foundation, Division of Science Resources Statistics. (2004). *Science and engineering indicators* (NSB 04–01). Arlington: National Science Board.

National Science Foundation. (2012). *Grant proposal guide* (nsf130010, Retrieved 31 Jan 2013).

Noble, D. (1977). *America by design: Science, technology, and the rise of corporate capitalism*. New York: Alfred A. Knopf.

Ollis, D. F., Neeley, K. A., & Luegenbiehl, H. (Eds.). (2004). *Liberal education in twenty-first century engineering: Responses to ABET/EC 2000 Criteria*. New York: Peter Lang.

Schyfter, P., Calvert, J., & Frow, E. (2013). Editorial introduction: Synthetic biology: Making biology into an engineering discipline. *Engineering Studies, 5*(1), 1–5.

Talbot, H. (1911). The engineering school graduate: His strength and his weakness. In H. Talbot (Ed.), *Technology and industrial efficiency* (pp. 114–123). New York: McGraw-Hill Book Company.

Williams, R. H. (2002). *Retooling: A historian confronts technological change*. Cambridge, MA: MIT Press.

Winsor, D. A. (1996). *Writing like an engineer: A rhetorical education*. Mahwah: Lawrence Erlbaum Associates.

Wulf, W. A. (2004, October 3). Annual meeting – president's remarks. *National Academy of Engineering Annual Meeting*. http://www.nae.edu/News/SpeechesandRemarks/page2004 AnnualMeeting-PresidentsRemarks.aspx

Gary Lee Downey Alumni Distinguished Professor, STS, Virginia Tech. B.S. in Mechanical Engineering, B.A. in Social Relations, Lehigh University. M.A. and Ph.D. in Cultural Anthropology, University of Chicago. Author of *The Machine in Me: An Anthropologist Sits among Computer Engineers* (Routledge 1998), co-author of *Engineers for Korea* (Morgan & Claypool Press 2014), co-editor of *Cyborgs and Citadels* (SAR Press 1997) and *What Is Global Engineering Education For?* (Morgan & Claypool 2011). Founder of *Engineering Cultures* course (https://www.youtube.com/user/downeygary). Founding Organizer of International Network for Engineering Studies (INES). Editor of *Engineering Studies* journal (Routledge), the *Engineering Studies* book series (MIT Press) and the G*lobal Engineering* book series (Morgan & Claypool). President, Society for Social Studies of Science (2013–2015). Main interest in engineering knowledge and personhood. (www.downey.sts.vt.edu).

Chapter 22
Implementing Social Awareness into Engineering Curricula

Javier Cañavate, Manuel José Lis Arias, and Josep Maria Casasús

Abstract The inclusion of competences related to social context in engineers' education has been recommended by several organizations and authors. Among these abilities appears the development of reflective and behavioral skills or the awareness of public debate in engineering. In order to create an engineer profile with expanded social commitment, some criteria have been implemented by accreditation boards.

In this chapter, we will review these criteria and requirements, analyze how they contribute to the future engineering profile, how universities have approached the inclusion of the new contents in their syllabi and identify the difficulties that this may present in real practice. A reflective digression on the need of going beyond the ethics, the habitus of the engineering profession, the implicit values in education and the existing gap between engineers and society is also included. Finally, the subject is illustrated with the description of engineering in Spain and the results of interviews with the presidents of professional engineering associations.

Keywords Engineering curricula • Social awareness • Social context • Accreditation criteria

Introduction

Within the last decade, a significant change in the vision of the engineer has been observed. Engineering education has always been a field of discussion among authors who envisage a humanist engineer, a figure that goes beyond his tasks and

J. Cañavate (✉) • M.J. Lis Arias
Department of Chemical Engineering, Universitat Politècnica de Catalunya BarcelonaTech, Jordi Girona, 31, Barcelona 08034, Spain
e-mail: francisco.javier.canavate@upc.edu; manuel-jose.lis@upc.edu

J.M. Casasús
GEP-21, Carrer Pla de l'Atmetllera 135 3 A, Terrassa 08225, Spain
e-mail: jmcasasus@gmail.com

S.H. Christensen et al. (eds.), *International Perspectives on Engineering Education*, Philosophy of Engineering and Technology 20, DOI 10.1007/978-3-319-16169-3_22

can influence and induce changes in society, versus the conception of an especially qualified technical performer.

Engineering programs, for most of the last century, have been consequent with technical advances and have promoted an engineer's profile that considers science as a founding factor and technology as a tool to develop the processes that industry requests. This notion, which is strongly related to the public image of engineers, has been challenged by changes in society that have taken place alongside the globalized world. The ability to engage in public debate and to link the techno-scientific world to the welfare of society, combined with other skills such as communication aptitude or capacity for transcending cross-professional cultures, is increasingly demanded.

As a consequence of these new requirements, engineering education has been enriched, including sets of competences that are outside what traditionally was considered the core of engineering knowledge. Environmental concern, soft skills, sustainability, ethics or integration of the social context in engineering practice are some of these demanded topics. Some universities have engaged actively, on their own initiative, in designing study plans that aim to develop these competences while others tend to continue providing a classic engineering profile. Eventually, especially during the last decades of the twentieth century, the accreditation organizations have been aware of the importance of effectively implementing these public requests in the education of engineers. Accreditation organizations have also designed some criteria aiming to fulfil the new needs.

However, the task is not completely straightforward, the choice of criteria that must be included in the core of engineering programs, the suitability of the proposed criteria to the real demands of society, the adoption of these criteria by universities may be arguable; their actual implementation by universities also presents some serious issues.

Accreditation Criteria

In engineering studies, accreditation is a process in which certification of quality is achieved. More precisely, it is an evidence of the ability of the programs to provide graduates with the competences necessary for the development of their future professions.

In order to monitor the set of competences that configure the future engineering profile, there are studies promoted by professional associations, such as the accrediting boards, or by institutions related to education. Some of these well-known documents are the series published periodically by the National Academy of Engineers of the USA. These institutional studies have been complemented by several authors such as Marie Paretti and Christine Burogyne (2005), James Duderstadt (2008) and publications as the *Journal of Engineering Education* (2008). A key feature of the most recent works is the discovery of new demands from the Society for Engineers and the subsequent proposal of a change in the engineers' education including an increase of competences situated outside the classic core of engineering programs. Progressively, the concern about environmental issues, sustainable development and the request for soft skills and ethical commitment have increased. The result may be a new paradigm that considers engineering as a social activity, dedicated to the "promotion of public good" and goes beyond applying expertise to solve technical problems or performing the assigned tasks.

An important starting point of this discussion was set by the Accreditation Board for Engineering and Technology in the USA (ABET) Criteria for Accreditation, published in 2000. The urge for departments to develop communication ability, cultural awareness, human interaction or ethics is a seminal event that can be considered as a reference point in the subject.

With these precedents, the situation nowadays is defined at least partially by the accreditation criteria that are defined by the respective agencies. In this sense, every institution has developed a set of guidelines that can be used as a reference for the development of engineering curricula.

The criteria proposed by ABET (2011) include a list of "student outcomes" that should be articulated by the engineering program: an ability to apply knowledge of mathematics, science and engineering; an ability to design and conduct experiments; an ability to design a system, component, or process to meet desired needs within realistic constraints such as economic, environmental, social, political, ethical issues, as well as health and safety, manufacturability, and sustainability; an ability to function in multidisciplinary teams; an ability to identify, formulate, and solve engineering problems; an understanding of professional and ethical responsibility; an ability to communicate effectively; the broad education necessary to understand the impact of engineering solutions in a global, economic, environmental, and societal context; lifelong learning; a knowledge of contemporary issues; an ability to use the techniques, skills, and modern engineering tools necessary for engineering practice. Following the trend started in previous years, these outcomes combine skills related to the specific engineering practices with a broader vision of the engineer. Some of these points outline the need for providing students with a perception of the ethical and social constraints that must be included in every engineering approach. Even in the curriculum description, there is a recommendation to impart to students a general educational component that "complements the technical content of the curriculum and is consistent with the program and institution objectives".

The European Network for Accreditation of European Engineering Programmes (EUR-ACE) (2008) proposes a set of skills based on six main outcomes: knowledge and understanding; engineering analysis; engineering design; investigations; engineering practice; transferable skills. These points aim to classify the results of the engineering studies in specific fields. Every field is then specified in terms of the essential competences that students should have acquired when they graduate. Analyzing these competences, most of them can be considered as basic or advanced sets of scientific and technical skills that are related to their branch of Engineering. Only in two of them some references to the non-technical awareness of the engineer are included. The first of them, knowledge and understanding, includes as a competence an awareness of the wider multidisciplinary context of engineering. The outcomes related to engineering practice also include a point specifying that engineers must be provided with "an awareness and knowledge of the non-technical implications of engineering practice". Some other references to the context of engineering practice are incorporated in the set titled "Transferable Skills", which includes soft skills like communication, team work, international perspective, and also "demonstrate awareness of the health, safety and legal issues and responsibilities of engineering practice, the impact of engineering solutions in a societal and environmental context, and commit to professional ethics, responsibilities and norms of engineering practice". The United Kingdom system of accreditation is quite similar to the European one and they are clearly inspired by each other.

The Canadian Engineering Accreditation Board (2011) describes "Graduate attributes" dividing the expected outcomes into sets that include the following headings: A knowledge base for engineering, Problem analysis, Investigation, Design, Use of Engineering Tools, Individual and Team work, Communication skills, Professionalism, Impact of engineering on society and the environment, Ethics and equity, Economics and project management, Life-long learning. Although different in arrangement, most of the attributes included in these sets are quite similar to those reflected by the EUR-ACE document. Nevertheless there is a remarkable difference that denotes a special sensitivity related to the fact that topics such as the impact of engineering in society or ethics are in an equal level of description compared to traditional competences associated to engineering. This conception is equally reflected in the description of the attributes that include such contents as "cultural and social awareness" in the engineering design, an "understanding of the roles and responsibilities of the professional engineer in society, especially the primary role of protection of the public and the public interest" under the headline of professionalism, or an ability that "includes an understanding of the interactions that engineering has with the economic, social, health, safety, legal, and cultural aspects of society, the uncertainties in the prediction of such interactions; and the concepts of sustainable design and development and environmental stewardship" to raise the conscience of the impact of engineering.

Engineers Australia (2008) includes in their recommendations for curricula some sets of expected outcomes labeled: enabling skills and knowledge development, in-depth technical competence, personal and professional skills development, engineering application experience, practical and 'hands-on' experience. These outcomes are in essence quite similar to the previous associations' guidelines. An emphasis on practical experience can be observed. An in-depth reading of these headlines show some points related to the social integration of the engineer. Points like "an understanding of and commitment to ethical and professional responsibilities" or "a commitment to safe and sustainable practices" can be considered as an intention of directing future engineers to a commitment to social welfare. Basically, the notion of context awareness is perhaps less intense than in other accreditation criteria. The more traditional vision of engineering offered by these publications may be related to the critical point of view of engineers expressed recurrently by Sharon Beder (1999) and other Australian authors.

In order to provide an input about criteria used in Eastern countries, we can analyze Malaysian Engineering Accreditation Council (EAC) (2008) Engineering Program Accreditation Manual. A change of point of view is observed. The concept of accreditation is more based on the process for obtaining accreditation than on the outcomes of the programs. The references to competences in the curricula seem more ambiguous. Nonetheless, there are also allusions to the denial of accreditation to programs that graduate students "not able to express themselves and unaware of the importance of sustainability, safety, and professional involvement etc., which reflect the lack in generic attributes expected of them".

The reading of these accreditation documents, even considering differences among them, implies that the need to include a broader perspective in engineering studies is present. From a critical point of view, it would be licit to discuss if these criteria are appropriate enough to the vision of the engineer that the society pursues. The direct application of the criteria may not be enough to induce a greater role of engineers of the future as regards the need of a public implication of engineers expressed by many authors and institutions (Revel 2007). The way in which these criteria are accepted and implemented by universities or even the justification in the existing differences among countries is also quite diverse. At the same time, some other initiatives including engineering programs based on on-line contents (Peercy and Cramer 2011), new conceptions of education that rate self-learning over official degrees may not be taken into consideration for accreditation criteria.

Universities have tried to adapt their programs to these demands. One way or the other, they have implemented systems to provide students with the evolving sets of competences that are considered necessary. The basic demands have evolved from environmental concern, to soft skills, sustainability, ethics and social integration of the profession. The ways of teaching these subjects have been diversified and will be covered below. However, every subject presents distinctive features that exert an influence on their consideration.

The Broader Perspective of the Engineer

According to the studies cited above and the criteria proposed by engineering agencies, we can divide the set of engineering competences into two big groups. First, the contents related to the development of the profession that are specific to engineering and have been taught in their actual form or as an evolution of other subjects present since the beginning of the engineering education program, which would include scientific and technical subjects, as well as technologies that have been progressively added according to their development and use. The second would include a field that is more varied and subject to interpretation. It would basically include five basic concepts: environmental concern, soft skills, ethics, sustainability and social context perception.

The environmental concern as part of engineering programs is a feature that can be nowadays considered as having been achieved. The protection of the environment, the calculations of environmental risks and the awareness of the potential damage to the planet are integral parts of engineering projects. This point has been assumed as part of technical restrictions and is based on legislation that acts as a framework for engineering practice. The necessary knowledge for applying environmental criteria was already present in technology and its application was more related to a change of perspective than to a matter of expertise. Environmental concern is now a technical subject and is deeply rooted in engineering tradition. The unfortunate ecological disasters of the past have also acted as a catalyst to promote a better system of environmental control. Even the ethical aspects related to the concepts of pollution, risk or health seem to be present in every lesson taught at universities.

The inclusion of soft skills has resulted from the need of engineers to communicate with other professional cultures or the public in general. This interaction has been especially intense in the case of engineers-managers (Christensen 2003) or in the case of engineers communicating the consequences, risks, results or planning of their projects to public and private stakeholders. The practice of engineering has also evolved and the relationship with clients is more necessary in most activities. The need for soft skills in engineering practice is nowadays undeniable and has been recognized for many years by professionals and employers (The Royal Academy of Engineering 2009). Analyzing the accreditation criteria or study plans from several universities, we can conclude that engineering programs already include some competences included in this group. However, there is a slight difference compared with the subjects related to environmental concern. Educators, especially in technical universities, are officially not specialists in these fields and, unlike the previous case, these skills were not taught traditionally in engineering schools. However, even when they are not trained to teach these competences, the engineering faculty could use their personal development to transmit the requested abilities to graduates. For example, a quite extensive skills list would be: strong work commitment, positive attitude, good communication, time management abilities, problem-solving, team work, self-confidence, ability to accept and learn from criticism, adaptability, etc. Most of these competences are common to the education of university professors and it is reasonably expected that educators may instill these features in their students by programming activities where those skills are developed.

Ethics is a word that has increased its popularity among engineering stakeholders. Sometimes, the word is related to values. Paradoxically, both concepts have generally been excluded from engineering programs. Values in engineering are mainly related to ethics and both terms are included in discussions about liability, legal responsibility, etc. Some of the concerns of this important subject were recognized by the Engineering Associations, who provided engineering codes and regulations of engineering practice. Even in this case, studies as the ones performed by George Catalano (2006) in the USA show that they mainly focus on reliability and integrity at work. As Yamun Nahar et al. (2009) point out, in most cases, engineering programs are limited to microethics and questions like whistleblowing or individual concerns. Macro ethics, or issues of the broader society are usually not dealt with. An important issue that arises in this sense is the lack of expertise in the subject by engineers' teachers. Compared to the two topics exposed previously, the lack of proper training of educators is more evident. In many cases, the vision of ethics for engineers is then quite restricted to concrete case studies and common practice. Deep philosophical thought is generally omitted.

The case of sustainability is considerably different. Sustainability beyond environmental concern is a complex subject that involves a humanist perspective. Humanities were not taught at all in engineering faculties, and social perspective is quite remote from engineering education. The approach followed in many engineering schools is to take on board sustainability from the point of view of "technology for the sustainability" which basically involves techniques and scientific methods to evaluate risks, environment or in some cases social statistics. Further social concern has been only considered and partially accepted from the emerging idea of corporate social responsibility. The convenience of including these concepts and the adequate teaching systems is still a subject of discussion.

University Approaches to Non-traditional Sets of Skills

Many universities have faced a serious issue when trying to implement the non-traditional sets of skills that include knowledge and methodologies beyond their typical core of contents. The problem is especially significant in the case of technical universities that are only devoted to technical knowledge. As mentioned above, the incorporation of the environmental concern in engineering studies or other related pseudotechnological subjects was relatively easy in comparison. After all, the addition of these topics was related to the technical contents of engineering and only a different change of perspective was required (Jamison et al. 2011).

When the incorporation of wider concepts like soft skills, ethics, sustainability or the implication of engineering in public debate was intended, implementation was more complicated. Some institutions did not have the experience, knowledge or the strong belief that were needed to fulfill the requirements. The main approaches adopted have been:

(a) Incorporate courses into the engineers' curricula:
This option has been widely spread among engineering programs since the 1980s (Atman and Nair 1996). Most syllabi nowadays include subjects that aim to provide either soft skills, sustainability concepts, ethics or social context to the graduates. Some of these courses are often labeled as science, technology and society. This straightforward approach collides with serious issues and visions of the academic world. Some common arguments put forward against these courses are as diverse as:

- The new courses imply a reduction of the technical/scientific time in already condensed programs and in a technical environment that requires everyday further extension to follow the continuous advances of technology.

- The scientific level of students is declining and they need more intensive teaching in fundamental courses.
- There is no reciprocity in humanist studies to promote interest for technology.
- Engineering educators are not interested in humanism or in subjects beyond their own scientific/technical fields.
- The engineers are more involved in research than before and need more basic preparation.
- Employers do not require humanist concepts.

These assertions are clearly arguable from several points of view (Mitcham 1998; Prausnitz 2003) but they are used anyway to undermine the effective implementation.

There are also some practical issues that have negative influences on the implementation of these courses:

- Some universities do not have departments or professors with the appropriate background for these courses. In some countries, including Spain, universities teaching engineering considered as top-ranking do not have departments related to non-technical subjects.
- When courses are taught by professors with a humanist background, they are often criticized for not having sufficient overlap with the engineering practice. They are seen as unrealistic and their professors tend to be isolated from the rest of the engineering faculty.
- When courses are taught by engineers, they are developed in a biased concept of "teaching what is useful for an engineer" and too much centered on practical aspects, lacking a deeper perspective. The humanist perspective is often lost because of the conception and tools.
- The approach of the "compulsory course" in order to achieve graduation is perceived as an inane requirement and as a less important subject not integrated in the rest of the curriculum.

Nowadays, the efforts for providing adequate humanist education have produced generations of engineers that have formed themselves in these subjects and are qualified to appropriately teach interesting courses. These professors may face also some adverse influences (Jones et al. 2011) that challenge their humanist interest:

- They may be considered as less prestigious because their interests and research do not involve scientific knowledge.
- Their professional development may be more limited because their work is not measured with the standards created to evaluate scientific or technical careers. This is the case in Spain where regulations about professors' promotion or salary increase in technical universities are basically linked to scientific technical research; other aspects or even publications not related to the specific scientific area of knowledge are not equally important.
- Sometimes they have to combine their regular work as professors/researchers in technical fields with their interest in social subjects.

(b) Integration of concepts related to the social perspective in regular courses:
This conception works along the curriculum, spreading through many engineering subjects. It is an approach used widely for ethics, soft skills, etc. This approach provides teachers with a better opportunity to introduce ethics in small doses in their lessons. There are many engineering subjects that could provide excellent examples where not only the technical side of the profession is involved (Brady and Lawson 2011). Nevertheless, this approach also presents downsides. First, it depends strongly on the knowledge and willingness of the teacher. Second, the evaluation of the outcome tends to be ambiguous because the social perspective is usually considered as less important than the technical one. As a consequence, students' task is devoted basically to solve the technical problem with a less intense focus on the social part. At the same time, it is difficult to coordinate a huge number of courses in a way that provides quality education in all social aspects of the engineering program and have a clear indicator showing that expected levels are achieved. Sometimes, the excessive dispersion of the knowledge results in a superficial, non-effective approach.

An example of this approach was the competence mapping in engineering curriculum at the School of Design Engineering (ETSID) at the Polytechnic University of Valencia (Spain). The experience is described by Edwards et al. (2009) and was essentially related to competence-based education without special incidence on social awareness. The School of Engineering Terrassa (Spain) also developed a system designed to assign competences at various levels to the subjects taught in the engineering curricula (EET 2013). The main issue in this situation was the detached assessment of the competence separately from the technical content. The regulations of the university were not intended to apply to that methodology and constituted a really difficult obstacle to overcome in order to provide qualifications and final grades to the students.

(c) Implementing social perspective by methodological aspects of teaching:
This approach has been used typically to develop soft skills. It consists in the use of a teaching method involving tasks that develop students' communication abilities, teamwork and may also be used for cultivating ethical, sustainable or societal concern (Riley 2011). The weaknesses of this system are similar to the previous one.

From the pedagogical point of view, Joseph Herkert (2002) has described several trends applied to ethics that can be extended to most of the other subjects. These trends can be applied in the several models of curricula explained above:

- Providing general frameworks: teaching codes and theories. The usefulness of this method is a classic subject of discussion.
- The case study method: a very popular approach that attracts students and tends to encourage them into reflection. Their detractors argue that sometimes the cases used commonly as examples because of their intrinsic interest are quite far from the real engineering practice.

To a certain extent, all these difficulties lead to some frustration among educators and institutions that are willing to expand the vision of engineering and their implication, using their privileged role, in promoting public welfare.

Implementation in Context: Some Critical Reflections

As stated above, in our educational environment, it is possible to distinguish five basic demands (environmental concern, soft skills, sustainability, ethics and social integration) related to the broadening of the engineer's perspective, and three strategies to introduce them to engineering studies (introducing specific courses into the curricula, integrating concepts in regular courses or changing some teaching methodological aspects). Before further describing common practices, it could be useful to introduce a critical reflection about the context in which the implementation takes place.

The five basic demands and the three strategies involve an exchange of explicit information like data, theories, concepts, cases, etc. The context of this exchange is the formal relationship between the teacher and the students inside the classroom. The objective, rather than transmission of knowledge, should be to provoke a reflective moment and the final goal is that the reflection eventually induces a change in practice. Does that kind of context and exchange guarantee a real internalization of concepts, perspectives, and especially, practices?

Richard Sennett (1998) refers to the case of an executive who very often changed companies and geographical locations. The professional was concerned with that situation because he wanted to educate his sons in constant values. There was a contradiction between what he experienced in real life (constant change of routines and values) and what he thought should be good for the education and life of his children (a constant foundation based on reliable values). That kind of conflict could also be experienced by an engineering student. What is explained and debated in one determinate course on any kind of subject related to the above-mentioned five basic demands can provide a contradiction with what is taught in other courses. For instance, after an intense debate on pollution, a chemical engineering student can find himself, in the following class, being taught just how to comply with the legal norms on water pollution, which contradicts his interest in learning not to pollute at all. In this situation, what one teacher is trying to communicate to the students (the dangers of all contamination) enters in contradiction with the approach used in some other subjects (concern about the levels of water contamination allowed by law). Likewise, the gap between the contents aiming to provide a broader engineer perspective and the real practices in companies can be quite important. Those cases illustrate the need of going beyond the realm of ethics in order to achieve a more socially engaged engineering. In many cases, the issues are not only ethical dilemmas based on values and individual decisions, but also a matter of politics that imply changing the laws through a collective movement independent of *de facto* powers.

Through the concept of *habitus*, Pierre Bourdieu provides a deeper insight into the conflict between what is intended and the real practices of engineering. *Habitus* is a central concept of Bourdieu's dispositional theory of action. The former Aristotelian notion of *hexis*, translated and enriched later on by Thomas Aquinas as *habitus*, is also a sociological conceptual tool. *Habitus* defines a practical competence that operates beneath the level of consciousness. It is a competence which is not acquired by the formal and explicit exchange of information, but through the practices. *Habitus* is what we acquire *in* action and *for* the action. It encompasses simple acts like our way of walking down the street or looking at the face of others, as well as much more complex actions and behaviors. *Habitus* is "a system of durable and transposable dispositions", and "functions as a matrix of perceptions, appreciations and actions" (Bourdieu 1977, p. 261). It is a latent practical knowledge ready to be activated by a new situation. The *habitus* is socially constituted and individually embodied, as schemata of perception and appreciation. It includes thinking patterns ready to guide our actions.

The engineering field has its own *habitus*. We should wonder if, in some way, broadening engineering in terms of social awareness is opposed to the actual *habitus* of engineering and if it would be possible to change the engineering studies at the deeper level of its *habitus*.

In order to provide a provisional answer to those questions, we can imagine a pedagogical case. Let us picture a lesson in which the students have entered in contact with some polemical ideas of sociologist Ulrich Beck (Beck 2009): the scientist and technologist really do not know completely and cannot predict the practical consequences of their projects and theories. The concepts are illustrated with examples like the catastrophe of Fukushima or the unpredicted earthquakes in Italy. As a result of the discussion, the class is aware of the risks existing in our society and the notion of the twenty-first Century world becoming some kind of testing ground. The objective could be that the students become aware that: science and engineering can have unpredictable consequences related to the environment and/or the society, a technological project may not have a closed or definitive resolution and the consequences of a project cannot always be calculated. These statements would constitute the explicit information critically discussed in the context of the formal relationship between the teacher and the students inside the classroom. However, some questions are still unanswered:

- While the students after the lesson may understand the theory explicitly and could be able to apply it to some real cases, is it really internalized?
- Is the *habitus* in contradiction with the topic taught in the class?
- Has the transmitted information permeated the *habitus*? Will the session change the practices?

Students learn in action through the resolution of class problems, cases or projects. They learn that all problems have a solution. The projects that they develop as students have a controlled result. Practice and action create their *habitus*. Then, their implicit habitus contradicts and nullifies the explicit content of the session on Beck.

The ideas that were intended to broaden the engineering perspective are directly opposed to the practice that they are taught in other subjects.

To engage deeply in this digression is out of the scope of this chapter, but, in our opinion, the concept of *habitus* would help us to raise some questions relevant to the contents of the implementation of social awareness, and to provide a framework to that objective.

We can apply a similar approach to examine the proposal of contents to include in engineering curricula. In many cases, the use of concepts such as "risk" or "values" implies accepting uncritically some aspects of the engineering *habitus*. As Langdon Winner wisely remarked some years ago (Winner 2010), the concept of "risk" leads to think (as a *habitus*, we would add) that it is always valuable to "take risks". The concept of "value", "this one amorphous category" (Winner 1986, p. 156) tends to mean something external to the real practice of engineering, like some kind of ornamental and humanistic floating aura. Insisting on "values" probably will induce the habitus of the student to guide his mind to classify that concept as some external item, unconnected to the rest of the curricular contents.

Also institutionally, the universities play, by their own nature, an important part on the development of habitus. By their use of tests, formal and socially established requirements, universities generate, at the level of *habitus*, the self-perception of the engineer as an "expert" above the non-expert citizens. In our societies, there is a hierarchical gap between experts like engineers, doctors or those citizens that have passed specific and socially established tests, and the rest of citizens (Illich 1973). This situation also creates a paradox: Is it possible that the very institution that creates that gap could bring closer both sides of the gap?

Applying Bourdieu's concept of *habitus*, we realize that we would need to change engineering studies at a deeper level. This understanding does not imply necessarily a pessimistic conclusion. *Habitus* not only helps to explain the perpetuation of the social world, it is equally present through the process of change and crisis. When social and environmental conditions evolve, the *habitus* becomes unadapted. The discovery and the diffusion of negative environmental consequences, for example, change the social perceptions and *habitus* undergoes a subsequent transformation. In any case, it is important to be conscious of the several stages that guide our decisions and actions, from the level of explicit knowledge to the more hidden, implicit and, in a certain way, unchosen level of habitus. It is also a due task to discover the several ways those levels enter in contradiction. And above everything, there is a need to be aware of the conflict required for a change of behavior or mentality, especially the internal conflict of the individual with the socially inherited preconceptions and *habitus*.

Engineering Programs in Spain

The experience in engineering programmes that place competences or learning outcomes at the core of the academic activity is lower in Spain than in other European countries. Traditional conceptions are based on curriculum packaged formats. Last decade, following the big social transformation of these years, university transformed its formative processes in order to adapt to the new situation.

In 2002, the Spanish Ministry of Education founded the National Agency for Quality Assessment and Accreditation (ANECA). Its first commandment was to impulse the conversion of the existing syllabus into the requirements defined by the European High Education Area (EHEA). After the Bergen Communiqué (2005), Ministers responsible for Higher Education committed themselves to implement the standards of the European Association for Quality Assurance in Higher Education (2005).

The pursued integration of the Spanish university system caused some problems with some academic activities and worried professional engineers and employers (Suarez et al. 2011). In order to provide guidelines for the task, a first step was to map the required competences from the point of view of different stakeholders: academics, graduates, employers and other agents. The results were included in documents that defined the learning outcomes of the official programmes and are related to the subjects included in the curricula. Some studies and projects describing the design of Spanish degrees based on this new scheme have been published by several authors (Edwards et al. 2009). Key issues in the process, related to some of the topics covered in this chapter, were validity, reliability of the competences, transportability of competence assessment or credentials.

After the transformation of the higher education system in Spain, the regulations for the universities teaching engineering degrees are now defined officially. The government publishes a list of official degrees with their names and the description of the minimum compulsory subjects that must be included in every engineering program. As in most engineering programs, these specifications include foundations of science and technology or management. Every field is developed in competences that the program should guarantee to the student. Significantly, soft skills, ethical values or social implication are not clearly included. Only the general law (RD1393/2007 2007) that describes the objectives of the university degrees mentions two objectives related to ethics and risk communication (our translation):

> Students must have the ability of gathering and interpreting relevant data (usually in their field of study) in order to emit judgments that include a reflection on relevant topics of social, scientific or ethical nature.
>
> Students must be able to transmit information, ideas, problems and solutions to specialized and non-specialized public.

However, the compulsory contents are considered only as a minimum. Universities complete the programs with their own contributions according to their idea of the competences required by a practicing engineer. When developing their study plans, universities must follow the basic regulations provided by the ministry and may also include some regulations in order to ensure that all their programs are conveniently providing what is considered important. For example, some universities include English language, soft skills competences, sustainability foundations, etc. The objectives mentioned above are supposed to be included in this category.

The approach followed by Universitat Politecnica de Catalunya-Barcelona Tech, one of the most prestigious technical universities in Spain, is a blend of the systems included in the previous point. As is the case in many other universities in Spain, the programs of Barcelona Tech are composed of the core of competences designated by the government, a set of competences designed by the University itself and a number of subjects elected by the school where the program is taught. As a result of this approach, nowadays the main engineering programs include competences in a set of soft skills, sustainability and second language. Humanistic studies, ethics or social considerations are not specified.

The system used to guarantee these outcomes is based on two methodologies:

- Courses: Sustainability is included as a compulsory course, including the social perspective.
- Other competences: Every regular subject taught in the program must include in its contents one or more of the competences established by the university. In order to provide the desired outcome, teaching methods and activities are implemented in a "regular" subject to achieve the desired level. The whole of the engineering program is designed in order to achieve the set of soft skills previously established, gradually increasing the student's competence through three levels.

This system presents its own pros and cons. Some of them are related to the deficiencies outlined above. We would point out to the less emphatic approach to the subjects that are not related to the "core" conception of the engineer, and to the lack of attention to the pedagogical work of the teachers who develop them. Other issues are derived from the conception of engineering itself and the application of the programs. Some new additions to the traditional engineering curricula have been quite integrated. That is the case of environmental studies, cooperation and sustainability. Also the education in soft skills is nowadays considered as important, as an answer to the requests of employers. As for other competences, excluding perhaps the social contents related to sustainability, most of the knowledge related to macro ethics or even micro ethics is not considered as part of the engineering programs. Some universities like Universidad Nacional Educación a Distancia (UNED) highlight their compromise with ethical values, but they are usually not clearly specified and considered as part of the pack of soft skills and general knowledge provided to the students.

The Role of Professional Associations in Spain

As previously stated, the situation related to ethics or social awareness for engineers is not well defined in Spain. Similar trends are observed in other countries (Brumsen 2005). Similarly to other professions, professional associations are sometimes the main depositories of ethics responsibility, usually by the adoption of professional codes. In order to do some research on this subject, we have interviewed two presidents of professional associations related to Industrial Engineering which includes Mechanical, Chemical, Electric, etc. (CETIB) and Telecommunications (COETTC).

Our interview covered several subjects that have been reviewed in this chapter. First, we asked the presidents about the role of accreditation criteria and their suitability to the profession of engineer. Both answers were quite restricted to the need of defining clearly engineering profiles and competences. The concrete set of criteria or its relevance was not mentioned.

The second question was related to the foundation values underlying the engineering profession. In this sense, both answers were quite representative of the conceptions of engineering that, in our experience, are quite common (our translation):

> Professional dignity, respect to the citizen's rights, caring for professional activity that adjusts to the citizen's interest.

This answer denotes the social conception of the engineer's profession and how professionals see themselves. The statement seems quite aligned with the idea of engineer as a social agent. Two main concepts are remarkable: *professional commitment,* which is associated with the main values of every occupational culture, and *citizens, people;* the statement is that engineers work for the people, not for companies or government and they must tend to activities that are required by public demand. The next answer is also quite representative of the general self-conception of engineers:

> Guarantee safety of any project beyond economic considerations, promoting safety to people and society and a higher comfort.

The conception of the engineer contained in this sentence is more technical. Engineers take care that everything is done in a safe and generally correct way, but they are "outside" the decisional group conducting the projects that are implemented and uniquely committed to promote that the projects they are involved in provide value to the society.

Our next question was related to the existence of an ethics code and its promotion among professionals. The answers were clearly affirmative, there are ethical and deontological codes, and the association has a committee devoted to this commitment that provides assessment, judgement and recommendations to the professionals.

Our next set of questions were related to the implications of engineers in society. We asked if engineers have a social value, if they see themselves as an executive profession or if they also have a social responsibility in the outcome and the conception of the project. The answers tend to consider that the social component must be present in engineers' profession. However, some comments are more related to the concept of civil or penal responsibility than to the broad macro-ethic conception. Thus, the presidents put the emphasis on legislation, and allusions to promote the welfare described before seem restricted to the safety and liability of a determinate project. So engineers seem circumscribed to their working environment and not in contact with the public. This conception was also related to the answer to public communication skills and ability required for engineers. In this case, the answer was (our translation):

> Implication in the public assessment and communication of engineers is restricted to the professional forums. Public communication is deserved to professionals.

From our point of view, it is quite clear that communication is better carried out by professionals. Nonetheless, in cases of technical projects with social implications, an engineer could provide an excellent outlook to observe the general situation. Instead the opinion presented above seems to be related to an engineer that is merely a technician.

Finally, when asked about the convenience of including ethical or humanistic concepts in engineering, all answers were affirmative. The general opinion was that engineers must have a close connection with people and that philosophical or ethical values were completely necessary in order to ensure the correct development of the profession.

Conclusion

The profession of engineer has evolved during the last decade. The changes in technology, globalization, intense communication with other professions, widening of the field of operation and need of assessment of the society in a changing world, have required new abilities from practicing engineers.

In order to satisfy society's needs, engineering programs have made an effort to adapt to the new engineer vision. The scope and intensity of these efforts are arguable to a certain point. One of the principal ways to implement these requirements have been through accreditation criteria.

Accreditation criteria have modified engineering programs under the influence of the accrediting institutions. The application of the criteria has been adopted quite directly by universities. Most of the accreditation criteria include competences related to environmental concern, sustainability and development of soft skills. Others are more ambitious and include special mentions to ethics, social awareness and social implications. However, allusions to these contents are quite diffuse in general.

The convenience of the desired outcomes, the adopted competences and ways of implementation are sometimes confronted with the possibilities that universities, teachers and students find in their respective environments. In practice, providing engineers with a broader perspective faces some serious issues. Some of them derive from the very nature of the knowledge that is being transmitted and the expertise of the institutions where engineering programs are taught. There are also difficulties in implementing the ideas in a way that is consequent with the actual programs and that do not collide with universities' conception of engineering. The approach followed in most cases has been teaching special courses or integrating the concepts along the existing subjects. The methodology is based mainly in case study method and providing theoretical background.

Our educational system and our society bring implicit values to the engineer's profession. The way the competences are acquired, through the formal and explicit exchange of information, through the context in which this exchange is developed and through the practice of engineering, can modify or confront the *habitus* of engineering. A deep analysis of the effect of these options, adjustments and antagonisms is required. These reflections can provide useful indications in order to integrate a broader vision of engineering into real practice and to really modify the *habitus* of the profession.

In Spain, there are general references to ethics or social values in the legal frameworks that regulate engineering programs, but the way of implementation or specific requirements is missing. Most universities teach these contents integrated in the set of soft skills that are provided to the graduates.

Following what we consider a general trend in the engineering world, professional associations are quite acquainted with the need to provide ethical assessment to practicing engineers, but mostly in the field of micro-ethics. Macro ethics, the general conception beyond the framework of a project or the strictly professional commitment is not envisioned. However these associations are aware of the importance of the formation in those subjects for future graduates.

Will engineers in the future really perform as society experts? Will they work for the society and promote public welfare? The prospects of achieving a real social implication, a public assessment of technology, an outlook to communicate change and provide vision should be important points in educational programs. The proposed changes in engineering programs seem quite subtle as to provide this change of paradigm.

References

ABET Engineering Accreditation Commission. (2011). *Criteria for accrediting engineering programs.* http://www.abet.org

Atman, C., & Nair, I. (1996). Engineering in context: An empirical study of freshmen students' conceptual frameworks. *Journal of Engineering Education, 85*(4), 317–326.

Beck, U. (2009). How to think about science, episode 5 – Ulrich Beck and Bruno Latour. In *How to think about science*. Ideas with Paul Kennedy, CBCRadio. http://www.cbc.ca/ideas/episodes/2009/01/02/how-to-think-about-science-part-1---24-listen/#episode5

Beder, S. (1999). Beyond technicalities: Expanding engineering thinking. *Journal of Professional Issues in Engineering Education and Practice, 125*(1), American Society of Civil Engineers.

Bergen Communiqué. (2005). *The European higher education area – Achieving the goals.* Communiqué of the Conference of European Ministers Responsible for Higher Education. http://www.ehea.info/Uploads/Declarations/Bergen_Communique1.pdf

Bourdieu, P. ([1972] 1977). *Outline of a theory of practice.* Cambridge: Cambridge University Press.

Brady, P. A., & Lawson, J. W. (2011). *Using case studies to characterize the broader meaning of engineering design for today's student architectural engineering education.* Architectural engineering conference proceedings: Oakland, CA. Available at: http://works.bepress.com/jwlawson/8

Brumsen, M. (2005). Ethics in engineering in the Netherlands: The role of professional associations, universities, and law. *International Journal of Engineering Education, 21*(3), 391–401.

Canadian Engineering Accreditation Board. (2011). *Accreditation criteria and procedures*. http://www.engineerscanada.ca/e/index.cfm

Catalano, G. D. (2006). *Engineering ethics: Peace, justice, and the earth* (Synthesis lectures on engineering, technology and society). San Rafael: Morgan and Claypool.

Christensen, S. H. (2003). Two cultures – Engineering and business: Towards a theory of occupational culture. In S. H. Christensen et al. (Eds.), *Profession, culture and communication: An interdisciplinary challenge to business and engineering* (pp. 3–92). Herning: Institute of Business and Technology Press.

Duderstadt, J. (2008). *Engineering for a changing world. A roadmap to the future of engineering practice, research, and education*. Ann Arbor: The Millennium Project, University of Michigan. http://milproj.ummu.umich.edu/publications/EngFlex_report/download/EngFlex%20Report.pdf

Edwards, M., Sánchez-Ruiz, L. M., & Sánchez-Díaz, C. (2009). Achieving competence-based curriculum in engineering education in Spain. *Proceedings of the IEEE, 97*(10), 1727–1736.

EET. (2013). Document available at http://www.eet.upc.edu/courses/bachelors-degree-programmes/information-on-courses/development-and-assessment-of-generic-competencies?set_language=en

Engineering Accreditation Council (EAC) Malaysia. (2008). *Determining accreditation decision*. http://www.eac.org.my/web/document/Determining%20Accreditation%20Decision.pdf

Engineers Australia Accreditation Board. (2008). *Accreditation management system education programs at the level of professional engineer.* http://www.engineersaustralia.org.au/nerb/board

European Association for Quality Assurance in Higher Education (ENQA). (2005). *Standards and guidelines for quality assurance in the European Higher Education Area*. http://www.enqa.eu/index.php/home/esg/

European Network for Accreditation of Engineering Education ENAEE. (2008). *Framework standards for the accreditation of engineering programmes*. http://www.enaee.eu/

Herkert, J. R. (2002). Continuing and emerging issues in engineering ethics education. *The Bridge (National Academy of Engineering), 32*(3), 8–13.

Illich, I. (1973). *Tools for conviviality*. New York: Harper and Row.

Jamison, A., Christensen, S. H., & Botin, L. (2011). *A hybrid imagination: Science and technology in cultural perspective*. San Rafael: Morgan & Claypool.

Jones, C., Saha, K., Evans S., & Pfister, T. (2011). *A provocation. The next 20 and beyond: Provocations from within the field. STS 20+20 conference*. Harvard University. http://stsnext20.org/conference/a-provocation/

Mitcham, C. (1998). The importance of philosophy to engineering. *Teorema, 17*(3), 27–47.

Nahar, Y., Baillie, C., Catalano, G., & Feinblatt, E. (2009). *Engineering values: An approach to explore values in education and practice.* http://rees2009.pbworks.com/f/rees2009_submission_3.pdf

Paretti, M. C., & Burgoyne, C. B. (2005). Work in progress – An integrated engineering communications curriculum for the 21st century. *Proceedings 35th annual conference frontiers in education (FIE '05)* (pp. F3F-5). http://ieeexplore.ieee.org/stamp/stamp.jsp?tp=&arnumber=1612098&isnumber=33854

Peercy, P. S., & Cramer, S. M. (2011). Redefining quality in engineering education through hybrid instruction. *Journal of Engineering Education, 100*(4), 625–629.

Prausnitz, J. M. (2003). *Towards breaking the silence between the two cultures: Engineering and the other humanities*. Lawrence Berkeley National Laboratory. Retrieved from http://escholarship.org/uc/item/30831547. June 2012.

RD1393/2007 REAL DECRETO. (2007). de 29 de octubre, por el que se establece la ordenación de las enseñanzas universitarias oficiales. http://www.boe.es/boe/dias/2007/10/30/pdfs/A44037-44048.pdf

Revel, M. (2007). The necessity for engineers to engage in public issues of global importance. In S. H. Christensen, M. Meganck, & B. Delahousse (Eds.), *Philosophy in engineering*. Aarhus: Academica. Chapter 14.
Riley, D. (2011). *Engineering thermodynamics and 21st century energy problems: A textbook companion for student engagement* (Synthesis lectures on engineering). San Rafael: Morgan and Claypool.
Sennett, R. (1998). *The corrosion of character. The personal consequences of work in the new capitalism*. London/New York: W.W. Norton & Company.
Suarez, B., Revilla, J. A., & Galan, L. (2011). Quality assessment in engineering education in Spain: Towards a new accreditation agency. *First EUCEET association conference: "New trends and challenges in civil engineering education"*, Patras.
The Royal Academy of Engineering. (2009). *Engineering values in IT: A joint study by The Royal Academy of Engineering, the Institution of Engineering and Technology, and the British Computer Society*. www.raeng.org.uk/engineeringIT
UNED (Escuela de Ingenieros Industriales). (s.d.). *Ética Profesional*. http://portal.uned.es/portal/page?_pageid=93,5972357&_dad=portal&_schema=PORTAL
Winner, L. (1986). *The whale and the reactor*. Chicago: The University of Chicago Press.
Winner, L. (2010). *Local citizens against global, corporate power. The role of neighborhood science between local community and global corporations*. Medialab Prado. Lecture available at http://medialab-prado.es/mmedia/5179/view

Javier Cañavate B.Sc. in Chemical Engineering, M.Sc. and Ph.D. in Industrial Engineering from Technical University of Catalonia, Barcelona Tech (UPC), Spain. College Professor and Dean of School of Engineering in Terrassa (EET), a school of the Technical University of Catalonia Barcelona Tech (UPC), Spain. He has worked as International coordinator and participated in several European educational projects. His teaching areas are Chemistry and Polymers and his scientific research areas Polymers and Composites. He is author of about 30 articles and book chapters in materials research, an official guide to the development and implementation of engineering programs and also participated as co-author in the books *Philosophy in Engineering* (Academica 2007), and *Engineering in Context* (Academica 2009).

Manuel José Lis Arias B.S. in Chemical Engineering, M.Sc. in Industrial Engineering, and Ph.D. in Industrial Engineering from Technological University of Catalonia (UPC). Since 1984, teaching in the Chemical Engineering Degree Program at UPC, Engineering School of Terrassa (EET). As Assistant Director of International Relations of EET, he was the coordinator of an EU Intensive Program on *Sustainable Development applied to Engineering*. Co-author of *Philosophy in Engineering* (Academica 2007) and *Engineering in Context* (Academica 2009). His research areas are: Social and Ethical considerations on Engineering Design and Smart Materials for Biomedical applications.

Josep Maria Casasús M.A. in Philosophy, Universitat Autònoma de Barcelona (UAB). Teacher in secondary schools and in the Art School of Terrassa, Spain. His main publications include articles on the relationship between art, design, cooperation and philosophy. He is a founder member of Gep21 (Grup d'Estètica i Política – segle XXI), a multidisciplinary group of philosophers, geographers, sociologists that intends to approach contemporary thought on aesthetics and politics to all kind of audiences outside the academic field. He is involved as activist in several campaigns. He has participated as co-author in the book *Engineering in Context* (Academica 2009) and *Aesthetics and Policy* (Gep21, 2010).

Chapter 23
Engineering as a Socio-technical Process: Case-Based Learning from the Example of Wind Technology Development

Matthias Heymann

Abstract This chapter describes an example of teaching engineering students social perspectives of engineering by using the case example of wind technology. It is part of the philosophy of engineering course taught to undergraduate engineering students at Aarhus University. The case of wind technology development is suited to discuss a large number of different social issues related to engineering, such as engineering approaches (science-based versus practice-oriented), the role of engineering styles and traditions, forms of learning and interaction in engineering, requirements and problems of engineering communication, innovation strategies, research policies, market structures and ideologies. The case of wind technology shows that engineering is more than developing technical artifacts. It is a way of "mixing with the world" in a much broader sense than reflected in many engineering curricula.

Keywords Engineering education • Case-based learning • Non-technical elements in engineering education • Engineering communication • Wind technology

Introduction

Since 2004, it has been mandated by law that all Danish undergraduate university programs have to include a compulsory course on the philosophy of science for that particular program. At the Faculty of Science and Technology, Aarhus University, the Centre for Science Studies has taken over the responsibility to offer these courses for most science and technology bachelor programs including mathematics, IT and computer science, physics, chemistry, medical chemistry, biology, molecular biology, molecular medicine, geology, nanotechnology, and, since 2011,

M. Heymann (✉)
Centre for Science Studies, Department of Mathematics, Aarhus University,
Ny Munkegade 118, Aarhus C 8000, Denmark
e-mail: Matthias.heymann@css.au.dk

S.H. Christensen et al. (eds.), *International Perspectives on Engineering Education*,
Philosophy of Engineering and Technology 20, DOI 10.1007/978-3-319-16169-3_23

engineering. The Centre adopted a teaching philosophy of using historical and contemporary case studies to anchor broader philosophical discussions in the particular subject discipline under consideration. With this approach it aims at tailoring courses to the interests of the students (avoiding purely theoretical lectures) and, at the same time, making important philosophical issues and themes related to their disciplines interesting and accessible to the students (Sørensen 2012; Andersen et al. 2009).

The design of the philosophy of engineering course takes a similar case based approach like the other philosophy of science courses. In contrast to most of these courses, however, a number of cases is taken up more than once to serve for the discussion of very different themes. These cases represent threads through the course rather than vehicles for addressing singular points of interest. This design has been adopted purposely to account for the complexity of engineering processes and of technology-society relations. A variety of very different issues in engineering play out in single cases; and, at the same time, a variety of very different issues in engineering are inherently linked, as single cases can highlight. Furthermore, students learn to study examples of engineering and technology from very different perspectives. A particularly rich case to discuss many of the issues in engineering to be covered in the course is the case of wind technology development since the 1970s. This chapter will present this case and draw conclusions about its educational value. The structure of the chapter follows the structure of teaching in class. Starting out with developing the question "Why were the Danes best?" subsequent sections – taught in subsequent classes – will contribute to solving this riddle.

Why Were the Danes Best? The Riddle of Modern Wind Technology

The energy crises of the 1970s increased interest in alternative energies and led many countries to launch wind power research programs. Governmental expenses for wind technology research in the period 1975–1988 reached $427.4 million in the USA and $103.3 million in the Federal Republic of Germany. The Danish government, in contrast, representing a much smaller country, still spent $19.1 million in the same period (Heymann 1995, p. 345). Some of the leading high-tech companies, such as General Electric, Boeing, Lockheed or Westinghouse in the USA or MAN, MBB and Dornier in Germany were involved in wind technology research. The intensive research and development efforts, however, showed surprising results. Danish wind turbine manufacturers produced the best turbines and sold by far the largest number of turbines in the 1980s. Danish wind turbines covered about 45 % of the total wind turbine capacity installed worldwide in 1990 (Heymann 1998, p. 642). As it turned out, Danish producers came up with a design of modern wind turbines, which proved superior compared to others and set the standard for international wind turbine design since the late 1980s: a three-blade rotor running upwind (on the weather side), medium rotor velocity and a robust mechanical structure. In the early 1980s, most Danish turbines were smaller in size and power, but about twice as heavy as American wind turbines. Turbines of US producers typically had a specific turbine weight (weight per power) around 30 kg/kW, some were even much lighter. Danish wind turbines, in contrast, typically ranged around 75 kg/kW, with some turbines even significantly heavier (Heymann 1998, p. 647).

As it turned out, most of the US-produced wind turbines failed rather miserably, many already after short operation time. Danish turbines, in contrast, proved surprisingly reliable. When in the early 1980s a huge wind turbine boom set off in California, Danish turbines showed superior performance and boosted Danish exports. In 1982, Denmark exported the first 30 wind turbines to Californian wind farms, which in that year installed a total number of about 1,200 turbines. In 1985, Danish companies sold 3,812 turbines to California, a share of 62 % of all newly installed turbines in California in that year; 1 year later they reached a share of even 70.8 % (Heymann 1995, pp. 397–399, and Heymann 1998, p. 646, 664). The total number of wind turbines operating in California reached about 15,000 in 1987 and stagnated in subsequent years. The performance of many turbines in California was poor. In autumn 1985, out of 8,460 turbines only 4,400 (48 %) proved operational. Out of 1976 installed Danish turbines, however, 1932 (98 %) were in operation. On average, Danish turbines could be operated 98 % of the time, while all other turbines only operated 38 % of the time (Stoddard 1986, p. 94). Among the top ten wind turbine producers with regard to turbine performance in California ranked five Danish producers (ranks 1, 2, 3, 5, 7) mostly outclassing four US producers on that list (ranks 4, 8, 9, 10; see California Energy Commission 1988, pp. 18–19). Technically, Danish turbines were clearly superior. The major problem of Danish turbine manufacturers in the early 1980s was not technology, but rather the poor command of English of their managers and engineers.

So, why proved Danish wind technology clearly better compared to US or German wind technology, even though the United States and Germany put about 20 times, respectively five times, as much money into it? There is no single answer to this question. A number of crucial issues in engineering need to be considered like pieces of a puzzle which only when put together give the full picture. These engineering issues are related to: (1) knowledge bases and engineering mentalities, (2) engineering communication and the social character of technical development, (3) power relations, technology policies and ideologies. These pieces of the puzzle will be revealed in subsequent sections in this chapter – like in subsequent classes in teaching. All these parts have insights to offer, which lie beyond traditional engineering perceptions and – to a large extent – outside of most engineering curricula.

Knowledge Bases and Engineering Mentalities

Wind technology involves many fields of scientific and engineering knowledge, such as aerodynamics, mechanical engineering and electrical engineering. Not surprisingly, well-established companies with a strong background in aerodynamics, mechanical and electrical engineering – companies such as Boeing, Lockheed, Westinghouse, General Electrics, MBB and MAN – dominated wind technology R&D programs in the 1970s and 1980s in countries like Sweden, Germany and the USA. These companies failed and all left the field of wind technology development after some 10–15 years of engagement. Similarly, most producers for the emerging huge market in California failed. A fundamental problem in both cases was over-ambition. Producers for the Californian market adopted highly ambitious strategies of light-weight design and maximizing efficiency. Most national research programs focused on innovative new designs and very ambitious large-scale turbines with a power output in the megawatt range. The German Growian, the largest wind turbine at its time, is a striking example. Built from 1981 to 1983, it had a height of 100 m, a rotor diameter of 100 m and a power output of 3 MW. Within 4 years it only operated about 500 h (instead of planned 15,000 h) due to enormous technical problems (Heymann 1995, pp. 369–382). All these research and development efforts proved utterly unsuccessful not to the least, because they aimed at either large-scale and/or highly ambitious new technologies from the outset, even though all actors totally lacked experience with wind technology and widely underestimated its complexity and demands (Heymann 1998; Garud and Karnøe 2003).

In Denmark, a different group of actors, which built on a completely different knowledge base, proved much more successful. Craftsmen with an interest in wind power like the carpenter Christian Riisager and the blacksmith Karl-Erik Jørgensen built on a historical example. About 20 years earlier, in 1957, the Danish engineer Johannes Juul had erected a 200-kW wind turbine at Gedser, the so-called Gedser turbine. Juul had experimented in the 1950s over about 10 years with various designs and elaborated a turbine design that proved to be the most reliable: a turbine with a robust, three-blade upwind rotor. This historical example proved of invaluable help (Heymann 1998). Riisager and Jørgensen rebuilt versions of the Gedser turbine in much smaller size with a power of about 20 kW. These turbines, in contrast to other designs in other countries, worked surprisingly well. The historical example proved a crucial starting point for wind turbine development in Denmark. A second feature of the Danish knowledge base relates to development approach and style. Juul like his successors in the 1970s were practice minded tinkerers who trusted more in testing ideas practically than applying sophisticated theoretical approaches. Riisager and others worked with simple and robust strategies, focusing on mechanical strength rather than lightweight design and efficiency. Copying an existing design was part of that strategy, as was the decision to build a small version of the turbine and use cheap and robust standard parts. Technical problems were solved through experimentation rather than theoretical design, calculation and computer simulation. Practical experience turned out to be a key advantage. It gave rise to a rich base of personal 'tacit' knowledge, a feeling for forces and loads and for the performance and limitations of technical components.

What can students learn about the role of knowledge bases from the case of wind technology? Certainly not that craftsmen are the better engineers. Academic engineering knowledge proved crucial in later stages of the Danish wind turbine development (see below); and craftsmen would have been fully inadequate in many sophisticated technological developments like the transistor, radar technology or nuclear power. Instead, three messages deserve attention. First, the knowledge generated through *practical experience* is of crucial importance to technical design. This type of knowledge has been described by Michael Polanyi as "personal" or "tacit" knowledge. It is a type of knowledge that cannot be replaced by or transformed into theoretical knowledge. Consequently, it cannot be taught or conveyed by one person to the next, but needs to be acquired in practical work (Polanyi 1966; Collins 2010). The role of tacit knowledge is not limited to practical tinkering and experimentation in the material world, like in the case of the Danish craftsmen that developed wind turbines. It is neither limited to tinkerers and craftsmen. The effective and successful application of engineering theories or computer-based design systems relies on experience-based tacit or personal knowledge as well. As tacit knowledge cannot be taught in classes, it tends to be undervalued or even ignored in engineering curricula. Reflecting on cases like wind turbine development helps to make it more visible.

Second, different knowledge bases in engineering contribute to shaping different engineering approaches, styles and strategies. The practical tradition and knowledge of craftsmen involved a preference for practical trial-and-error approaches. Theoretical education and the knowledge of academic engineers, in contrast, created a focus on theoretical or theory-based approaches. Peter Karnøe has described the different approaches to wind turbine development as "bottom-up" and "top-down" (Karnøe 1991). While craftsmen adopted a bottom-up approach to wind technology development, engineers and planners in ministries and companies developed top-down visions of ambitious technical solutions.

Third, different knowledge bases do not only consist of different sets of knowledge and shape different approaches. Different knowledge bases also create different orientations, values, mentalities and ideologies (Heymann 1996). A craftsmen knowledge base characteristically relates to a conservative rather than ambitious and innovative engineering style. This conservatism involves an appreciation and persistence of existing and proven solutions, an inclination to security-oriented construction and reluctance against high technical or economic risks. Craftsmen like Riisager focused on practicability and reliability rather than novelty and sophistication. The successful reconstruction of a small and cheap version of a historical turbine as an initial step fully satisfied their ambitions. Engineers, in contrast, are socialized from early on in their education with the expectation to be innovative and create novel solutions. This expectation is a part of engineering identity and self-confidence. High recognition of science and theoretical knowledge in twentieth century western culture underpinned reliance in science-based (in contrast to practical) approaches and strengthened the confidence in science-based engineering. Andrew Jamison and Michael Hård referred to this state of mind as "hubris" (Hård and Jamison 2005; also Jamison et al. 2011). Hubris facilitated technological breakthroughs and rapid technological change. Hubris, on the other hand, also caused the underestimation of technical problems and overestimation of engineering capabilities, as the case of wind power has shown (Heymann 1998).

Engineering Communication and the Social Character of Technical Development

Knowledge bases and concepts like tacit knowledge, bottom-up versus top-down approaches and over-ambition and hubris represent first important pieces for solving the riddle of modern wind technology; but they don't provide a full answer. A second direction of explanation is related to the social character of engineering. Modern wind turbines "embody the steady accretion of inputs from many actors" Garud and Karnøe emphasized (2003, p. 282). Communication and cooperation proved instrumental in wind technology development (the following argument is developed in greater detail in Nielsen and Heymann 2012). Danish craftsmen and tinkerers who built wind turbines in the 1970s were not alone in their endeavors. They would likely have failed without collaborators, supporters, idealistic buyers, combatants and fellow campaigners. In fact, wind turbine development in Denmark was embedded in strong social movements campaigning against nuclear power and for renewable energy (Jamison 2001). This context proved to be of crucial importance for the creation of networks of communication and cooperation. The antinuclear and alternative energy movement(s) carried common values and pursued common social ambitions like environmental protection, democratic, self-controlled technology and social justice. They shaped common ideals and identities and created a common language and effective platforms for intensive communication across social and spatial boundaries and disciplinary and professional demarcations. People with different background, students and professionals, workers and academics, producers and consumers met and pursued common goals.

The social movements provided motivation, visibility and support to wind turbine developers in various ways. Notably, these movements pursued not only political goals, but also practical interests. Wind turbine builders and other practical-minded people with an interest in wind power constituted a notable fraction in these movements. A number of examples will help to understand how social movements helped to create communication platforms, bring different people together and facilitate interaction and exchange. A first unique feature in the emergence of Danish wind technology were so-called "wind meetings", in which people interested in wind power, engineers, technicians, users, buyers, activists and others regularly met to discuss relevant issues, exchange knowledge and experience and enter collaboration. Second, an example of such type of collaboration was related to the construction of turbine blades. Most turbine builders had little knowledge about the aerodynamics of blades they needed. Two companies, LM and Okjær, recognized this niche, specialized in blade production and supported significantly early wind turbine development (Karnøe 1991, p. 193). This way of sharing work to bring in the best expertise needed was unique and emerged only in Denmark.

Third, the newly-founded journal Naturlig Energi, a product of the alternative energy movement, represented an important platform for communication. The journal published detailed performance data of all wind turbines in use in Denmark. These data indicated failures and problems of wind turbines. It created a learning base for wind turbine developers and, at the same time, a base of information for turbine buyers. Yet another crucial institution facilitating exchange and learning was the Test Station for Small Wind Turbines, which was established as part of the Danish wind power research program in 1979. The Test Station carried out mandatory system assessments of wind turbines to enable quality control of wind turbine systems eligible for government subsidy. The Test Station, staffed with professional engineers who subscribed to the social and technical values of the renewable energy movement, proved instrumental in spreading knowledge and creating acceptance of best practices and common technical solutions, for example of standard design principles which proved most successful and became known worldwide as "Danish Design" (Stoddard 1986, p. 83).

Fourth, cooperation proved also crucial in the transition from technology development (building first wind turbines) to market success (setting up a production and sales of large numbers of turbines). Pioneers like Riisager did not have the technical competence to stand his ground in a field of quick and continuous technical improvement and competition. He neither had the capital and commercial expertise to build production facilities, launch marketing efforts and organize sales of wind turbines on emerging markets in Denmark and California. At this stage, small machine companies stepped in, acquired licenses and patents from the pioneers, professionalized wind turbine development based on larger technological resources, built production facilities for the mass production of turbines and succeeded in entering, and soon dominating, profitable markets. Vestas, later one of the leading wind turbine producers, was a machine company with 120 employees producing hydraulic cranes for light trucks. In 1978, it tested and took over Karl-Erik Jørgensen's wind turbine design. Vestas had employed the first academic engineer in 1971 and possessed technical knowledge and resources for the advancement and stepwise enlargement of the simple first turbines. Other small turbine builders drew on the service of the company Tripod, which was founded by two engineers to offer calculations and measurements for the improvement of wind turbine design (Karnøe 1991, p. 214). By the mid-1980s all Danish wind turbine producers had employed academic engineers.

The important role of communication and collaboration becomes particular visible in the case of a lack of it. Failures in wind technology development in Germany had much to do with the social structure and organization of turbine construction and the different styles, habits and scopes of communication it facilitated. Established development styles in Germany both in small companies and in large corporations included a strong element of competition and confidentiality, which hampered and limited communication and reduced collaboration and the sharing of knowledge. Competition was an explicit element in the development strategy of the government's renewable energy research program launched in the 1970s. Large engineering companies involved in wind technology research and development were used to operating in contexts of international competition. They kept the long-standing practice of official corporate communication to keep knowledge in technical development, particularly development problems, secret. In the context of competition, engineering communication meant something very different from knowledge-sharing and collaboration. Only selected and manipulated information –

mostly controlled by professional communication departments – crossed corporate borders to the outside world for the purpose of public relations and marketing. It served for concealing rather than revealing, technical and operational details and problems.

In wind technology development these communication habits caused insulation and fragmentation and an exclusion of collaborative approaches and synergies. An international comparison of wind technology development in the 1980s and 1990s concluded: "The statement that 'competition vitalizes the economy' does not fit for the pioneering phase of wind energy [technology]" (Neukirch 2010, p. 214). Quite to the contrary, success depended on protected spaces and close regional interaction of wind turbine manufacturers and users, as was the case in Denmark. In their comparison of Danish and Dutch wind turbine development Linda Kamp et al. (2004) emphasized the role of learning and interaction as important elements in the Danish success story. While Dutch wind engineers relied on traditional learning-by-searching mechanisms, Danish actors proved more open to learning-by-interacting. Across a wide range of interests and professional expertise Danish wind turbine builders engaged with other wind actors, learning from each and negotiating designs, technologies and policies.

This part of the story teaches students that engineering is more than technical work. A crucial part of any engineering work is social interaction, namely communication, collaborative learning and cooperation. In short, engineering is a social process. It is hardly reflected in engineering programs at universities that engineering is a social process that requires social skills as well as platforms and infrastructures facilitating communication and cooperation. Communication is not limited to technical communication with engineering peers to exchange technical knowledge. Engineering projects usually involve heterogeneous actors with different background, expertise, interests and social commitments. In the case of wind turbine development in Denmark, shared social values in the renewable energy movement of the 1970s and institutions like regular "wind meetings", the journal Naturlig Energi and the Test Station for Small Windmills facilitated communication and learning across the borders of disciplinary, professional or social demarcation. This case and its social context were certainly special and hardly provide a blueprint for other instances of technical development and innovation. But the message it carries holds more generally. First, social interaction and communication play a crucial role in engineering; and second, social interaction and communication are not simple tools at hand if needed. They demand favorable social conditions and appropriate institutions. Engineering students may profit from learning this lesson.

Power Relations, Technology Policies and Ideologies

Power relations and political, economic and cultural conditions strongly affected wind technology development. Many governments had to learn (and did learn) that wind power use was not only (and maybe not in the first place) a problem of wind technology. It was also a question of power and interests, markets and monopolies, political and legal constraints. Making available a working technology (as most national research programs aimed at) was not enough to produce a success on the market (even if the technology was technically successful). At first sight, wind turbines in most countries failed technologically, proved unreliable and were much too expensive to sell. At second sight, a more complicated logic becomes apparent: wind turbines shouldn't sell. Energy markets were controlled by powerful corporations and trusts which had monopolized energy production and sale (Lucas and Papaconstantinou 1985). Why should they be interested in an alternative energy technology that gives competitors access to the market? Wind turbines could be owned by any citizen, farmers for example, who would produce their own electricity and feed surplus electricity in the grid, thus curtailing markets, monopolies and power of large energy companies. Consequently, the power industry inhibited market access of wind turbines by creating economic, technical and legal barriers. In Germany, a large number of legal cases emerged in the 1980s, because power companies declined or hampered access of wind turbines to their electricity nets (Heymann 1995, pp. 420–426).

The problem, however, was not only vested corporate interests. The structure of the energy production system proved highly incompatible to wind power. Power companies operated highly centralized systems of power production, which were based on large power plants. Wind turbines with a much smaller power output, in contrast, represented a highly decentral energy source. Furthermore, wind turbines provided only varying power output depending on variable wind velocities. The integration of decentral and variable electricity production in the electricity system demanded extended net infrastructures and compensating power capacity in the case the wind didn't blow. Such adjustments, which represent an effective hybridization of the market structure to accommodate both central and decentral elements, were blocked by the power industry – at least as long as legislators did not push them by force (Heymann 1999). California attempted to make wind power compatible to centralized energy production by building huge wind farms with thousands of wind turbines – an approach that demanded very large areas (and investments) and was hardly conferrable to more densely populated European countries.

Most governmental wind power programs launched in the 1970s and early 1980s did not take power relations, market conditions and the structure of the energy system into account. Governmental programs in the USA, Germany, the Netherlands or Sweden focused on the support of wind technology development only. Their rationale went as follows: Wind technologies, once they had sufficiently matured, had to compete on the market against the established technologies. This innovation strategy failed miserably. It was guided by a free market ideology for a sector in which a free market did not exist. Furthermore, it took a centralized structure of power production for granted, a given constraint to which alternative technologies were expected to adjust. It was not to the least this condition that tempted administrators and pushed engineers to pursue the development of much larger wind turbines than successfully achieved by any historical examples. In hindsight, this technological jump from windmills or small turbines to megawatt-size machines (a jump by a factor of about 10 in terms of size and by a factor of about 100 in terms of power) and from proven technology to new technological solutions under the condition of a total lack of experience with wind technology seems a particularly striking example of hubris.

A different picture offered California. California only supported market development by granting lucrative tax reductions and subsidies. The Californian tax schemes were put into effect uncoordinated with and totally independently of the federal US wind technology research program. This strategy also failed. While huge wind parks emerged within very few years the largest fraction of turbines never worked reliably and rather produced losses than electricity and revenue. A (subsidized) market had been created, whereas a mature technology did not exist. An exception to these cases proved wind power research and development policies in Denmark. The Danish government supported both technology and market, not to the least as a response to demands of a strong renewable energy movement. This innovation strategy linked well to the (independent) technical development of small wind turbines. Technology and market emerged concurrently and could mutually be stabilized. Turbines were sold and operated, invaluable experience gained. As a result, Danish wind turbines improved quickly in reliability, size and competitiveness and helped the expansion of wind turbine markets. The approach proved so successful that other countries, most notably Germany, copied a combined strategy of technology and market support about 10 years later. Market subsidies, appropriate legal conditions and Danish Design as technological basis let Germany surpass Denmark in 1997 in terms of installed wind power capacity and electricity production (Ibenholt 2002).

Technology development and energy policies do not take place in a cultural vacuum. They are embedded in and shaped by sets of beliefs, values and norms. The political, technological and industrial elite in most countries conceived of the centralized power system based on large nuclear and fossil power plants as an indispensable backbone of sufficient, reliable, safe and cheap energy supply. It was believed to be precondition to industrial and economic progress, social development and wealth. In this mindset any compromise to the established power infrastructure was conceived as an inacceptable risk to energy security, reliability and cost effectiveness. Consequently, the expansion of nuclear energy, which was structurally compatible to the centralized power infrastructure, represented the most promising development path. The use of alternative renewable energies, in contrast, appeared to be a great risk and – given their small power capacity and limited performance – were hardly taken seriously. The American physicist and environmentalist Amory Lovins called this set of convictions and beliefs the "hard energy path". He contrasted it with a fundamentally different view that had emerged during the 1970s, the "soft energy path" (Lovins 1977). Large parts of the growing environmental movement shared alternative visions of clean, sustainable, small-size, decentral, more democratic energy production. These visions entailed very different beliefs, values and norms. The demand for alternative, clean energy technologies like wind power was linked to much broader social goals like the creation of a better, more human, more just and more environment-sensitive and sustainable society

(Schumacher 1973). The "hard" and the "soft energy path" were more than alternative energy strategies. They represented a polarized society, a split of ideas and ideologies which ran much deeper than conflicting visions of future energy.

This intellectual and ideological substratum affected technology policies significantly. A successful innovation of wind turbines, the development and maturation of wind technology and political decisions for the support of it on the market depended on values, ideologies and cultural commitments and their persistence and political influence. In the countries in which a commitment to the "hard energy path" and its inherent values and beliefs proved influential, effective policies for wind power had a much lower chance, even in the case of a strong environmental and anti-nuclear movement, as the example of Germany in the 1970s and 1980s shows. Denmark, in contrast, was characterized by a political tradition of liberalism and decentralism which allowed for and fostered social spaces of emerging alternative visions and imposed much lower barriers to political participation (Lucas and Papaconstantinou 1985). This cultural context was a foundation upon which the anti-nuclear and renewable energy movements could build and gain political influence.

This part of the story sets engineering in broader perspective and context. It serves to convey a number of messages to the students. Technology development and innovation transcend the domain of technology and its technical and social elements. Technology development takes place in and makes part of a larger context of power relations, market structures and policies as well as beliefs, values and ideologies. Many engineers still tend to see engineering as a highly specialized and more or less contained practice, which is taking place in the protected spaces of the company or the laboratory and only guided by technical goals and constraints. In this view, engineers are advised to keep to their technical competence and keep out of politics and public debate. The case of wind technology helps to show that this position has a number of serious flaws. First, engineers are not acting free of political and social values, but are part of social groups and carriers of convictions and ideologies. Second, technologies carry non-technical values, commitments and goals (which may be perceived differently by different actors and social groups). Langdon Winner has famously explained this condition with the concise phrase that "artifacts have politics" (Winner 1980). Third, engineers have to serve customers and markets. Ignoring market conditions and social interests and trends reduces the chance of economic success. Companies like Apple have learned the lesson and made it their competitive advantage. They sell emotional experience and identity rather than only technology. Wengenroth (2001) drew the conclusion that engineers and innovators need to abandon their focus on technology and strengthen cultural competence and communication.

Engineering curricula with a strong focus on science and technology rather conceal these economic, political and cultural settings, of which engineers are an influential part. If students learn to develop awareness for this condition, if they learn to perceive themselves as a part of a larger culture with influential and conflicting values and goals, then they may more easily develop the political and cultural sensitivity required in technological development and innovation. There is no best technology or best technological path based on engineering (or whatever other) criteria. There are different social groups giving technologies different meanings and appropriating technologies for different social and cultural ends.

Elements of Innovation: Conclusions from the Case of Wind Technology

Can students learn anything from the case of wind technology and wind power use? The success story of Danish wind technology does not provide any blueprint to successful innovation. It is a remarkable, but unusual case of innovation, a type of development which occurred in a specific period of time with unique conditions and contexts. The role of craftsmen in this case is probably highly uncommon in late twentieth century cases of innovation. The role of social movements likewise represents a unique feature that occurred in a limited time window in the 1970s and 1980s. Communication and collaboration across such wide range of different social actors has likely been rather rare in processes of innovation. Contested power relations, social conflict and incompatible ideologies of proponents of the "hard energy path" and the "soft energy path" culminated in the 1970s. In spite of these special circumstances and unique characteristics, the case of Danish wind technology provides important lessons.

Modern wind technology, which by many engineers and politicians was mistaken for a relatively old-fashioned and simple technology, turned out to be a very complex case of innovation. Its historical investigation helps to make crucial elements of innovation visible, which may appear in different forms at other times and places, but are of general importance and nowadays still valid. In this chapter, these elements have been subsumed under the three headings knowledge bases and engineering mentalities, engineering communication and the social character of technical development and power relations, technology policies and ideologies. I will refer to these headings in short as (1) knowledge base, (2) social interactions and (3) techno-political settings, which represent major categories in a model of elements of innovation, which I will discuss below (Fig. 23.1, see next page).

Knowledge Base

The case of wind technology does not indicate that knowledge and skills of craftsmen are superior to the knowledge and skills of academic engineers. It rather provides the message to engineering students that knowledge bases are important and can shape development paths in different ways. First, historical knowledge and experience matter. Without Juul's blueprint for a working wind turbine, Danish

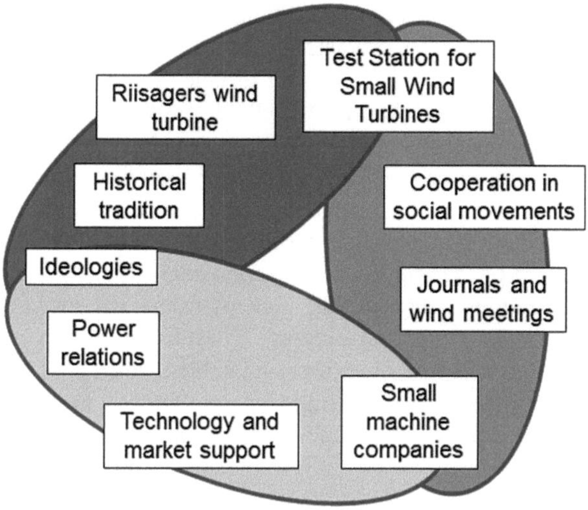

Fig. 23.1 A model of elements of innovation in Danish wind technology development based on the three categories knowledge base (*dark*), social interactions (*medium*) and techno-political settings (*light*)

developments would most likely have been less focused and successful. Second, and most importantly, the role of craftsmen reminds us of the importance of practical experience and tacit knowledge. It is a crucial point to learn that "engineering theory and technical skill are two irreducibly distinct components of all technological knowledge" and that "no technical praxis is completely reducible to abstract theory" (Staudenmaier 1985, pp. 115–116). Engineering curricula set the focus on theoretical knowledge. All the more important it is to discuss with students the role of tacit forms of technical knowledge. Both types of knowledge, theoretical-scientific and practical-personal (or "tacit"), are crucial in technical work. There is no general rule when one or the other type of knowledge is more likely to be effective in concrete development tasks. The historical record in engineering design in the twentieth century shows that theoretical knowledge has tended to be overestimated, whereas practical knowledge was underestimated (Ferguson 1992). Numerous ambitions by academic engineers to make engineering design a *science*, which can only be based on established theoretical knowledge proved misled and failed (Heymann 2005, 2009).

Third, the reflection of engineering mentalities and identities is another crucial point to be considered in engineering education. How do engineers attack a problem? Empirical design research has shown that unconsciously academic engineers tend to attack design problems differently than practice-minded engineers without theoretical education: they do it more theoretically, more systematically and with higher ambitions (Ehrlenspiel 2003, pp. 112–114). The case of wind technology has clearly shown that academic engineers (mostly in settings of large companies with strong theoretical knowledge base) proved to be much more ambitious than Danish wind turbine builders and awfully underestimated the technical problems, whereas Danish 'low-tech' developers appreciated the role of a slow pace of progress based on small, incremental steps. While Danish turbine builders started with building 20-kW turbines and simple proven technologies, some engineering companies aimed at sophisticated new technological solutions and started with developing turbines in the megawatt-range. While this case may represent an extreme example, there is a general point to it. The technological ambition to be *innovative*, develop *new* and *better* technology than in the past, push the front of technology as far as possible, makes part of the identity of academic engineers – and it usually makes part of the socialization of engineering students from early on in their university education. This identity along with a record of past technical successes contributes to creating a mentality of self-assuredness and hubris, which contrasts notably with the conservatism of practice-minded engineers (particularly in settings of the small workshop with limited material and knowledge resources).

Social Interactions

Another misleading feature of engineering education is the general focus on knowledge, mostly scientific and technical, while the role of social interaction in engineering is almost ignored. The case of Danish wind technology has shown the crucial importance of interaction particularly clearly. Communication and cooperation among many diverse actors in the frame of social movements, technical cooperation with the Test Station for Small Wind Turbines, knowledge exchange through detailed operation records in the journal Naturlig Energi and many other examples make this point (Nielsen and Heymann 2012). Communication and social interaction, however, don't come by themselves (as some engineering teachers may tend to believe). They require appropriate languages, appropriate institutions, appropriate skills and the establishment of stable and trustful relations. In technical development projects the range and network of interaction certainly varies. In some cases interaction may be limited to more or less homogeneous actors in a corporate laboratory. In others – like wind technology – it involves numerous heterogeneous actors spread in the whole society. In some cases of development only communication for collaborative learning and technical improvement will be required. In other cases, intensive interactions across social boundaries may be crucial.

Engineering students usually learn to some extent *technical* communication and cooperation with their teachers and peers. Engineering science provides a defined language for technical communication; and the common background and identity in the discipline and at the university serve for trustful relations. But how well are engineering students prepared for communication with heterogeneous actors outside university? Which skills do they acquire at university for selling their message or creating trustful relations to managers, marketing specialists, controllers, customers, shareholders, politicians, journalists etc.? The teaching of the role of social interaction in engineering will not provide the skills required for these types of interaction. It may contribute to it by defining relevant assignments for the students and running appropriate tutorials. In any case, it can help to make students aware of the importance of technical and non-technical interaction, of the skills it requires and of the fact, that these skills usually don't come by themselves, but need to be learned and practiced.

Technopolitical Settings

The case of wind technology, finally, helps engineering students to understand the role of broader settings, in which engineering and development take place and of which they are a part. Engineering involves more than scientific and technical knowledge, more than just technical activity and more than just a R&D laboratory in which development takes place. Power relations, political interests, technology policies, market structures, legal conditions, ideologies and cultural dispositions and trends are reflected in it. Disregarding any of these elements increases the likelihood of failure – if not technologically then on the market. The failure of early wind technology in many countries was clearly related to unfavorable settings such as monopolized markets, strong interests in nuclear power, short-sighted technology policies and "hard energy path" ideologies. The success of wind technology in Denmark, on the other hand, decisively rooted in differences of power relations and policies as well as in successful efforts by wind power supporters to shape relevant techno-political settings in constructive ways (e.g., negotiate an accord with the power industry, establish institutions to support wind power, campaign for a positive image of wind power).

Historians of technology commonly agree that broader "contexts", the "cultural ambience", as Staudenmaier calls it (1985), matter to innovation in crucial ways. True as it is, there is a downside to concepts like "contexts", "ambience" and "settings". They suggest portraying engineering as an activity that is surrounded by a set of given conditions, which has influence on research and development. While this perspective may have its value, it draws artificial boundaries between engineering and the surrounding culture. I suggest inviting engineers and engineering students to consider engineering a part of that very culture. Engineers, in fact, contribute significantly to shaping techno-political settings. They profit from conceiving themselves actors in broader culture rather than technical experts subject to it. This argument is related (though not equal) to attempts of grasping the complexities of innovation by describing "innovations systems" or, more recently, "innovation cultures" (e. g., Godin 2010; Ulijn and Weggeman 2001). While the term innovation system implies a somewhat static character of system elements, the term innovation culture has received grown attention, because it is broader, more flexible, less static and better suited to account for cultural interests. Figure 23.1 could be reconceived as a depiction of elements of the innovation culture in Danish wind technology development. While the case of wind technology does not provide a blueprint for innovation, it offers rich potential to learn about engineering and innovation.

Acknowledgements The writing of this chapter was made possible by a grant from the The Danish Council for Strategic Research (DSF) to the Program of Research on Opportunities and Challenges in Engineering Education in Denmark (PROCEED).

References

Andersen, H., Klostergaard, L., Knudsen, H., Kragh, H., Nielsen, K., Pedersen, K., Møller, S., & Kragh, H. (2009). Vedkommende videnskabsteori. *Aktuel Naturvidenskab, 1*, 32–35.

California Energy Commission. (1988). *Results from the wind project performance reporting system*. 1986 Annual report. Sacramento: California Energy Commission.

Collins, H. (2010). *Tacit and explicit knowledge*. Chicago: University of Chicago Press.

Ehrlenspiel, K. (2003). *Integrierte Produktentwicklung. Methoden für Prozeßorganisation, Produkterstellung und Konstruktion*. München: Carl Hanser Verlag.

Ferguson, E. S. (1992). *Engineering and the mind's eye*. Cambridge, MA: MIT Press.

Garud, R., & Karnøe, P. (2003). Bricolage versus breakthrough: Distributed and embedded agency in technology entrepreneurship. *Research Policy, 32*(2), 277–300.

Godin, B. (2010). National innovation system: The system approach in historical perspective. *Science, Technology and Human Values, 35*, 476–501.

Hård, M., & Jamison, A. (2005). *Hubris and hybrids: A cultural history of technology and science*. New York: Routledge.

Heymann, M. (1995). *Die Geschichte der Windenergienutzung 1890–1990*. Campus: Frankfurt am Main.

Heymann, M. (1996). Technisches Wissen, Orientierungen und Mentalitäten: Hintergründe zur Mißerfolgsgeschichte der Windenergietechnik im 20. Jahrhundert. *Technikgeschichte, 63*(3), 237–254.

Heymann, M. (1998). Signs of Hubris – The shaping of wind technology styles in Germany, Denmark, and the United States, 1940–1990. *Technology and Culture, 39*(4), 641–670.

Heymann, M. (1999). A fight of systems? Wind power and electric power systems in Denmark, Germany, and the USA. *Centaurus, 41*(1–2), 112–136.

Heymann, M. (2005). *Kunst und Wissenschaft in der Technik des 20. Jahrhunderts. Zur Geschichte der Konstruktionswissenschaften*. Zürich: Chronos.

Heymann, M. (2009). "Art" or science? Competing claims in the history of engineering design. In S. H. Christensen, M. Meganck, & B. Delahousse (Eds.), *Engineering in context* (pp. 227–244). Aarhus: Academica.

Ibenholt, K. (2002). Explaining learning curves for wind power. *Energy Policy, 30*, 1181–1189.

Jamison, A. (2001). *The making of green knowledge: Environmental politics and cultural transformation*. New York: Cambridge University Press.

Jamison, A., Christensen, S. H., & Botin, L. (2011). *A hybrid imagination: Science and technology in cultural perspective*. Morgan and Claypool. Available online at http://www.morganclaypool.com/doi/abs/10.2200/S00339ED1V01Y201104ETS016

Kamp, L. M., Smits, R. E. H. M., & Andriesse, C. D. (2004). Notions on learning applied to wind turbine development in the Netherlands and Denmark. *Energy Policy, 32*(14), 1625–1637.

Karnøe, P. (1991). *Dansk Vindmølleindustri – en overraskende international succes. Om innovationer, industriudvikling og teknologipolitik*. Frederiksberg: Samfundslitteratur.

Lovins, A. B. (1977). *Soft energy paths: Toward a durable peace*. New York: Harper Colophon Books.

Lucas, N., & Papaconstantinou, D. (1985). *Western European energy policies: A comparative study of the influence of institutional structure on technical change*. Oxford: Clarendon.

Neukirch, M. (2010). *Die internationale Pionierphase der Windenergienutzung*. Dissertation Georg-August University of Göttingen. http://webdoc.sub.gwdg.de/diss/2010/neukirch/neukirch.pdf. Accessed 12 Feb 2013.

Nielsen, K. H., & Heymann, M. (2012). Winds of change: Communication and wind power technology development in Denmark and Germany from 1973 to ca. 1985. *Engineering Studies, 4*(1), 11–31.

Polanyi, M. (1966). *The tacit dimension*. Chicago: University of Chicago Press.

Schumacher, E.-F. (1973). *Small is beautiful. A study of economics as if people mattered*. New York: Harper & Row.

Sørensen, H. K. (2012). Making philosophy of science relevant for science students. *Research publications on science studies*, No. 18, Centre for Science Studies, Aarhus University. http://css.au.dk/fileadmin/reposs/reposs-018.pdf. Accessed 12 Feb 2012.

Staudenmaier, J. M. (1985). *Technology's storytellers. Reweaving the human fabric*. Cambridge, MA: MIT Press.

Stoddard, F. S. (1986). The California experience. *Proceedings of the DANWEA conference 1986*, (pp. 83–101). Copenhagen: Centec Business Consultants.

Ulijn, J. M., & Weggeman, M. (2001). Towards an innovation culture: What are its national, corporate, marketing and engineering aspects: Some experimental evidence. In C. L. Cooper, S. Cartwright, & P. C. Earley (Eds.), *The international handbook of organizational culture and climate* (pp. 487–517). London: Wiley.

Wengenroth, U. (2001). Vom Innovationssystem zur Innovationskultur. Perspektivwechsel in der Innovationsforschung. In J. Abele, G. Barkleit, & T. Hänseroth (Eds.), *Innovationskulturen und Fortschrittserwartungen im geteilten Deutschland* (pp. 23–32). Köln: Böhlau.

Winner, L. (1980). Do artifacts have politics? *Daedalus, 109*(1), 121–136. Republished 1986 in: Winner, L. (Ed.). *The whale and the reactor, a search for limits in an age of high technology* (pp. 19–39). Chicago: Chicago University Press.

Matthias Heymann Diploma in Physics, University of Hamburg, Ph.D. in History of Technology, University of Technology Munich. Associate Professor for the History of Technology at Aarhus University, Denmark. Research interest in the history of engineering, energy and environmental science and technology. He is author of *Die Geschichte der Windenergienutzung 1890–1990* (Campus 1995), *Kunst und Wissenschaft in der Technik des 20. Jahrhunderts. Zur Geschichte der Konstruktionswissenschaften* (Chronos 2005), *Engineers, Markets and Visions: The Turbulent History of Natural-gas Liquefaction* (Piper 2006), and *Scientists, Pioneers, Visionaries: Hydrogen as Energy Carrier* (Piper 2009) and serves as Associate Editor of Centaurus and Domain Editor of Wiley Interdisciplinary Review Climate Change, Domain History, Society, Culture.

Chapter 24
Getting Context Back in Engineering Education

Anders Buch and Louis L. Bucciarelli

When I heard the learn'd astronomer;
When the proofs, the figures, were ranged in columns before me;
When I was shown the charts and the diagrams, to add, divide, and measure them;
When I, sitting, heard the astronomer, where he lectured with much applause in the lecture-room;
How soon, unaccountable, I became tired and sick;
Till rising and gliding out, I wander'd off by myself,
In the mystical moist night-air, and from time to time,
Look'd up in perfect silence at the stars

Walt Whitman (1819–1892). Leaves of Grass, 1900

Abstract Discussions about reform in engineering education have mainly centered on issues of curriculum and didactics but these discussions rarely address fundamental questions about the nature and character of knowledge and learning. This neglect has led the discussions down the wrong track and failed to critique implicit and inadequate conceptions of knowledge and learning. Our discussion will draw upon John Dewey's philosophy of human experience and inquiry as a resource that can remedy the neglect. This chapter thus focuses on learning and by example proposes ways that engineering knowledge and skills can be contextualized, taught – and learned.

Keywords Engineering education • Learning • John Dewey • Contextualization of knowledge

A. Buch (✉)
Department of Philosophy and Learning, Aalborg University Copenhagen,
A.C. Meyers Vænge 15, Copenhagen, SV 2450, Denmark
e-mail: buch@learning.aau.dk

L.L. Bucciarelli
School of Engineering & School of Humanities, Art & Social Science,
Massachusetts Institute of Technology, 77 Massachusetts Av., Cambridge,
MA 02139-4307, USA
e-mail: llbjr@mit.edu

S.H. Christensen et al. (eds.), *International Perspectives on Engineering Education*,
Philosophy of Engineering and Technology 20, DOI 10.1007/978-3-319-16169-3_24

Introduction

Whitman's poem, written more than a century ago, depicts some of the challenges facing education in science, technology, engineering and mathematics – then and now. Today, science and technology are shaping our very human existence in countless ways and we marvel, delight and despair when we reflect on the changes that science and technology have brought to our world. But, equally, we feel that science and technology may be lacking in some respects – not being able to encompass qualitative, aesthetic, ethical, existential and spiritual dimensions among others. Thus, educationalists have turned to the liberal arts and other domains to supplement the education of children and adolescents. In engineering, more specifically, a number of academically inclined engineers and others have found engineering education lacking. The critics support their claim with diverse arguments and propose different reform initiatives (Buch 2012). Some critics suggest that engineering education should pay more attention to business; others suggest that the education should develop a sense of professional responsibility within the students and still others stress the importance of contextualizing technical skills and knowledge. Though the motivation varies the critics seems to agree on broadening engineering education and thus supplementing the technical and scientific curriculum with elements from the humanities and the social sciences.

No one questions the importance of teaching science and technology to engineering students, but there seems to be disagreement about the dominant role of science education in engineering and indeed about how engineering education should in fact be broadened. The discussion centers round the question: *What are the fundamentals of engineering education?* As engineers specialize in engineering subfields like mechanical, chemical, electrical, etc., engineering they are supposed to have a basis for their specialization. Traditionally this basis has been considered to derive from courses in physics, chemistry, calculus and other science disciplines. But as engineering has moved into new domains such as e.g., biotechnology the list of fundamentals becomes still longer and broader. Should it also include biology? And engineers are often occupied with the development of new artifacts, products and services within enterprises. Should the list then be expanded to include management, economics, marketing, etc.? In fact it is quite difficult to delimit the field of engineering and thus equally difficult to answer the question of what the fundamentals of engineer education should be.

Discussions about engineering education have also to a large extent focused on another fundamental question: *how should engineering be taught to students?* Whereas the first question focus on *curriculum*, this second questions address issues of *didactics*. If we can agree on a curriculum in engineering education, how can engineering then be taught most effectively? Various methods have been suggested, implemented and tested. At our own institutions – MIT and Aalborg University – two models have been influential: Professor Ed Crawley, of MIT's Department of Aeronautics and Astronautics in collaboration with Swedish universities has developed the CDIO-model (Conceive – Design – Implement – Operate) and Aalborg

University have developed the PBL-model (Problem Based Learning). Both models have spread to other universities and have been implemented in engineering programs around the world.

Although we recognize the importance of discussions about *curriculum* and *didactics* we will, however, approach engineering education from another vantage point – in order for us to return to these important issues from another angle and with another perspective. John Amos Comenius contrasted *didactics* – the science of teaching – to *mathetics* – the science of learning – in his *Spicilegium didacticum* published in 1680. He argued that we need to have a clear vision of how we learn before we can develop our teaching methods. We will follow Comenius and argue along similar lines that we need to understand how we learn in order to address questions about curriculum and didactics in engineering education. Thus we address a third question, namely: *how is engineering learned?* – or in more general terms *How do we learn (engineering)?* in order to return to the other two questions: *What are the fundamentals of engineering education? How should engineering be taught?*

In this chapter we will thus focus on learning and by example propose ways that engineering knowledge and skills can be taught – and learned. It might be objected that a broader discussion on learning will defocus the discussion on how engineering education can in fact become more adequate. We tend to disagree. Discussions about reform in engineering education have mainly centered on issues of curriculum and didactics but these discussions rarely address fundamental questions about the nature and character of knowledge and learning. This neglect has led the discussions down the wrong track and failed to critique implicit and inadequate conceptions of knowledge and learning. Our discussion will draw upon John Dewey's philosophy of human experience and inquiry as a resource that can remedy the neglect.[1]

Traditional Ways of Teaching Engineering: Decontextualizing

In order to get our argument started it is necessary briefly to review traditional ways of teaching engineering. Here we will see how engineering knowledge is traditionally conceived.

Prerequisites to enrolling in engineering education are courses in the calculus, physics, chemistry, now biology. These are seen as fundamental. And they are in the way that they provide the student with the vocabulary, the tools, the concepts and principles upon which the engineering sciences base their development of more concepts and principles and methods – enabling students to solve the problems assigned in mechanics, thermodynamics, electronics, etc. The engineering sciences in turn are held as fundamental to practice, e.g., engineering design and to further study and research at the graduate level.

[1] This chapter draws on and includes sections from Louis Bucciarelli's (2012) *Bachelor of Arts in Engineering – the Full Proposal.*

The purpose of assignments and exercises within engineering education, whether mechanical, electrical, civil, chemical, etc., is to convey a well established body of instrumental, disciplinary knowledge from faculty to the student. The abilities stressed are problem solving within the discipline's paradigm using its concepts and principles alone. Here, for example, is an excerpt from a well-know textbook in engineering mechanics.

> The main objective of a basic mechanics course should be to develop in the engineering student the ability to analyze a given problem in a simple and logical manner and to apply to its solution a few fundamental and well-understood principles. (Beer et al. 2006, p. xiii)

The mechanics problem is given – not to be formulated by the student; it demands a simple and logical analysis – not a conjectural, inferential thinking up and about – and is to be solved using a few fundamental and well-understood principles, not by trying several, alternative, perhaps conflicting, approaches and perspectives. The work-life of an engineering student, hence graduate, from this perspective is neat, well posed, deductive and principled. Solving well posed, single-answer problems is the dominant learning experience of the undergraduate. It is a solitary activity, engaged in competition with one's peers. It is an essential activity in engineering practice, but it is not all. Solving problems is a necessary part of engineering work – work within a bounded discipline, e.g., structures, electronics, etc., with its own particular resources including concepts, principles, heuristics, metaphors, methods, codes, standards, supplier catalogues, instruments, techniques of fabrication, and more. It is necessary work but it does not suffice. For example, in design, many with whom engineers must work may not see the world in the way they do. The (over) emphasis on solving well-posed, single-answer problems with its reductionist, deterministic ideology works against taking the social and constitutive features of engineering seriously.

Instrumental rationality is evident in the ways a 'problem' is described to be 'solved' and what is deemed a legitimate 'solution'. The exercise of instrumental rationality requires abstraction and simplification. This is key to methods for problem solving in all fields of engineering.

> [Abstraction requires] Simplification of a complex problem by breaking it down into manageable components. Specifically modeling in *quantitative terms* critical aspects of the physical and human world, and necessarily simplifying or *eliminating* [our emphasis] less important elements for the sake of problem analysis and design… (MIT, Engineering Council for Undergraduate Education 2005, p. 4)

For a problem to be treated as an engineering problem it must be expressed in quantitative terms. Only factors, aspects and features of the 'real' world that can be construed as measurable and quantified matter. Numerical measures of inputs, outputs, parameters, variables, behavior and performance, costs and benefits are the essential ingredients of a problem. One might wonder what criteria are used in eliminating, (or deforming), more qualitative elements for the sake of problem analysis and design. Is it perhaps the case that only those 'elements' that can be quantified are considered at all? Anything that can't be measured is, ipso facto, irrelevant, not of interest or significance?

This way of thinking is evident in the desire to make engineering design into an engineering science. And it's this way of thinking that leads us to write and speak of knowledge as if it were some kind of material stuff or well-defined entity; e.g., We gain knowledge, store it away somewhere in our head; transfer it to our students; students claim that my course is 'like taking a drink from a fire hydrant'. Our research contributes to the body of knowledge and we measure this in large part by the number of our publications. We know more now than before. The 'knowledge as stuff' metaphor leads us astray – in our case, down the path of curriculum reform that constrains our discussion to what material we must cover, what we must leave out, what we should keep in – what the 'fundamentals' of engineering are. We refuse to go that way. Instrumental rationality is essential to engineering but it fails when taken too far, when presumed a basis all of our thoughts, for dealing with events and features, phenomena and people, beliefs and values, that cannot, ought not, be reduced to quantitative measure.

Scott D.N. Cook and John Seely Brown (1999) have called this dominant position 'the epistemology of possession': knowledge being something that is well-defined, compartmentalized, mental, value-free, something that individuals can either possess or lack and something that individuals can instrumentally put to use – regardless of context.[2] This epistemology is flawed in two important respects. It fails to recognize that knowledge is always produced in specific settings where individuals interact with their concrete environment and community in a process of learning – it fails to recognize the importance of context in the process of learning. Secondly, it fails to recognize what is going on in a process of learning. We will draw upon the philosophy of John Dewey to elicit how context matters in the course of experiencing the world and how the process of learning unfolds.

Experience: Contextualizing Knowledge

John Dewey was a strong critic of both empiricism and rationalism in the philosophical tradition. Empiricism and rationalism have both profoundly affected our conceptions of knowledge. But knowledge, Dewey claimed, cannot be adequately understood as either accumulated sensory impressions of individuals or ideas structured by a priori categories. Instead, it is necessary to understand the process and structure of the acquisition of *experience*. He departed from both empiricism and rationalism by rejecting the epistemological approach to the understanding of the production and acquisition of knowledge. Instead he gave an ontological account of human experience in naturalistic terms (Dewey 1948). He saw humans as biological organisms situated in and constantly interacting with the environment in the process of life. Biological evolution is not just a process of passive adaptation to an unchangeable environment: through our interactions we – and the environment – are both transformed. Dewey in his later writings (Dewey and Bentley 1991)

[2] The critique of this epistemology is elaborated in Bucciarelli 2003, p. 43 ff.

actually preferred the term 'transaction' to 'interaction' because the former stresses the fact that both man and environment are transformed in the interaction.

> In the first place, the interaction of organism and environment, resulting in some adaptation which secures utilization of the latter, is the primary fact, the basic category. Knowledge is relegated to a derived position, secondary in origin, even if its importance, when once it is established, is overshadowing. Knowledge is not something separate and self-sufficing, but is involved in the process by which life is sustained and evolved. The senses lose their place as gateways of knowing to take their rightful place as stimuli to action. (Dewey 1948, p. 67)

Experience thus occurs in the course of action when our doings are interrupted. Experience is thus not a cognitive state, but "a clue in behavior" that can alter our habits and challenge us in an act of inquiry. It is important to note, that Dewey not only disentangles *experience* from the mentalistic connotations layered upon the term by empiricism, but he also demonstrates how experience is inextricably linked with context. It is not possible to experience out of context: an experience is always situated in time and in relation to a specific environment. Likewise, experience is not an isolated phenomenon – it is always linked to and builds on previous experiences and it has consequences for succeeding experiences. The demonstration of the continuity of experience is a fundamental cornerstone in Dewey's contextualism. Another important element is related to the social mediation of experience. It is not the case that humans face the environment alone. The upbringing and education of humans is mediated by social meanings and categories that shape experience: "Things come to him clothed in language, not in physical nakedness, and this garb of communication make him a sharer in the beliefs of those about him" (Dewey 1948, p. 92). Our social categories, perspectives and interests thus pervasively mediate the line of experiences.

The continuity and mediated character of experience implies that experience does not evolve in a mechanical and automatic way. In Dewey's thinking experience is fundamentally historical in the sense that it has a narrative structure that raises particular events above the otherwise continuous flow of experienced moments in time. Dewey's concept of experience is thus different from *an* experience in the common sense meaning of the term. Humans actively group experiences of moments that stand out from the rest and thereby form 'consummatory' experiences that structure and forms prior and subsequent experiences. The highlighting of particular significant moments in the stream of experiences is guided by our selective interests and contributes to the establishment of a background for the acquisition of further experiences. The organism – that is the human – is thus shaping and reshaping a context of interpretation in encountering and interacting with the world. Humans are not solitary beings though, but fundamentally social creatures. The social mediation of experiences afforded by communication among humans transforms this context from a personal level to a collective sphere – which Dewey in his late writings labels 'Culture'. In an unfinished revision of *Experience and Nature* Dewey writes:

> The name 'culture' in its anthropological…sense designates the vast range of things experienced in an indefinite variety of ways. It possesses as a name just that body of substantial references which 'experience' as a name has lost. (Dewey 2008, p. 363)

Culture, in Dewey's naturalistic sense, can thus be seen as the collective formation of our selective interests that shapes experience. Culture thus provides a context for our engagement with the environment.

Inquiry: The Method of Learning

For Dewey knowledge acquisition and learning forms a mode of experiencing the world – and a very important mode. Organisms (humans) are involved in interactions with the environment and form habits when the interaction is successful, i.e., when we are successful in dealing with and solving the problems we encounter. Sometimes, however, our habits do not suffice. The environment may have changed or we may be confronted with new phenomena and aspects of the environment. When this happens an opportunity for learning appears. Dewey preferred the terms 'pattern of inquiry' or 'reflective thought and learning' to signify the process that takes place in a situation where an organism is confronted with a problem (Dewey 2005). For Dewey 'reflective thinking' (and learning) is synonymous with 'the scientific method'. Reflective thinking is a fundamental human approach of 'active, persistent and careful' investigation of beliefs and presuppositions in the light of the evidence that can support them and consequences that can spring from them. Inquiry is not a passive 'contemplative' reflection of formal modes of inferences from premises but an active engagement with the environment through experiments and tests. A characteristic of reflective thinking is the ability and willingness to refrain from jumping to conclusions and suspend judgment until sufficient evidence has appeared through examination and experiment.

It is important to notice that Dewey's equation between 'inquiry', 'reflective thought and learning' and 'the scientific method' does not give special priority to the methods most commonly used within the *natural* sciences – as opposed to the humanities or the social sciences. For Dewey the methods of the natural sciences are a subset of a more general and encompassing scientific method that is also at play in a cogent investigation of human and social affairs. The scientific method is thus a method that has led to the establishment of the sciences as bodies of knowledge – not the other way around. Reflective thinking and learning is a fundamental logical mode of inference founded in our biological existence (Dewey 1938). The method consists of five subsequent and related steps. The first step of the method is unfolding when an organism encounters a problem that blocks ongoing activities and reveals the inadequacy of the previous habits. In this situation humans are confronted with uncertainty and hesitation: the course of actions is interrupted. The second stage forces the individual to reflect on the location and character of the problem, thus trying to develop a definition of the problem in order to direct inquiry towards a solution of the problem. Thirdly, investigations and research into the problem must be conducted in order to establish evidence that can contribute to the solution of the problem. Fourthly, hypotheses are constructed and entertained. It is considered whether the hypotheses suggest actions that potentially could lead to the solution of the problem, and the consequences implied by the hypotheses are considered. Lastly, the preferred hypotheses are selected and tested. A successful hypothesis will solve the problem – an unsuccessful will not. If this line of inquiry does not lead to solve the problem it must be reiterated, additional evidence must be collected and new hypotheses must be tested.

'Knowledge' is not a predominant concept in Dewey's vocabulary. In fact he hesitated to use the concept because it easily leads to a-contextual and reified understandings of what is in fact a deeply contextual, contingent and dynamic process of coming to know (something). He preferred to describe the processes and situations in which we 'think', 'reason' and 'learn'. Thus, Dewey, like Comenius, makes the argument that we must understand the process of learning before we engage with questions about curriculum and didactics. Let us, therefore, turn to learning in engineering practice.

Learning in Engineering Practice

It is easy to conclude that thought and practice within engineering is mundane, done mechanically (looking up in tables, deciphering numbers arranged in columns, etc. as Wittman's poem depicts…), routine or uninteresting. That's the case for some tasks but not true in general. Quite the contrary: The challenges engineers face are never so neatly defined as problems to be solved (they first must be constructed), nor bounded so narrowly (defining interfaces requires more than a look-up table), nor devoid of opportunity for creativity (even in the smallest item) as the general public might presume. Engineers derive great pleasure and satisfaction from getting things to work right in accord with their conjectured solutions, their proposals, their designs. Finding an elegant solution to a problem, or going from ideas, words on paper, a statement of specifications, to a device that actually does what the boss or a client says it should do is quite an amazing achievement – and it is sensed that way, for there is no rulebook for doing such. Engineering work is immensely satisfying, albeit constrained and instrumental, quantitative and material.

The narrowness of the domain, the instrumental nature of engineering work, frees the engineer from social concerns. Working uncontaminated by human foibles, varying opinion, subjective judgment – or this is the way it seems – one can dream of reinventing and saving the world, through the miracle of modern technology – oblivious to what goes on in the world.

This fascination with technology in and of itself alone is characteristic of the exciting part of engineering work. And it is what sustains the energy and engagement of faculty in their teaching of undergraduate as well as graduate students. Here lie the roots of the value system fundamental to engineering education.

But the system is deficient. It is deficient because it ignores context – the context of practice, the context of use, the context of the individual psyche and the context that our culture provides – barely acknowledged in the teaching of engineering. We rarely explore or show how social and political interests contribute in important ways to the forms of technologies we produce. We assume that engineering knowledge and know-how is universally accessible and understood by all in the same way – free of cultural variety or individual expression.

With blinders on, what is seen is only the 'hard' stuff; what is discussed in earnest is limited to how to get from a well-posed problem statement to the unique solution, from a list of functional requirements in design to specifications of the product. It is a value system that glorifies the material to the extent that the system will not allow any serious discussion of values and visions other than those co-opted within itself. The way we structure our curriculum and teach our subjects conspire to instill in the student the idea that engineering work is value-free. Engineering work as instrumental problem-solving might be perceived to be so, but that is but one part of engineering competence. While teaching the 'fundamentals' of science and mathematics and the engineering sciences remains necessary, we must do so in more authentic contexts, showing how other than quantifiable constraints, costs, and benefits contribute in important ways to the forms of technologies we produce. We ought not as faculty imply, as we do, that solving single answer problems or finding optimum designs alone, uncontaminated by the legitimate interests of others who see the world in other ways than we, is what engineers do all of the time.

The fundamental problem and barrier to broadening then is the instrumental focus of our traditional curriculum, the reified conception of the 'content' of engineering knowledge. Teaching engineering as value-free, or what is much the same thing, teaching as though technological innovation in itself is always good and ignoring the complexities of cultural context – including who sets the ends – severely constrains opportunities for curriculum reform. The almost entire attention to means and methods to solve problems is selling students short, is doctrinaire, stifles questioning, creativity, reflection and debate, can be hazardous to society and borders on the irresponsible.

Letting Context Back in and Restoring Reflection in Engineering Education

Gary Downey (2005) has proposed, "to adapt pedagogies in engineering science courses to emphasize the limitations of the knowledge they convey along with their strengths". He asks "How can one teach engineering science courses so that students come to understand what they are not learning?" What are the boundaries? What is a legitimate question in thermodynamics? How does thermodynamics connect up with heat transfer? What is different about how thermo is taught in Chem E and in Mech E? And why are they different? Questions about the scope, validity, authority and relevance of engineering knowledge situate the 'hard' content of engineering science as 'text' within a historical, social and societal frame. Contextualizing engineering in this way enables consideration of the fundamentals of an engineering science above and beyond the instrumental analysis one finds in the text-book, e.g., the complex relation between theory and technique throughout history; the critical role of language, especially mathematical expression, in instrumental reasoning; the fundamental importance of rhetoric, of narrative, in establishing meaning and a connection with the reader. Along the way we see what's left out, what's kept in; what's a laughing matter, what's a legitimate question; what's not a solution, what's a good outcome.

In what follows we will give examples that can illustrate ways engineering might be contextualized in engineering education. We organize the examples in accord with three kinds of learning experiences in the current curriculum – the engineering sciences, laboratories, and design projects.[3]

Engineering Science

Engineering beam theory is one of the important topics of a mechanics course in engineering education. In this respect it is as fundamental as the mathematics and physics upon which it builds.

Think of what follows as describing the possible contents of an engineering science module that contextualizes engineering knowledge. The main purpose is to show the way historical context can be brought to bear in engineering education, in this case, engineering beam theory and practice. The objective is to engage the student in the critical evaluation of historical precedents that are bound to appear 'foreign' yet familiar in the sense that the questions posed and addressed are the same as found in today's mechanic text-book.

The module would begin with Galileo's analysis of the 'resistance' of beams to fracture, an analysis found in the 'Second Day' of his "Dialogue concerning Two New Sciences". Allowing the anachronism, Galileo is very much in the mold of contemporary engineering science thought and practice in his attention to the different materials of the world of construction, in his logical analysis, in the overall flow of his treatment of different kinds of beams, in the generality of his results obtained proceeding from but a few fundamental concepts and principles.

His text might strike the reader as bizarre; there are no equations; his analysis is throughout in terms of ratio and proportion. In contrast to the box of the scaling rules for beams taken from the Mechanics Text-Book published two centuries down the road, Galileo provides a derivation of the resistance of a cantilever to fracture due to an end load. His analysis relies on the principle of equilibrium of the lever; his result expresses the fracture load in terms of the ratio of the length of the beam to its thickness and a property of the material – its resistance when subject to tension.

[3] This tri-partite structure is not meant as a rigid template for contextualizing engineering education. Our organization is patterned on Sheppard et al. (2009).

But, when compared to today's theory, his result is wrong! The constant of proportionality in the relationship of end-load to the dimensions of the beam is wrong. He goes on to deduce results for beams supported and loaded differently. And these scaling laws are correct! In fact, we can claim that Galileo provides justification for the rule governing the "Resistance to lateral pressure, or transverse action" found in a Mechanics Text-Book of the nineteenth century (Kelt 1866)! We can ask the student to explain all this. How can a theory in the engineering sciences be both wrong and correct?

At this point we go through the analysis of beam behavior as found in today's text-book – i.e., the consideration of displacements and deformation, the concepts of stress and strain, the principles of equilibrium, continuity, how the properties of the material out of which the beam is made enter the picture, etc. We compare assumptions, concepts, principles and methods of analysis with Galileo's treatment. We assign a single answer problem – a cantilever, a steel I-Beam, so students, upon finding the solution, get a taste of the pleasures of engineering. The reality of today's steel I-beam – uniform in dimensions and properties – provides a way into comparing the technological infrastructure of the late Renaissance and today.

We cannot leave Galileo aside until we say something about the dialogical form of his treatise. Who are these three discussants? What different roles do they play? Who does Simplicius stand in for? We ask the students to recast the text-book derivation of engineering beam theory in dialogue form where the wrongly intuited student takes on the role of Simplicius.

Galileo is better known for his defense of Copernicus. So the students can read selections from the "Two World Systems" – for the power of his scientific narrative. We ask the students to compare how Simplicius is treated in the two treatises.

Clearly, we could go on with a study of the period, starting with Galileo's trials and tribulations with ecclesiastical authority (Santillana 1955). Or a tangential connection to Descartes and his style of writing, of reasoning in natural philosophy, can be made reading his letter to Mersenne in which he 'reviews' Galileo's "Two New Sciences" (Descartes 2012). Philosophy and history and literature – and engineering science conjoined.

Engineering Design

There are other approaches to the fundamentals for designing. Rather than keying off a particular design exercise, we can look across disciplines to search out and explore themes common across design domains and engage students in illustrative design tasks only when they are prepared to recognize and deal with the other-than-technical ingredients of the task. We consider three possible themes: the organization of work, methods for making of decisions, and representations of users.

All designing requires the organization of work, the setting of channels for communication both informal and formal, the acceptance of a particular hierarchy – who is an authority, who a serf – and good designing requires a healthy esprit, among all participants who, despite their different interests and responsibilities, must work together in harmony. Schools of business, in recognition of this need, have anthropologists/sociologist on their faculty – 'ethnography' is not a bad word in an MBA program. If we are truly interested in what drives a 'culture of innovation' – in architecture, in policy making, in urban planning, as well as in the domains of engineering – then the kinds of questions scholars of anthropology/sociology address, their field of view, the methods they use, and the insights they prompt, provide another collection of fundamentals for design studies.[4]

The particular prescriptions and methods for making of decisions in designing in different domains, methods intended to provide confidence in the integrity of the process as well as the final product, are also worthy of study. Here philosophy is fundamental. Take, for example, the methods promoted in engineering design textbooks for choosing among a set of design options in accord with a given set of design criteria. The methods that seek a best design option in accord with a set of criteria, relying on an aggregation of weighted criteria, have been shown to lack rigor (Franssen 2005). Here Arrow's impossibility theorem is fundamental; here the distinctions among different kinds of scales of measurement must be understood if the methods are to be applied 'correctly'. Yet the fundamental basis, or lack thereof,

[4] Such learning is essential to moving beyond simplistic analyses of failure as well. To understand events, to move beyond myth-making about whistle-blowing, to prepare students for recognizing the antecedents of, and sociology of mistakes, one might start with Vaughan's summary analysis, contained in her final chapter (Vaughan 1997). There she talks about "paradigm" and "structural" (not engineering structure, but sociological), and "script" and "social construction" and "culture" and how to do good history (ethically), and calls upon the insights of authors like Kuhn, Latour, Geertz – so they need to be read and evaluated if one is to grasp the full force of her analysis.

of these methods is rarely addressed in the engineering design classroom where the rush toward quantification provides an unwarranted confidence in the scoring of options.

While these methods have a generality in that they might be applied to the problem of (social) choice in dramatically different domains, their grounding in anything that might be labeled 'fundamental' is fragile. This is not to say they ought not be taught and used but if so, then their limitations ought to be part of the lesson – and prerequisite to teaching design. Otherwise, in the words of Burr (1893), our student is but following the "handbook method of construction" leaving him "… defenseless against his own ignorance".

There are designers – then there are users, clients, customers, inhabitants, …. society and culture. The different representations, images and treatments of users implicit and explicit in the work of participants in design in different domains is another feature worthy of study. The impacts of technology can be considered as part of this, though we are uncomfortable with the phrase.[5] Literature and history can provide fundamental insight into the understanding of the relationship of the products of design to society whether it be contemporary fiction or cultural studies. The study of narratives of designers themselves when "speaking" of their users requires a broader schooling in literature than is customary, even an understanding of how 'deconstruction' might help in this regard (horrors!!). Sociology and psychology are relevant too. Study in these domains can provide a solid basis for consideration of user images and impacts – whether ergonomic models, open source ideologies, focus groups, marketing methods, modes of social interaction – or lack thereof (e.g., Turkle 2011) – and extend to include the meaning of movements and ideas bearing the labels 'environment', 'ecology', 'sustainability', 'mitigation' and the like.

Whatever array of fundamentals and/or domains of knowledge are deemed relevant to designing, learning the basics in this broad sense would be prerequisite to engaging in specific design exercises. The latter would provide a grounding and vehicle for critical reflection on the fundamentals in themselves.

In the Laboratory

In the ordinary engineering laboratory course, as in a course in engineering science or capstone design project, the focus is on the technical ingredients in a well defined, bounded, and constrained setting. In the laboratory course, the primary foci are learning to design the experiment, the testing of the relationship of theory to experimental outcomes, the analysis of data, and the learning of principles of operation and uses of instrumentation.

[5] The phrase suggests the technology is made independent of culture, put out on the market, then does its work, has impacts – like it had a life of its own. A "softer" vision of the interaction of technology and culture is required.

To further our discussion of fundamentals, we ask what knowledge and know-how is prerequisite to developing these competencies in our students, irrespective of professional domain.

For example, in almost all domains data analysis relies upon statistical methods. Immediately probability and statistics appear fundamental. The methods for testing significance, setting confidence limits, etc. need to be learned. But much more than learning the use of these 'tools' is required if we claim to be teaching the basics. Here history again comes to our aid.

The histories of the development of probabilistic and statistical thinking themselves, as in the work of Ian Hacking, reveal how cultural attitudes mix with mathematical quest mix with application in measuring society and predicting the future (Hacking 1975, 1990). Even in engineering applications, the acceptance of probabilistic ways of viewing and talking about, say, the success of a mass production process and what quality of the process means have not been above debate (Bayard 2000). The liberal study of the historical development and cultures of probabilistic thinking provides a rich context for the discussion of questions the use of such methods provoke when the results of such thinking are made the basis of policy and practice in, for example, medicine and health policy. Think of the question regarding the costs and benefits of periodic screening for breast cancer.

Constructing experiments that require probabilistic reasoning and statistical analysis is, obviously, not a realistic option if we strive to replicate a professional evaluation of the effectiveness of a new pharmaceutical product, say. But it is possible, in all domains, to mimic what's done in practice with regard to data analysis and, in this way, make students sensitive to the issues the use of such methods raise.

In a similar way, we can explore the nature and ways-of-use of instruments and systems of instrumentation in different domains. Here we have a rich resource in the work of scholars allied with programs in STS.[6] Study ought not be restricted to the apparatus of scientists and engineers but extend out to include the survey as instrument, e.g., to 'measure' the beliefs, values, attitudes of heterogeneous groups of persons. Such diversity of concerns can increase the sensitivity of students to issues like 'bias' – e.g., is bias is only a real concern in the writing of a survey but not important when using a generic, standardized instrument for the measurement of a physical quantity? We can ask how anomaly is treated in different domains. We can ask how data processing defines the 'picture' we see as 'results' of an experiment.

Again, we can construct experiments to bring these issues to the fore in a way that generates critical reflection on the characteristics of instruments, whether survey or sensor – their transparency, robustness, reliability and the like.

What is a good (or bad) experiment design? The criteria for answering this question are different in different domains. A control group is essential in some; not so when you only have one artifact to put to the test. Who sets the rules that state when you have taken enough data? Who sets the specifications of the test for behavior of the material under such-and-such extreme conditions? What qualifies as a 'failed'

[6] Latour and Woolgar, Peter Galison, Terry Shinn, Davis Baird all tell different stories but in each case the instrument is an agent of more than measurement.

experiment? What do you learn from failure? How does the cost of an experiment 'trade-off' against the reliability of the results? The STS literature addressing questions such as these is sparse compared with that concerned with measurement and instrumentation. The questions themselves are not well formed, perhaps, suggesting that they might be worthy of research from a perspective that sees beyond the scope of the dictates of the engineering textbook to include the social constituents of experiment design. The development of experimental tasks that would bring these kinds of questions to the fore should not be difficult.

In all cases of laboratory learning, we construct tasks that serve as a vehicle for critical reflection by leading the student, in the course of designing an experiment, taking measurements, analyzing the data, to think about and respond to questions about the soundness of method, the transparency of the instrument, the relation of experiment to theory, the justification of claims made on the basis of results. Questions about ethical behavior should arise naturally, prompted by this broadened view of the context of experimental work.

Conclusion

The premise of our argument is that engineering education should be broadened in the sense that technical skills and knowledge should be contextualized. The technical curriculum of engineering education should be substantiated, supported, supplemented and critiqued by including historical, social, philosophical and ethical reflections of scientific and technological results and procedures. There are many reasons that support this premise, but we do not pretend to cover them all here. For example, we have not considered the *benefits* to students, industry or society of broadening curriculum or the *effectiveness* of adjusting didactics in engineering education by considering contextualizing teaching methods. Instead we have chosen to reflect on engineering education from the perspective of learning.

Going back to John Dewey's theory of experiential learning it becomes clear that the possibility of human experience presupposes context. As biological beings our organisms are in a continual process of interaction with the environment – not only to adapt to, but also in order to transform the environment. In this process of transactions we experience. One experience building upon the previous and conditioning to succeeding. The continuity of experience entails that knowledge – the 'product' of experiences – cannot be compartmentalized into technical knowledge and 'the rest'. Technical knowledge is intrinsically interwoven with other types of knowledges in our stream of experiences. Furthermore, as we interact with our fellow human beings in social relations, our experiences are mediated. Through social interaction humans transforms their individual experiences from a personal level to a collective sphere – context becomes culture. Dewey thus shows us that context and culture matters: it is a fundamental prerequisite in experiencing and interacting with our environment. To delimit historical, social and cultural phenomena from the curriculum in engineering education decontextualizes technical knowledge and dramatically confines the room for experiencing and interacting with the environment.

Secondly, Dewey helps us understand that learning is better understood as a process of inquiry in relation to problems or disturbances in our flow of experiences than as acquisition of knowledge. Inquiry is in fact what the scientific method is all about. The scientific method is thus a method that has led to the establishment of the sciences as bodies of knowledge – not the other way around. This insight from Dewey ought to direct us away from the obsessive focus on what should be included and what should be excluded from engineering *curriculum*. This line of reasoning puts the cart before the horse. Instead, we ought to focus on installing opportunities for learning in engineering education. This will start by 'interrupting' experience and identifying problems to be solved. This is where learning starts and this is where engineering education should start. The 'tricky' part has to do with how this can be done – it has to do with *didactics*. The didactics of problem based learning (PBL) provides some cues, but we must be cautious not to equate working with problems with the process of inquiry. Dewey warns us:

> Qualification of a situation as problematic does not, however, carry inquiry far. It is but an initial step in institution of a problem. A problem is not a task to be performed which a person puts upon himself or that is placed upon him by others – like a so called arithmetical "problem" in school work. (…) to set up a problem that does not grow out of an actual situation is to start on a course of dead work, nonetheless dead because the work is "busy work". Problems that are self-set are mere excuses for seeming to do something intellectual, something that has the semblance but not the substance of scientific activity. (Dewey 1938, p. 108)

To set the process of inquiry in motion problems must be founded in real life – the historical, social and cultural elements in the situation must be brought out, investigated and critically examined – alongside the technical elements. To provide conditions for 'problematic situations' to appear in education is hard work – it requires close attention to the prerequisites of students, their interests and the 'situation' of teaching.

All of this means more work for faculty and, no doubt, less time for research. How then to justify this effort? One way is to portray the teaching as a form of research. That is, the scenarios we set out may be construed as opportunities for research by faculty– not just occasions for learning with students. As we have described these learning events in engineering science, design, and laboratory, their construction would require the collaboration of faculty from different disciplines. Admittedly, it is a challenge for faculty to engage in teaching more than "problems that are self-set" in their own discipline, none the less in collaboration with faculty from other disciplines, most of whom would not see the world in the same way. But a case may be made that the preparation for, and actual interaction with students in the inquiry and reflective modes Dewey prescribes can be seen as a kind of research – a research project as demanding as the specialized kinds of research that place such demands on faculty time. Furthermore the idea that classroom learning should "grow out of an actual situation" implies that the learning experiences we construct for (and with) our students will have a freshness about them each time they are initiated – a freshness like at the start of a new research project. Each time Galileo is "taught", it would be different – even if taught by the same faculty team. There would be different probings of the mathematics, different connections to seventeenth century thought, different philosophical questions, depending upon the interests and competencies of faculty as well as students. Teaching in this way then becomes research but a form of research intimately connected with the preparation of students for reflective, professional practice.

Finally, teaching engineering also requires close attention to the actual practice of engineering activities – be it in design practice, in the laboratory or wherever engineers are solving problems. We have gestured at some of these practices and advocated that engineering education should take its point of departure in authentic engineering situations where problems emerge. These situations are mostly complex in nature – comprised by technical, organizational, social, political, etc. elements and the students will need to acquire knowledge about the problem situation in order to frame and solve the problem. In engineering education the technical part is the 'text' and the historical, social, political, etc. elements are the 'con-text'. In principle this could have been the other way around, but we recognize that engineering education must give priority to solutions that foregrounds material and 'practical' aspects. On the other hand we must emphasize that the instrumental and technical approach to problem solving must be balanced – context has to be considered.

References

Bayard, D. (2000). How to make chance manageable: Statistical thinking and cognitive devices in manufacturing control. In M. R. Levin (Ed.), *Cultures of control* (pp. 153–176). Amsterdam: Harwood Academic.

Beer, F. P., Johnston, E. R., Jr., & DeWolf, J. T. (2006). *Mechanics of materials* (4th ed.). London: McGraw-Hill.

Bucciarelli, L. (2003). *Engineering philosophy*. Delft: Delft University Press.

Bucciarelli, L. (2012). *Bachelor of arts in engineering – The full proposal*. http://dspace.mit.edu/handle/1721.1/71008

Buch, A. (2012). Governing engineering. In S. H. Christensen, C. Mitcham, L. Bocong, & Y. An (Eds.), *Engineering, development and philosophy: American, Chinese, and European perspectives*. Dordrecht: Springer.

Burr, W. H. (1893). The ideal engineering education. *Proceedings of the society for the promotion of engineering education* (Vol. 1, Section E). Chicago, 31 July–5 Aug 1893.

Cook, S. D. N., & Brown, J. S. (1999). Bridging epistemologies: The generative dance between organizational knowledge and organizational knowing. *Organization Science, 10*(4), 381–400.

De Santillana, G. (1955). *The crime of Galileo*. Chicago: University of Chicago Press.

Descartes, R. (2012). Oeuvres Complètes de René Descartes. Electronic edition, Correspondence 1619–1650, Descartes to Mersenne: Oct 11 1638.

Dewey, J. (1938). *Logic: The theory of inquiry*. New York: Henry Holt and Company.

Dewey, J. (1948). *Reconstructions in philosophy*. Boston: Beacon.

Dewey, J. (2005). *How we think*. New York: Barnes and Noble.

Dewey, J. (2008). Experience and nature. In J. A. Boydston (Ed.), *The later works of John Dewey* (Vol. 1). Carbondale: Southern Illinois University Press.
Dewey, J., & Bentley, A. (1991). Knowing and the known. In J. A. Boydston (Ed.), *The later works of John Dewey* (Vol. 16, pp. 1–294). Carbondale: Southern Illinois University Press.
Downey, G. (2005). Are engineers losing control of technology? *Chemical Engineering Research and Design, 83*(A6), 583–595.
Engineering Council for Undergraduate Education. (2005). *From useful abstractions to useful designs – Thoughts on the foundations of the engineering method.* Part I, Draft, 7 May.
Franssen, M. (2005). Arrow's theorem, multi-criteria decision problems and multi-attribute design problems in engineering design. *Research in Engineering Design, 16*, 42–56.
Hacking, I. (1975). *Emergence of probability*. Cambridge: Cambridge University Press.
Hacking, I. (1990). *The taming of chance*. Cambridge: Cambridge University Press.
Kelt, T. (1866). *The mechanistic text-book and engineer's practical guide.* http://books.google.com/books?id=LiwIAAAAIAAJ&printsec=frontcover&dq=thomas+kelt&lr=&cd=1#v=onepage&q=&f=false
Sheppard, S. D., Macatanguay, K., Colby, A., & Sullivan, W. M. (2009). *Educating engineers. Designing for the future of the field. A report of the Carnegie foundation for the advancement of teaching*. San Francisco: Jossey Bass.
Turkle, S. (2011). *Alone together, why we expect more from technology and less from each other.* New York: Basic Books.
Vaughan, D. (1997). *The challenger launch decision: Risky technology culture and deviance at NASA*. Chicago: University of Chicago Press.

Anders Buch M.A. in Philosophy, University of Copenhagen, and Ph.D. in Educational Studies, Roskilde University. He holds an associate professorship in Techno-Anthropology at Aalborg University Copenhagen at the Department for Learning and Philosophy and he is affiliated to the Centre for Design, Innovation and Sustainable Transitions (DIST). He has published articles and books on knowledge, learning, education, and the professional development of engineers. He is presently involved in the strategic research alliance PROCEED: "Program of Research on Opportunities and challenges in engineering education in Denmark."

Louis L. Bucciarelli B.S. in Mechanical Engineering. M.S. in Aeronautical Engineering from Cornell University, and Ph.D. in Aeronautics and Astronautics from MIT. Professor Emeritus in Engineering & Technology Studies. Current research concerns innovation in engineering education.

Chapter 25
Techno-anthropology and Engineering Education: Between Hybridity and Social Responsibility

Lars Botin and Tom Børsen

Abstract This chapter investigates the relationship between Techno-Anthropology and engineering, and indicates how emphases upon hybridity/hybridization and social responsibility can beneficially supplement engineering education and practice in a post-normal world of constant change and globalization. The authors describe what they consider the core competences of Techno-Anthropology – hybrid imagination, meta-reflections on technological practices, social responsibility in relation to technological design and development, and interactional expertise – and compare these competences to conventional engineering skills. The chapter describes how Techno-Anthropology competences might materialize (i.e., how one can give advice or participate in ethical and sustainable problem-solving, and in value-based or anthropology-driven design), and it links such abilities to conventional engineering. With reference to this comparison, the chapter concludes that Techno-Anthropology will not replace engineering, but will supplement and complement it. The chapter suggests that collaborative teams (with anthropologists, philosophers, scientists, engineers, and Techno-Anthropologists) be established to face today's manifold technological challenges. This proposal will only function if team-members are prepared for such interdisciplinary collaboration. Techno-Anthropologists can facilitate these preparation processes. Only if different disciplines team up can technological hubris, technological determinism/substantivism, and essentialism/instrumentalism be avoided.

Keywords Techno-Anthropology • Hybridity • Interactional expertise • Social responsibility • Science and engineering education

---

L. Botin (✉)
Department of Development and Planning, Aalborg University,
Vestre Havnepromenade 5, Aalborg DK-9000, Denmark
e-mail: botin@plan.aau.dk

T. Børsen
Department of Learning and Philosophy, Aalborg University Copenhagen,
A.C. Meyers Vænge 15, Copenhagen 2450, Denmark
e-mail: tom@learning.aau.dk

S.H. Christensen et al. (eds.), *International Perspectives on Engineering Education*,
Philosophy of Engineering and Technology 20, DOI 10.1007/978-3-319-16169-3_25

Introduction

During the past three decades the educational programs within the institutional framework of the university have undergone substantial changes. The changes that can be observed have to do with, among other things, disciplines and disciplinarity, as particular interests put the classical virtues of science under pressure (Lyotard 1979). The dissolution of the monistic belief in science as the ultimate and absolute explanatory force was, before 1980, on its way through the critique of idealistic social-constructivists like Peter L. Berger and Thomas Luckman (1966). But it was not until the economic and societal conditions changed in the beginning of the 1980s that we saw how multi-, trans-, and inter-disciplinarity manifested itself in new university programs in engineering and science.

The American historian of technology and science Rosalind Williams described, in *Re-tooling: A Historian confronts Technological Change* (2002), how this took place at MIT while she was dean at the faculty of science and engineering from 1995 to 2000. There she experienced how new educational hybrids came into being at the various departments within the faculty. She writes in dramatic terms about the substantial changes that took place within a fairly short period in the 1990s at MIT:

> There is no "end to engineering" in the sense that it is disappearing. If anything, engineering-like activities are expanding. What is disappearing is engineering as a coherent and *independent* profession that is defined by well-understood relationships with industrial and other social organizations, with the material world, and with guiding principles such as functionality. Engineering is "ending" only in the sense that nature is ending: as a *distinct and separate* realm. The two processes of disintegration are linked. Engineering emerged in a world in which its mission was the control of non-human nature and in which that mission was defined by strong institutional authorities. Now it exists in a *hybrid world* in which there is no longer a clear boundary between autonomous, non-human nature and human-generated processes. Institutional authorities are also losing their boundaries and autonomy. (Williams 2002, p. 31)

Williams indicates how engineering has lost its ties and bonds to nature and the natural sciences, because these entities have lost their precise contours and contents. This existential loss has been addressed by a diverse range of researchers and philosophers of technology, such as Bruno Latour (1993), Don Ihde (1990), Carl Mitcham (1994), Albert Borgmann (1984) and Andrew Feenberg (1991), just to mention a few. According to Williams, this loss has been substituted for by a new belief in the forces of commerce, finance, and market. Williams writes, "In the 1990s, the trend toward hybridizing engineering and management only became more pronounced" (Williams 2002, p. 60). Further down the line she underscores the new condition of engineering, writing, "…engineering and management are the 'hot mix'" (Williams 2002, p. 61).

In a more recent context, Andrew Jamison, Steen Hyldgaard Christensen, and Lars Botin claim, in *A Hybrid Imagination: Science and Technology in Cultural Perspective* (2011), that the engineering/managerial mix will not adequately address what they see as three epochal-typical challenges, nor will it detect potentials, new solutions to those challenges that transcend the engineering/management hybrids of the 1980s and 1990s.

They portray thc three challenges as follows. First, engineering and engineering education need to address the challenges of sustainability in relation to climate change and scarce resources. Second, there is a need to frame the techno-scientific development in a discourse of responsibility and ethics. Third, contemporary engineering has to address new ways of thinking about the relationship between science and technology; e.g., developing techno-science with a cultural awareness (Jamison et al. 2011, p. 7).

In order to meet these challenges, Jamison, Christensen, and Botin show that there are three overall strategies, which, inspired by Gibbons et al. (1994), are classified as modes. We take the concept *modes* to mean stylistic and paradigmatic ways of viewing knowledge production. Modes can, in this reading, be compared to the 'thought collectives' and styles identified by Polish physician, biologist, and philosopher of science Ludwik Fleck (1935) in the sense that they reflect different attitudes/holdings of individuals or collective subjects.

Mode 1 is described as the residual, classical and traditional scientific procedure, where in-depth disciplinary knowledge and practice are conceived as comprising true science. Many of the ideals of this specific kind of knowledge production are rooted in positivist science and can also be inscribed in Robert K. Merton's ethos of science, CUDOS (1942). Mode 2, described as the current dominant way of dealing with science and technology in institutional frameworks, reflects in many ways what Williams described as the condition at MIT in the 1990s. The strong influence of business on knowledge production and research at universities has resulted in commodification and commercialization, where the outcome of research and education is measured in financial terms.

Jamison et al. (2011) are highly critical of this commodification of knowledge, research, and education, and they advocate for a third mode that takes into account the complexity of the challenges, while also adopting a caring and concerned attitude towards the humans and non-humans involved in engineering processes. The strategy of Mode 3 is coined as 'emergent or hybrid', and is characterized by a quest for "contextualization, engagement and cross-disciplinarity" (Jamison et al. 2011, p. 7). The strategy is finally described as *hybrid imagination*:

> In order to meet the challenges facing science and engineering in the world today it is not sufficient to reaffirm a traditional faith in reason and truth and reassert the importance of a largely outmoded form of imagined academic community. There is instead a need to foster a hybrid imagination, connecting science, technology and society in new ways, by combining scientific knowledge and technical skills with cultural understanding, or empathy. (Jamison et al. 2011, p. 11)

Ole Ravn Christensen and Tom Børsen (2009) observe that new university programs have emerged as reactions to the challenges identified by Jamison et al., and hence see a need for Mode 3 university education. They also articulate what they see as central traits of such programs, and they question whether these traits are actually present in the new so-called hybrid university programs.

One proposal for a program that fulfills these criteria is the "Bachelor of Arts in Engineering" put forth by Louis Bucciarelli (2011), professor of Engineering & Technology Studies at MIT. Bucciarelli characterizes the core of conventional

(Mode 1 and Mode 2) engineering as "problem-solving within [a] discipline's paradigm using its concepts and principles alone" (Bucciarelli 2011, p. 18). The problem to be solved, according to Bucciarelli, …

> … is given – not to be formulated by the students; it demands a *simple and logical analysis* – not a conjectural, inferential thinking up and about; and is to be solved using a *few fundamental and well-understood principles* – not by trying several, alternative, perhaps conflicting, approaches and perspectives. (Bucciarelli 2011, p. 18)

Conventional engineering is split up into majors – mechanical, electrical, chemical, etc. – and prerequisite courses, such as calculus, physics, chemistry, and biology, that the student needs to pass before enrolling in a graduate engineering program (a major). The prerequisites …

> … provide the student with the vocabulary, the tools, the concepts and principles and methods – enabling students to solve the problems assigned in mechanics, thermodynamics, electronics, etc. (Bucciarelli 2011, p. 18)

Hence, "traditional" engineering education is dominated by "an instrumental rationality". Therefore Bucciarelli finds conventional engineering education to be one-dimensional: "Instrumental rationality is but one mode of thinking in engineering; we must allow that much more goes in the classroom than learning how to solve well-posed, single-answer problems" (Bucciarelli 2011, p. 21).

Engineering students need also to reflect on the fact that, "Norms and beliefs – about what is a 'robust' design, about the capabilities of the user, of citizens [–] matter. Ethics matters. Culture matters" (Bucciarelli 2011, p. 22). To bring this focus into engineering education Bucciarelli suggests the establishment of a liberal arts program in engineering:

> The guiding principle [for this program] is that the humanities and social sciences should dominate – it's a bachelor of arts degree – in perspective, class-room ambience, choice of content, as well as credit hours. Restricting attention to a semester, three fourths of the semester hours would be devoted to the liberal studies of engineering/technology. The remaining one fourth being designated "free electives" – though we might consider requiring the student to "take" 25 % of his or her total credit hours in the sciences. (Bucciarelli 2011, p. 24)

The core content of a B.A. in engineering will analyze three central areas of engineering using analytical tools from the humanities and social sciences (e.g., ethnographic, ethical, historical, and sociological approaches):

> Take exemplary, substantive content of the "traditional" under-graduate engineering program – the engineering sciences, the laboratory tests, the design projects – and subject this to study from the perspective of humanities, arts, and the social sciences. (Bucciarelli 2011, p. 1)

The B.A. program will generate competences related to …

> … a more *cross-disciplinary*, whole-systems approach to engineering that emphasizes *contextualized* problem formulation, the ability to lead team-centered projects, the skill to *communicate across disciplines*, and the desire for life-long learning of the engineering craft in a rapidly changing world. (Bucciarelli 2011, p. 2, italics added for emphasis)

Mode 1 and Mode 2 Engineering: Or 'Conventional Engineering and Engineering Education'

When we look at engineering in a historical perspective, we see two phases of both temporal and conceptual character. One type of engineer is taught, and performs, in close relation to an empirical-analytical framework – cf. Jürgen Habermas' knowledge and human interests (1968), where engineering, interpreted by Bent Flyvbjerg (2001), is seen as applied science. This type of engineering-as-applied-science has been classified as Mode 1 engineering and rests upon sound and reliable knowledge that in turn is based on consistent and coherent criterions for truth (Gibbons et al. 1994). In a historical perspective we see how engineers were taught in that mode for centuries, and to some extent we still meet engineers that perceive themselves as mediators of Big Science in the practical field (Jamison et al. 2011). And often engineers are called upon in order to perform and act in this specific mode, as when architects collaborate with engineers. Architects involve engineers to find out if their buildings will hold and stand, in order to make their projects technically and scientifically reliable (Whitbeck 1998).

Jamison et al. (2011) trace the roots of this linkage between science and technology/engineering back to the seventeenth century, but emphasize that the actual application of scientific laws, rules, and procedures did not take place until the industrial revolution in the beginning of the nineteenth century:

> Science provided ways of knowing – mathematical logic and calculus, systematic experimentation, mechanical models, chemical theories – that separated knowledge from everyday life. Scientific knowledge was abstract and codified, and it was communicated in writing. By means of systematic experimentation, for instance, the traditional techniques of energy production, mining and metallurgy could be made much more efficient. Using quantitative methods and mechanical models could provide much greater precision and, not least, possibilities for control and management over processes of both primary and secondary production. It would not be until the mid-19th century that scientific knowledge would be used directly in industry. More important in the early period than any particular uses of science was the change in attitude. (Jamison et al. 2011, pp. 55–56)

Another type of engineer emerging in more recent times can, according to the classification made by Michael Gibbons et al. (1994), be called Mode 2 engineers. This approach is highly entrepreneurial and management practice-oriented, with theoretical frameworks stemming from system theory, organization theory, and institution theory. This of course gives a different type of knowledge and practice, with a focus on optimization of competences and skills related to work procedures, and to the performance of systems and technologies, as seen in an overall economics perspective.

Mode 2 engineering differs from Mode 1 on both the theoretical and the practical levels, but when viewed from a deeper epistemological perspective we find that the differences are fewer than the similarities. Mode 1 and Mode 2 engineers are both conceptualizing and perceiving from a technical, quantitative point of view, which makes the performance of engineering in both modes mechanical, reductionist, and structured for control, either of natural or organizational processes and ends, and thereby they both reflect instrumental rationality:

> The relevant contextual knowledge is thus almost exclusively economic and managerial, with a focus on identifying and analyzing the "entrepreneurial" skills that are considered crucial for bringing scientific discoveries or technical inventions to market. (Jamison et al. 2011, p. 14)

According to Habermas (1987), this fusion, or hybridization, resulted in a monster, an 'unholy' alliance of science, technology, and capitalism that lead to a dehumanizing instrumental rationality of dominance and control.

The Dutch pragmatist philosopher of technology Martijntje Smits has developed a theory of monsters (2006), which defines monsters as hybridizations of opposites in dynamic cultural environments. She detects four types of approaches to cope with monsters of technology: exorcism, adaption, embracement, and assimilation. The exorcism strategy demonizes the monsters and hence expels them from, for example, engineering education. The adaption strategy reduces the monsters to rational models, whereby the monster character disappears and dissolves. According to the embracement strategy we fully accept the monsters as part of reality and are engulfed. The assimilation approach portrays the technological monsters in their cultural context and in that way reveals the opposite as uniting rather than absolute. Conventional Mode 1 and Mode 2 engineering would turn towards the three first categories whereas, according to Smits, appropriate engagement with technological development and innovation would try to assimilate. She writes,

> Technological innovation is a rich source of new phenomena. These phenomena have to be appropriated to make them fit into our lives and practices. The appropriation process has various aspects, because new technology has to fit into diverse existing orders: social, technical, organizational and others…However, new technology also has to be attuned to cultural order, since our perception of technology is mediated by our cultural categories and contemporary myths regarding nature and what it is to be human. Domestication of new technology is a process in which *cultural imagination and technological change are intertwined.* (Smits 2006, p. 499)

It is this intertwinement of science, technology, and cultural imagination/appropriation that characterizes how we should deal, in a post-normal reality, with responsible design and development, because either way we look there are no categorical or absolute answers to be found about scientific and technological evolution:

> Because scientific consensus about the truth of complex environmental risks is unlikely to be achieved given the post-normal situation (facts uncertain, values in dispute, high decision stakes), we will have to drop our demand for a single certain truth and strive instead for transparency of the various positions and learn to live with ambiguity and pluralism in risk assessment. (van der Sluijs 2005, p. 91)

In Smits' perspective, Walter G. Vincenti, American engineer and historian of engineering, steps forward as a *monster adapter.* In his *What Engineers Know and How They Know It: Analytical Studies from Aeronautical History* (1990), Vincenti tries to cope in a rational and analytical way with the uncertainties, values, interests, and beliefs of the social and cultural context.

Based on five cases from the aeronautical industry from the first part of the twentieth century, of which he had substantial personal experience, he claims that engineering is more than applied science. It is a science in itself with a specific method that is particular to engineering. He calls this method 'variation and selection', which epistemologically can be classified as evolutionary in its content. It must be said that Vincenti is of the opinion that engineering needs to broaden its perspective on the world and reality, and needs to relate better to society and ethics. He was also co-founder of Stanford University's Department of Values, Technology, and Society (in 1971, and later renamed Science, Technology and Society). Yet, it is striking that his argumentation and conceptualizations remain analytical and scientific.

It is in the institutions where engineers are taught, and in the companies and organizations where engineers work, that this picture for how engineering and engineering education are conceptualized and maintained. This common perception is fertilized and preserved by influential professional associations and institutions like, for instance, IDA (Danish Associations of Engineers), which safeguards the disciplinary boundaries of engineers and engineering through interventions in relation to curricula in engineering education at universities. A recent example of such an intervention was seen in the demands made by IDA (2013) in relation to re-accreditation of the Architecture and Design civil-engineering program at Aalborg University (DK), where specific demands were made in relation to upgrading skills in algebra and calculus, at the expense of historical sentiment of architecture, academic writing skills and social contextualization abilities and aesthetical judgments.

In order to further exemplify how Mode 1 engineering is still alive we turn to the Bio-Medical Engineering and Informatics (BMEI) Master's program at AAU, which we see as an exemplary sample of how Mode 1 science is present in contemporary engineering education. An additional reason why we have chosen to focus on the BMEI program is that it is related to Techno-Anthropology by means of overlapping teaching faculty. BMEI faculty delivers a substantial part of the teaching at the Techno-Anthropology bachelor program in Aalborg. Of the overall nine learning objectives of the BMEI study program, only three objectives partly address elements that are not strictly science-proper or applied science. One objective concerns communication and dialogue "with fellow medical engineers, with health care personnel, including medical specialists, as well as with non-specialists" (AAU 2012c). Another objective issues a fairly vague call for disciplinary and interdisciplinary professional responsibility, and a third objective deals with management of complex and unpredictable situations.

Despite the apparent negligence of contextual knowledge in the program, we nevertheless witness an opening of, and an understanding of, communication and collaboration between experts and non-experts in interactional settings, which should be addressed in the student projects.

If we look into the bachelor study-program of BMEI, we see the same tendency materializing. Even the required philosophy of science course is purely based on cases from the empirical-analytical sciences (AAU 2012b). Hence, the BMEI programs reflect predominantly a Mode 1 approach to engineering, and, to a lesser degree,

Mode 2 hybridization between applied science and (stakeholder) management. The question is whether Mode 1 and Mode 2 engineering approaches are appropriate in a world that urges and needs different types of solutions?

The BMEI curricula hint at the challenge when it calls for management of complex and unpredictable situations. Our point of view is that only Mode 3 engineering can deliver responses to such challenges. In the following section we address Techno-Anthropology, as we seek to define the contours of Mode 3 engineering.

Techno-anthropology: Mode 3 Engineering?

The German poet and scientist Johann Wolfgang Goethe wrote in a poem, "Transform what has been created/Not allowing it to harden in defense/that is eternal living deed" (Goethe 1821/1992). Techno-Anthropology looks at intersections and interfaces in technological processes in order to direct movements and representations toward sustainable and responsible solutions, not letting things "harden in defense" (Botin 2013). In this section we portray the central competences of Techno-Anthropology (as reflected in the poem of Goethe), and investigate to what extent these competences can constitute the Mode 3 engineer.[1]

In May 2010 Aalborg University applied to the Danish accreditation authorities (ACE Denmark) for permission to set up two new interdisciplinary study-programs in Techno-Anthropology on, respectively, the bachelor's and master's level. Following the requirements of the Bologna process, Danish universities need to get the accreditation authorities' approval to set up new study programs. According to the guidelines of ACE Denmark, a successful application to set up a new study program must show that: (1) There is a need for the program on the labor market, (2) The program is research based (i.e., is taught by researchers), (3) The program's title and overall competence profile correspond, (4) The program's modules, taken together, must reflect and satisfy the overall competence profile of the program, and (5) The quality of the program must be continuously and satisfactory evaluated.

The application was approved in November 2010, and a Study Board was set up, with the responsibility of turning the ideas and visions presented in the accreditation application into concrete study programs. In September 2011 the first students enrolled at Aalborg University's new bachelor program in Techno-Anthropology, and a Master's program in Techno-Anthropology started in September 2012.

No empirical studies of the students' perceptions of Techno-Anthropology have so far been published, though the programs' teaching modules and semesters have been continuously evaluated and modified. Techno-Anthropology is still a program in the making, and insight into how students perceive Techno-Anthropology is welcomed in the educational development.

[1] A more thorough introduction to the formal elements of Techno-Anthropology study-programs is found in (Børsen 2013a). Part of this section is based on that work.

The Danish Globalization council – a body set up in April 2005 by the Danish Government, comprising representatives of all sections of society, with the task of advising the Government on a strategy for Denmark in the global economy – stated that there is a need to attract more students to natural scientific and technical areas (Globaliseringsrådet 2005).

The designers of Techno-Anthropology at Aalborg University agree with this general recommendation, though they believe that the understanding of natural and technical sciences needs to be specified. Natural and technical sciences do not only cover "hard" scientific and technical knowledge about the material world; they also cover the "social, cultural, organizational, institutional and ethical assumptions and implications" underpinning the production and use of scientific knowledge and technology, to quote and paraphrase from the curriculum of the Master's program (AAU 2012a, p. 6).

With these remarks in mind, the Techno-Anthropology programs are hosted by the School of Engineering and Science at Aalborg University, and graduates of Techno-Anthropology formally become either Bachelors or Masters of Science (B.Sc. and M.Sc.), and do not earn B.A. or M.A. degrees (as was the case, for example in Bucciarelli's proposed liberal education program in engineering). Many of the courses and project modules in the B.Sc. and M.Sc. programs include terms like "Science", "Technology", "Knowledge production", and "Innovation". A full list of course titles and descriptions can be found in the curricula plans (AAU 2011, 2012a). These courses are classified as science, engineering, and technology courses, and thereby reflect a broad and contextual perception of science and engineering education.

Hence, the Techno-Anthropology programs are not conventional science or engineering programs. They are interdisciplinary endeavors integrating different disciplinary approaches: anthropology and social studies; philosophy and ethics; and natural and technical sciences of instrumental character. Indeed it is this mix of different disciplines that the designers of Techno-Anthropology consider truly scientific.

A literary search on the terms "inter-/trans-/multi-/cross-disciplinary", "university/tertiary/higher/graduate" and "education/study/learning" reveals that limited educational research explicitly focused on cross-, multi-, inter-, and trans-disciplinary university study-programs has been published, though a number of analyses of health care study programs are available (e.g., Hall and Weaver 2001; Baldwin 2007). These analyses typically argue for, or document, a need for handling complexities of proper patient care and treatment, and they show a need for health professionals and stakeholders to collaborate, typically suggesting interdisciplinary teamwork or fieldwork as possible educational approaches to managing such challenges.

Scholarly treatments of interdisciplinary study programs that combine elements of the humanities and social sciences on the one hand, and the natural and technical sciences on the other, are rarely found. However, we suspect that similar approaches might prepare Techno-Anthropology students to face the complexities emerging at the human–technology intersection.

However, the research published on inter- and trans-disciplinarity does reveal a rich theoretical vocabulary, and we take that vocabulary as one of our theoretical perspectives. Peter Weingart sees the emergence of inter- and trans-disciplinarity as a reaction against the hegemony of disciplines. He claims that inter- or trans-disciplinary knowledge production has to some extent replaced disciplinary science, and …

> …can be summarized as follows: the university has lost its monopoly as the institution of knowledge production since many other organizations are also performing that function. Transitory networks and contexts are formed which replace traditional disciplines. Knowledge production outside disciplines is no longer the search for basic laws (fundamental research) but takes place in contexts of application. (Weingart 2010, p. 12)

Inter- and trans-disciplinary knowledge production must generate solutions to concrete and contextualized problems situated in "society". Weingart states that the claimed shift from disciplinary science to inter- or trans-disciplinary knowledge production "is based on impressionistic evidence only" (ibid.). He argues that it has not been proven that mainstream knowledge production is inter- or trans-disciplinary, rather than disciplinary, though it takes place outside the university. Applied research, that tries to solve locally situated societal problems, can and might well be mono-, cross-, or multi-disciplinary rather than inter- or trans-disciplinary.

What *is* shown by a number of ethnographic science and technology studies (e.g., Gusterson 1996; Traweek 1988; Collins and Pinch 1998a, b), along with science and technology related journalistic pieces (Law 2006; Moynihan and Cassels 2005; Rampton and Stauber 2002; Turney 2000; Washburn 2005), is that knowledge production has become applied and contextualized. It remains to be seen if it has also become inter- or trans-disciplinary.

> In cultural terms, interdisciplinarity is most often a highly personal process, or series of processes of self-transformation, whereas trans-disciplinarity more often involves the seeking of niches in a competitive market, in a process or series of processes of furthering self-interest. (Jamison 2013 chapter four, p. 15)

Christensen and Børsen (2009) argue that the legitimation of new university programs is based on performativity (the ability to provide solutions to potential challenges as perceived by the takers and users) or paralogy (the ability to relate the challenges to new ways of thinking in local contexts). It is argued that new study programs need to be, (1) in constant flux (they are always in the making), and (2) interwoven with the demands of the local context (e.g., potential takers and users). Hence, Techno-Anthropology as new study-programs needs to liaise to the requirements of the surrounding society by performativity (instrumentally solving external stakeholders' challenges) or paralogy (reformulating challenges by new ways of thinking, so that they take a manageable form).

Following the recommendation of Christensen and Børsen, the study program designers and the study board for Techno-Anthropology have interacted with a number of potential so-called takers of Techno-Anthropology candidates. These takers include design firms, public and public/private innovation units, consultancies of various kinds, NGOs, and technological production firms (a list of central takers can be found in AAU 2013). The on-going interaction has so far resulted in the following list of challenges that Techno-Anthropologists can solve or re-think. The list is a slightly modified version of a list presented by Børsen (2013a), and is constantly developing:

- **Incommensurability between different professions and expert groups**. The hospital is an iconic example of such a Techno-Anthropological challenge, where different professions and expert cultures (e.g., doctors with different specialties, radiographs, nurses, public health specialists, management, the administrative and political layers, and management) fail to interact properly. Lack of understanding between hospital staff, patients, and relatives increases the complexity of such challenges and makes optimal healthcare services difficult.
- **Cultural clashes between users of techno-scientific products and the technical experts.** Genetically modified as well as radiated foods, geo-engineering and different forms of technical enhancement illustrate this cluster of challenges: new technology often becomes the focus of controversy and conflict, rather than problem solving endeavors.
- **Problems facing technology users when they try use the technology.** Technology is not also ways user-friendly. One example is the difficulty of young people using the Danish tax authorities web resources; another difficulty is the understandability and transparency of manuals of technical artifacts for domestic purposes.
- **Unintended (and undesirable) cultural and ethical consequences of new technology.** The introduction of a new technology will lead to unintended uses and consequences for the users and society as such (Ash et al. 2004). An illustrative example is the introduction of information and communication technology in schools that should have resulted in richer learning outcomes, but in many cases have had the opposite effect when the ICT equipment is used for other purposes, such as social media or sending text messages to friends, and thereby diverts attention from the teaching.
- **Dysfunctional technology**. The list of technologies that do not function is long, and includes a number of mega-projects in the West, along with developing projects in the South. Attempts to solve problems are often technological fixes, with technological solutions offered without assurance that the problem at stake is really amenable to being solved by such technology. The use of drones in Afghanistan is one example, as drones will not solve the root causes of that conflict.

According to Børsen (2013a), the overall research domain of Techno-Anthropology is *Technology*. This is a term with many facets and must be addressed from different angles. Three of the facets are technical products (designed artifacts and procedures), technical experts, and users/stakeholders. Techno-Anthropology focuses on these facets and their relationships, cf. Fig. 25.1 (ibid.).

Figure 25.1 shows the various components in the Techno-Anthropological research domain: technical experts/procedures + artifacts/users + stakeholders. It also shows the components of hybrid imagination: interactional expertise/

Fig. 25.1 The techno-anthropological field

anthropology-driven innovation and social responsibility. The central Techno-Anthropological competencies can be seen as reactions to the listed technological and techno-scientific challenges, and are found in between the three corners of Fig. 25.1:

1. Expert-user interface: interactional expertise is a competence that can "repair" a lack of understanding between experts and lay people, cf. public understanding of science. It might also manage what C. P. Snow (1990) identified as the clash of the two expert cultures: the humanities and the sciences. This competence is partly about mapping different groups' horizons of understanding and cultural codes. Interactional expertise is also about translating between them, so that the two cultures can generate an understanding of other perspectives that can be incorporated into their own cultural schemes.
2. Expert-artifact interface: here we argue that social responsibility competence is central. This quality is about ethically sensitizing technological expert cultures so that they are able to make informed, robust, and committed ethical judgments about their scientific and technological production. This requires, in part, the formation of ethical value systems that can be used to evaluate situated science and technology projects. This is a reciprocal activity; technology and value systems interaction is bidirectional. Not all situations activate the same ethical principles; knowledge of the specific contextual elements surrounding a particular technology is important for ethical judgment. Social responsibility competences are also about identifying appropriate reactions to situations where a technology project violates ethical judgments. This endeavor is not only about forecasting and evaluating potential or likely consequences of a project, it is also about not over-selling such forecasting by presenting scenarios as more certain than they actually are. Likewise, it is not about over-emphasizing uncertainty and thereby delaying preventive efforts.

3. User-artifact interface: this interface we denote with the terms anthropology-driven and value sensitive design. Techno-Anthropology is action-oriented. Hence, it is also the intention of the Techno-Anthropology programs to enable students to actively take part in bridging opposing perspectives on concrete science, engineering, and technology (SET) projects by initiating value-sensitive design or anthropological-driven innovation, cf. technology monsters. On the one hand, the education is not classified as an engineering education, which could create "trouble" in relation to both design and innovation, because, as Vincenti wrote in 1990, these are the core competences of the engineer. On the other hand, Techno-Anthropology is hosted in The School of Engineering and Science, which means that students are required to work with experiments, observations, models, and, as we see it, with proposals for design solutions. This means that Techno-Anthropology transcends classical anthropology approaches with respect to both affiliation (anthropology is generally placed under humanities or social sciences) and the aim of provoking change through action-oriented research, value sensitive design solutions, and anthropology-driven innovation.

In this section we set out asking whether Techno-Anthropology constitutes Mode 3 engineering. In the following we shall provide a response.

The challenges and strategies of hybrid imagination and Mode 3 (Jamison et al. 2011, p. 7) pointed towards sustainability, responsibility, ethics, and cultural awareness in engineering and science. In addition to those, a number of scientific organizations argue that the proximity between the wider society SET requires ethical and social responsibility competences. In UNESCO's medium term strategy for 2008–2013 it is stated that:

> The ethical dimensions of the current scientific and technological evolution must be fully addressed. Ensuring the world remains secure for everyone means that scientific and technological progress must be placed in a context of ethical reflection rooted in the cultural, legal, philosophical and religious heritage of all our communities. (UNESCO 2002)

Similar thoughts are represented in the Pugwash Conferences on Science and World Affairs (Børsen 2013b) and the World Science Forum.

Our short introduction of interactional expertise misses a focus on the potential consequences of humanity's biological and cultural constitution of given technologies. To make such assessments, all legitimate voices need to be synthesized into future projections. Such projections are by nature uncertain, but this does not mean that they are worthless. What is also missing is a discussion of the responsibility of the experts involved.

Social responsibility brings attention to these neglected points. It is crucial in this regard not to neglect important perspectives or uncertainty issues. Experts are responsible for not overselling their results (and neglecting patterns of ignorance), and they can be held responsible for doing so.

Social responsibility of SET has to do with the correspondence between the wider consequences of SET projects and *ethical* assumptions embedded in different social, cultural, institutional, and organizational domains. But it is difficult to evaluate the social responsibility of SET projects:

- It should be possible to assess the potential environmental, health, and societal consequences of SET projects. Although the future is not exactly predictable, a prognosis of future scenarios – within given uncertainty levels – is possible. If it is impossible to assess the consequences then one needs to decide how to handle this uncertainty.
- In order to evaluate the social responsibility of a SET project, one needs to map, criticise, and communicate the ethical assumptions embedded in different social, cultural, institutional, and organisational domains involved in, or affected by, the project.

Techno-Anthropology is directed towards both of these points.

With respect to how users interact with technologies, we stressed that for technological design to be sustainable and responsible (and democratic), values and interests of users (and developers) should be taken seriously. Value sensitive design (VSD) is means of doing this within Techno-Anthropology. VSD places itself in between the poles of endogenous/internal and exogenous/external theories on values in relation to technology. This underpins the hybridity of VSD, because "people and social systems affect technological development, and new technologies shape (but do not rigidly determine) individual behavior and social systems" (Friedman and Freier 2005, p. 369).

If technology both mediates and constitutes, it is imperative for VSD to set up regulations and requirements in relation to design-processes. Mary L. Cummings indicates 12 human values that should be attended to in the design:

> …human welfare, ownership and property, privacy, freedom from bias, universal usability, trust, autonomy, informed consent, accountability, calmness, identity, and environmental sustainability. (Cummings 2006, p. 702)

Cummings describes how technology projects direct our focus towards two or three of these values, hence generating an interdependent focus that merge the context and the technology. The actual design-process is divided into three phases: conceptual investigation, empirical investigation, and technical investigation. The conceptual investigation is characterized by enquiry of philosophical and theoretical character, which according to Cummings is fairly distant from conventional engineering design practices. The empirical investigation is based on quantitative and qualitative analyses of the social context. The technical investigation is a classical engineering practice where the designer focuses on the technical performance of the design through experiments and tests. VSD is, according to the Dutch philosopher of technology Jeroen van den Hoven, an appropriate answer to the challenges that engineering and technology are facing right now. He writes,

> If I am not mistaken we are now entering a third phase in the development of ICT, namely one where the needs of human users, the values of citizens, patients, and some of our social questions are considered in their own right and are starting to drive research and development of ICT. (van den Hoven 2007, p. 71)

Anthropology-driven design is derived from the Scandinavian model on participatory design or/and user-driven innovation (Bødker et al. 2004). The specific anthropological approach is that the Techno-Anthropologist observes all actors, whether involved directly or indirectly, in the innovational process through intensive and extensive field studies, and thereby draws heavily on classical anthropology.

The Techno-Anthropologist observes and interacts with the end users, as well as the technical experts in the lab, and tries to bridge gaps and make connections as she moves back and forth in the field, hence facing the challenges and strategies of hybrid imagination and Mode 3 engineering (Jamison et al. 2011). Techno-anthropological core competencies can, in other words, supplement Mode 1 and Mode 2 engineering.

Conclusion

As with all other study programs at Aalborg University, Techno-Anthropology applies the Aalborg Model of Problem Based Learning (Barge 2010). This means that each semester contains approximately 50 % course work and 50 % project work that is driven by the students under supervision of one or two supervisors. In the course modules, tools to identify and analyze the technological challenges mentioned above are presented and exemplified, while it is in the project modules that the Techno-Anthropology students engage first-hand with these challenges. As we see it, all of this requires a *hybrid imagination*. Andrew Jamison is very eloquent on this point, writing,

> Hybridity, or a hybrid imagination, is something that has to come from within; it requires a student who is interested in obtaining what might be best characterized as a dual competence. But it requires something else, as well: a motivation, a commitment, a sense of engagement in the broader process of social and cultural change that is sustainable development. (Jamison 2013 chapter 5, p. 23)

In this chapter, we have identified three components of the hybrid imagination: (1) interactional expertise/social responsibility competences/abilities to (2) enable anthropology-driven innovation, and (3) foster value sensitive design.

The techno-anthropologist is, in other words, a socially responsible interactional expert that carries out anthropology-driven innovation or value sensitive design. The purpose of Techno-Anthropology is not to replace science or engineering. Rather, the Techno-Anthropologist can team up with scientists and engineers in identifying and addressing the wider technological challenges that one profession cannot manage alone. The overall picture of the technological challenges, which the instrumental and reductionist engineering/management hybrid do not address at all, reflects concerns regarding science, technology, politics, ethics, society, and culture.

We have pointed towards fruitful and appropriate approaches that are present in contemporary theorizing and practice, where Jamison's *The Making of Green*

Engineers (2013), Bucciarelli's proposal for a Bachelors of Arts program in Engineering (2011), and Smits' pragmatic concept of "monster assimilation" foreground more cautious and conservative openings that are to be found in the writings of, for instance, Vincenti (1990). We think that Techno-Anthropology is part of this avant-garde of new ways of conceiving and practicing engineering. We think that for Techno-Anthropologists to have an impact on, and relevance in, engineering and engineering education, this *greening, artistry, and assimilation* has to be present in the actual technology domain. It is not something that can be impressed or imposed from the outside by anthropologists or philosophers with an interest in technology. It has to be present, nurtured, and fostered from within engineering, which also means that as we *green, create, and assimilate* we apply the modes and styles of inter-disciplinarity with the aim of promoting and provoking cross-fertilization. Techno-Anthropology places itself in between culture/humanities and techno-science/engineering and searches to bridge the gap and create gateways and paths for dialogue and interaction. The main focus is to enhance and support this dialogue and interaction as a way of fostering and nurturing appropriate processes and solutions in a complex post-normal reality.

The conclusion here is that Techno-Anthropology fuels hybrid imagination and can create platforms, programs, projects, courses, and supervision that nurtures and fosters empathy, cultural understanding, and social responsibility in science, engineering, and technology – provided that representatives of these endeavors embrace and collaborate with Techno-Anthropologists.

References

Ash, J. S., Berg, M., & Coiera, E. (2004). Unintended consequences of information technology in health care: The nature of patient care information system-related errors. *Journal of American Medical Informatics Association (JAMIA), 11*, 104–112.

Baldwin, D. W. C., Jr. (2007). Some historical notes on interdisciplinary and professional education and practice in health care in the USA. *Journal of Interprofessional Care, 21*(1), 23–37.

Barge, S. (2010). *Principles of problem and project based learning. The Aalborg PBL model.* Aalborg: Aalborg University.

Berger, P. L., & Luckmann, T. (1966). *The social construction of reality: A treatise on the sociology of knowledge*. Garden City: Anchor Books.

Bødker, K., Kensing, F., & Simonsen, J. (2004). *Participatory IT design: Designing for business and workplace realities*. Cambridge, MA: MIT Press.

Borgmann, A. (1984). *Technology and the character of contemporary life: A philosophical enquiry*. Chicago: The Chicago University Press.

Børsen, T. (2013a). Identifying core competencies in techno-anthropology: Interactional expertise, social responsibility, and anthropology-driven design. In T. Børsen & L. Botin (Eds.), *What is techno-anthropology?* Aalborg: Aalborg University Press.

Børsen, T. (2013b). Extended report from working group 5: Social responsibility of scientists at the 59th Pugwash conference on science and world affairs in Berlin, 1–4 July 2011. *Science and Engineering Ethics, 19*(1), 299–308.

Botin, L. (2013). Techno-anthropology. Betweeness and hybridization. In T. Børsen & L. Botin (Eds.), *What is techno-anthropology?* Aalborg: Aalborg University Press.

Bucciarelli, L. (2011). *Bachelor of arts in engineering*. Cambridge: MIT Press.
Christensen, O. R., & Børsen, T. (2009). From anomaly to paralogy. The post-modern condition and its consequences for University Science Education. In O. Skovsmose, P. Valero, & O. R. Christensen (Eds.), *University science and mathematics education in transition* (pp. 283–299). New York: Springer.
Collins, H., & Pinch, T. (1998a). *The Golem. What you should know about science*. Cambridge: Cambridge University Press.
Collins, H., & Pinch, T. (1998b). *The Golem at large. What you should know about technology*. Cambridge: Cambridge University Press.
Cummings, M. L. (2006). Integrating ethics in design through the value sensitive design approach. *Science and Engineering Ethics, 12*(4), 701–715.
Feenberg, A. (1991). *The critical theory of technology*. Oxford: Oxford University Press.
Fleck, L. (1935/1979). *The genesis and development of a scientific fact*. Chicago: University of Chicago Press.
Flyvbjerg, B. (2001). *Making social science matter. Why social enquiry fails and how it can succeed again*. Cambridge: Cambridge University Press.
Friedman, B., & Freier, N. G. (2005). Value sensitive design. In K. E. Fisher, S. Erdelez, & L. McKechnie (Eds.), *Theories of information behavior: A researcher's guide*. Medford: Information Today.
Gibbons, M., Limoges, C., Nowotny, H., Schwartzman, S., Scott, P., & Trow, M. (1994). *The new production of knowledge – The dynamics of science and research in contemporary societies*. London: Sage.
Goethe, J. W. (1821/1992). Eins und Alles. In: K. O. Conrady (Ed.). *Das Grosse deutsche Gedichtbuch*. Berlin: Artemis & Winkler.
Gusterson, H. (1996). *Nuclear rites. A weapons laboratory at the end of the cold war*. Berkeley: University of California Press.
Habermas, J. (1968/1971). *Knowledge and human interests*. Boston: Beacon Press.
Habermas, J. (1985/1987). *The philosophical discourse of modernity*. Cambridge, MA: MIT Press.
Hall, P., & Weaver, L. (2001). Interdisciplinary education and teamwork: A long and winding road. *Medical Education, 35*(9), 867–875.
Ihde, D. (1990). *Technology and the lifeworld: From garden to earth*. Bloomington: Indiana University Press.
Jamison, A. (2013). *The making of green engineers. Sustainable development and hybrid imagination*. London: Morgan & Claypool.
Jamison, A., Christensen, S. H., & Botin, L. (2011). *A hybrid imagination. Science and technology in cultural perspective*. London: Morgan & Claypool.
Latour, B. (1993). *We have never been modern*. New York: Harvester Wheatsheaf.
Law, J. (2006). *Big pharma. Exposing the global healthcare agenda*. New York: Carroll and Graf.
Lyotard, J. F. (1979/1984). *The post-modern condition: A report on knowledge*. Minneapolis: University of Minnesota Press.
Merton, R. K. (1942/1979). The normative structure of science. In: R. K. Merton (Ed.), *The sociology of science: theoretical and empirical investigations*. Chicago: Chicago University Press.
Mitcham, C. (1994). *Thinking through technology. The path between engineering and philosophy*. Chicago: Chicago University Press.
Moynihan, R., & Cassels, A. (2005). *Selling sickness: How the world's biggest pharmaceutical companies are turning us all into patients*. New York: Nation Books.
Rampton, S., & Stauber, J. (2002). *Trust us we're experts! How industry manipulates science and gambles with your future*. New York: Putnam Books.
Smits, M. (2006). Taming monsters: The cultural domestication of new technology. *Technology in Society, 28*(4), 489–504.
Snow, C. P. (1990). The two cultures. In *Leonardo* (Vol. 23, 2/3, pp. 169–173), New foundations: Classroom lessons in Art/Science/Technology for the 1990s. Cambridge: MIT Press.

Traweek, S. (1988). *Beamtimes and lifetimes. The world of high energy physicists*. Cambridge, MA: Harvard University Press.
Turney, J. (2000). *Frankenstein's footsteps: Science, genetics and popular culture*. New Haven: Yale University Press.
van den Hoven, J. (2007). ICT and value sensitive design. In P. Goujon, S. Lavelle, P. Duquenoy, K. Kimppa, & V. Laurent (Eds.), *The information society: Innovations, legitimacy, ethics and democracy*. Boston: Springer.
van der Sluijs, J. (2005). Uncertainty as a monster in the science-policy interface. Four coping strategies. *Water Science and Technology, 52*(6), 87–92.
Vincenti, W. G. (1990). *What engineers know and how they know it. Analytical studies from aeronautical history*. Baltimore: John Hopkins University Press.
Washburn, J. (2005). *University Inc. The corporate corruption of American higher education*. New York: Basic Books.
Weingart, P. (2010). A short history of knowledge formations. In R. Frodeman, J. Thompson Klein, & C. Mitcham (Eds.), *The Oxford handbook of interdisciplinarity*. Oxford: Oxford University Press.
Whitbeck, C. (1998). *Ethics in engineering practice and research*. Cambridge: Cambridge University Press.
Williams, R. (2002). *Re-tooling. A historian confronts technological change*. Cambridge, MA: MIT Press.

Reports, Curricular and Study Guides

AAU. (2011). Curriculum for B.Sc. in Techno-Anthropology.
AAU. (2012a). Curriculum for M.Sc. in Techno-Anthropology.
AAU. (2012b). Curriculum for B.Sc. in Bio Medical Engineering and Informatics.
AAU. (2012c). Curriculum for M.Sc. in Bio Medical Engineering and Informatics.
AAU. (2013). The AAU study guide.
Globaliseringsrådet. (2005). (Danish Globalization council) report.
IDA. (2013). The Danish Accreditation Council Report.
UNESCO. (2002). Report on medium term strategy 2008–2013.

Lars Botin M.A. in Art History, Aarhus University, Ph.D. in Art, Technology and Science, Aalborg University. He holds an associate professorship in Participation and Technology at the Department of Development and Planning at Aalborg University. He is affiliated to the research-group of Techno-Anthropology. He has published books and articles on philosophy of science and technology, architecture and health informatics. Besides he has co-authored *A Hybrid Imagination – Science and Technology in Cultural Perspective* (Morgan & Claypool Publishers 2011) together with Andrew Jamison (main-author) and Steen Hyldgaard Christensen. He is involved in the research alliance PROCEED: "Program of Research on Opportunities and Challenges in Engineering Education in Denmark."

Tom Børsen M.Sc. in Chemistry and Philosophy from University of Copenhagen, and Ph.D. in University Science Education from The Danish University for Pharmaceutical Sciences. He is an Associate Professor at the Department of Learning and Philosophy, Aalborg University Copenhagen, and former Chairman of the Study Board for Techno-Anthropology under the School of Engineering and Science, Aalborg University. His research is focused on academic, ethical and social responsibilities of experts and in expert cultures as well as on inter- and trans-disciplinary university education.

MIX
Papier aus verantwortungsvollen Quellen
Paper from responsible sources
FSC® C105338
FSC
www.fsc.org

If you have any concerns about our products,
you can contact us on
ProductSafety@springernature.com

In case Publisher is established outside the EU,
the EU authorized representative is:
Springer Nature Customer Service Center GmbH
Europaplatz 3, 69115 Heidelberg, Germany

Printed by Libri Plureos GmbH
in Hamburg, Germany